teaching primary mathematics

edition three

In memory of Jack Briggs
(1932 – 2001)
Teacher, scholar, warm friend who embraced
everything he did with enthusiasm and joy.

teaching primary mathematics

GEORGE BOOKER   DENISE BOND   LEN SPARROW   PAUL SWAN

Copyright © Pearson Education Australia 2004 (a division of Pearson Australia Group Pty Ltd)

Pearson Education Australia
Unit 4, Level 2
14 Aquatic Drive
Frenchs Forest NSW 2086

www.pearsoned.com.au

The Copyright Act 1968 of Australia allows a maximum of one chapter or 10% of this book, whichever is the greater, to be copied by any educational institution for its educational purposes provided that that educational institution (or the body that administers it) has given a remuneration notice to Copyright Agency Limited (CAL) under the Act. For details of the CAL licence for educational institutions contact: Copyright Agency Limited, telephone: (02) 9394 7600, email: info@copyright.com.au

All rights reserved. Except under the conditions described in the Copyright Act 1968 of Australia and subsequent amendments, no part of this publication may be reproduced, stored in a retrieval system or transmitted in any form or by any means, electronic, mechanical, photocopying, recording or otherwise, without the prior written permission of the copyright owner.

All material from Principles and Standards for School Mathematics is reprinted with permission from the NCTM Principles and Standards for School Mathematics, (c) 2000 by the National Council of Teachers of Mathematics. All material from A National Statement on Mathematics for Australian Schools 1991 and Numeracy Benchmarks Years 3, 5 & 7 2000 is reprinted with permission of the publisher Curriculum Corporation, PO Box 177, Carlton South, Vic. 3053 http://www.curriculum.edu.au
Email: sales@curriculum.edu.au Tel: (03) 9207 9600 FAX: 1300 780 545 (within Australia)
#61 3 9639 1616 (outside Australia).

Senior Acquisitions Editor: Nicole Meehan
Project Editor: Jane Roy/ Rebecca Pomponio
Editorial Coordinator: Jill Gillies
Copy Editor: Jennifer Coombs
Proofreader: Ron Buck
Indexer: JIS Indexing Services
Cover and internal design by DiZign Pty Ltd
Typeset by Midland Typesetters, Maryborough, Vic.

Printed in Malaysia

2 3 4 5 07 06 05 04

National Library of Australia
Cataloguing-in-Publication Data

Teaching Primary Mathematics

3rd ed.
Bibliography.
Includes index.
ISBN 1 74103 098 6.

1. Mathematics – study and teaching (Primary).
I. Booker, George.

372.7044

Every effort has been made to trace and acknowledge copyright. However, should any infringement have occurred, the publishers tender their apologies and invite copyright owners to contact them.

An imprint of Pearson Education Australia (a division of Pearson Australia Group Pty Ltd)

Teaching Primary Mathematics
Detailed Table of Contents

Preface to the first edition	viii
Preface to the second edition	x
Preface to the third edition	xii

PART ONE
Mathematics and mathematics education   1

CHAPTER ONE Approaches to mathematics teaching and learning   5

Numeracy	7
Constructing mathematical concepts and processes	10
Using materials to develop mathematical thinking	14
Patterns in mathematics	17
Language and mathematics	20
Symbols in mathematics	23
The social context of mathematical development	25
Gender and mathematics	28
Mathematics and technology	31
Mathematics and education	35

CHAPTER TWO Thinking mathematically and problem solving   37

Making sense in mathematics	38
Problem solving	42
Difficulties with problem solving	51
Problem solving as a process	56
Reflecting on the process of problem solving	62
Problem writing	65

PART TWO Content and processes in the primary mathematics curriculum   67

CHAPTER THREE Numeration for whole numbers and fraction ideas   71

Section I: Whole number numeration	71
One-digit numbers: the basis for all numbers	81
Two-digit numbers: establishing a system of place value	94
Three-digit numbers: consolidating place value and establishing renaming	107
Thousands—four, five and six-digit numbers: consolidating, renaming and extending place value	115
Rounding	120
Large numbers: extending the system for renaming numbers	125
Section II: Fraction ideas	133
Developing the fraction concept	139

Recording fractional amounts	145	Learning the multiplication basic facts	266
Ratios	167	The multiplication algorithm	275
Patterning—moving to algebra	170	Estimation with multiplication	288
		Multiplying numbers mentally	290

CHAPTER FOUR Computation and estimation for whole numbers, decimal fractions, common fractions and per cents — 175

Teaching computation	175	Using calculators in multiplication	292
Learning computation	178	Extending multiplication to fractions	294
Concepts for the operations	179	Section IV: Overview to division	302
Basic facts	180	The division concept	304
Algorithms for computation with larger numbers	182	Introducing remainders	307
		Extending the division concept	309
Section 1: Overview to addition	190	Linking division to the nearest multiplication fact	311
The addition concept	191	The division algorithm	316
Learning the addition basic facts	195	Estimation with division	331
Developing the addition algorithm	204	Dividing numbers mentally	333
Estimation with addition	213	Using calculators in division	335
Adding numbers mentally	216	Division with fractions	337
Using calculators in addition	217	The relationship between division and the fraction concept	340
Addition with decimal fractions and like common fractions	220	Section V: Errors and misconceptions with computation	348
Section II: Overview to subtraction	225	Origins of errors in mathematics	350
The subtraction concepts	226	Sources of learned difficulties in mathematics and its teachings	352
Learning the subtraction basic facts	234	Intervention in mathematics	356
Developing the subtraction algorithm	238	Teaching computation: concluding remarks	359
Estimation with subtraction	246	Moving to algebra	363
Subtracting numbers mentally	248		
Using calculators in subtraction	249		
Subtraction with decimal fractions and common fractions	251	CHAPTER FIVE Shape and space	367
Subtraction and addition with unlike common fractions	254	Section I: What is geometry?	368
		Historical development	368
Section III: Overview to multiplication	258	The language and vocabulary of shape and space	369
The multiplication concepts	259	Forms of geometry	370

Section II: Why shape and space are important in the primary school — 371
Learning shape and space in the primary school — 372
Ways of working in geometry — 374
How children learn shape and space concepts — 378

Section III: How to teach shape and space — 384
Shape and space teaching and constructivism — 384
Hoffer and van Hiele — 385
The inquiry model — 386
Principles for teaching shape and space — 386
A development for teaching shape and space — 392

Section IV: Shape and space in the primary school — 396
Shape and structure — 396
Two-dimensional (2D) shapes — 397
Three-dimensional (3D) shapes — 417
Two dimensions to three dimensions and back again — 422
Transformation and symmetry — 427
Location and arrangement — 444
Postscript — 449

CHAPTER SIX Measurement — 451
Why teach measurement? — 451
Measurement across the primary years — 459
Measurement topics in primary school mathematics — 461
Teaching measurement in primary schools — 469
Likely difficulties and misconceptions in measurement — 495
Measurement units recommended for Australian use — 496

CHAPTER SEVEN Chance and data — 500
Chance — 501
Data handling — 506
Data analysis — 509
Teaching chance and data — 521
Data sense and numeracy — 530

PART THREE
Implementing effective mathematics learning 531

CHAPTER EIGHT Planning mathematics — 537
Planning a mathematics program — 538
Planning learning — 542
The learning environment — 545
Beliefs and attitudes in mathematics learning — 553
Teaching and mathematics — 555

CHAPTER NINE Assessment in mathematics — 556
The process of assessment — 558
Alternatives to written tests — 565
Formal testing — 571
Assessing problem solving — 579
Using and communicating assessment information — 581
Assessment, learning and teaching — 583

References — 584
Index — 593

Preface to the first edition

In the primary school, mathematics has traditionally involved computational arithmetic largely focussed on paper and pencil skills and the solution of routine exercises. At a time when computation outside of the classroom makes substantial use of electronic calculating devices, the continued emphasis of a traditional syllabus and approach is being questioned. This has led to a broadening of the curriculum to include more geometric notions and the development at an earlier stage of topics pertinent to modern life such as probability. It has also led to the view that problem-solving should be central to the curriculum as the focus shifts from learning standard procedures to applying a variety of processes to real-life needs. This book both accepts and questions these changes. It covers an extensive set of topics and approaches to reflect the broader scope of mathematics teaching and learning in the primary school and it discusses many topics long considered to be central to the mathematics curriculum with a different emphasis in terms of understanding and applications. The approach is consistent with contemporary research on how children can be led to construct their own mathematical knowledge while allowing for a diversity of organisational and instructional means of achieving this.

Part 1 of the book discusses a number of issues central to the development of mathematics. Part 2, which is the focus of the book, outlines the content and methods for developing both understanding and skills. Specific sequences of development are advanced and particular attention is paid to the use of materials to develop ideas and the role of language in formulating descriptions for the various processes and concepts. The symbolic and graphical expressions that grow out of these uses of models and their expression are then founded on a full and meaningful background of experiences leading to a pool of processes that can be brought to bear on problem situations. But problem solving does not just happen as a consequence of having appropriate skills and understanding; it needs to be developed in a systematic way. While the development of the various topics has been embedded in a problem-solving approach, there is still a need to build up specific strategies and a general orientation to problem solving.

Preface to the first edition

Writing and putting together such a book is not an easy task, nor is it simply achieved in the process of creating the text. The material owes much to the many students and teachers we have interacted with over the years, to children we have worked with and observed, and to the opportunities we have had to test or advance new ideas and to discuss their merits with colleagues at conferences and in staffrooms and corridors of the institutions in which we have taught and studied. Particular thanks must go to Beth Muller who contributed much to the initial discussions which set the framework for the book, to those students who reacted to our ideas as the final manuscript was put together, to the Graduate Diploma of Special Education students of 1990, and to Denise Bond especially who carefully proofread and reacted to much of the final manuscript. Several others were involved in the framing of the project: Peter Langford as series editor; John Monro and Tony Jones who contributed ideas to the initial drafts the flavour of which we maintained through to the final product; and Ron Harper who displayed patience in waiting for the manuscript in finished form.

It is the children in primary schools now and in the future who are the real audience for this book. If the ideas can be integrated into the approach to teaching mathematics of those students taking education courses now and be utilised by teachers who have already been teaching successfully for some time, we believe that children will develop an interest in mathematics, an ability to achieve in primary school mathematics and a readiness to apply and use their mathematical knowledge to a greater extent than before. We wish all who use this text, who take up its ideas and challenges, success in the important task of building a more numerate society for the twenty-first century.

George Booker

Preface to the second edition

Writing a text such as *Teaching Primary Mathematics* is an interesting experience. On the one hand, it allows, even forces, you to pull together all your ideas on the teaching and learning of mathematics into a comprehensive, well-organised form. On the other hand, it fixes these thoughts and allows you to reflect on them as facts so as to re-think fundamental concepts, approaches and issues. Mathematics education itself is also evolving to reflect the changing nature of society and the increasing impact of technology. This new world requires people who make sense of the mathematics they use, the problem situations they meet and the results they determine, apply and interpret. For this to be achieved, the various topics that have often been taught in isolation-space, measurement, numeration, number operations, chance and data-need to be related to one another through connections that are brought out in the initial teaching and called upon in using mathematics. In this way, mathematics can be seen as a way of thinking, not just a collection of procedures to be learned and used in standardised ways.

These changes are reflected in the second edition. Chance and Data now forms a complete chapter with suggestions of how it might be taught from the beginning of school; estimation and mental strategies have been brought to the fore in the teaching of computation, although emphasis is still given to the use of materials to build up ways of thinking and appropriate recording to keep track of what is happening and allow reflection on the processes that enable learners to take control of the thinking for themselves. Problem solving has been subsumed under the broader heading of Thinking Mathematically, which also includes seeing mathematics as a sense-making activity through an emphasis on number sense and spatial sense and proposes a problem-solving approach to the development of all mathematical ideas. In order to bring about these new emphases, assessment practices also need to be re-examined and re-focussed and the final chapter investigates a range of new assessment and diagnostic procedures.

Many people have helped us to frame this new edition; students who have studied from the first edition, teachers who have used it to form their mathematics programs, lecturers who have used it in their own courses and the authors, collectively and individually, who have reflected on their initial work and its impact. One new author has joined the team. Denise Bond who was a critical reader of the first edition wrote the chapter on mathematical thinking. Charles Lovitt contributed many of the activities and ideas for the Chance and Data chapter. Prominent mathematics educators from around Australia and New Zealand—Andy Begg, Janette Bobis, Ken Clements, Joanne Mulligan, Steven Nisbet and Dianne Siemon—were invited to put forward their own points of view on crucial aspects of mathematics learning and teaching. Very strong support was also given by Peter van Vliet of Addison Wesley Longman to the project as a whole, and this was backed up by fine editorial co-ordination by Carmen Riorden and attracive and pertinent design of the finished product by Melissa Fraser. To all of these I offer sincere thanks and know that the real rewards will come in seeing the ideas and challanges the text presents picked up by students and teachers in order to increase the mathematical enjoyment and ability of generations of children.

George Booker

Preface to the third edition

Preparing a third edition of *Teaching Primary Mathematics* has been a challenge and a pleasure. It has provided an opportunity to reflect on and refine the way experiences with materials, language and the patterns in thinking that emerge from them can be used to construct mathematical ideas. The need for more numerate citizens in an increasingly technological society has refocussed attention on the understanding of fundamental mathematical concepts and processes. Suggestions for sequencing learning and organising teaching have been designed to enable children and their teachers to value meaning in the mathematics that they learn and apply as much skill as possible in ensuring that these processes are carried out fluently and accurately. Care has also been taken to connect the various facets of mathematics rather than simply provide a series of isolated techniques and approaches. In this way, a range of mathematical ideas can be brought together to solve problems in everyday life, study or work, and generalise what is known to further aspects of mathematics.

Many of the new ideas and changes have originated in work with children in a variety of contexts from regular classrooms, through focussed investigations at the Mathematics Assistance Centre of Griffith University, to research settings examining particular curriculum or social issues. Teachers, lecturers and students in both pre-service and in-service courses have influenced the new developments through their questions, enthusiasm and feedback on the uses and adaptations they have made of the previous editions. Two new authors, Paul Swan and Len Sparrow have joined the writing team, bringing new perspectives from their work with children and teachers in their home state of Western Australia as well as more broadly through participation in conferences and projects in and outside of Australia.

Specifically, there have been refinements to the teaching sequences and activities for numeration and computation, with a particular emphasis on likely difficulties with each set of concepts and processes. The chapter on mathematical thinking has been extended, based on a deeper awareness of the manner in which number sense and problem solving can be fostered. The treatment of geometry has been reorganised to more closely reflect the way in which children's thinking about shape and space develops, with many practical activities that can be used directly with children. There are more investigation and teaching activities with measurement and chance and data along with discussion of the types of misconceptions that are frequently observed among children. The chapters on planning and assessment have also been extended to incorporate contemporary curriculum expectations such as outcome based learning, benchmarks and the assessment of problem solving.

It has been especially pleasing to see how *Teaching Primary Mathematics* has been used to frame school and class programs throughout Australia and further afield. The wish expressed with the second edition that the ideas and challenges presented would increase the mathematical confidence, enjoyment and ability of children and their teachers has well and truly been met. This new edition is sure to continue that expectation.

To that end, both the authors and publishers wish to express their gratitude to the Editorial Reviewer Board who provided encouragement and suggestions during the writing process. Apart from the many anonymous reviewers, we would also like to extend our thanks to Mr Leo Crameri, Lecturer, Faculty of Education, University of Southern Queensland; Dr Carmen Diezmann, Senior Lecturer, School of Mathematics, Science and Technology Education, Queensland University of Technology; Ms Wendy Hastings, Lecturer, School of Teacher Education, Charles Sturt University, Bathurst; Mr Vivekanand (V.) Mohan-Ram, Senior Lecturer, School of Education, Northern Territory University and Mr Barry Squire, Honorary Fellow, School of Education, University of New England for their valuable contribution and time. The support given by the editorial and production team at Pearson Education has ensured a fresh look while the enthusiasm and close collaboration of the authors have provided a deeper and more consistent presentation of ideas. My thanks to all who have contributed to this epochal third edition and best wishes to all those who use the book to frame their mathematical thinking and develop the qualities needed to participate fully in a world shaped by spatial and quantitative reasoning.

George Booker

PART ONE

Mathematics and mathematics education

> Tomorrow's citizens need a predisposition to look at the world through mathematical eyes, to see the benefits (and risks) of thinking quantitatively about commonplace issues, and to approach complex problems with confidence in the value of careful reasoning. Numeracy empowers people by giving them tools to think for themselves, to ask intelligent questions of experts, and to confront authority confidently. (Steen, 2001, p. 2)

THE WORLD we now live in is changing rapidly. As society has become more complex, the mathematical reasoning required for society and individuals to function appropriately have also become more complex. Once, it was sufficient to learn procedures to add, subtract, multiply and divide. Demonstrating that proficiency assured the outside world that one was an educated person while those were precisely the skills needed to function well in the work and leisure that society afforded. Now these skills can be carried out by the machines we work, communicate and play with and it is simply taken for granted that they will be available. More advanced behaviours are expected and the emphasis has shifted from simply learning skills to understanding the thinking on which processes are built so that they might be adapted to new and novel applications.

While arithmetic was sufficient to encompass the mathematics needed by most people in the past, as those needs have broadened, the term *numeracy* has emerged to provide a more satisfactory description of the extended mathematical processes and understanding that are now required in everyday situations. At the same time, technological change, societal change and the impact of change on individuals is making the world a lot less certain. These changes demand people who make sense of the mathematics they use, the problem situations they meet and the results they determine, apply and interpret. This requires mathematically literate adults who can analyse the pronouncements of political and business leaders when they use arguments and data to their own ends, and who can also communicate their concerns in persuasive ways using quantitative data and reasoning.

If such a broad understanding and use of mathematical ideas is to be gained and extended across all years of schooling, it is essential that the various topics that have often been taught in isolation—space, measurement, numeration, number operations, statistics, probability and logic—are related to one another and that these links are both brought out in the initial teaching and called upon in using mathematics. As Ritchhart (1994, p. 19) puts it, 'too often they are taught in a discrete and separate way when teachers attempt to cover the curriculum rather than uncover it'. Mathematics needs to be viewed more as a way of thinking than as a subject in which the products of someone's thoughts are learned and used without reference to the problems which gave rise to them or the forms in which they evolved. In order to promote numeracy, it is necessary to engage students in authentic mathematical tasks, games and investigations that require thinking and understanding rather than the memorisation of facts, procedures and techniques.

Understanding has always been a critical outcome of mathematical education but it has come to be seen as much more crucial as problem solving has been identified as being central to mathematics. It is understanding that allows ideas and techniques to be adapted to new ends. A skill can only be applied in the manner in which it was practised. Non-routine problems, however, usually require that existing knowledge be transformed to meet the needs of new situations; or that the problem as stated be transformed to match the mathematical processes that have been developed previously.

At the same time, understanding is invaluable in allowing ideas to be developed more efficiently and effectively. For example, multiplication facts used to be introduced solely through the repetition of 'tables', yet when the first set of facts, the twos facts, are met there is already knowledge of the answers to be drawn on.

$$\begin{array}{c} 7 \\ \times 2 \end{array} \qquad \begin{array}{c} \bullet\bullet\bullet\bullet\bullet\bullet\bullet \\ \bullet\bullet\bullet\bullet\bullet\bullet\bullet \\ \text{2 sevens are 14} \end{array} \qquad \begin{array}{c} 7 \\ +7 \\ \hline 14 \end{array}$$

This thinking then allows other related facts to be constructed from those that are already known:

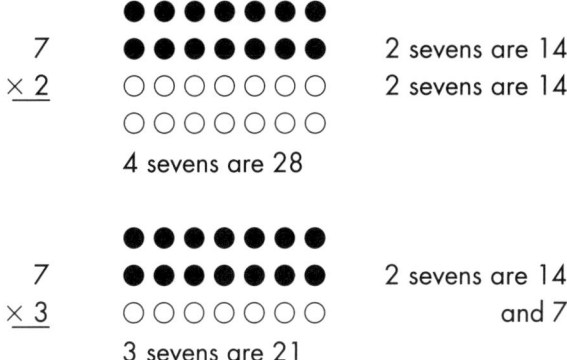

Using secure addition and numeration knowledge allows learners to build up these new facts more efficiently. At the same time they develop confidence in their ability and see the computational skills as a cohesive body of knowledge rather than as a series of fragmented ideas.

Similarly, when underlying geometrical ideas are developed in a meaningful way so that a high degree of spatial sense results, measurement can be viewed as a bringing together of spatial knowledge with the numeration and computation understanding that readily allows measurement concepts to be quantified. For example, a cylinder may be viewed as a series of discs:

The volume of the cylinder can be given by summing the volume of the discs. Each disc has an area of πr, and its volume is given by multiplying this area by its thickness. But the height of all of the discs must be the same as the height of the cylinder (h). So the volume of the cylinder is given by multiplying the area of a disc by the total height of all of the discs; $\pi r^2 h$.

If all new material can be linked to existing knowledge in this way not only will this mean the new concepts and strategies can be efficiently built out of established ways of thinking and proceeding, but the learning will also be more effective as an integrated view of the subject is formed. Concepts and processes built up in this way will not only be differentiated from each other but will also be retrieved more readily when needed. Attached to a meaning that is seen as part of a whole view of the subject, accessing them is a matter of sensing the fundamental meaning then retrieving that particular aspect of mathematical knowledge rather than floundering through a vast array of individual facts and techniques. Understanding can then be seen as central to mathematical knowing:

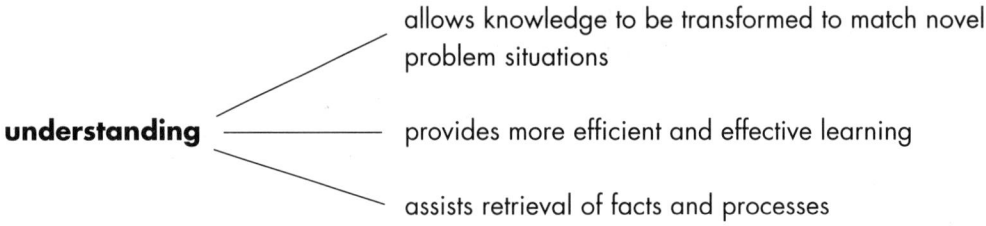

Just as mathematical ways of thinking that will be needed to deal with this new world will be different, so too the teaching that will help to prepare for it will also require changes. An ability to adapt to these new ways requires understanding over skill, and teaching must move from instruction on how to perform and apply techniques to focus on building the meaning and understanding that will enable new situations to be examined and new processes to be applied. A teacher's role will be to assist learning, recognising that students construct their own ways of knowing, rather than telling students what to learn. Classrooms will focus on the types of cooperative learning and doing that matches the world of business and science rather than encouraging solely individual learning and assessment. Problem solving will be central to learning at all stages from the beginning concepts through the development of sophisticated processes, instead of being seen as the end-point of instruction, to be applied to problems resembling the way the content has been taught.

CHAPTER ONE

Approaches to mathematics teaching and learning

The goal of teachers should be to help people understand mathematics and to encourage them to believe it is both natural and enjoyable to continue using and learning mathematics. It is essential that we teach in such a way that students see mathematics as a sensible, natural and enjoyable part of their environment.

Willoughby, 2000, p. 8

CONCEPTIONS of how mathematics is learned have changed. Once, the way in which the content was organised was considered paramount and good teaching focused on ways of transmitting this preformed knowledge from teacher to learner. Material was categorised into a detailed syllabus, suggesting that an existing mathematics simply needed to be conveyed to children. Examples of each new idea were produced by the teacher, explained to the class and followed up with practice in the form of worksheets or textbook pages. The role of the learner was to practise what was provided until it could be readily reproduced. Only then would the (successful) learner be shown and given practice in ways of applying this knowledge to different situations. In turn, the degree to which these procedures could be acquired and used determined the mathematical status of the individual learner and revealed the mathematical aptitudes with which he or she had been endowed.

While an appropriately organised curriculum is always important, teaching is now more learner centred than content driven. In the first instance, this means that the organisation of materials to be taught needs to derive as much from an understanding of how children learn as from the structure of the knowledge to be gained. New ideas and ways of thinking need to be linked to well-understood existing knowledge and their development needs to proceed in ways that reflect the manner in which the learner sees and makes sense of them rather than in the order seen by someone who already has this understanding. It also reflects a realisation that children do not simply take in mathematical knowledge that is merely transmitted to them, no matter how well organised and justified it is. Children are frequently observed to build their own ways of doing mathematics despite material or procedures introduced by a teacher. Sometimes this has given alternative ways of coming to terms with mathematics and of using it to solve problems:

Asked how many counters in this row, a young child might answer

○ ○ ○ ○ ○ ○ ○ ○ ○ ○ ○

5 and 5 is 10 and 1 more is 11 rather than simply count 1, 2, 3, . . ., 11

At other times it has produced consistent patterns of errors or misconceptions when place value is overlooked, renaming is neglected, and the significance of zero is not appreciated:

$$\begin{array}{c} 47 \\ +39 \\ \hline 716 \end{array} \qquad \begin{array}{c} 64 \\ -39 \\ \hline 35 \end{array} \qquad \begin{array}{c} 49 \\ \times 47 \\ \hline 1663 \end{array} \qquad \begin{array}{r} 63 \\ 9)\overline{5427} \end{array}$$

Learning that builds on the needs and knowledge of individual students also parallels the way in which mathematics evolved as people tried to come to terms with and make sense of problems in their everyday lives. In the early years of school, concepts in number, space, measurement, chance and data have always been developed in a similar manner through story situations from the children's own lives. Thus addition and subtraction concepts and facts have grown from realistic embodiments rather than through exercises in acquiring the addition and subtraction symbols + and − and their use with number symbols. This emphasis on problem situations out of which mathematics can grow is essential all the way through a student's schooling. As problems are understood and reconciled, the mathematics that is needed and that can develop from making sense of the solution process is personally developed and owned.

Learning mathematics then is necessarily an active process; the concepts and processes are too complex and the ideas often too abstract to allow them to be simply accepted through reading or telling. Children need to be involved in the formation of these new ways of thinking if they are to be personally meaningful and able to be used in different settings and formats. Experiences with problematic situations are fundamental to the way in which concepts and processes are built up or acquired, and resources for assisting learning need to incorporate play, games, everyday situations and objects from the child's world as well as specialised materials that might be seen to embody mathematical ideas.

Learning is also a social activity and both the mathematics and the manner in which it is learned are influenced by the way children interact with each other and with their teachers.

Children not only construct meaning through the experiences they have with materials and problems, but also through examining and reflecting on their own reasoning and the reasoning of others. Talking about ideas that are being generated, ways of attacking tasks, resolving difficulties and describing outcomes that arise are integral to learning mathematics. Rather than working in isolation from other learners, children need to work cooperatively so as to encourage mathematical discussion and resolution of the various interpretations that emerge. The learning of mathematics is now viewed as an active, problem-solving process in which social interactions help promote understanding and reconcile the various interpretations and ways of thinking and acting that can arise.

Numeracy

The nature of what is to be studied has also changed. As we move through ever more technological times there are increasing calls for a more numerate population. It is no longer considered sufficient for children to study mathematics; they need to be able to use their mathematical knowledge in an ever broadening range of activities. As Orrill (2001) remarks, we are living in a society 'awash in numbers' and 'drenched in data'. For those comfortable and competent in thinking about numbers, this provides a basis to evaluate such issues as the benefits and risks of medication, estimates for budgets that will allow or disallow access to education and transport, and many other concerns that were once only available to specialists and those in the know. Conversely, individuals who lack an ability to think numerically will be disadvantaged and at the mercy of other peoples' interpretation and manipulation of numbers. Indeed, 'an innumerate citizen today is as vulnerable as the illiterate peasant of Gutenberg's time' (Steen, 1997). In contrast to the study of further mathematics, numeracy is concerned with applying elementary ideas in sophisticated settings rather than generalising these ideas to more abstract concepts and more complex processes.

> Mathematics thrives as a discipline and as a school subject because it was (and still is) the tool par excellence for comprehending ideas of the scientific age. Numeracy will thrive similarly because it is the natural tool for comprehending information in the computer age. (Steen, 2001, p. 111)

The term *numeracy,* concerned with using, communicating and making sense of mathematics in a range of everyday applications, emerged to provide a more satisfactory description of these extended mathematical processes and ways of thinking. Initially proposed as a 'mirror image' to literacy (Crowther, 1959) related to relatively advanced mathematics, comparison with 'reading and writing' has frequently led to its interpretation as the skills associated with simple number computation. However, the mathematics used in everyday life has always meant more than the manipulation of numbers. Just as the meaning of literacy has broadened to incorporate the integration of reading, writing, listening, speaking and critical thinking, and to recognise the importance of context in the making of meaning, its correlate numeracy (sometimes referred to as mathematical or quantitative literacy) now extends to an ability to explore, conjecture and reason logically and to use a variety of mathematical methods to solve problems (National Council of Teachers of Mathematics, 1989, 2000).

Sense making, such as number sense which was first referred to in the context of numeracy as 'at homeness with numbers', is also fundamental along with the need for 'an appreciation and understanding of information presented in mathematical terms in order to use and apply mathematical skills and to communicate mathematically' (Cockcroft, 1982). Positive attitudes towards *involvement* in mathematics, problem solving and applications as well as a capacity to work systematically and logically and communicate with and about mathematics (*A National Statement on Mathematics for Australian Schools*, 1991) are also central to being numerate. Mathematical communication abilities are essential in understanding and assessing the proposals of others, to convey arguments and justifications to a broader audience, and to analyse and interpret information. Discussing and writing about the mathematics that has been completed, focusing on the thinking processes that were followed, the attempts that were made along the way and the justification that allows particular approaches and solutions can be used to build up an ability to read, write and speak with and about mathematics.

As the certainties of the past have given way to the uncertainties of the present and future, this also means that the formal skills and techniques of arithmetic and geometry that gave exact and unalterable results must make room for ways to examine and explore less certain situations using statistics and probability and include a range of estimation and approximation processes. Thus, numeracy should be seen to include the content of mathematics, particularly number sense, spatial sense and chance and data, together with a focus on problem solving and the uses of mathematics in communication:

Numeracy

content	+ problem solving	+ sense making	+ communication
• number • space • measure • chance and data • algebra	• processes • analyse the problem • explore possible strategies	*number sense* • understanding of number and computation *spatial sense* • visualisation of properties and relationships	• discuss and write about mathematics • reflection • present arguments in mathematical form
New emphases • technology– calculators computers as tools to aid thinking • estimation vs exact • mental as well as pencil and paper	• select and try a solution process • analyse solution and possible answers in problem context	• inclination to use flexibility • personal strategies • ability to interpret	• interpret data presented graphically and statistically

The Australian Association of Mathematics Teachers (1997 p. 15) was among the first to attempt to provide a working definition of numeracy to assist teachers in their planning and teaching:

To be numerate is to use mathematics effectively to meet the general demands of life at home, in paid [sic] work, and for participation in community and civic life.

In school education, numeracy is a fundamental component of learning, performance, discourse and critique across all areas of the curriculum. It involves the disposition to use, in context, a combination of:

- underpinning mathematical concepts and skills from across the discipline (numerical, spatial, graphical, statistical and algebraic)
- mathematical thinking and strategies
- general thinking skills
- grounded appreciation of context

Building on the initial notions of numeracy that first evolved in the United Kingdom (Crowther, 1959; Cockcroft, 1982), the Department for Education and Employment (DfEE, 1998, p. 11) formulated a concise statement of what was required as part of the new national curriculum:

An understanding of the number system, a repertoire of computational skills and an inclination and ability to solve number problems in a variety of contexts. Numeracy also demands practical understanding of the ways in which information is gathered by counting, measuring, and is presented in graphs, diagrams, charts and tables.

Further general statements were proposed for Europe (OECD, 1999) and Australia (DETYA, 2000):

Mathematical literacy is an individual's capacity to identify and understand the role that mathematics plays in the world, to make well-founded mathematical judgements and to engage in mathematics, in ways that meet the needs of the individual's current and future life as a constructive, concerned and reflective citizen (OECD, 1999, p. 41).

Numeracy provides key enabling skills essential for achieving success in schooling. Sound numeracy skills acquired in schooling support effective participation in personal, economic and civic contexts. The increase of globalisation and the use of technology have generated increased demands for a more numerate Australia (DETYA, 2000, p. 1).

The national *Numeracy Benchmarks* which articulate nationally agreed minimum acceptable standards for year levels 3, 5 and 7 were also published in 2000. Although they do not attempt to describe the whole of numeracy learning, nor the full range of what students will be taught, they represent important and essential elements of numeracy at a minimum acceptable level. They further note that numeracy must be seen as only a part of mathematics, and that students should attain high standards of knowledge, skills and understanding through a comprehensive and balanced curriculum (Curriculum Corporation, 2000).

Such calls, of course, are not new. For instance, in the preface to a series of school arithmetic books, *Living Arithmetic*, first published in 1938, William Brownell wrote:

> These books are designed to implement a new purpose in the teaching of arithmetic; namely, to develop an ability to see meaning in number, to understand computational processes, in addition to using them skilfully, and to sense the quantitative significance in everyday social situations. Proponents of this new purpose in arithmetic consider it as important to understand the significance of quantitative statements encountered in reading as to compute with speed and accuracy. They are as much interested in arithmetical thinking as in arithmetical skill.

Constructing mathematical concepts and processes

While the new emphasis on teaching the mathematics that has traditionally been the mainstay of the primary school has been on the development of numeracy, it has also been argued (Ma, 1999) that an understanding of elementary mathematical ideas essentially underpins the development of all mathematics. She argues that the early mathematics of number and space is *fundamental* in that all of the new branches, whether pure or applied, from measurement, through trigonometry to calculus and beyond have developed from basic concepts and processes established with numerical and spatial reasoning. Second, it is *primary*, containing the rudiments of many important concepts needed in more advanced topics. This has profound implications for teaching, as she notes in quoting Chinese teachers' wisdom on the development of early mathematical ideas: 'if students learn a concept thoroughly the first time it is introduced, one will get twice the result from half the effort. Otherwise, one will get half the result with twice the effort' (Ma, 1999, p. 115). Third, it is *elementary* because it is the beginning of students' learning and appears straightforward and easy, yet provides a basis for later generalisations. Rather than merely being a simple collection of disconnected number facts and computational algorithms, numeracy establishes the basis on which future mathematical thinking is constructed.

This perspective, that mathematics is learnt by individuals constructing ideas, processes and understanding for themselves rather than through the transmission of preformed knowledge from teacher to learner, now dominates conceptions of mathematics learning (Malone and Taylor, 1993; Goldin, 2002, p. 204). The view of learning known as *constructivism* considers that:

- knowledge is actively created or invented, not passively received
- new ways of knowing are built through reflection on physical and mental actions

- learning is a social process requiring engagement in dialogue, discussion, argumentation and negotiation of meanings

Evidence for a constructivist epistemology began to surface as errors that children made were analysed and the misconceptions that they had formed were seen to fall into common patterns rather than reflect individual inconsistencies. Writing teen numbers back to front, for example 41 for fourteen, occurs because children generalise the pattern for other two-digit numbers and write the digits in the order in which they hear them. Misinterpreting numbers with zeros, such as seeing 3005 as three hundred and five based on its symbolic writing rather than three thousand and five as place value notions suggest, carrying out computational procedures from left to right in the same way that numbers are read, subtracting the smaller number from the larger or dividing the larger number by the smaller (so that $3 \div 6$ is given as 2) all show how children are able to construct their own ways of interpreting and carrying out mathematical processes.

The pervasiveness of these patterns of thinking from child to child and from situation to situation despite well-structured and well-presented teaching forced an awareness that there was something compelling individual children to build their own view of mathematics. It was not so much that the children were mistaken but that their alternative was incomplete—reasonable from their perspective but inappropriate or inconsistent from the viewpoint of mathematics. Their errors did not arise carelessly, but deliberately, as they set out to obtain answers using thinking they felt would get them there. This realisation that children were able to construct their own set of *explanations* for their mathematics led to a reconception of what was happening as mathematical ideas and processes were learnt. If children were able to invent their own methods and explanations that showed up in patterns of errors, they would also be building their own understanding of appropriate ways of thinking rather than simply taking in a teacher's explanations. Thus students came to be viewed as inevitably performing constructions, 'some flimsy and indistinguishable from rote learning, some powerful and highly generative' but leaving teachers unsure of 'the kinds of constructions students made when [they] demonstrate a technique by showing or produce a solution by telling' (Noddings, 1993, p. 38).

From these observations, the approach to teaching and learning mathematics known as constructivism was proposed. In essence, this acknowledges that mathematical knowledge is a product of an individual's mental acts. It is a way of seeing, of organising experience; a set of mental constructs which are used to view the world (Confrey, 1985). Such a view suggests that the major role of the teacher is to help children to create more powerful constructions and that developing autonomy and self-motivation is vital (Clements, 1997). In this way, the classroom is transformed from a place where teachers attempt to gain the desired performance by children to a place where children are seeking out challenges, attempting to solve them, negotiating paths to those solutions with the teacher and other children, and then reflecting on what has happened.

An essential feature of this view is that existing conceptions, whether gained from everyday experiences or previous learning, guide the understanding and interpretation of any new information or situation that is met. This often results in resistance to adopting new forms of knowledge or an unwillingness to give up or adapt previously successful thinking. Indeed, old ways of thinking are not usually given up without resistance and their replacement by, or extension to, new conceptions is guided by those that already exist. Consequently, the intuitive beliefs and methods of children may appear very different from accepted mathematical practice and may also be very resistant to change (Brousseau, 1997). Teachers will need to ask questions that challenge ill-formed ideas and inappropriate generalisations and pose new problems that will require the revision of old constructions or ways of thinking. The role of a teacher in a constructivist classroom is thus more one of a guide or mentor than a director of what needs to be done in order to become proficient, although there will still be a place for showing how things are to be done. Then the challenge will be to lead children to come to understand and accept this as a method of their own rather than simply practising and learning by rote another person's way of doing something. Evidence from children who have experienced difficulty in learning mathematics has also shown that those who simply acquired teacher-taught techniques by rote were often unable to apply this knowledge and frequently forgot, or at least were unable to recall when needed, knowledge that had been earlier assumed to have been learnt (Booker, 2000). In contrast, those who participate actively in their own learning are more able to use this knowledge and tended to maintain it for future use and adaptation. As the novelist Isabel Allende in *Portrait in Sepia* (2001, p. 151) put it:

> Every time I asked a question, that magnificent teacher, instead of giving the answer, showed me how to find it. She taught me to organise my thoughts, to do research, to read and listen, to seek alternatives, to resolve old problems with new solutions, to argue logically. Above all, she taught me not to believe anything blindly, to doubt, and to question even what seemed irrefutably true.

A constructivist perspective on the teaching and learning of mathematics, then, focuses on the learner, setting out to guide her or him in the construction of mathematical ways of knowing and operating based on existing knowledge. This requires three phases (Herscovics and Bergeron, 1984):

- determine the form of knowledge that may be used as a foundation for building the intended concept or process
- ascertain whether such a basis is present for each learner
- ensure that each step in the proposed construction is accessible to each individual

Under the guidance of a competent teacher, mathematics learning can then be viewed as a process of reconstructing particular ways of thinking rather than reinventing pre-existing mathematical thoughts. When children construct their own mathematics, that knowledge is both personal and owned; something over which they have control so that their learning experiences empower them rather than leave them relying on procedures that have been developed by someone unknown, in response to problems that are no longer remembered from a time and situation no one can recall.

This contrasts with more traditional approaches to teaching using a 'good definition' to transmit a given concept and a suitable standardised procedure to provide a mathematical process. This form of teaching is still paramount in many classrooms—not necessarily by design or even intent, but through the premature introduction of concise definitions, formal symbolism or technical terminology. Pressures to cover a specific program in a given time may prompt a teacher to choose such an approach as it seems more efficient in the short term. However, teachers' familiarity with the mathematics and their own formal ways of thinking often cause them to overlook the obstacles that the child must overcome in building this cognition. If the mathematics to be introduced cannot be related to the child's experiences, it simply will not make sense and the child will be reduced to manipulating meaningless symbols using rules that are not understood.

Many difficulties in learning mathematics can thus be ascribed to discrepancies in the type of activities engaged in or the timing of their introduction. Consequently, an essential part of teaching mathematics is to provide meaningful experiences at appropriate points out of which appreciation and understanding of concepts and ways of thinking can be built. Such activities can be meaningful in the teaching of mathematics only when two interrelated aspects are present. The meaning inherent in the situation, materials, patterns, language or symbols through which the new notions are being expressed must match the mathematical meaning that is intended to be built up. These meanings must also match the children's level of development and ability to take in information, to generalise and construct a reasonable view of the underlying mathematics. In other words, experiences need to be meaningful for both the mathematics and the child.

Materials in use today to represent numbers, such as ice-cream sticks bundled to show tens and ones, not only provide a structure that the children can see, but also match identically the behaviour of the number system itself. Such materials have a long history, from the use of bead frames to show tens and hundreds popularised by Maria Montessori in the early 1900s and the use of blocks to show the ones, tens, hundreds and thousands promoted in arithmetic books of the nineteenth century.

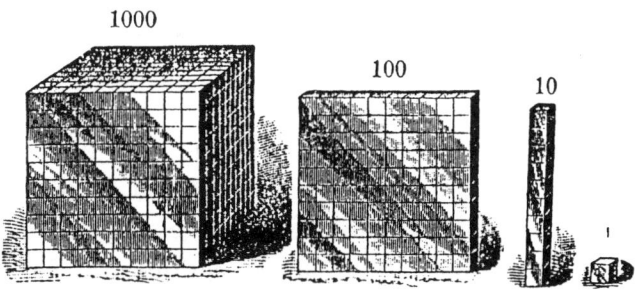

Fish's New Arithmetical Series 1883

Nonetheless, as Cobb *et al.* observed, 'manipulative materials can play a central role if we wish students to learn with understanding, but the way the materials are interpreted and acted upon must necessarily be negotiated by the teacher and students' (Cobb, Yackel and Wood, 1992, p. 7). The materials are not 'transparent' representations of a readily apprehensible mathematics, but instead are vehicles for the potential meanings that children might construct.

Using materials to develop mathematical thinking

The need for materials is fundamental in teaching mathematics because so many of the ideas that have to be learned are not intrinsically obvious. They were generalised and developed from diverse and obscure situations over a long period of time, usually by mature thinkers who had particular social or intellectual needs. If young children are to be assisted to develop the same forms of thinking, then situations in which these ideas can be discerned and discussed are essential. It is for this reason that the teaching of mathematics has had a long tradition of using structured materials, materials through which the underlying mathematical ideas might come to be perceived and appreciated. For instance, ten frames and Bundling sticks are used to establish early numbers, Base 10 materials show the place value patterns for larger numbers, and region models give meaning to fraction ideas. Even comparatively simple notions such as the initial number concepts are abstractions rather than something that can be seen in the immediate environment. The number four does not exist in the real world, but is a representation that the child needs to construct for him- or herself by linking objects with language and symbols within meaningful experiences with materials that sometimes show and sometimes do not show the concept of *fourness*. Similarly, the basic spatial forms can only take on meaning through seeing and making representations of the general notion of what will come to be called a rectangle or a triangle and learning to distinguish one from the other.

At the same time, it must be borne in mind that materials by themselves do not literally carry mathematical meaning. While they might assist in the initial building of understanding, it is reflection on the actions of the materials and the situations that they represent that allows the generalisation to a mathematical way of thinking rather than a rote learning of the results of these actions. If children focus solely on the outcomes, it is possible that they will simply learn at a surface level how to manipulate the materials rather than the deeper, fundamental mathematical ideas. This risk becomes even larger if the materials come to be seen as ends in themselves rather than links to the mathematical concepts and processes they are intended to represent.

For instance, many children are introduced to a number line early in their learning of number concepts and then use it when *counting on* is introduced. When asked to find 5 and 3 more, 5 is identified on the number line, 3 more spacings are counted, and then the number is read off. This is not counting on at all, but is simply counting 1, 2, 3 and then finding the number corresponding to the 3 that has been counted, in this case 8.

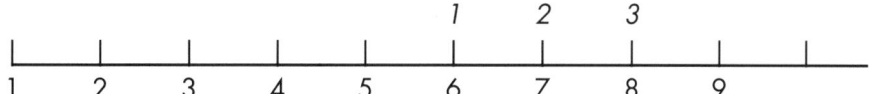

Counting on requires that the counting be a mental operation, counting 6, 7, 8 while keeping track of the count by identifying 6 with 1, 7 with 2 and 8 with 3. It is hardly surprising that later, when the number line is unavailable, such children use their fingers in counting the answers to simple addition combinations; the material used has not enabled a link to be made to the underlying mathematical idea but has instead provided a means of obtaining answers in an unrelated, and less sophisticated, way. Use of the number line in this way with a focus on the counting numbers 1, 2, 3, . . . also leads to children's difficulties with initial measurement techniques. If zero is not integrated into their conception of numbers from the beginning, they assume that the initial mark to be used on a ruler is the 1 and then find that all of their measurements are one unit too short.

Another common example in which children experience difficulties because of the way they see manipulating material as an end in itself is when Bundling sticks or Base 10 materials are used to introduce the addition algorithm. For good reasons, the first examples given usually involve simple combinations such as 43 + 35. A correct answer can be given by simply putting the tens and then the ones together. Yet, later on, it will be necessary to join the ones before the tens in order to cope with additional tens that might result. With the materials, this can be accounted for by rearranging the sticks that result; in the written form, it needs to occur as the algorithm proceeds or else there is a danger that an example like 47 + 39 is answered with 716 rather than 86. Thus, materials by themselves will not be sufficient and their use needs to actively engage children's thinking with guidance from teachers and other users as they build towards long-term outcomes of which the learner may initially have little awareness.

Materials are not something to be used only at the beginning of mathematics teaching and learning. Concepts and processes need to be introduced through realistic problem settings at all levels and materials offer a very feasible way of portraying both the problem and its possible solutions, leading to the generalisations which constitute higher mathematics. At the same time, there is a need to guard against the use of *ad hoc* exemplars, that is, materials that seem to assist the task at hand but that in the longer run prove to be inadequate for further developments. Rather, materials that can be used for many different purposes can assist in building an integrated view of mathematics. There would seem to be advantages in using the same representation in different situations rather than using different representations in the same situation (Clements and McMillen, 1996). The teaching of fraction ideas is perhaps the most compelling topic where this has been overlooked. Models involving regions of a circle show the parts within a whole but do not allow children readily to show for themselves the basic part/whole conception. Indeed, many children choose the length across the diameter of the circle on which to measure off equal parts rather than determine same-sized portions of the area of the circle.

A circle partitioned to show 8 (equal) parts

Another difficulty is that these circular models cannot readily be extended to decimal fractions or per cents yet these are all the same fundamental fraction idea. Similarly, when materials based on regional models are used to portray addition a child may see that

3 parts out of 4 parts + 3 parts out of 4 parts is 6 parts out of 8 parts

It is even more difficult to envisage how these models would be able to show subtraction, let alone multiplication or division.

While some materials would seem to suit certain purposes well, other situations demand different models. Yet, if there are too many different models for a given concept, children will need to be operating at a formal level of cognition to bring the diverse representations together into the one understanding. It is more likely that they will see each material or model as a distinct entity and try to learn many fragmented rules. Thus children who have been introduced to multiplication as *groups of*, *steps* on a number line and in terms of *area* tend to view each separately with its own ways of proceeding. Consequently, they end up with no firm basis of multiplication at all. In contrast, if the use of materials at first focuses on one strong model or material, not only will this give security to the initial conception, but it will provide a basis for future development to all of the possible representations by reference back to a secure, fundamental idea.

It is rather analogous to the manner in which a plant grows; at first a strong central stem is produced, and only when this is secure do other limbs branch off. Where many stems emerge at the same time, competition ensures that not one is strong enough to last on its own and frequently the whole plant simply withers and dies. Thus, wherever possible the material or model that will assist the end-point of development should be the one that introduces the initial idea and be the backbone of the development from beginning to end. If the same materials can also be used for different but related aspects of mathematics, so much the better. This will be of invaluable assistance in allowing a cohesive, connected view of mathematics and its applications to be developed.

A further difficulty with many of the materials proposed for classroom use is that they actually demand knowledgeable users. It is the experience and understanding of the proponent that carries the mathematics rather than the materials themselves. The use of circular regions for common fractions mentioned earlier are an instance of this. In reality it is the angular measure about the centre that determines the number of parts yet children often meet and use these models some time before they come to terms with angles as such. Another is when the values of the Base 10 materials used to show hundreds, tens and ones are suddenly altered to show decimal fractions. It is important that the materials themselves are able to portray the underlying ideas so that constant interjection and explanation by the teacher who is using them is not necessary. Otherwise, the materials may simply be a distracter and the children resort to learning how to manipulate them at a surface level instead of coming to terms with the ideas that are supposed to be portrayed.

Materials are fundamental to learning mathematics in all forms and at almost all levels although their nature and the ways in which they are referred to may change. At some point they may need to be very concrete, able to be picked up, pulled apart or manipulated. The use of such materials can build in from the outset that mathematics is fundamentally an experimental science; conjectures need to be made and evaluated in place of simply acquiring a mass of seemingly immutable facts and rules. Mathematics at all levels is a developing body of knowledge. Activities inherent in the use of materials and models can introduce that way of thinking from the very beginnings of its development in each child. Such materials will include Base 10 blocks to represent numbers and calculations; clocks, money, rulers, balances and other instruments for measurement; solid objects, drawing instruments and street directories for geometry; physical graphs and the use of dice or spinners for chance and data.

At a later time, a pictorial reference may suffice. Actual objects may give way to pictures or diagrams of them to represent numbers and computations; computer programs may simulate measurement, geometry and chance and data; and diagrams formed *in one's head* may replace pencil and paper drawings for problem solving across all areas of mathematics. Indeed, it may be that only the reference to the experiences that were produced with materials will be needed to allow the development of further ideas.

Patterns in mathematics

One further reason for the use of materials to introduce many early ideas in mathematics is that they readily show the patterns that will later need to be applied to larger numbers, fraction ideas or complex ideas in geometry and measurement when materials can no longer model the situations directly. Patterning is fundamental to mathematical thinking—in fact, mathematics has been defined as the study of patterns (Klaebe, 1986) in number, shape and arrangements. Number patterns allow the extension of the place value ideas readily shown with materials to provide a means of recording large numbers. They also allow the development of decimal fractions as a system of recording small numbers and the infinite number of fractions between any two whole numbers. In time, they gave rise to a consistent extension of the number concept itself to allow *negative numbers*, at first sight a contradiction since they clearly do not relate to any known quantity at all. In the tenth century, Arab mathematicians proposed a number line as a model to make sense of the negative amounts that occurred in linear and quadratic equations, suggesting that numbers could be viewed as distances from zero:

A number line gave meaning to the concept of negative numbers

Since quadratic equations involved multiplication, soon the notion of multiplying negative with positive numbers and negative with negative numbers arose. This was a stumbling block for mathematicians and students alike until patterns consistent with earlier multiplication were proposed:

$4 \times 3 = 12$ As the number multiplying $-3 \times 4 = -12$ As the number multiplied
$3 \times 3 = 9$ decreases by 1, the $-3 \times 3 = -9$ decreases by 1, the
$2 \times 3 = 6$ product decreases by 3. $-3 \times 2 = -6$ product increases by 3.
$1 \times 3 = 3$ Thus -1×3 must be -3 $-3 \times 1 = -3$ Thus -3×-1 must be 3
$0 \times 3 = 0$ $-3 \times 0 = 0$
$-1 \times 3 = -3$ $-3 \times -1 = 3$

In this way, it was shown that when a negative and a positive number are multiplied, the result must be negative, while two negative numbers must multiply to give a positive result. A new mathematical way of thinking was built by extending the patterns shown for (positive) whole numbers in a consistent manner and this was to lead to the solution for all algebraic equations and pave the way for calculus and its use in coming to terms with the physical world during the seventeenth century. Later, a similar use of patterns led to the development of *imaginary numbers* involving the square root of -1, $i = \sqrt{-1}$.

Children are usually introduced to patterns in the arrangement of shapes and colours from their earliest days in school. Such thinking is critical to making sense of their world, but care needs to be taken that patterns formed with materials do not supplant the more useful patterns shown in building the numbers to ten using a ten frame, the place value patterns for two-digit, three-digit and four-digit numbers that lead to naming and recording larger numbers meaningfully. There are also number patterns based on sequences such as those shown by the arrangement of blocks in the same way the ancient Greeks used pebbles to form triangular and square numbers:

1   1 + 2 = **3**   1 + 2 + 3 = **6**   1 + 2 + 3 + 4 = **10**   **15, 21, 28, 36 . . .**

Forming numbers in triangle patterns shows how they simply increase by the next counting number each time

Another famous number pattern is that given by Fibonacci, the mathematician Leonardo of Pisa, who first translated the Arab forms of computation into Latin to give the European world the Hindu-Arabic number system. This pattern has been found to underpin the Greek golden ratio by which they designed buildings pleasing to the eye, and even in nature, where the Nautilus shell and many plants seem to involve spirals based on this pattern:

1   1 + 1 = **2**   1 + 2 = **3**   2 + 3 = **5**   3 + 5 = **8**   5 + 8 = **13**   **21, 34, 55 . . .**

After the first two numbers, each new number is the sum of the two numbers that precede it

Patterns also underlie many of the thinking strategies needed to learn basic facts and bring them to the point of automatic response. For instance, in addition the patterns first shown as objects on ten frames build numbers in terms of doubles and one more can be called on to provide answers for facts such as 4 + 5 or 4 and 3, and then extended to use similar thinking for 6 and 7 or 7 and 8. Realising that adding zero or multiplying by one leaves numbers unchanged also reduces the number of individual facts to be learned and assists in aspects of the algorithms for larger numbers which often cause children difficulties. These ways of thinking are readily built up using materials and the patterns extended to all other numbers. Base 10 materials also give rise to other number patterns that assist in multiplication such as numbers resulting from multiplication with nine have a 'digit sum' of 9. Multiplying by a teen number is simply a result of multiplying by ten and adding a basic fact—for example 14 × 8 is 80 + 32 or 112. Indeed, it is the consistent patterns followed in the algorithms for all numbers based on the ideas of place value and renaming that are the key to computational fluency.

Nonetheless, using pattern blocks, colour tiles or attribute blocks to form patterns is essential to building general thinking skills, spatial sense and the arrangements that underpin much of chance (see Chapters 3, 5 and 7). They are also important in leading into the development of algebra (Bay-Williams, 2001), and this strand of the mathematics curriculum across primary, middle and high school is now universally called 'patterning and algebra'. Children in primary schools enjoy studying patterns and determining how they work, whether the patterns are spatial or numerical, repeating or growing, additive or multiplicative.

As they learn to model and describe patterns in their world, children are building the thinking that is 'an important precursor to the more formalised study of algebra in the middle and secondary schools' (*Principles and Standards for School Mathematics*, 2000, p. 159).

Language and mathematics

While the role of materials and the patterns they develop is fundamental, materials by themselves do not literally carry meaning. Their most important use is to provide experiences that can be discussed and reflected on to allow the effects of the embodied actions to emerge as mathematical ways of thinking. It is language that communicates ideas, not only describing concepts but also helping them to take shape in each learner's mind (Usiskin, 1996). Discussion among the participants is needed to bring out 'the explicit construction of links between understood actions on the objects and related symbol processes' (Ma, 1999, p. 20). In building problem-solving abilities, students need to report, display, explain and argue for their own solutions to see that 'getting the right answer is only the beginning rather the end . . . the ability to communicate thinking convincingly is equally important' (Schoenfeld, 2001, p. 53).

Language is the key to all aspects of mathematics learning, from the formation of concepts and processes, through problem solving, to the development of numerate students who will thrive in the technology-dominated world in which they will live. However, as teachers and students attempt to come tō terms with their mathematical understandings, the meanings given and taken may differ not only from one another but also from the intended mathematics (Pirie, 1998). Care is needed that the language used matches the materials and experiences provided, the language levels of the learners and the mathematical concepts and processes being addressed.

Yet, mathematics classrooms were once envisaged as silent places and communication from one child to another about the mathematics they were trying to come to terms with was discouraged, seen as a form of cheating. Now talking about the ideas that arise is encouraged and cooperation and communication are desired behaviours rather than forbidden practices. When mathematical ideas are communicated, particular care is taken in formulating language to keep track of what is happening with materials or representations which will eventually allow formal symbolic recording, mental operations or approximations to be conducted confidently.

In Part Two of the book, this aspect of the development of mathematics is fundamental, with particular attention paid to the use of materials and patterns to engender an appropriate way of governing the more formal symbolic expressions of mathematical processes through the use of a language that at first describes and then directs the thinking that is involved. Care has been taken in establishing uses of materials to lay the foundation for particular processes and in using a form of language at a level appropriate to both the learner and the mathematics. The same remarks about meaning that were applied to the use of materials are also pertinent to the language that can be encouraged to emerge from the concrete experiences. As mathematical ideas are constructed, so too must be the language which enables them to become the individual learner's own. In turn, this leads to the thinking which ensures that control over the processes is possessed by each child. Consistency in this use of language is extremely important. This does not necessarily mean that individual words need to be used in preference to others. Rather, that the meaning that they have is available to the particular group of children and is consistent with the use of materials, the patterns that emerge, and from one situation to the next.

Mathematics has often been called a symbolic language, reflecting the importance given to the concise symbols used to stand for concepts that in time allow these ideas to be manipulated easily. Yet, mathematics is both oral and written and language is needed to communicate the meaning of symbolic statements. For example, while the symbol π provides a concise way of stating the ratio of circumference to diameter of a circle, unless the symbol can be verbalised it is unlikely that any meaning will be attached to it or the various formulae, such as πr^2 for the area of a circle or $2\pi r$ for circumference, in which it is used. Indeed, it is through using lengthy informal statements that sufficient understanding gives rise to a need for more succinct specialised words as well as symbols to express mathematical ideas.

This specialised vocabulary of mathematics can be as concise and dense as the mathematical symbols with which we most readily associate the writing of ideas in mathematics and should be left until the underlying meaning is quite clear. There needs to be a careful building up of experiences and descriptions on which new mathematical words are based and care needs to be taken in introducing these words by reference to the underlying meaning. Since many of these words have entered mathematics from Greek, Arabic or Latin sources, investigating the original meanings of the words that are used can also be helpful to children. For instance, many children have difficulties when asked to determine the perimeter of a shape, but can readily measure the distance all the way around the outside of the shape. There is often a distinction between knowing the meaning of a concept and being able to recall and use the corresponding mathematical vocabulary. In fact, *perimeter* is made up of Greek words that literally mean to measure around the outside. It is helpful to children to realise that we use the Latin expression with the same meaning, *circumference*, when we measure the distance around the outside of a circle.

Other words in measurement also have origins that can help children move from an informal description to adopt the more formal mathematical names. Some shapes are named in terms of the angles they contain, using Greek number words for prefixes together with part of the word *gonia*, which originally meant corner and expanded to indicate angle, as in pentagon (5 angles), hexagon (6 angles) and so on. Others are named using Latin prefixes, as in triangle (3 angles), rectangle (from words that mean right angle, hence 4 right angles), while others are named from Latin words that refer to sides as in quadrilateral or equilateral. Interestingly, a triangle in ancient Greek was trigon, which gave its name to the branch of mathematics called trigonometry, while a four-sided shape was called a tetragon before it was called a quadrilateral, thus giving rise to the quadratic equation which was concerned with 'squared' amounts.

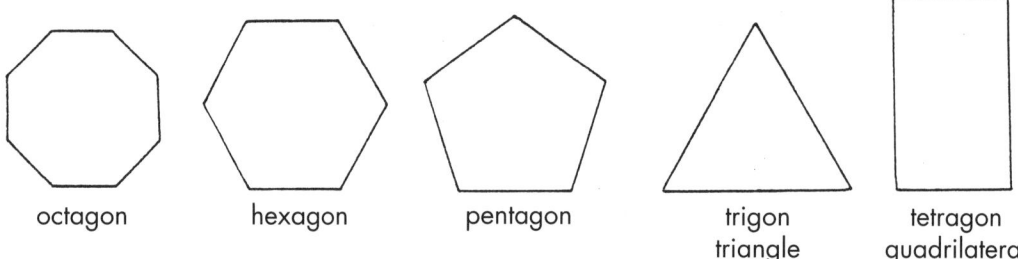

Having some knowledge of the counting words in Latin and Greek can help with many of the seemingly more complex words that are used in mathematics to name shapes, large numbers and even properties such as bisect or trisect. Knowing that *dia* means across can help with the meaning of diameter (a measure across a circle) and diagonal (a line from one vertex to another), while understanding that *isos* means same can help fix the meaning of an isosceles triangle. Treating the words of mathematics as parts of a specialised language can lead to building up the meanings for children in the same way that word meanings are built up in a first language, rather than assuming that they can be readily tagged to quite complex mathematical ideas. A very good source for this information is a book published by the Mathematical Association of America on the etymology of mathematical terms (Schwartzman, 1994).

Another way in which the notion of language enters mathematics is through the use of words that have subtly different meanings from those exhibited in everyday use. For instance, some words have their meaning expanded; in everyday use the word *more* means to increase but in many mathematics situations it means to find the difference as in 'I have 3 oranges, you have 5 oranges. How many more oranges do you have than me?'. Similarly, in everyday use a rectangle refers to a shape in which one side is longer than the other while a square has all sides of equal length and a curve is a line in the form of a bend, with no straight part. In mathematics, the concept of rectangle includes squares since each contain four right angles and the notion of a curve includes straight lines. Other words have their meaning contracted: the word *product* is used to refer to the consequence of many actions in everyday life, but in mathematics it is the result we get when multiplying; distance in everyday experience takes into account the route followed while in mathematics it means the shortest distance. Other terms have a meaning that has no bearing at all on the understanding that may be used in other situations. In mathematics, *average* means that a lot of complicated adding is required before a difficult division is undertaken to provide an answer, whereas in everyday life average indicates nothing at all out of the ordinary. Perhaps these specialised uses of words would also not be so difficult if their origins were considered. The word average originates in the Arabic word *awariyah* which translates as 'goods damaged in shipping' because of the process of sharing losses proportionally when goods were lost, stolen or damaged while being shipped from one port to another (Rubenstein, 1996).

Even supposedly helpful phrases used to define mathematical meanings can be difficult for children. There are rarely any repeated words to ensure that the meaning is clear; often learning is confused simply with a facility to recall a succinct definition; and the form of expression used is more complex than that used in everyday speech or other learning situations. For instance, the use of the word *not* is crucial in mathematics, defining precisely what can be allowed. Elsewhere such a use is less common and tends to convey other than positive aspects of meaning. As an example of the complex nature of written statements in mathematics, consider the understanding and ability to analyse English expressions needed to really know what a prime number is: 'a prime number is only divisible by one and itself'. Or how to understand the meaning of the word exponent—'the exponent tells how many times the number is a factor'—which immediately requires some mental image of the number as well as the meaning of the word factor. It is this need to recall specialised meanings of words inherent in one definition before it makes sense in itself that makes the language of mathematics quite awesome to many children. On top of this, they have to read complex descriptions before they can even begin to solve problems that allow them to apply the knowledge they have so carefully built up or allow them to generalise and build further processes.

Symbols in mathematics

Language is crucial to the learning of mathematics because it is through discussion that learners can come to terms with mathematical ideas, develop ways of expressing concepts and processes and take on the ways of thinking as their own. In due course, this verbal description and explanation of what is happening needs to give way to the symbolic expressions that are often seen to characterise mathematics. Although symbols are in many ways the most distinctive aspect of mathematics, the notion of a symbol needs to be analysed before its place in mathematics can be determined. There is frequently confusion about whether a sign or a symbol is being used; often + is referred to as the plus sign, whereas mathematicians write about the addition symbol. The issue of whether the operation of addition should be addressed using the word 'plus' is one that will be taken up later, but it is necessary at first to consider whether sign or symbol is the appropriate label. In life outside of the classroom, the distinction is clear:

Could all patrons please ensure that their mobile phones are switched off during the performance

No mobile phone sign   No mobile phone symbol

A sign is something that can be read; its meaning is inherent in the words that make up its message. On the other hand, a symbol is a representation of an idea. The idea has to be internalised before the symbol can be recognised. In mathematics, it is the same. The symbol system represents concepts and processes that have been built up through an appropriate range of experiences and discussions about what is occurring. There is no way that the meaning even of elementary symbols such as >, +, ÷ or % can be inferred directly, and it is not possible to read them. Consequently, mathematics involves symbols not signs. More importantly, this implies that the mathematical meaning must be acquired before any form of symbolism is introduced. This is as true of the first number symbols as it is for the representations of more complex operations at higher levels of mathematics. After all, if the use of the zero symbol took many centuries before it was universally accepted as a means of recording nothing, why should a young child be expected to read the meaning of zero as indicating that there are none of something almost immediately?

The essence of mathematics is in the experiences with materials and patterns or with ideas and the way in which they are to be described. If young children are exposed to sufficient activities before being asked to put pen to paper, many of the difficulties still observed today when mathematics is treated solely as a symbol system with attached rules will simply not emerge. They will know what it is they are trying to represent and appreciate the need and power of a concise form of expression. As has been said, 'mathematicians are not symbol-minded!' reported at the 4th International Congress on Mathematics Education, Berkeley, 1980. It is simplistic indeed to think of mathematics as existing in the symbols through which it is represented and no less so to believe that teaching mathematics as a set of symbols simply to be manipulated has much to do with mathematics at all. Rather, mathematical thinking is based not on the symbols that are used, but on the meaning on which they are based and come to represent so that, as Davis put it, each symbol comes to be regarded 'as a window through which one looks at a particular reality' (1993, p. 54).

Yet it is the precision and conciseness of expression that is simultaneously the power and the weakness of mathematical symbolism. Power in that it divorces mathematics from its referents so enabling it to be applied to many, different and new situations. Weakness in that, as it renders mathematics context free, what results is seemingly applicable to nothing. Only if children are led to a full understanding of the underlying mathematical ideas, then to see the use and potency of the symbolic forms, will mathematics be reconstructed as a personal and powerful way of thinking. As Polya said in *How to Solve It*:

> Intelligent students are justified in their aversion to mathematics if not given ample opportunity to convince themselves by their own experience that the language of mathematical symbols assists the mind. To help them to such an experience is one of the most important tasks of the teacher (1990, p. 141).

Mathematics then is fundamentally a way of thinking, to be subsumed gradually and to be essentially constructed anew by each learner. It exists in the blend of meaningful experiences that incorporate materials, patterns, language and symbols. In general, a suitable building up of this view of mathematics proceeds from realistic problem situations through their representation in concrete form to a use of language that describes and organises what is occurring before being recorded in some powerful, concise, symbolic form when appropriate. In turn, this will allow the mathematics constructed by each learner to be used in applications to problems in their own world and to create further and perhaps deeper ways of conceptualising mathematical understanding.

The social context of mathematical development

Children develop mathematical concepts as they engage in this cycle of mathematical activity, with the social interactions in the classroom as crucial to learning as the interactions with ideas bound up in materials or patterns and their uses. Cooperative learning is one framework for organising the social situation to foster this development. At one level, this means an accent on group activities and the organisation of tasks so that children work to accomplish shared goals. Thus, instead of having children perform individually on problems, worksheets, even pages from a text, they can work together in pairs or groups of 3 or 4 on shared tasks, taking turns to record any working or observations in order to discuss them with the larger group later. Of course, it is likely that at first one child may do more of the thinking and activity than the others. Using the task to generate whole class discussion about the mathematics of the situations on which they are working can then focus the need of all children to be able to talk about the activity. Indeed, judicious questioning of the one who watched more than participated will quickly draw that child's attention to the need to attend to all aspects of the task at hand and to voice uncertainties as they occur rather than leave it to the more capable or dominant child. It will also allow a variety of ways of thinking about a particular situation to arise and add to the richness of the learning for all participants.

In this way, learning cooperatively can go beyond merely working together on set tasks and an atmosphere can be created in which children construct their own mathematical conceptions. The goals of learning, the discussions about the means to those goals and the individual paths taken need to be at the centre of classroom learning. An attitude that each individual will reach his or her learning goal only if the others in the group also reach their goals is as important as the goals themselves. This is in distinct contrast to a classroom where learning is individualistic or competitive with children working by themselves at their own pace to achieve goals unrelated to those of their classmates.

Cooperative learning is also crucial in promoting children's ability to communicate and reason mathematically. The interactions between teachers and children, and especially among children, influence what is learned and how it is learned. In particular, attempts to communicate their thinking help to develop their understanding that mathematics is conjectural in nature—that mathematical activity is concerned with reasoning about possibilities rather than learning the results presented by others. Indeed, trying to make sense of methods and explanations they see or hear from others is fundamental to constructing mathematical meanings (Yachel et al, 1990). Within a framework of learning cooperatively, each child can be viewed both as an active reorganiser of his or her personal mathematical experiences and as a member of a community or group in which he or she actively participates in the continual regeneration of 'taken-as-shared' ways of doing mathematics (Cobb and Bauersfeld, 1995). Institutionalised practices such as using tens and ones in a place-value sense or following a particular method for measuring the area of a circle can then emerge anew for each child, yet conform to accepted norms of mathematical behaviour.

Expectations concerning these interactions are just as important. For a child's goal in the classroom is not necessarily to learn *per se*, but to complete tasks in ways that are acceptable within the particular environment. Their actions in the classroom are fundamentally influenced by their beliefs about their own role and the teacher's role (Brousseau, 1997). Consequently, the role of a teacher must change from being an organiser and imparter of knowledge to one whose task is to facilitate learning through assisting children to build up concepts and processes for themselves. Not only by providing experiences out of which mathematics can grow, or even by focusing discussion on the ideas that are emerging, but by establishing and maintaining a classroom environment that will maximise productive social interactions. Children need to be led to value persistence in working at a challenging task in contrast to mere repetition of similar exercises; to engage in meaningful activity in preference to procedures learned by rote; and to see that cooperation and negotiation are productive at both a personal and social level.

Instructional games and mathematics learning

Outside the classroom, the effort applied in mastery of play activities focused attention on the potential of games for involving children in mathematics. First, because children 'construct much of their reality through playing' (Steffe and Wiegel, 1994, p. 117) and their games almost always involve sustained attention, high-level thinking and collective as well as individual effort. Second, the rule-governed behaviour in these games is suggestive of the actions envisaged in the teaching and learning of mathematics. The match between the expectations for involvement in mathematical learning and the behaviours freely committed to game playing led to the development of instructional games which could be included in mathematics programs. Initially this concentrated on practice aspects such as that required for basic fact learning, to provide motivation or to reward children for progress they have made. However, observations of the games in use led to their extension to a wider range of concepts and processes (Larouche, Bergeron and Herscovics, 1984) and use can also be made of the interest generated by games to assist children to generalise to the more abstract recorded forms and higher level mathematical ideas. Vygotsky (1978) has argued that 'the influence of play on a child's development is enormous' in that action and meaning can become separated and abstract thinking can thereby begin.

As learners participate in the playing of instructional games, the manipulation of materials and verbalisation of actions, thoughts and interpretations assist in the construction of mathematical concepts. An element of chance ensures that each player has an opportunity to win and build self-esteem. Games themselves are seen as fun, not only providing motivation but also ensuring the full engagement on which constructive learning depends (Blum and Yocom, 1996). Often this means that while children may not engage in learning to please a parent or teacher, and can rarely accept that mathematics will be useful in later life, they will willingly learn in order to participate with their peers in socially rewarding activities. There are many computer games that provide this type of learning situation and motivation (Booker, 2002).

Games should not be regarded merely as a useful activity when teaching and learning have been accomplished, but should be seen as an integral part of a balanced program. They are most effective when structured around mathematical ideas and when playing is dependent on mathematical understanding. In this way, they provide a context that is real to children as they become fully engaged in something in which the outcome matters, leading to a realisation of the value of the underlying mathematical processes. These include more general notions such as predicting, testing, conjecturing, generalising, justifying and checking (Ainley, 1990), which are often integral to the purpose of winning a game. This same purpose has children willingly trying out new ideas which they must justify in a meaningful way to those against whom they are playing (Ernest, 1986). There is also a strong incentive to check or challenge the mathematics of other players during games when there is a chance of winning or being defeated, providing a meaningful context for discussion and a need for clear communication.

When they are integral to teaching, games also more readily allow a teacher to make judgements about student understanding. Ways of thinking become much more apparent during playing, not only as ideas are made public, but also through the actions which reveal underlying thinking. Within a game, children are not endeavouring to provide an answer or reason which they think will match a teacher's expectations, but focus on those that make sense to them. Close observation of small groups provides much greater insight into students' ways of knowing and allows the development of more appropriate and effective mathematics programs. Feedback from peers is provided immediately compared to that with teacher-directed instruction and worksheets where the responses to a misconception are often so delayed that the thinking that resulted in a particular answer can no longer be remembered. In these ways, games also have an important contribution to make to the shift in assessment from judgement of student output to portrayal of capabilities.

Because of these many contributions to the development of knowledge and positive influences on the effective components of learning situations, instructional games have always featured in mathematics teaching and learning in the early years. In particular, they have been used to develop and construct concepts, to build, maintain and consolidate skills such as basic fact mastery, and to improve problem-solving abilities and strategic thinking. However, much of this use has been to provide additional practice on mathematical ideas or processes introduced in prior teaching. Games should also be used to put forward new notions, to lay the foundations for concepts and processes that would be formalised later, and to allow discussion on possible generalisations from earlier knowledge prior to any formal introduction of new concepts or processes.

For these social and cognitive reasons, games have an important place in mathematics education (Booker, 2000). They contribute to teaching and learning by providing a background in which mathematical concepts can be developed and constructed. Problem-solving ability is improved when the discovery and use of strategies is required and previously acquired skills are maintained through motivating practice. Reference will be made to each of these aspects of game use in Part Two of the book, where a variety of instructional games will be introduced. At the same time, the social interactions conducive to learning have also been borne in mind so that methods of play requiring cooperation have been deliberately invoked. In this way, children learn that without cooperation a game may not proceed and there will certainly not be any chance of winning. Listening to other players, talking about what is happening and even assisting others to understand and complete the tasks involved in the game need to be seen to be critical playing behaviours. Children then learn from one another as much as from the structured activities through sharing of the method of play, consequences and needs of the game. In turn, the act of winning, or even almost winning, produces a desire for further play and thus further experience with the underlying concept or skill.

Above all, involvement in instructional games induces children to make sense of their ideas and the interpretations of others. The dialogue engaged in while playing facilitates the construction of mathematical knowledge, allowing the articulation and manipulation of each player's thinking. Such communication helps to extend a conceptual framework through a process of reflection and points to the central role of language, as the social interaction gives rise to genuine mathematical issues. In turn, these problems engender an exchange of ideas with children striving to make sense of their mathematical activity and lead them to see mathematics as a social process of sense making requiring the construction of consensual mathematical understandings.

Gender and mathematics

All students, regardless of their personal characteristics and backgrounds, must have opportunities to study—and support to learn—mathematics. Equity does not mean that every student should receive identical instruction; instead, it demands that reasonable and appropriate accommodations be made as needed to promote access and attainment for all students . . . Low expectations are especially problematic because students who live in poverty, students who are not native speakers of English, females [and other disadvantaged groups] have traditionally been far more likely to be the victims of low expectations. High expectations for mathematics learning must be communicated in words and deeds to all students (*Principles and Standards for School Mathematics*, 2000, pp. 12–13).

Attitudes to mathematics are influenced not only by approaches to teaching or by the materials or activities designed to bring about learning. The social context outside as well as inside the classroom has as much or even more bearing. This issue has came to light with a growing awareness of the need to encourage girls in particular to participate more in mathematics and science learning. But what is true for girls in general may often be as true for individual children or particular minority groups regardless of gender. Nonetheless, the issue of mathematics and gender is one that all teachers need to address. Willis (1989, p. 3) summed up the extent of awareness of this 'problem of girls and mathematics':

> It is perhaps only a slight oversimplification to suggest that the commonsense conception of girls and mathematics thirty years ago was that there was no problem, it was 'only natural' that girls could not, did not and would not want to do mathematics. Twenty years ago, however, that girls could not do as well as boys in mathematics had come to be regarded as a problem worthy of attention of psychologists and educators. Ten years later, fuelled in part by the increasing popularity of 'social' explanations of educational achievement, but also by the re-emergence of feminism amidst social, political and economic change, the problem had also changed. We no longer asked why girls could not but rather why they did not do as well as boys at mathematics. Now, the problem has become why girls will not do as well as boys, why they choose to participate less . . . the solution is seen to be to require that more girls be encouraged to undertake more mathematics.

Her analysis of the reasons for the differences that have been observed showed some evidence that girls are less likely to experience play activities outside of school conducive to mathematical development. However, patterns of student–teacher interaction within schools differing along gender lines were more plausible causes. In particular, teachers tended to interact more frequently with boys than with girls in mathematics, especially with high-achieving and highly confident boys. While boys initiated many more interactions than girls, boys also received more disciplinary interactions encouraging them to be task oriented and more process questions demanding higher-order thinking and analysis. Girls' requests for assistance tended to be more often ignored while their 'called out' answers were less likely to be accepted than those of the boys. Overt encouragement of girls' participation and persistence on tasks and acknowledgement of their successes were suggested to ensure a more equal opportunity for girls in mathematics, in particular fostering the 'risk taking' on which much problem solving and generalisation of higher-order mathematics skills depends.

Of more importance was the observation that girls tended to attribute their failures in mathematics to lack of ability and their successes to external features, such as luck or an easy test. In contrast, boys attributed success to internal aspects such as natural ability or working hard, while their failures they put down to confusing questions or lack of teaching of certain material. Consequently, girls were more likely to doubt their capacity to avoid failure or achieve success. Added to this, they were less likely to exhibit confidence in their continued achievement in mathematics and were less convinced that mathematics would be useful in their future lives. The notion of 'fear of success' where there is a feeling that success in mathematics is not a desirable facet of femaleness is a further social factor that enters the development of mathematical ability by the time girls are reaching the end of primary school. External issues such as the wishes of parents, the advice of school councillors and the like were seen to be less encouraging of girls staying on at mathematics, leading to lower participation and restricted career and life opportunities.

During the 1990s, much effort was applied to reverse this situation and achieve gender equity in education, with mathematics and science a priority. The extent to which mathematics is perceived to be a 'male domain' seems to be waning and, particularly among younger students, girls' mathematical potential is no longer regarded as inferior to boys (Forgasz and Leder, 2001). Consequently, there is now a growing acceptance that there is no longer a 'problem' for girls and mathematics and the focus has switched to a perceived problem for boys. However, more males than females continue to study the most demanding mathematics courses offered and males still dominate in mathematically related careers. Burton (2001) calls this 'the fable of the underachieving boy' and draws attention to the continuing need to encourage girls to participate in mathematics at all levels, and to engage with the science and technology courses that lead to more high status and influential careers.

Some of these continuing difficulties arise in the way the curriculum is framed in general and in the particular learning activities that are chosen. One way in which a teacher can minimise these disadvantages, not only for girls but also for other underachieving groups, is to focus on the extent and nature of the early experiences that each individual needs and receives. Praising results along lines that attribute each person's success to their own ability or hard work and suggest that failure will be overcome by further work or alternative activities will also assist. Indeed, if all children were to believe that success is largely a consequence of application to appropriate learning activities and failure to insufficient work, they would be in a position to see how to achieve even more no matter what their level on assessment surveys. Talking about those who work hard at mathematics as opposed to those who do not work hard enough would be much more profitable than the current acceptance that some people, the 'smart' ones, are good at mathematics while most, the 'dumb' ones, are not. If someone believes that mathematics ability is something one is born with, there really is little they can do to change the situation. Mathematics is something that should be available to all; success lies in the approaches taken to teaching and learning and is not simply something for which there is a predisposition within the individual learner.

Nonetheless, community reactions to girls' improved performance need to be borne in mind and efforts made to show the advantages to each individual as well as society as a whole in improving the mathematical capabilities of all students regardless of gender. As Coupland and Wood (1998, p. 243) noted in a review of newspaper coverage of the changing success rates for male and female students:

> People thought that there was something to be concerned about now that girls were doing better than boys, while for years the status quo of boys doing better was simply accepted as the natural order of things . . . Very few reports actually said anything congratulatory about the females' performance!

Mathematics and technology

One other outside influence is particularly important to the development of mathematics. Information and communication technology (ICT) is changing the nature of what is needed as well as the ways in which things are to be done. Calculators can be used to develop concepts as well as carry out readily the many calculations that adults use them for (Battista, 2002; Huinker, 2002; Swan and Sparrow, 1998). Software is available for computers that promotes mathematical thinking as children discover patterns and solve problems (Kerrigan, 2002). The Internet not only provides information for children and their teachers to use in developing problem solving and applications, it also makes available teaching and learning materials for most mathematical topics (Rich and Jotner, 2002). Attitudes to mathematics are changing with these technological advances. Why should so much time and effort be given to learning mathematical processes when machines can do almost everything more quickly? The first answer is that this is not true: a calculator or computer is only as good as the person who operates it—or, more accurately, the understanding of the person who uses the device and interprets its outputs. It is more reasonable to expect that more mathematics is necessary rather than less, focused on understanding of concepts and operations, the uses of these in solving problems, and seeing the processes and outcomes in terms of sense making. Since publishing on the Internet is easy, material that is retrieved has to be carefully checked to see that the results it purports to show are indeed accurate and reliable.

Nonetheless, some aspects of computational mathematics will diminish in importance. Some parents, and even some teachers, will decry the use of calculators in schools on the grounds that they will reduce a child's ability to think or to carry out mathematics quickly and accurately. Yet, as Wheatley points out, throughout history humans have invented and used tools that have been instrumental in cultural advance and 'a calculator is simply another tool that has been created to facilitate our activities' (1995, p. 116). Calculators allow us to do more things, more quickly and differently in response to the needs of a changing world. While teachers, parents and employers acknowledge the constant change in the world in which we live, they also seem to expect school to remain unchanging, wanting all of the methods and topics that they experienced to remain in the curriculum. One of the roles a teacher must now take on board is to show how education too is responding to the changing world—that it is using new ways to achieve the real goals always expected of education: students who leave school with ways of knowing that will be adaptable to new problems, receptive of new ideas, and who can take advantage of the new technologies that will be available to them. Calculators and computers are part of this world and teachers are using them to promote deeper and more effective mathematical thinking rather than mere button pushers who can only cope with routine problems from the past.

Because a calculator or computer can do things at a greater speed, it will also reveal someone's misunderstandings more quickly. It is much easier to make more mistakes, more frequently on a calculator when you do not know what you are doing than it is with pencil and paper computations. On the other hand, use of a calculator can allow a teacher to introduce more realistic problems, sooner and more frequently than when only traditional computational skills are available. Consequently, the use of calculators is universally advocated both to teach and use mathematical ideas from the beginning of schooling. 'It should be taken for granted that a calculator is available whenever it can be used, from years 1–12' (*A National Statement on Mathematics for Australian Schools*, 1991, p. 109), accompanied by 'explicit teaching of the conceptual and technical skills which underlie the correct use of a calculator'. Since the world in which they will live is governed by the use of calculating and computing devices, children need to grow up with this technology as much a part of their classroom experiences as it is part of their world outside of school, seeing their use in developing ways of thinking and solving problems as a natural part of learning and using mathematics.

However, the manner in which particular operations can be carried out most efficiently on a calculator is not obvious. Steps that with other methods need to be carried out one at a time and one after the other can be combined. Answers can be seen intuitively rather than only at the end of a large number of separate steps. Methods can be quickly explored with the data of the problem rather than analysed through logical or algebraic expressions and answers can be quickly put back into the problem to see if they make sense.

Consider this problem:

When Tosca got on the bus, there were already some people on it. At the next stop, 8 people got on and 4 got off. At the stop after that, 9 people got on, and at the terminus all 21 people on the bus got off. How many were on the bus when it stopped to pick up Tosca?

In the past, this type of problem would have been answered by a guess and check method or using algebra. By focusing on the meaning of the situation, the calculator can be used to explore other ways of solving it, in a sense more intuitive, but also bringing into play fundamental understandings of the underlying mathematics without being encumbered by the mechanics of the computation. Working backwards using a calculator is simply a matter of reversing each change: $21 - 9 + 4 - 8 = 8$. Then there is Tosca! So the number on the bus prior to her getting on was 1 less than this, or 7. Or should the bus driver also be considered?

But calculators are no more a simple replacement for other ways of doing things than they are obvious to use. Most adults who use calculators at work, home or for their leisure activities were self-taught and consequently feel less than adequately competent with their machines, unable to use certain functions or do certain computations efficiently. Children need to learn effective ways of using this technology from their earliest days in school, seeing it as simply one way among many to determine answers, another way of expressing results, and, above all, as a tool to which they can apply their understanding and reasoning. In order to use a calculator in these ways, they will need a number of mathematical understandings and skills. Recognition of the various symbols and the actions they represent ($+$, $-$, $\times$, $\div$, $=$, $\sqrt{}$, %) is essential, as is knowing the concepts for these operations and the variety of meanings and situations in which they might be used. Familiarity with the operations in other contexts will assist in knowing when to use them, give some awareness of likely outcomes and provide a measure of the reasonableness of results. There will also be a need for a plan to 'see' what is happening when the calculator does not show it and for recording intermediate results that might be helpful later. Thus, it would be helpful to learn to use the non-writing hand to operate the calculator, to the point of using several fingers akin to touch-typing, so that the writing hand will be available to write any outcome or important point along the way. However, it should be noted that today's children and young adults are much more adept at holding a calculator in one hand, using only a thumb to key in numbers and operations in the same way that they have used electronic games and mobile phones! In time, an awareness of how the various memory functions operate, including the constant keys, will allow more efficient computation.

These points will be addressed in Part Two of the book when number concepts are built up, suggesting how calculator use can be an integral part of the learning of numeration and computation from the outset. It is important for children and their teachers to realise that a calculator needs to be seen as just one way of finding solutions among many. At times it will be easier to find an answer using mental calculations, at other times an approximate result would suffice while an accurate solution may just as readily be found using pencil and paper as with a calculator. The most important end result is that each individual child should be in a position to choose the form of calculation best suited to a particular situation or task. The activities developed throughout this book refer to suitable calculator activities but the difficulty that always arises relates to the different key formats, operating procedures and capabilities. Perhaps it is wisest to assume that familiarity with calculators will be built up individually by each reader, and then explored jointly with the children for whose mathematics education they will take responsibility. As Dick (1988, p. 41) noted:

> Although calculators are not the villains portrayed by some, it would be just as erroneous to view them as a panacea for mathematics instruction. Regardless of whether calculators or paper and pencil are used, students learn by actively thinking about and reflecting on what they're doing . . . the use of calculators *per se* does not preclude our students from thinking,

but they must be presented with activities that require them to think! Often the calculator can supply more and varied opportunities for learning.

These statements are of course even more true for computers although one can question to what extent a computer is a mathematics-related tool at all, especially at the primary level. Their use in recording data, expressing and displaying results is unquestionable. Programs such as *Shape Makers*, the *Geometer's Sketchpad* or *Cabri Geometry* allow children to investigate spatial concepts in a dynamic way using problem-solving and problem-posing situations, rather than learning the results of static, hand-drawn ideal representations of results that a teacher knows and simply wishes children to memorise. Chance and data packages such as *Data Explorer*, *Graphers* or the *Graph Club* allow real data to be analysed and manipulated. Programs that foster mathematical thinking range from those which build on early number learning such as *Win with Maths!*, *I Love Maths* or *Numeracy Connections* to those which require sophisticated problem solving such as *Zoombinis*, *Building Perspectives Deluxe* or *Factory Deluxe*.

On the other hand, much of high school and tertiary mathematics, such as sketching and interpreting all kinds of graphs, manipulating algebraic expressions, using calculus or statistical analysis, can now be done on a relatively inexpensive, enhanced calculator. When they are coupled with data probes to measure and record a range of physical activities, the analysis of many real situations can come within the reach of young children. Many other new technologies such as handheld computers, for example the *Palm* range, are also coming into general use and will soon be within school budgets.

Perhaps the computer is more useful across the curriculum than when it is reserved for any one area such as mathematics. Teachers and children can use programs such as *Publisher* to prepare and present reports, make posters and develop classroom games and activities. Spreadsheets can be used to present data, investigate changes to the situations that the data represent, and readily draw fraction shapes or tables of values. The Internet has many sites where teaching materials can be accessed, clip art to illustrate projects and problems, and ideas for units of work. Typing the area of mathematics for which help is needed into a search engine such as Google quickly locates helpful materials. For example, simply typing in 'box plots' gave rise to several tools for constructing them, some discussion on their advantage in analysing agricultural data, and ways of drawing them on graphical calculators. Nonetheless, it is ironic that primary classrooms have adopted computer technology for a wide range of activities across many subject areas, while the calculator has been decried as something that might restrict children's mathematical development rather than enhance it. Perhaps now that calculators with graphic and other advanced features are becoming so powerful that they are increasingly referred to as handheld computers, there will be a greater focus on their use as a tool to aid the thinking and development of mathematically literate students. Technology will then be seen as a natural part of the domain of learning and using mathematics, as essential as the pencil and paper in earlier, less technologically sophisticated times.

Mathematics and education

Although mathematics education is a comparatively recent discipline, only beginning to develop from the end of the nineteenth century (Kilpatrick, 1992), and really taking form in the 1970s (Moon, 1986), mathematics itself has been an essential aspect of education for thousands of years. From the outset, the teaching and learning of mathematics has involved a balance between learning processes developed by others and coming to terms with the patterns of thinking on which they depend. But it was the nature of the mathematical ideas that were seen to be discovered or invented that gave rise to the form that this education was to take. While attention might have been given to the representations used to exhibit the ideas or the techniques that were to be used in the applications, the principal focus was on handing on self-evident truths to the next generation.

Today there are many issues to consider when teaching mathematics. How to encourage children to construct their own mathematics is clearly one of them. But then how do we account for the time-honoured processes that are in common use? Should all children recreate their mathematics anew? What of the mathematical achievements of the past? There is surely a need to include a historical and cultural perspective on the development of mathematics, and problems from the past can provide a setting in which to allow children to develop their ability to think mathematically.

> From a certain field I harvest 4 wagons of grain per unit area. From a second field, I harvest 3 units of grain per unit area. The yield of the first field was 50 wagons more than that of the second. The areas of both fields is known to be 30 units. How large is each field? (Babylonia, 1500 BC—cited in Swetz, 1996, p. 203).

A second issue from the history of mathematical ideas and forms of expression is that obstacles encountered in their development often recur as misconceptions for individual learners. Notions of zero as a number, negative numbers, the use of place value for decimal fractions, the concept of an angle, the laws of chance and the use of coordinates in geometry have all been difficult ideas for mathematics and for the many students who subsequently had to learn them. Such epistemological obstacles are generally strong and durable, continuing to appear even when there is a belief that they have been eradicated. Teaching cannot escape from them entirely; they form compulsory passages on the road to knowledge (Artigue, 1992).

A very readable book on the origins of much of the mathematics that is taught in the primary school and early secondary years is *The Crest of the Peacock* (Joseph, 1991), which not only gives many interesting examples but shows that mathematics is not a European invention as so many people now believe. Mathematics has its origins in the needs of many cultures, from Africa to the Americas to the Middle East, and many of our concepts and procedures come from Babylonian, Egyptian, Chinese and Indian mathematicians. It is notable that not only do our pencil and paper techniques for both whole number and common fraction computations originate with Hindu mathematicians, but that as early as the fifth century they saw the implications of this for teaching: that understanding of the number system together with a few powerful processes should allow all people access to the power of mathematics.

> If the sun performs one revolution in a year, how much does he accomplish in a given number of days? Does not even an ignorant person calculate the sun in such problems simply by scribbling with a piece of chalk? (Varahamihira, 505, in Datta and Singh, 1962).

Unfortunately, many centuries of teaching children the thinking behind such 'scribbling' has shown us that it is not always easy to achieve. It is the hope of the authors of this book that the ways of developing mathematical understanding promoted here will allow many more children to develop mathematical thinking abilities that will permit such problems to be solved in a confident and straightforward manner. Then, perhaps, students leaving school will view mathematics in the manner of these early Indian mathematicians, following the quotation that gave rise to Joseph's book: 'Like the crest of a peacock, like the gem on the head of a snake, so is mathematics at the head of all knowledge' (Vedanga Jyotisa, *c*. 500 BC, cited in Joseph, 1991).

A full background to the history of mathematics is beyond the scope of this book, although indications of where and why certain mathematical forms and ideas came into being will be indicated. General histories of mathematics have been published by Smith (Dover reprints), National Council of Teachers of Mathematics (1991), Katz (1993) and Mankiewicz (2000). Information about the various number systems that have evolved in different civilisations is readily available through books such as *From one to zero* (Ifrah, 1987), *Numbers: Their History and Meaning* (Flegg, 1984) and *The Nothing That Is* (Kaplan, 2000). There are also teaching materials using the history of mathematics in the form of posters and learning activities available from most educational suppliers (e.g. *Mathematical History: Activities, Puzzles Stories and Games* from the NCTM, 2002). Recently, a particular focus has been given to *ethnomathematics* (d'Ambrosio, 1985, <www.rpi.edu/~eglash/isgem.htm>) that brings together various cultural, social and historical factors involved in the development of mathematics.

The material that follows suggests that certain ways of proceeding can be followed to build up children's mathematical ability. They have been found to be successful over a long period of time and in a variety of situations. They do not constitute a 'fail-proof' or 'fool-proof' prescription for teaching. Instead, they should be seen as an essential background to teaching, a framework of ideas and understanding on which learning can be based. It is essential that new ideas be posed in the form of interesting and motivating experiences, that realistic problems be set as the beginning point for developing new ideas, and that through discussion and activities children are led to construct for themselves the strategies, techniques and ways of thinking that constitute mathematics. In this way, we might bring together this constructive aspect of mathematical learning with the acculturation of ways of thinking that arose in response to particular problems within a range of cultures and times, and 'begin to understand how it is possible for students to construct for themselves the mathematical processes that, historically, took several thousand years to evolve' (Cobb, Yackel and Wood, 1992, p. 2).

CHAPTER TWO
Thinking mathematically and problem solving

A mathematical way of thinking is becoming ever more necessary in a world that is being shaped by the impact of new technologies and the society changes that these are bringing to every community. Attitudes and ways of knowing that enable new or unfamiliar tasks to be dealt with are now as essential as the procedures that have always been used to handle familiar tasks easily and efficiently.

Booker and Bond, 2001, p. 43

As the nature of the mathematics studied in primary and middle school has broadened from arithmetic, measurement and simple geometry of the past to incorporate numeracy many changes in the way people perceive mathematics have occurred. An ability to think with and about mathematics has come to be the dominant feature of what has to be learned rather than the set procedures and directed solving of straightforward problems that occurred in the past.

Mathematics can be thought of in many ways. It can be thought of in terms of its content such as number, space and measurement; the process for computing, constructing or measuring; or its uses in applications across a diverse range of situations. It is, however, mathematical thinking that allows concepts, processes and their uses to be built up, problems to be explored and solved, conjectures to be made and examined, and complex ideas about the world to be communicated in precise and concise ways.

In everyday life people find it necessary to use their ability to think mathematically in order to make decisions. For example, which box of cereal is more economical—the small, medium or large box? In many cases the large box is not always the most economical. As such, mathematical thinking is concerned with using, communicating and making sense of mathematics. The *Principles and Standards for School Mathematics* (2000) stress the importance of mathematical competence in mathematical thinking and problem solving:

> We live in a mathematical world. Whenever we decide on a purchase, choose an insurance or health plan or use a spreadsheet we rely on mathematical understanding. The WWW, CD ROMs and other media disseminate vast quantities of quantitative information. The level of mathematical thinking and problem solving needed in the workplace has increased dramatically. In such a world those who understand and can do mathematics will have opportunities that others do not. Mathematical competence opens doors to productive futures. A lack of mathematical competence closes those doors.

Making sense in mathematics

For students to really think mathematically they need to make sense of the ways numbers are used in everyday life. They need to be able to use and make sense of numbers in order to make judgements, interpret data and communicate effectively. Reys (1995, p. 1) states that:

> Number sense refers to a person's general understanding of numbers and operations along with the ability and inclination to use this understanding in flexible ways to make mathematical judgments and to develop useful strategies for solving complex problems. It reflects an inclination and an ability to use numbers and quantitative methods as a means of communicating, processing and interpreting information. It results in an expectation that numbers are useful and that mathematics has a certain regularity.

Being able to work with numbers comfortably and competently is important for both children and adults. Number sense involves aspects such as the meaning of numbers, ways of representing numbers, relationships between numbers and the size of numbers. It means being able to estimate, to do mental arithmetic, to computate and to use pencil and paper and calculators as necessary. It is not something that is taught in discrete lessons but rather is something that evolves over time and indeed often begins before children come to school. The following problem highlights the importance of having a sound sense of numbers.

There were 237 people at the square dance party. If each dance set needed 5 pairs of dancers, how many dance sets could there be for each dance and how many people would not be able to dance?

Reading the problem carefully shows that each set involves 5 pairs, or 10 people. At first glance this problem might be solved using division; however, division will result in a decimal fraction, which is not useful when dealing with people and pairs of dancers.

```
      23.7
10)237.0    23.7 people in a set
    20
    ‾‾
    37
    30
    ‾‾
    70
    70
    ‾‾
```

In contrast a sense of numbers enables us to know that 237 has 23 tens and 7 ones. Thus there can be 23 sets of dancers with 7 people not in a set.

Number sense and its development is closely related to problem solving and the approaches undertaken by the teacher. Developing a mathematics program with problem solving and mathematical thinking as a focus requires a particular approach to teaching and learning. It is not enough simply to provide a range of problem solving strategies. Means of analysing the problem in order to be able to get started on it and ways of exploring mathematical knowledge and understanding in order to come up with possible paths to a solution are crucial. These need to be coupled with an inclination to try one or more of the possibilities to produce a result along with a willingness to return to the problem question and context to determine whether an answer is reasonable and has in fact provided a full solution to the problem.

Making sense of the mathematics being undertaken also needs to be seen as a key component, replacing speed in solving exercises and the external reward of a teacher's mark or approval as measures of success. This is important not only in the interpretation of problems and their solutions, but also in terms of *number sense*, which is based on a full understanding of concepts such as place value and the renaming of numbers in equivalent forms. Knowing basic facts will also underpin number sense—not as an end in itself, but rather as a resource that can be readily extended to lead to a variety of personally invented strategies to complete computations mentally or with a calculator in ways that take advantage of mathematical understanding to provide more efficient and more manageable methods. Equally important is an inclination to use this understanding and facility in flexible ways to make mathematical judgements and interpret information (McIntosh, Reys and Reys, 1998).

For example, how would you decide which country has the larger population using information presented on the United Nations online population database:

Denmark: 5.32×10^6 people
The Netherlands: 1.586×10^7 people

It is necessary to consider both the size of the initial number, 5.32 and 1.586, and each multiple of ten, 10^6 and 10^7. Although 5.32 is larger, if each is shown as a number with 10^6 as a factor, then the comparison is really between 5.32 and 15.86. Thus The Netherlands has a larger population, approximately three times that of Denmark.

Similarly, spatial sense, which builds on an ability to visualise shapes, their properties and the relationships from one to another, is also crucial in supporting problem solving. It underpins a sense of measurement and provides a basis for much of the thinking involved with ratios, fraction ideas and the presentation and interpretation of data and measures of probability. The use of diagrams and the ability to visualise what is happening can be very powerful tools when solving problems. The following problem about a farmer fencing a paddock can be readily seen using a diagram.

A farmer wants to make a new rectangular paddock which will be 26 m longer than it is wide. If she has 980 m of fencing, how long and wide should she make the paddock?

We know the paddock is to be rectangular in shape and we can show this with a diagram.

We also know it is to be 26 m longer than it is wide which we can show by drawing a line to show a square and the extra 26 m.

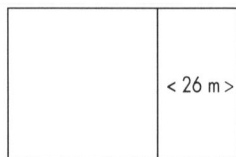

At this point we have a square and the additional 26 m on the length. If we take the 980 m of fencing and divide it by 4 we can get measurements for a paddock with a square shape.

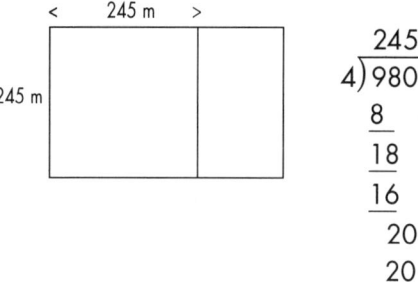

We now know the farmer could make a square paddock with each side 245 m; however, she would like the paddock to be rectangular with the length 26 m longer than the width.

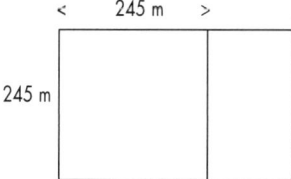

Therefore in order to get the correct measurements 26 m needs to be divided by 2 to give 13 m which is then added to one side measurement to give the length and taken away from the other side measurement to give the width.

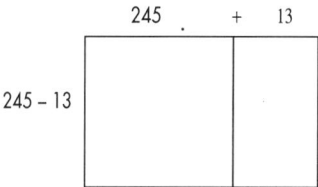

We can now see that the length is 258 m and the width is 232 m.

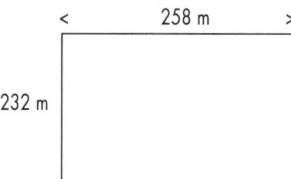

Having completed the problem and come up with a possible solution we need to refer back to the problem and check a number of points. Is the length 26 m longer that the width and do the measurements add to a total perimeter of 980 m? As we can answer yes to each of these questions we can assume our possible solution is correct. Therefore in this case being able to use a diagram and visualise the paddock and the fencing assisted with the problem solving process.

When there is an accent on problem solving and sense making, mathematical ideas are built up through describing, questioning, predicting, arguing and justifying ways of thinking.

Mathematical communication abilities are needed to understand and assess the proposals of other learners, to convey arguments and justifications to a broader audience, and to analyse and interpret information. One way of building up an ability to read, write and speak with and about mathematics is through the use of journal writing, where students are asked to write about their mathematical experiences, knowledge, doubts and the types of learning experiences they would like to follow. Another is through the use of classroom discussion about the mathematics that has been completed, focusing on the thinking processes that were followed, the attempts that were made along the way and the justification that allows a particular approach and set of answers.

In this way, the essence of mathematics as a way of thinking that involves conjectures, argumentation, justification and negotiation of meanings can come to the fore and equip students with the tools to work mathematically in a range of contexts in and outside of mathematics. Thinking mathematically needs to pervade all facets of mathematics learning and, as such, is a process that comes over the entire curriculum. It is a way of thinking to be applied over the program of mathematics.

Number sense and mathematical thinking often evolve out of problem solving. Solving problems enables students to think about and use mathematics processes and enables ways of thinking and sense making to be developed, applied and communicated. The *Principles and Standards for School Mathematics* (2000, p. 52) states:

> by learning problem solving in mathematics, students should acquire ways of thinking, habits of persistence and curiosity, and confidence in unfamiliar situations that will serve them well outside the mathematics classroom. In everyday life and in the workplace, being a good problem solver can lead to great advantages.

Problem solving

The terms 'problems' and 'problem solving' occur in many disciplines but are perhaps more closely related to mathematics than any other. Over the years much has been written about 'problems' and 'problem solving' giving rise to various schools of thought.

In mathematics education, problem solving has been emphasised since Polya's work in the 1940s. Polya, who is often considered 'the father of problem solving', describes it as follows (Polya, 1990; Polya, 1965, p. 117):

> Solving a problem is finding the unknown means to a distinctly conceived end . . . to find a way where no way is known off-hand. For a question to be a problem, it must present a challenge that cannot be resolved by some routine procedure. Problem solving is a process of accepting a challenge and striving to resolve it.

Polya believed that in order to solve a problem a student had first to come to terms with what the problem was really about. Once he or she had gained an insight into the problem only then could a plan for solving it be devised. When the plan had been carried out Polya emphasised the need to look back over the problem in terms of the solution. His four-step problem solving model, which has been used as the basis for many subsequent frameworks, can be summarised as:

1. understand the problem
2. devise a plan
3. carry out the plan
4. look back

Schoenfeld (1989) discussed problem solving in terms of 'tasks to be solved'. He believes that for problem solving to occur a student must first be motivated to solve the problem and to have no obvious ways to do so. He states that, for any student, a mathematical problem is a task:

- in which the student is interested and engaged and for which they wish to obtain a resolution; and
- for which the student does not have readily accessible mathematical means by which to achieve that resolution.

Siemon and Booker (1990) have a similar view of problem solving, highlighting the need for the student to want or need to solve the problem individually or as a group and having no immediate means to do so. They go on to describe problem solving as a process of achieving the solution to a problem, often with identifiable beginning, middle and end phases. They state that a problem is a task or situation:

- that you want to or need to solve;
- that you believe you have some reasonable chance of solving, either individually or in a group; but
- for which you or the group have no immediately available solution strategy.

These views on problem solving highlight that a *problem* is a *task* for which there is no immediate or obvious solution and *problem solving* is the process students undertake when engaging with this task. Problem solving involves engaging in tasks for which the solution strategy is not immediately obvious. In order to discover the possible solution students need to draw on their current knowledge and processes and will often develop new knowledge and understandings as they progress towards a solution.

A key issue in problem solving is the idea of no immediately available or obvious solution. A task that is a problem for students in Year 1 would not necessarily be a problem for students in Year 7. Indeed a task that would be a problem for some students in a particular year level would not necessarily be a problem for all students in that year level.

Three dogs each have five puppies. How many puppies are there altogether?

This task would not be a problem for students in Year 7 as most students would be able to think of an immediate solution. Their mathematical knowledge and understanding of the concept of multiplication and multiplication basic facts would lead them immediately to think about 3 fives and come up with the solution of 15 puppies. This task therefore could not be considered a problem for these students.

However, for students in Year 1 who would in most cases not have this mathematical knowledge and understanding, it could well be conceived as a problem for which there is no immediate or obvious solution. In most situations a Year 1 student would need to think about how to solve this problem and would perhaps come up with the possibility of using materials and counting or drawing a diagram and counting.

When selecting problems for a class program it is important to keep in mind the concept of problem solving as a task or situation for which there is no immediate or obvious solution. At times in a classroom students may be 'problem solving' when in fact they are reading 'problems' and immediately knowing what to do to solve this 'problem'. In these situations the activity being undertaken by the students could not in fact be considered to be problem solving.

Strategies and problem solving

In 1965 Polya observed that students needed techniques to help them plan for solutions. This observation provided the catalyst for over two decades of research into the identification and utilisation of problem solving strategies. The outcome of this research ultimately led to strategy driven problem solving programs in schools. These programs centred around the teaching of particular strategies such as *make a list*, *work backwards*, *guess and check*, where the strategy rather than the problem was the focus. A strategy was introduced and then the class would solve a variety of problems using the identified strategy.

For example, the strategy of 'make a list' would be taught and then the class would spend time solving problems by making a list. Students did not need to understand or come to terms with problems as they knew immediately each one could be solved using the particular strategy. This method of teaching problem solving continued for some time until researchers began to notice that students didn't necessarily become more proficient at problem solving in situations outside of the specific lessons. This thinking was not generalised into different situations where the students had to identify the appropriate strategy for themselves.

Research began to focus on problem solving and cognition and the methodology of problem solving. This emphasis led to changes in the nature of the mathematics curriculum itself and strengthened the importance of problem solving in school mathematics. Throughout the 1970s the mathematics community expressed the need for clearer guidelines and a more concise sense of direction. The National Council of Teachers of Mathematics (NCTM, 1980a) responded to these concerns with a document titled *Agenda for Action: Recommendations for School Mathematics of the 1980s*, which outlined eight explicit recommendations, the first of which was that problem solving should be the focus of school mathematics.

By 1989 the NCTM had taken its earlier recommendation even further and was now stating that problem solving must be integral to all mathematical activities. Problem solving was to be viewed not as a separate topic but as a process that should permeate the entire mathematics program from beginning to end. Viewing problem solving in this way would provide the context in which concepts and processes could be learned. This approach enables mathematical constructs to be grounded in and emerge from students' own solutions to problems that are, to them, real and genuine. Hence, as problem solving as such is an individualised endeavour, mathematics becomes both functional and meaningful to each individual.

Similar calls were made in Australia. State and territory education departments began to interpret problem solving as a process, placing importance on the procedures and strategies used by the students rather than their answers. Problem solving was often viewed as the central focus of the curriculum and integrated across all mathematical areas.

In 1991 the Australian Education Council published *A National Statement on Mathematics for Australian Schools*. The purpose of this statement was to provide a framework around which states and territories and thus schools could build their mathematics curriculum. It identifies important components of mathematics education and states that experiences with problems should be provided to enable students to use a wide range of problem solving strategies across all topics in mathematics. This document is still the central framework for the various syllabuses that have evolved.

Today many educators believe that the most important goal of the study of mathematics is fostering and developing students' abilities to solve problems. Yet, as mentioned, adherence to traditional styles of teaching leads to difficulties with problem solving. For problem solving to be worthwhile it is essential that teachers view it as a valuable, motivating and pedagogically sound approach for introducing, developing and applying concepts and processes.

Small-group instruction, team teaching, learning centres and technology such as computers and calculators have become more common in classrooms. However, this style of teaching is often only conducted after the 'real work' is completed—after the content involving rules and procedures has been taught. It is usually not used as a means of teaching a concept but rather as consolidation or reinforcement. Activities where students are seen to be talking, interacting and even enjoying themselves are not always accepted as pedagogically sound. Yet, this is often how students learn best—in environments where they can engage in activities that allow exploration, language and socialisation from which they can make sense of complex ideas.

Worthwhile problems and building new knowledge

For students to really develop mathematical ways of thinking and number sense it is essential for good worthwhile problems to be selected for the class program. A teacher needs not only to select problems for which there are no immediate or obvious solutions but also to select problems which will consolidate, extend and stimulate mathematical knowledge and understandings.

When choosing problems for a particular mathematics classroom a teacher needs to thoroughly explore the problem and the possible mathematical ideas which can be brought by the students when working through the problem. In some cases it may be necessary to modify the problem to ensure particular mathematical ideas are enhanced or consolidated. There are many problems which are interesting and fun but at the end of the day the teacher needs to ask what mathematical thinking is being developed, stimulated, extended or consolidated by this problem.

Problem solving which is integral to the mathematics program and not an isolated activity should assist students to consolidate and extend their knowledge and understandings. Well-chosen problems will give students the opportunity to clarify and come to terms with mathematical processes and understanding. For example, a problem regarding pet ownership could involve quite different mathematical processes and understandings depending on the age of the students.

Who in the class has pets? How many pets do they have? What type of pets do they own?

In an early year's classroom where the students are beginning to learn mathematical knowledge and understandings the problem might take the following focus:

- *gathering information*—students might discuss, learn about and try some ways to gather information, for example asking people individually, a survey with pictures of animals or asking groups with a show of hands
- *recording data*—students might discuss, learn about and try a number of ways to record the information, for example counting and writing the number, using ten frames, tally marks, using tables to record the data
- *displaying the results*—students might learn what a picture graph is, how to use the data recorded and how to construct a picture graph

In a middle year's classroom where students have more established mathematical knowledge and understandings the problem might take a quite different focus where previous work is revised and new learning is undertaken:

- *gathering information*—consolidation of how to gather information where surveys and interviews are revised
- *recording data*—consolidation of how to gather information where tables and tally marks are revised
- *analysing data*—students might discuss, learn about and try a number of ways to analyse data, for example finding averages and percentages on the results
- *displaying the results*—students might discuss and learn what a bar graph and line graphs are and how to construct bar graphs and line graphs

Problems can integrate multiple topics and involve complex mathematics. The teacher's role in selecting worthwhile problems is important and at the centre of successful problem solving. Thinking about the mathematical ideas that can be consolidated and learnt through a particular problem helps build a balanced program where problem solving is at the fore.

When selecting problems a teacher needs to bear in mind what mathematical ideas will not only be consolidated but also enhanced. There are many problems that may be interesting and fun but may lead to very little new learning. While there is a place for these problems it is important that a wide range and variety of problems are solved.

Organising a problem solving program

One idea to assist teachers selecting and modify problems is to use a problem structure table. During the course of the year as problems are selected and used they are entered into the table to help the teachers see what sort of problems are being used. If a particular problem was successful it can be modified and have the context changed and used again at another time of the year. This would encourage students to think about previous problems and how they are solved as an important strategy when solving problems.

The table shown below can be used to help teachers to see the type of problems they are using and how often they are used. It is not meant to be the only way problems can be thought of but rather one way in which a teacher might think about problems. As teachers come to terms with problem solving in their classrooms they may think of other ways of keeping track of the various types of problems and come up with their own recording and tracking tables.

	Operation obvious	Operation less obvious	Too much information	Too little information	Strategic thinking
1 or 2 steps					
Many steps					

An important aspect to developing a successful problem solving program is being able to establish a classroom environment in which problem solving is at the fore. Siemon (1998) has put together seven suggestion for creating a problem solving environment.

Suggestion 1: Problem solving should be integrated into the school mathematics program so that it becomes intimately associated with what it means to know, understand and use mathematics. For example, learning about place value can and should be as much a negotiated problem solving experience as an investigation involving the collection, interpretation and presentation of data.

Suggestion 2: Recognise that children have different goals, motivations and expectations that are as much shaped by the classroom culture in which the mathematics teaching and learning occurs. For example, many children believe that mathematics is about 'doing sums quickly to get answers . . . it doesn't have to make sense'. Such approaches need to be gently challenged and changed. The construction of meaning and the justification of procedures need to be emphasised and valued. One way of doing this is through assessment: 'one mark for the answer, nine for a sensible story which we can all understand'.

Suggestion 3: Develop a classroom culture in which students expect to elaborate, defend and amend or reject their ideas in a non-threatening, mutually supportive way. Value sense making and encourage students to recognise, develop and make connections between conceptual and procedural knowledge.

Suggestion 4: Recognise that whether a particular task is an exercise or a problem depends on the person(s) expected to engage with the task. Adopt an inquiry-based approach to everything that is considered in mathematics. Assume that everything is a problem until the whole group agrees that it is an exercise—a task where everyone knows what to do to arrive at a solution.

Suggestion 5: Establish classroom norms and processes which ensure that all students have the opportunity to contribute to a shared view about how an idea, exercise or problem might be modelled, described and recorded. Guide students to the most effective and generalisable forms of representation. Talk about how different problems are alike and how they are different. Keep a classroom record of these discussions.

Suggestion 6: Recognise that problem solving is not something which can be defined, packaged and delivered to students. It is an experience that needs to be shared with others, discussed and acknowledged for what it is—a cyclical process that requires conscious monitoring and direction in much the same way as the creative writing process.

Suggestion 7: Choose materials, problems and games that provide opportunities to discuss mathematical ideas and how they might be represented and applied. Model problem solving behaviour. Engage in think aloud solution attempts and ask children to check your reasoning. Make mistakes to focus discussion on the value of checking and justifying one's thinking.

Solving problems

To gain a full understanding of problems and problem solving it is necessary to read problems, look at their structure and of course to solve problems. The following problem is based on one by Polya.

> **The Hockey Herons**
> Sally is a member of the local hockey team, the Hockey Herons. One of her responsibilities is to cut out articles and stories about the Herons from magazines and newspapers and put them into a scrapbook.
>   One day she dropped the scrapbook and all the pages fell out. Luckily, by noting the date of the articles and stories, she was able to put the pages back in order.
>   Sally decided to number all the pages of the scrapbook. She had a box of stickers and on each sticker was one of the digits 0–9. She used the stickers to number each page of the scrapbook starting with 1.
>   When Sally had finished, she noticed she had used 537 stickers. How many pages are in the scrapbook?

Students and adults attempting to solve this problem do so in a multitude of ways. Many simply divide by 2; as each page needs a number and as there is a number on the front and back of a page, dividing by 2 would give you the number of pages. These people have overlooked the information of one sticker for each digit, meaning a page number like 346 would need three stickers. Some take this information and think of the one-digit numbers and divide by 10, reasoning that as there are 10 digits dividing by 10 would give you the number of digits, thus overlooking the fact that the question does not ask for how many digits but how many pages.

Others take a more organised approach and use strategies such as organising the information into tables or lists, which can help in monitoring a guess and check approach. It can also lead to a solution process that focuses on the number of digits used for the pages with one-digit, two-digit, three-digit or four-digit page numbers. For example:

number of digits	number of pages	number of stickers used
1	9	9
2		
3		

While identifying the response for the one-digit pages is easy, determining the number of pages containing two digits can be difficult. Three answers are commonly given, yet only one can be correct! Counting or examining a two-digit number board shows that there are 10 numbers for each decade (10–19, 20–29, 30–39, 40–49, . . .,) so there must be 90 numbers altogether and hence 180 stickers have been used.

An answer of 89 is frequently given because the student thinks 99 take away 10 while the other common answer of 91 comes from 100 take away 9. Reasoning about the problem suggests that since the last two-digit page number is 99 and there are 9 one-digit numbers then the number of two-digit numbers must be 90. Similar reasoning processes or considerations of each hundred (100–199, 200–299, 300–399, . . .,) shows that the number of three-digit numbers will be 900. If this information is added to the table it looks like this:

number of digits	number of pages	number of stickers used
1	9	9
2	90	180
3	900	2700

At this point students often recognise the pattern and would no longer need to count or check. Returning to the problem context shows that the number of three-digit pages must be less than 900 as only 537 stickers have been used. In fact since there are 189 stickers used for the one and two-digit numbers, there must be 537 take away 189 or 348 stickers for three-digit numbers. Dividing 348 by 3 shows there are 116 three-digit pages, giving a total of 116 add 90 add 9 or 215 pages altogether.

Reflecting on the process involved in coming to terms with solving a problem like this shows the complexity of understanding and bringing together various ways of thinking that problem solving demands. Several strategies may be involved in the one problem: making an organised list or table, reasoning logically about numbers or the situation, looking for and using patterns, or working backwards. At the same time, this problem called on understanding of numeration (what is a two-digit or a three-digit number, how do we use place value patterns to determine how many there are) and computation (multiplication, addition, subtraction and division were all involved in the solution). Problem solving is a process that demands understanding of mathematical context as connected knowledge along with an ability to think laterally and use strategies to bring diverse aspects of mathematics to focus on a solution.

A number of other questions could be asked using this same problem and the same problem could be used to create a different context. In this way a unit of work around the one problem could be developed and the students could also go on to think about and write problems around this concept. For example:

> How many pages had Sally numbered once she had used 195 stickers?
> Would Sally have enough stickers to number 270 pages if she had 650 stickers in total?
> When Sally finished numbering the pages of the scrapbook, her friend Carla said 'I once wrote the numbers on the pages of a book that had 536 pages, starting with page 1! There were nearly 2000 digits and my hand was very tired when I finished writing'. How many digits did Carla use to number the pages?
> Steve, Sally's brother, had a large, old dictionary. 'Boy, am I glad,' said Steve, 'that I did not have to write the numbers on the pages of my dictionary. I would have written a total of 7333 digits to number the pages!' How many pages are there in Steve's dictionary?

Difficulties with problem solving

Literature centred around problem solving reveals that the dominant emphasis in teaching problem solving focused on particular strategies that could apply to various classes of problems (Schoenfeld, 1992; Stanic & Kilpatrick, 1989). However, many students with good numeration and computation abilities were unable to access and use these strategies to solve problems. Successful problem solvers tended to use these strategies only after determining the intent of the problem; problem solving was found to depend more on an analysis of problem structure than on the use of strategies. Many researchers have found that students, when problem solving, simply manipulate numbers with little or no thought to the problem itself (Artzt and Armour-Thomas, 1990; Bond, 1996; Lester, 1985; Schoenfeld, 1992). In addition to monitoring a student's thinking while solving problems, researchers began to observe and record a variety of behaviours. Patterns and trends began to emerge and a number of observations were found to be common to students with poor problem solving processes. These students simply:

- located and manipulated numbers
- tried different operations

- looked for key words
- read problems quickly and cursorily
- gave no consideration to the problem context
- did not assess the reasonableness of answers
- showed little perseverance if answers could not be obtained from the first approaches tried
- did not access problem solving strategies, even though they could generally be named and described when discussed as entities in their own right

In order to better understand these limitations in problem solving capabilities each one is discussed below using classroom observations.

Locating and manipulating numbers

One loaf of bread and six rolls cost $1.80. At the same prices, two loaves of bread and four rolls cost $2.40. How much does one loaf of bread cost?

$$1 + 2 + 6 + 4 \Rightarrow \$1.80$$
$$1 + 2 + 6 + 4 + \$1.80 + \$2.40$$

Many students focused only on the surface level of a problem, viewing it as a sea of words in which numbers needed to be located and then manipulated. Here the student has located the numbers by underlining them in the problem and has written them in a horizontal number sentence to try to obtain an answer. Adding the amount of bread and bread rolls with the cost shows a lack of understanding of the problem on any level.

Trying different operations

In the garden there are 138 grape bunches on the grape vine and 63 passionfruit on the passionfruit vine. The gardener picks 79 grape bunches and 29 passionfruit. How much fruit is still on the vines?

When numbers were identified, often one particular operation was tried, then, if the response did not appear to be a likely answer, a different operation or combination of operations was tried. To begin with, the student tried multiplying the 79 grape bunches by the 29 passionfruit, which shows no understanding of the problem, as multiplying grapes by passionfruit would not help in any context! When an obviously large number was obtained the next operation tried was subtraction and 63 passionfruit was taken away from 138 grape bunches. Again this was not at all helpful so having tried multiplication and subtraction the next operation tried was addition and the 138 grape bunches were added to the 63 passionfruit. At this point the teacher intervened and told the student he was on the right track (was he?) and so he continued adding and added the 79 grape bunches to the 29 passionfruit. Having obtained these totals he was still unsure what to do with them until the teacher told him one total was the amount of fruit on the vines to begin with and the other total was the amount of fruit picked from the vines and he was able to subtract these totals correctly and achieve an answer. In addition to trying different operations it is common for students frequently to ask the teacher to verify whether the answer is reasonable or to provide further suggestions on how to proceed. The student attempting to solve the problem tends to have no means of determining when an answer might be at hand or what bearing the problem itself might have on the way the numbers should be interpreted and used.

Key words

A new dam was built and it took 24 days to fill with water. If it doubled itself in size each day, how long did it take for the dam to be half full?

24   12   6   3   1.5   0

12 days

½ of 24

Commonly students identified key words while overlooking key meanings. In this example the student has focused on the key word of half and divided by 2 and has missed the key meaning of doubling each day. If the dam doubles in size each day and is now full then for it to double itself it had to be half full the day before.

Problems were read quickly and cursorily

Pam earns $12.00 an hour at the plant nursery. If she works each weekend for four hours each day, how much will she earn over four weekends?

$$\begin{array}{r} \$12 \\ \times\ 4 \\ \hline \$48 \text{ hrs} \\ \times\ 4 \text{ (weekends)} \\ \hline \$192 \end{array}$$

Problems were read quickly, usually only once, with little or no thought to ensuring no important information was missed. In this problem the student has overlooked the 4 hours per day which means Pam is working 8 hours each weekend. The student has simply read the 4 hours and calculated the earnings on 4 hours per weekend rather than the correct total of 8 hours.

No consideration of the problem context

How much will it cost to transport by road 150 tonnes of sugar stored at the sugar mill for 3 weeks? There are 2 trucks and each truck holds 12 tonnes. It costs $48.00 for the first trip and $32.00 for every trip after that.

$$\begin{array}{r} 12.5 \\ 12\overline{)150.0} \\ 12 \\ \hline 30 \\ 24 \\ \hline 6.0 \\ \end{array}$$

2 truck.

12½ trips

When attempting to solve problems students frequently do not take into account the context of the problem. In this problem the student worked out that it would be necessary for the trucks to do twelve and a half trips to transport the sugar. Considering the problem even less, many students when doing this problem simply divided 12 by 2 and get 6 trips for each truck. The student in this case did not stop to think about the problem and consider that it would not be possible for a truck to do half a trip and that it would be necessary for one truck to do 6 trips and the other truck 7 trips. Had he considered all the conditions he may have noticed the cheapest cost of all is if only one truck is used for all the trips.

Reasonableness of answers was not assessed

Each morning starting at 6 o'clock, Sue and her father deliver milk to 129 homes. Each afternoon, starting at 4 o'clock, they deliver to 64 homes. How many deliveries do they make in one week?

$$\begin{array}{r} 129 \\ \times 64 \\ \hline 516 \\ 7560 \\ \hline 8076 \\ \times 7 \\ \hline 56532 \end{array}$$

When gaining an answer it is important to assess how reasonable the answer is in light of the problem. Here the student multiplied the deliveries per day instead of adding them. As a result she has ended up with an extremely large number of deliveries (56 532), which she simply recorded without question. Had she returned to the problem she may have realised that it is not reasonable for two people to make that many deliveries in one week and thus attempted to discover where an error may have occurred.

Little perseverance

The city council charges each household yearly rates. On Elke's last rates bill, she was charged $38.25 per quarter for water and $41.25 per quarter for rubbish collection. How much did she pay daily for water and rubbish collection?

$$365 \div 4 = 91.25$$

Can't do problem as can't have quarter of day !!

When faced with a difficulty many students give up rather than trying to think of ways of overcoming the uncertainty. In this problem the student has divided 365 (the number of days in a year) by 4 to work out how many days in a quarter. When the answer obtained was not a whole number the student wrote that the problem could not be done as the division does not turn out evenly and it was not possible to have a fraction of a day. No attempt was made to overcome this difficulty. For example, the student could have found the closest multiple of 4 (364) and worked out that there is an extra day so one-quarter may have had one additional day. Alternatively the student could have multiplied the water and rubbish collection totals by

4, added them and then divided by 365 to obtain a money cost per day. The actual cost per day needs to be a rounded amount—it works out to be about 87c per day.

Problem solving strategies not accessed

Sydney had half as many visitors in 2001 as it did in 2002. In 2003 there were 7492 more visitors than in 2000 and in 2002 there were 12 854 less than in 2003. If there were 98 460 visitors in 2000, how many were there in 2001?

[Handwritten notes:]
- make list
- tables
- work backwards
- materials
- guess and check
- look for patterns
- draw diagram

Often students are unable to start a problem. The student attempting to solve this problem was unable to begin. He read the problem several times and still could not commence. His teacher gave him the prompt to think of strategies which could be used to solve the problem. The student wrote a list of all the strategies he knew and was able to explain each one. He was, however, unable to identify which strategies might be useful for this problem and was unable to proceed any further.

Problem solving as a process

If students are to develop problem solving abilities then clearly collaborative rather than individual work, with discussion revolving around the methods of solution rather than answers, is essential. In particular, emphasis needs to be given to discussing the meaning of the problem before any plan is drawn up or work on a solution is begun. Initial emphasis needs to be focused on the value and construction of an overall plan to manage the problem solving process.

As mentioned earlier, many models or frameworks for managing problem solving are based on Polya's model. Often textbooks and resource books depict these steps in a linear way and do not cater for the processes of self-monitoring, self-regulation or self-assessment (Fernandez, 1994). These models encourage students to view problem solving as a procedure. Students tend to focus on completing each step and give little thought to monitoring processes. Siemon and Booker (1990) developed a model, which recognised the need to come back and reflect on the solution obtained in light of the original question and to check to see if the answer made sense. The model was designed to draw attention to the problem solving process and to encourage students to direct and monitor their own problem solving behaviour.

A number of questions that helped students to monitor the process and reflect on the solution in the context of the problem were identified at each stage to encourage the students to self-regulate the process of solving problems.

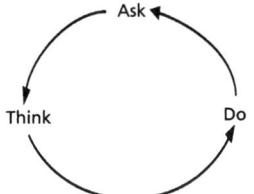

- What do I need to find out?
- What is the problem asking?
- What do I need to know?
- How can I find out?
- Did it work and does the answer make sense?

At this point it is important to look at a problem and work through the process of solving it while at the same time keeping in mind what is occurring to try to come to terms with problem solving-models and processes.

At the preschool there were 18 children playing in two sandpits. Two children from the first sandpit went to play on the swings and then four children from the first sandpit went to join their friends in the second sandpit. There were now the same number of children in each sandpit. How many children were in each sandpit at the beginning?

The first stage involves reading the problem and attempting to come to terms with what the problem is asking. It is necessary to delve beneath the surface level of the problem and begin to *analyse* its structure. The problem states that there are 18 children and that 2 went away and are no longer counted while 4 go over to the second sandpit and that at the end it is necessary to have the same number in each sandpit. All the required information is available in the problem and no additional information is needed. The question at the end is asking how many children were in each sandpit to begin with, but the real nature of the problem is that there were 18 children in 2 sandpits, the children numbers changed and then there was the same number of children in each sandpit. To put it more simply, there was a situation, something changed and then there was a different situation.

After analysing the problem and coming to terms with its structure, it is now necessary to *explore* ways in which it might be solved. Having looked below the surface level of the problem and really understood what the problem is asking it is now possible to start to see ways of solving it. The detailed analysis of the problem leads to seeing possible solutions. Possible ways that came to mind during the analysis were:

- materials—use blocks to represent each child and work through the problem from the beginning
- guess and check—we know there are 18 children and could make a guess to begin, check the guess and make adjustments as necessary
- work backwards—two went away leaving a total of 16, so there must have been eight in each sandpit at the end—work backwards from here
- think of a similar problem—think about the problem and see if it reminds you of another problem. If so think about how it was solved

Often when solving problems a combination of strategies is used in conjunction with each other. For example, when attempting to solve the sandpit problem it is possible to work backwards, guessing and checking while using materials. At other times each possibility will need to be worked through until one that makes sense and produces an answer is found.

Trying a solution

It is during this exploration stage that we start to think about and decide which strategy or combination of strategies we might use. Having selected an idea, we proceed with our decision and try our solution. If we return to the possibilities listed before we can think about which one to select. If we chose materials and guess and check as a combination we can go ahead and move into the stage of trying our selection.

We know there are 2 sandpits and 18 children and sandpit one loses more than sandpit two, so it makes sense to start with perhaps 12 and 6. Two go away from the first sandpit reducing its number to 10 and then 4 go over to the second sandpit leaving us with 6 and 10 respectively. If we look at these numbers as a total we can see that at the end of the problem there are 16 children which means there needs to be 8 in each sandpit. Knowing this we can then adjust our original guess and realise 2 children need to be removed from the second sandpit and added to the first. This would mean our first guess would need to be adjusted from 12 to 14 and from 6 to 4. Let us check to see if our reasoning is sound. There are 14 children in the first sandpit and 4 in the second. Two children from the first sandpit go away which reduces it to 12 and then 4 go over to the second sandpit which leaves us with 8 and 8 respectively.

Reflecting

At this point we need to go back to the problem and reflect on our solution to see if it makes sense in light of the problem and what it is asking. Did we start with 18 children, did 2 leave the first sandpit, did 4 go over to the second sandpit and did we end up with the same number in each sandpit? Luckily we can answer 'yes' to all these conditions. However, if we could not do so we would need to recheck our solution to try to discover where we went wrong or to think of our other possible solutions and try another approach.

Another way of solving the problem is to reason that after 2 children left the first sandpit, there would be 16 children and therefore 8 in each sandpit. Retracing the problem would mean that 4 of the children in sandpit two would go over to sandpit one and there would then be 14 children in sandpit one and 4 children in sandpit two. At this point many students might assume that they have 'the answer'. But reflection on the conditions shows them that a total of 16 children is not correct. There were 2 more children that left from the first sandpit. These need to be added to the number in the first sandpit and a correct solution would show that originally there were 14 children in sandpit one and 4 children in sandpit two.

If we reflect on how we solved this problem we can identify key elements within the process. To commence it was necessary for us to *analyse* the problem to unfold its layers and discover its structure. Next we *explored* possible ways of solving the problem before selecting one to *try*. Having tried an idea we checked our solution in light of the problem in case we needed to try another possibility.

Applying the problem process

If we look at another problem while keeping this process in mind we can start to come to terms with insights into approaches used by successful problem solvers.

On Tuesday Larry went to do his weekly shopping. He spent $\frac{3}{4}$ of his money at the supermarket. He then went and spent $\frac{1}{3}$ of what he had left at the butcher's shop. Afterwards he had $12 left in his wallet. How much money did he begin with?

Again it is important to read the problem and attempt to come to terms with what the problem is asking. It is necessary to delve beneath the surface level of the problem and begin to *analyse* its structure. When analysing the problem we can see that Larry took some money to do his shopping. He spent some at the supermarket some at the butcher's shop and then had $12 left. All of the information needed to solve this problem is available and no additional information is needed. The question at the end asks how much money did he start with but really the real question is how much did he spend at the supermarket and butcher's shop. To put it simply the final situation is known, something happened before this and the initial situation needs to be determined.

Having *analysed* the problem and come to terms with its structure possible ways in which it might be solved can now be *explored*. Having looked below the surface level of the problem and really understood what the problem is asking it is possible to see ways of solving it. Possible ways which come to mind during the analysis are:

- materials—counters could be used to represent each dollar spent and to work backwards through the problem from when Larry had $12 left
- guess and check—a series of guesses as to how much Larry took shopping could be made, checked and adjusted as needed
- work backwards—Larry had $12 at the end after going to the supermarket and butcher
- think of a similar problem—think about the problem and if it is similar to any other problem think about how it was solved
- use a diagram—show what Larry spent in pictorial form to show each fraction spent and how much was left over

As stated previously, when solving problems a combination of strategies might be used in conjunction with each other. For example, when attempting to solve this problem it is possible to work backwards while using materials.

Trying a possible solution

Now that we have analysed the problem and thought about possible strategies or combination of strategies it is time to decide which strategy or combination of strategies might be used. In this situation a number of strategies will be discussed to show the possibilities of solving problems different ways. There is no single best way to solve a problem but rather to use the strategy or combination of strategies which makes most sense to the person solving the problem.

Working backwards

We know Larry had $12 left at the end of his shopping after going to the butcher. This amount is $\frac{2}{3}$ of what he had left after the supermarket. From this we can see he must have spent $6 at the butcher and had $18 left after the supermarket. We know he spent $\frac{3}{4}$ of what he took with him at the supermarket therefore $18 is $\frac{1}{4}$ of the total. Therefore Larry must have taken $18 by 4 or $72.

At this point we need to reflect on our solution to see if it makes sense in light of the problem. If Larry started with $72 and spent $\frac{3}{4}$ at the supermarket he would have spent $54 leaving him with $18. If he spent $\frac{1}{3}$ of $18 at the butcher he would have spent $6 leaving him with $12. We can therefore see that our solution makes sense in light of the problem. If the solution didn't make sense we would need to return to our thinking to try and discover what went wrong.

Using materials

Materials could be used to solve this problem where counters could represent each dollar. Materials could be used in conjunction with working backwards or guess and check. If we work backwards we would need 12 counters to represent the $12 left at the end of the shopping. We know this is $\frac{2}{3}$ of what Larry had left after the supermarket so we would need to add another 6 counters for a total of 18 counters. We know that this is $\frac{1}{4}$ of what he took with him so we would need another 3 sets of 18 counters which could be added to reach a total of 72 counters or $72 dollars.

From our previous reflections we know that $72 makes sense in light of the problem.

Drawing a diagram

Another way to solve this problem is to use a diagram. If we use a rectangle to represent the total amount Larry took shopping we can show by shading how much he spent each time at the supermarket and the butcher quite clearly.

We know he spent $\frac{3}{4}$ of his money at the supermarket which we can show by dividing the rectangle into four equal parts and shading 3 of them.

After leaving the supermarket Larry then went to the butcher's shop where he spent $\frac{1}{3}$ of the money he had left. This can be shown by dividing the unshaded section into 3 equal parts and shading one of the parts.

We know that at this point he had $12 left which represented $\frac{2}{3}$ of what he had left after the supermarket shopping. Therefore the two unshaded parts must be worth $12 or $6 per part.

At this point we can see that each part is worth $6 and there are twelve parts so Larry must have started with $72.

Again we need to go back to the problem and look at the solution in light of the problem and see if this answer is reasonable or not. We believe Larry started with $72. He spent $\frac{3}{4}$ at the supermarket which means he would have spent $54 and have $18 left. He then spent $\frac{1}{3}$ of $18 at the butcher which means he spent $6 dollars and would have $12 left over. If we look at the problem we can see that this is correct and our diagram has assisted in solving the problem.

Reflecting on the process of problem solving

Thinking about how the various ways this problem was solved highlights the key elements within the problem solving process. This process, which has evolved over a number of years working with students in problem solving research projects (Bond, 1996), draws on the Siemon model. When starting the process it is necessary to *analyse* the problem to unfold its layers and discover its structure and what the problem was really asking. Next possible ways to solve the problem were *explored* before one or a combination of ways were selected to *try*. Finally once something was tried it was important to check the solution in light of the problem to see if the solution was reasonable. This process highlights the cyclic nature of problem solving and brings to the fore the importance of understanding a problem and its structure before proceeding. This process can be shown using the following model.

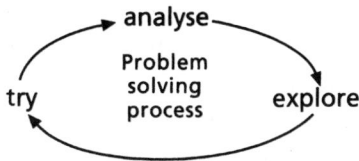

Analyse

During the analysis of the problem students are encouraged to analyse orally each problem and to think through and discuss what the problem is asking. It is in this step that students need to look below the surface level of the problem and come to terms with the problem structure. The real question the problem is asking is rarely found in the question at the end of the problem.

Students need to delve into the problem to discover what it is really about. During this stage of the process students may need to:

- read the problem aloud
- think of previous problems and other similar problems
- select important information that may be useful from the problem
- discuss what the problem is asking

Students who can list and describe strategies but are unable to identify which strategies might be helpful with a particular problem have overlooked or skipped the analysis stage of the problem solving process. They have not unfolded the problem and come to terms with what it is asking. It is only once students have really understood the problem and its different layers that they begin to see how it might be solved.

Explore

Once the nature and question of the problem is established students move to the next stage of exploring ways and ideas of how the problem might be solved. If the analysis stage has been thoroughly completed then students will begin to see ways in which it might be solved. Here strategies and how they might be useful in solving the problem can be discussed. Strategies include:

- drawing a diagram or graph
- using materials
- making a table or list
- working backwards
- looking for a pattern
- thinking of a similar problem
- guessing and checking
- acting out the problem
- using smaller numbers

Strategies are an important part of the problem solving process and students need to know a wide range of strategies and how to access them. However, they are only useful once the nature of the problem has been established.

Many problems can be solved in a variety of ways using different strategies. How a teacher might solve a problem can be contrary to how a student might solve the same problem. Imposing a teacher's strategy on the student usually only serves to teach that one problem as the strategy is not one which the student would have selected and therefore does not necessarily make sense to the student. Encouraging a class to solve a problem in a variety of ways and then discussing and justifying the different approaches gives access to many strategies and exposes students to a variety of different mathematical thinking.

Try

Having explored ways and ideas a possible solution strategy is then selected and tried. It is at this point that pencil and paper or calculators can be used. Work prior to this should centre around thinking and discussing and only after the decision to try a possible solution strategy has been made should the students turn to pencil and paper or calculators.

Many students, particularly when first using this process, are tempted to skip the discussion and explorations stages and go straight to manipulating numbers in the hope of finding a solution. To overcome this it is often necessary to restrict the use of pencil and paper and/or calculators and allow the students access to them only after it has been determined that the first two steps have been thoroughly completed.

Once an answer is obtained it is important to compare it with the original analysis of the problem to determine whether the solution obtained is reasonable given the context of the problem. It will also raise the question of whether other answers exist, and even whether there might be other solution strategies given the knowledge that the problem has been solved. In this way the process is cyclic and should the answer be unreasonable given the context of the problem then the process would need to begin again. The overall process for managing and discussing problem solutions could be summarised as follows:

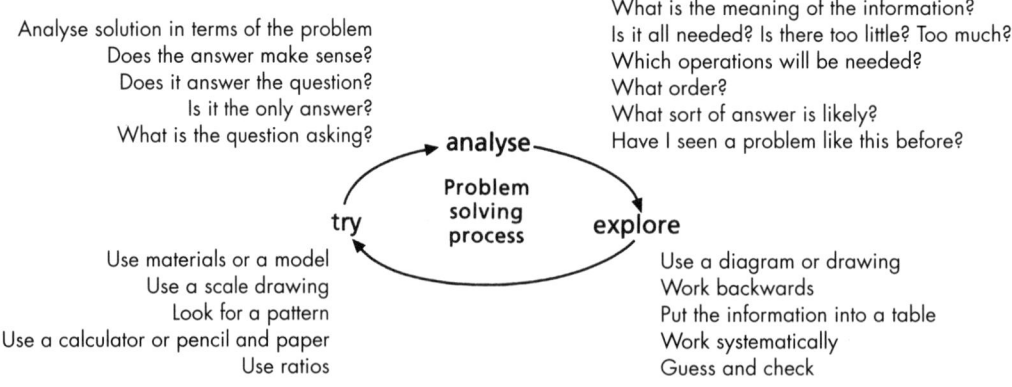

Reflection

Although not a formal part of the problem solving process, it is important to give time to reflecting on any plans drawn up, processes followed and strategies used so that the students can see the significance of coming to terms with the structure of the problem as well as the value and applicability of particular strategies that might then be used. Analysing problems in this way also leads to a realisation that a problem is not solved until any answer obtained can be justified. Learning to reflect on the whole process leads to the development of a deeper understanding of the entire process.

Problem writing

Problem writing is an area that is gaining prominence as a method of assisting students in coming to terms with problems and problem solving. Talking, listening, reading and writing about mathematics helps students clarify their own ideas and enables teachers to gain valuable insights into their students' thinking. Traditionally, communicating in mathematics involved the use of signs and symbols, and activities such as discussing, predicting, questioning, justifying a point of view and reflecting were not seen as important or even necessary in the learning of mathematics. *A National Statement on Mathematics for Australian Schools* (1991) highlights the point that an ability to communicate mathematically is essential to learning mathematics and is a part of numeracy.

If analysis of the problem is a crucial step then activities that involve students writing and discussing problems can be a powerful tool. The genre of writing problems requires the student to understand the structure of the problem and in order to do this a detailed analysis is required. At times it is only when students attempt to write a similar problem to one they have solved that they really come to see the structure of the problem and therefore come to terms with its nature. In addition, problems written by students can be used in the class program and provide not only a purpose for writing them but also a source of readily identified problems. Solving problems written by other class members is very motivating and provides an added dimension to the problem solving program.

Investigate and discuss

Listed below are six problems. Using a personal journal solve each problem keeping notes on:
- what you first thought when you read the problem
- how you attempted to come to terms with the problem
- what you did to solve the problem
- any difficulties you encountered and how you overcame them
- your reflections on the process

1. A camping barbeque and a gas bottle together cost $163. If the barbeque was $87 more than the gas bottle how much did the gas bottle cost?
2. Amanda went on 8 more rides at the show than Joseph. Ray went on twice as many as Joseph. Together Amanda, Joseph and Ray went on 36 rides. How many rides did each person go on?
3. The chicken farmer was sorting her chickens at the end of the laying season. She kept $\frac{1}{3}$ as breeding stock and half of the rest she sold. This left her with 30 chickens which she decided to keep for laying. How many chickens did the farmer have to start with?
4. Mary bought apples at the market at a cost of 6 for $1 which she sold at her fruit shop for 4 for $1. If she made $16 profit how many apples did she sell?
5. One chapter of a book contains 7 pages. The sum of all the page numbers in this chapter is 826. What are the page numbers of the chapter?
6. A professional photographer does not use 2 out of every 3 photos he takes. If he throws out 72 photos, how many has he taken?

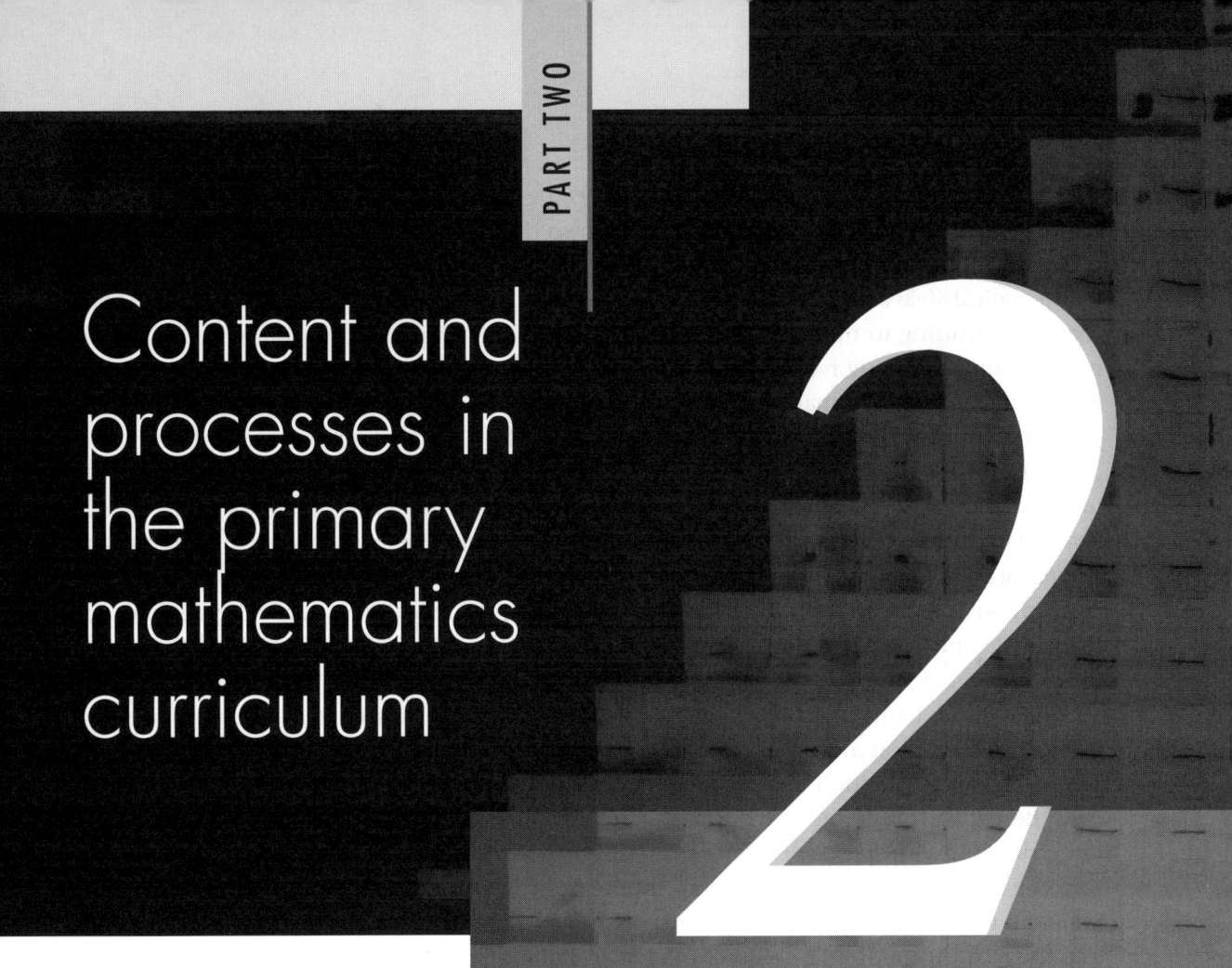

PART TWO

Content and processes in the primary mathematics curriculum

THE USEFULNESS of mathematics in solving problems in everyday life, work or further study has always been the justification for giving it a central place in the curriculum. As a consequence of these considerations, primary school mathematics programs have tended to devote most time and energy to the acquisition of number concepts and computational skills. Geometrical ideas have also been included, but rarely in a systematic or complete manner, and measurement has usually been taught as a series of *ad hoc* procedures rather than as an integrated topic. When aspects of graphing, statistical measures or simple ideas of probability have been included, these too have been one-off investigations rather than part of a developing awareness of the important role that chance and data play in a modern society. Breaking up the content into this large number of seemingly unrelated topics has often led children to learn them in isolation as a set of rotely learnt procedures. Yet, to use this knowledge in the unfamiliar settings that applications and problem solving demand, more than this is needed. Indeed, the degree to which material is acquired meaningfully is an important contributor to an ability to use it in new and unfamiliar tasks.

An understanding of number will always be fundamental to learning and using mathematics. Not only is it a major topic in its own right, it underlies all of the computational processes and in many ways sets up the patterns of thinking on which mathematics relies. If mathematical ideas are to be taught without confusion, it is important for the numeration ideas that give meaning to numbers to be well established. Further, misconceptions and gaps in number understanding have proven to represent the largest source of children's difficulties in arithmetic and are a major factor in their inability to solve other than routine, one-step problems. Even when children's errors appear in addition, subtraction, multiplication or division, the real source of the difficulty often lies in an inadequate understanding of the numbers they are working with. Difficulties children experience with computations involving decimal fractions or common fractions are as likely to lie with the fraction ideas underlying the recording and renaming of the fractions in equivalent forms as with the processes themselves. Similarly an inability to come to terms with problems and applications usually derives from an inadequate understanding of the concepts for the processes that will be involved, rather than difficulties in carrying out any calculations that are required.

On the other hand, in an age when calculators and computers seem able to carry out every possible computation more quickly and more accurately than the most competent humans, it seems obvious to question the need for traditional computational skills. Yet, to have control over these tools and to be aware of their strengths and weaknesses demands a high level of understanding of the processes that they are able to do so readily. This understanding is unlikely to come from the use of the machines by themselves. In the first place, there is a need for materials that reveal the underlying concepts and that also show the individual steps required and the relationship from one step to the next. Building up the pencil and paper algorithms themselves allows insight into the processes that seem to be so automatic. Understanding does not come from merely carrying out a task; it requires reflection and the overall view of what is occurring that can only be seen through seeing a process in its entirety. It is still important for children to develop computation with pencil and paper, but this needs to be seen as only one facet of a fully developed computational ability that includes a proper and full conceptualisation of each operation, facility with the basic facts on which each process is built, a capacity for mental computation and estimation and a knowledge of how and when to use calculators and other electronic assistance. Above all, it is the confidence to choose which aspect of computation is appropriate for a given situation or task that is most important in an age when routine computations can be done at the touch of a button. Without control and understanding, it is possible to be more wrong, more quickly and more frequently than ever before. With understanding, it should be possible to solve more complex and more demanding problems as a matter of course.

However, essential numeracy cannot be confined solely to number work and the concepts, facts and algorithms for the four operations. In the ever more technologically focused world in which today's primary school children will live their lives, measurement and spatial abilities and understanding will be essential; more emphasis will be required on the associated calculator and estimation abilities and on interpreting graphical and statistical information. Hence the inclusion of chance and data as a separate and full strand of mathematics at all levels of schooling as a means of dealing with uncertainty and complexity, rather than simply leaving it as a branch of measurement out of which it has grown. Within the measurement strand, children still need to understand the attributes of length and perimeter, area, volume and capacity, mass and weight, angle, temperature and time that underpin them. Only with a full conceptualisation of measurement will an ability to measure accurately, use formulas, make estimates and accurate calculations in everyday situations develop. Geometric knowledge, processes and insights, particularly a well-developed spatial sense, are also crucial in everyday life and most occupations. Without adequate knowledge of two-dimensional and three-dimensional shapes, their presentation in diagrammatic form and the relationships between these shapes, their representations and properties, it is not possible even to begin to solve many of the problems that will be faced in further mathematics, employment and real life situations.

As problem solving has become the main goal of the mathematics syllabus, the use and interpretation of number, data, chance, geometry, measurement and computation that this demands means that the need for a properly established background of proficiency and understanding has become even more essential. At the same time, the inclusion of particular problem-solving considerations has diminished the time available to develop this knowledge, even though the integration of calculator use from the earliest years and the availability of computers for other mathematical tasks have to some degree offset this loss of time by requiring less highly developed skills. This use of technology also increases the need for well-understood processes; not only the understanding that can direct the use of calculators and computers but also an ability to estimate results, to make approximate calculations and diagrams and to check the sensibility of particular answers. As *A National Statement on Mathematics for Australian Schools* highlights, it is important that children

> gain considerable experience in dealing with non-routine problems and unfamiliar situations. Choosing and using mathematical ideas to understand, to explain and to solve, can and should happen at every level of mathematical development. Considering whether mathematics might help to deal with situations which are not necessarily mathematical and judging whether a situation is one in which mathematics might appropriately be used are important. Students need to recognise when mathematics might be useful, choose the mathematics, do the mathematics and evaluate its effectiveness in the circumstances (1991, pp. 12–13).

Organising the development of this background demands an analysis of the requirements for each concept and process. This has always been recognised but, in the past, attention was largely given solely to the mathematical needs of the task. The basic components of each task were organised and presented in such a way that before a new level was encountered each subskill had already been met and learning proceeded from the simple to the more complex. However, while this type of analysis and organisation is necessary, it is not sufficient. Children do not simply receive knowledge as it is given and organised; they construct their own meanings for what is presented. If the order of presentation does not accord with their needs as novices, they will easily form misconceptions related to their individual prior knowledge and understanding. It is necessary in planning the introduction of number and spatial ideas, measurement, chance and data, and computational processes to consider not only the underlying mathematical understandings but how these are best developed by children. Materials and models are needed to provide a basis for the concepts. A meaningful language is needed to name numbers, geometrical and measurement notions, to direct computational processes and to lead to a system of recording numbers and computation that is understandable both in terms of the mathematical ideas and the children's developing mathematical worlds.

Children begin school with a large base of informal knowledge but this is often incomplete and needs to be fully formed and based on mathematical reasoning. Whereas past emphasis in mathematics learning stressed skills as the major aspect, there is now a need to ensure a balance between processes and applications, with understanding seen as the facilitator between the two. Consequently, a program for building up mathematics in the primary school initially needs to address the concepts and thinking that underpin applications but should move largely to a focus on applications and problem solving in the later years. This can only be achieved if understanding is seen to be fundamental so that the knowledge to be acquired is reduced to a manageable amount through building from one idea to another; to seeing the content as an interconnected set of concepts and processes rather than a collection of distinct topics or isolated procedures.

Initial ideas should be based on authentic problem situations—situations and solutions that are accessible to the new learners and that allow general patterns to be discerned. This knowledge can then be turned around and applied to the story situations that have traditionally been the basis of applications in the beginning years. At first, most applications will involve a fairly direct use of this knowledge, but as processes and understanding are consolidated, more complex problems can be introduced and emphasis can be given to developing the techniques and strategies on which such problem solving depends. If new processes and understanding are seen to grow out of existing knowledge, time will be available to allow new ways of thinking about problem solving to evolve. If mathematics is simply seen as a collection of distinct topics, either there will be no time to allow the development of particular problem solving approaches or the range of processes and understanding on which problem solving depends will not be available. Planning and implementing a sequence of learning activities to build up mathematics in this way is crucial if all children are to be able to achieve their potential and if society is to receive the mathematically literate members on which it depends.

CHAPTER THREE
Numeration for whole numbers and fraction ideas

In reading and saying numbers there has been a tendency merely to name the digits in their order. Thus, 2093 may be read 'Two, oh, nine, three'. This practice is of course all right for a telephone number but it does not tell *how many*—to omit the terms hundreds, thousands and the like is to rob numbers of their quantitative content . . . for [number] to be meaningful, children must understand the whole numbers in terms of ones, tens, hundreds and so on. They need abundant experience in constructing numbers, supplemented by experiences in recognising the totals of objects (using bundled sticks or similar) constructed by others. These experiences and the resulting understandings pay large dividends later on when the rational principles involved in computation are learned.

Brownell, 1945, pp. 484—485

Section I: Whole number numeration

NUMERATION is concerned with understanding numbers and their properties. This understanding underpins many of the uses of mathematics in everyday life, ranging from the way numbers are used to quantify items and situations, through the building of recording and retrieval systems, to their uses in computation and measurement. Within mathematics, numbers are used to organise diagrams and constructions in shape and space, to depict situations in chance and data concisely, and, through computer simulations, to represent the most complex modelling of engineering, economic and geophysical conditions.

The importance of numeration is not just related to its direct uses. Coming to terms with numbers establishes many of the thinking patterns on which mathematics relies. For example, the equivalent forms used for numbers suggest that other problems can be represented in diverse ways; the manner in which new numbers build on from earlier numbers shows the importance of prerequisite understanding in the construction of further mathematics; and a sense of numbers coupled with rounding is essential for estimation and approximation processes.

Further, misconceptions and gaps in number understanding have proven to represent the largest source of children's difficulties with numeracy. Underdeveloped meanings for the concepts of zero, place value or renaming are major factors in computational errors. Many of the strategies for building up basic facts also utilise numeration ideas, so a lack of these may create an inability to learn addition and multiplication facts or recall them for use within the computational algorithms. Consequently, although a child's errors may appear in addition, subtraction, multiplication or division, the source of the difficulty will often lie in an inadequate grasp of numeration. Difficulty in solving multi-step problems is also often due to an inability to conceptualise numbers and their relative importance, leading to numbers being manipulated without any relationship to the sense they take on in the problem context.

Consequently, if mathematical ideas are to be learned without confusion, it is important that the concepts and processes that give meaning to numbers be well established. A number itself is a representation. It can take the form of materials that directly or indirectly signify a quantity, a word which is ascribed to or describes that quantity, and symbols that are used to record the quantity succinctly. A number is named in each and all of these ways and is represented by the links among them:

Representation

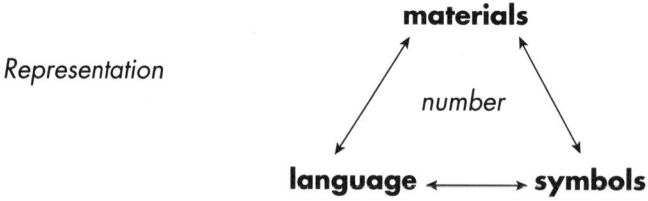

Bringing this notion of representation to the fore draws attention to the fact that numbers are necessarily abstract. They do not exist as such in the real world and have to be constructed anew by each learner. It also raises the idea that numbers can be represented in various ways, so that children will need to be able to recognise these equivalent forms and use this understanding as a basis for counting, rounding, comparing and determining relative magnitudes (*A National Statement on Mathematics for Australian Schools*, 1991). Numeration, then, is fundamentally concerned with the understandings and skills needed to name, rename and process numbers.

Naming numbers

The use of materials begins with counting objects but these are soon structured in some way to give a clearer picture. For instance, with numbers to ten a ten frame is used that highlights both a pattern that can be readily seen and the fact that 10 of the objects themselves form a new unit—that is, that 10 ones make 1 ten.

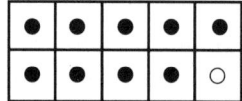

As numbers become larger, these materials give way to Bundling sticks that also show that 10 ones are 1 ten

and then to a more abstract use of Base 10 materials to show the relationship continuing to 10 tens forming 1 hundred and 10 hundreds forming 1 thousand:

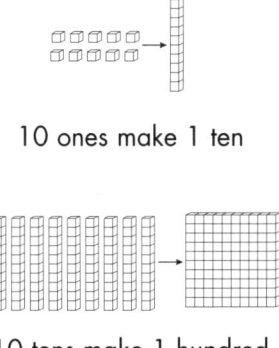

10 ones make 1 ten

10 tens make 1 hundred

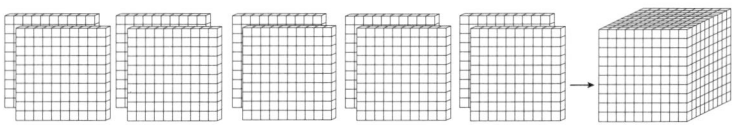

10 hundreds make 1 thousand

While the language used to name the initial numbers is quite arbitrary, later numbers have a relationship to these which reflects the base ten nature of the number system. For instance, the word six is simply assigned to the objects ::: but it is then more apparent why six tens are named sixty, six hundreds are named six hundred or six ten-thousands are named sixty thousand. Similarly, the initial, arbitrary number symbols or digits 0–9 are used repeatedly in a place value system whereby the position or place that a particular digit occupies is used to signify its value. Thus the digit 6 represents 6 ones by itself, 6 tens in 365, 6 hundreds in 8651, and 6 thousands in 6728.

However, after the three-digit and four-digit numbers, there is also a new way of writing and reading numbers that needs to be developed meaningfully. The use of materials is no longer possible, but the digits are grouped to reflect a new focus on the thousands, millions, billions and so on. Thus large numbers, for example 476 302 917, are written with a space separating each group of three digits and read as 476 million 302 thousand 917 rather than place by place as with the earlier numbers.

Naming numbers is the most fundamental concept needed in numeration. However, even when number names are known, it is not always an easy matter to read them or to write them. Most numbers are simply read from left to right like words, but there are exceptions. While 68 signifies sixty-eight or 6 tens 8 ones, 18 expresses eighteen or 1 ten 8 ones. The manner of reading is the opposite of the way in which the number is written. This difference between the reading and writing of teen numbers continues for larger numbers: 618 is six hundred and eighteen (six hundred and eight and one ten) while four thousand six hundred and eighteen (four thousands, six hundreds and eight and one ten) is written 4618, and so on.

68   18   618   4618

Numbers are not always read from left to right

Numbers that contain zeros pose further difficulties: while 60 represents sixty or 6 tens, 605 signifies 6 hundreds 0 tens 5 ones, or six hundred and five rather than the sixty-five that some children see. Similarly, although a number might be said 'six thousand and eight', it is not written 68 but 6008 where the zeros show that there is nothing in the hundreds and tens places. In order to learn to read, write and use numbers meaningfully, it is crucial that they have a basis in materials that show the place value system on which both symbols and language are structured.

thousands	hundreds	tens	ones

Zero shows there are none of a particular place

Renaming numbers

In everyday use, numbers often need to be renamed in a variety of ways rather than be simply understood in terms of counting or even place value. At its most fundamental, a number such as 68 can be viewed not only as 6 tens 8 ones but as 68 ones. Further possibilities arise with larger numbers, so that 268 can be interpreted as 2 hundreds 6 tens 8 ones or 26 tens 8 ones or 268 ones. It is particularly important in computation that a number such as 608 can be renamed as 60 tens 8 ones. These alternative representations can be developed through the use of Base 10 materials and then consolidated with a number expander which can be unfolded to see the various ways in which the number can be renamed:

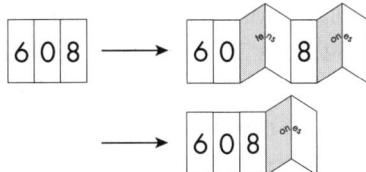

A number can be renamed in several equivalent ways

Within subtraction it is also important that 68 can be seen as both 6 tens 8 ones and 5 tens 18 ones, while in division, 6857 may need to be interpreted as 68 hundreds 5 tens 7 ones:

$$\begin{array}{r} 5\;18 \\ \cancel{6}\;\cancel{8} \\ -\;2\;9 \\ \hline 3\;9 \end{array}$$

68 is renamed as 5 tens 18 ones
to allow the subtraction to proceed

$$\begin{array}{r} 9 \\ 7\overline{)6857} \\ \underline{63} \\ 55 \end{array}$$

Since 6 thousands cannot be shared,
68 hundreds are shared among 7

Problem situations might require 689 to be viewed as one hundred more than 589 or twenty less than 709.

Number processes

There are also crucial number processes that build on an ability to name and rename numbers. Numbers need to be compared, ordered and sequenced using the materials with which they are represented and the place value with which they are written as a basis for developing meaningful processes. For instance, when a number 1 hundred less is called for, a child needs to see that a number such as 6825 has 8 in the hundreds place and simply decrease this digit by one to give 6725 rather than endeavour to count back or subtract 100. Similarly, numbers such as 8517, 8576, 8502 and 8525 all have 85 hundreds and can be ordered by simply examining the tens and ones places.

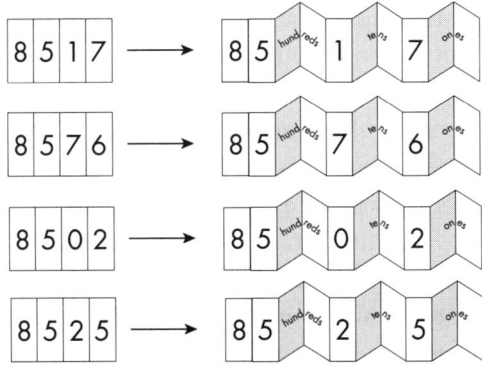

Counting on and back in ones, tens, hundreds and so on use the numbers themselves as counting objects. For example, since 342 has 34 tens, it is possible to count back in tens by thinking 34 tens 2 ones, 34 tens 1 one, 34 tens 0 ones, 33 tens 9 ones, 33 tens 8 ones to give 342, 341, 340, 339, 338. An ability to round numbers is essential for estimation, approximate calculations and in checking results. This too needs to be built on a sound basis of place value, renaming and the use of materials so as to become a process that can be used with full understanding in contrast to the rote procedure of 'rounding up' or 'rounding down' so often applied incorrectly. Since 4735 has 47 hundreds, it will round to 47 hundreds or 48 hundreds; 35 is closer to 47 hundreds, so 4735 rounds to 4700.

Teaching numeration

Numeration is the foundation for most of the thinking processes used in mathematics as well as being a crucial topic in its own right. Yet, the teaching of many aspects of numeration has frequently been inadequate. In particular, number symbols have often been the focus of instruction in place of the materials and patterns that allow insight into number meanings. Children first need to visualise numbers and then use materials as a basis for talking about them in order to develop the language used to name numbers. Only then will the symbols make sense and be used as a concise representation of ideas that are understood. Moving too quickly to the symbols leaves children at the mercy of memorised rules which have no sense to them. It is for this reason that such teaching is frequently referred to as 'symbol-minded'.

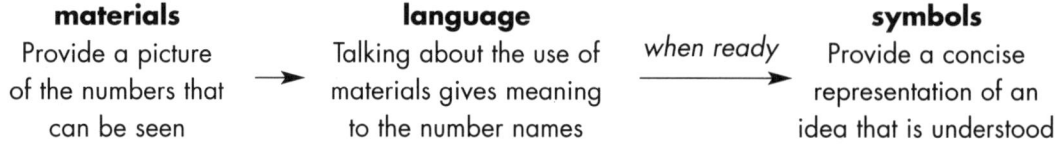

| **materials** | **language** | | **symbols** |
| Provide a picture of the numbers that can be seen | → Talking about the use of materials gives meaning to the number names | *when ready* → | Provide a concise representation of an idea that is understood |

Another reason for children's difficulties is that the sequence of steps used to establish number understanding has not always been in accord with their needs as they meet powerful ideas for the first time. To someone who already has a knowledge of numbers, the early ideas are simple and basic when in fact they are quite complex and provide the *basis* for all future number development. Concepts such as zero and the notion that ten is a new unit in the number system are essential for the development of place value but this is not at all obvious to children when they first meet them. Zero may be dismissed as nothing, while ten is simply seen as the next new number. If the teen numbers are then built up one at a time in the counting order—eleven, twelve, . . ., nineteen—children may have no awareness of the tens and ones structure of numbers. Building the new unit—10 ones make 1 ten, 10 tens make 1 hundred and so on—should be the first step in developing new numbers. This new place can then be linked to the earlier numbers, leaving the occurrence of zero until last so that its meaning is appreciated. Sequencing teaching in this way, so that new ideas grow out of those already built up, is far more productive than simply developing numbers in line with the counting sequence readily used by those who already know the numbers.

Often the teaching of number has not brought out the full set of meanings that will be needed later. When a teacher refers to a number in a 'house' or 'column' instead of the digit in a place, why should a child consider place value important? Rounding, counting on and counting back need to build on place value ideas rather than relate to numbers in a (number) line. Materials should be used to establish renaming as well as place value to provide children with a variety of interlinked ways of thinking about numbers. Not just in terms of the counting with which number meaning begins, or even with the system of place value that governs the construction and computation of numbers, but also with the various ways in which a number can be represented through renaming.

Learning numeration

A lot of number learning occurs informally, even before a child reaches school. Many children arrive at school with a substantial grasp of counting and the numbers to ten. It follows that a program limited only to these numbers will be neither stimulating nor motivating. Nor will it restrict a child's number development to these small numbers. Some children will develop their own 'understanding' of numbers to one hundred, but few are likely to build this on a basis of place value, rather than counting, without specific teaching. Mathematics programs need to take into account the time when new ideas are best introduced as well as the order in which numbers are best developed. Indeed, a major cause of children's difficulties with mathematics has been that their own ways of dealing with numbers have been able to develop when the school program has lagged behind their capacities and needs. For instance, when children learn the addition facts, teen numbers are essential for most of the answers, yet may only be introduced at the same time as they are needed. Children may then count on their fingers to find answers instead of thinking in terms of tens and ones. Similarly, if children are to have control over addition with two-digit numbers, they need three-digit numbers to make sense of their answers. Given the effects of inflation, not least in terms of children's expectations of what material goods they wish to have, there are social needs as well as mathematical ones for introducing larger numbers sooner than has been the case in the recent past.

On the other hand, there are other number ideas, such as the way that large numbers are written, that have been introduced *too quickly* for children to do anything but resort to 'rules' without meaning to cope with them. Children may then believe that thousand is related to the writing of three zeros and write 600002 for sixty thousand and two. Other ideas, such as rounding four-, five- and six-digit numbers, have been introduced *too closely together*, leading children to focus exclusively on the largest place to give an answer such as 800 000 when asked to round 825 217 to the nearest ten thousand. Renaming is an idea that for many was *not addressed at all*. Yet in many applications it is necessary to consider the number of tens, hundreds or whatever that a number has rather than simply focus on the digit in that place.

For example, the number 9804 has 9 in the thousands place, 8 in the hundreds place, 0 in the tens place and 4 in the ones place, but there are actually 98 hundreds, 980 tens and 9804 ones:

>9804 has 98 hundreds 0 tens 4 ones
> or 980 tens 4 ones
> or 9804 ones

For five-digit and larger numbers, there is also a second place value pattern, yet without specific teaching designed to highlight this system many children are unaware of it and become confused with these large numbers. They may be unsure of how large one billion is and not realise the pattern for larger groupings based on Latin counting numbers—*milli*on with its underlying meaning of thousand thousands is the first place after the thousands, *bi*llion is the second place, *tri*llion the third place, *quadri*llion the fourth place and so on.

Place value

While an ability to make full use of place value develops over many years, 'the foundations must be carefully laid during the early years through games, counting and grouping activities, and calculator explorations' (*A National Statement on Mathematics for Australian Schools*, 1991, p. 111). In fact, the concept of place value is developed along with the two-digit numbers and it is important to bring the full aspects of these numbers to children's attention soon after the one-digit numbers and the concept of ten have been built up. In contrast, a lot of the first year of school has traditionally been given to activities that are concerned with logical tasks such as ordering, sorting and classifying because of a belief that these are necessary prerequisites for the development of number. Research (Clements and Callahan, 1983; Cobb, 1987; Kamii and Housman, 2000) has thrown doubt on this supposition, highlighting the importance of counting-based cognitive tasks instead and revealing the stronger link between logical activities and the development of general problem-solving abilities.

Children first meet numbers to ten through patterning and counting activities, learning the sequence of number names and associating these arbitrary names and symbols with objects so as to give them meaning. For numbers greater than ten there are no new number symbols. Instead, there is a set of rules that generate the new numbers from those that have already been learnt. Larger numbers are based on notions of place value:

> 45 is 4 tens 5 ones
> 964 is 9 hundreds 6 tens 4 ones
> 3781 is 3 thousands 7 hundreds 8 tens 1 one

However, place value only refers to the written form of the number where the value of any digit in a whole number (or decimal fraction) is determined by its position in that number.

Materials that show grouping in tens and ones and highlight the nature of ten as a new unit (10 ones are 1 ten) are crucial in building up a knowledge of place value as the basis for number understanding and in establishing the two-digit numbers:

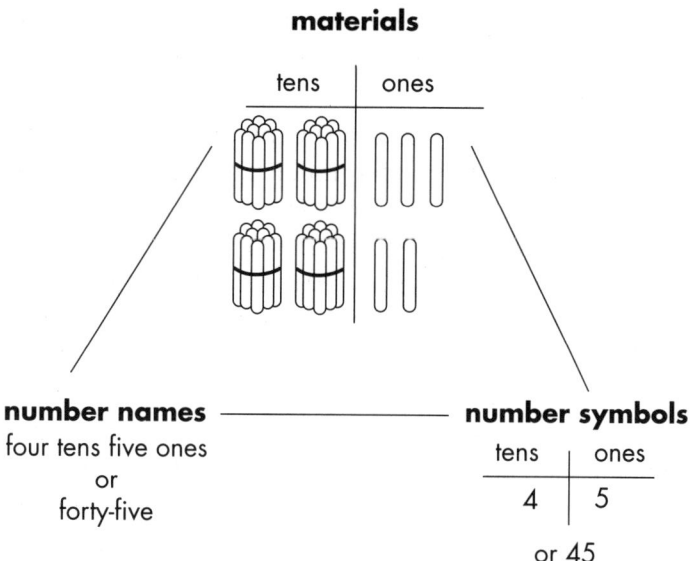

While teen numbers are written and represented with materials in exactly the same way as the numbers 20–99, the conflict between the pronunciation of the teen number words and the order of the symbols used to represent them has induced children to record them incorrectly, for example writing 81 for eighteen, or confusing eighteen with 80. Consequently, the development of the teen numbers is best left until last when the focus can be on the irregular language that is the source of children's difficulty.

This sequence of development for the two-digit numbers provides the foundation for numeration ideas, establishing the fundamental importance of place value and highlighting the need to proceed in an order and manner that allows children to construct meaning for new numbers from those they already know. It readily extends to three- and four-digit numbers, where Base 10 materials also focus on grouping 10 tens to make 1 hundred and 10 hundreds to make 1 thousand. However, larger numbers cannot readily be represented with materials so instead must depend on an extension of the Base 10 pattern to provide meaning through the associated language. In turn this language provides a meaningful link to the recording of the number symbols.

The concept of zero also causes difficulties for children, not only in reading and writing larger numbers but because the notion of 'writing something for nothing' itself seems odd. The reasons for writing a zero in a particular place as well as the significance of what the zero shows need to be carefully brought out when the numbers are developed. Materials do not necessarily assist because when there are no tens or hundreds there is nothing to see with the material either. Instead, care must be taken in sequencing the development of the number names and the number symbols to allow an appreciation of the lack of particular units and the use of the zero to signify that there is none of that place. It follows that the reference to zero as a place holder is also misplaced. Since the zero tells us that there is none of a particular place, it can hardly be 'holding the place' for any other digit! There are no place holders in numbers; either there is some digit already in the place, be it 0 or any of the nine other digits, or the number is incorrectly written.

Rather than passively taking in numbers in the manner and order presented by adults, children actively construct their own number understanding. Much of this construction is independent of formal instruction, based on language understanding, but active intervention in the form of schooling is necessary to provide meaning through an awareness of place value ideas. Materials are needed to bring out the Base 10 nature of the number system, reasons can be provided for the names used for larger numbers, and the sequence of development must reflect this constructive process. While the size of the numbers and their order within the number sequence might seem an adequate basis for organising their teaching, it is even more necessary to consider the development from one new idea to the next, the regularity of the patterns in naming the numbers and the difficulties that could arise in recording them. This has produced a different sequence from that traditionally used but one that experience has shown to be particularly smoothly effected with young children. But, while the specific teaching of numeration ideas is a very necessary aspect of the development of number understanding, it is important to realise that the topic is not self-contained. Just as numeration ideas are crucial for the development of computation, so the use of these processes also plays a critical role in bringing about a full realisation of the significance of numeration.

SEQUENCE FOR DEVELOPING NUMERATION

1. Introduce the numbers 0–9, linking objects to number names and number symbols and building in counting on and back, comparison, sequencing and ordinal numbers.
2. Establish an understanding of ten as a new unit composed of 10 of the ones: 10 ones are 1 ten. Then introduce the multiples of ten: ninety, eighty, . . ., twenty.
3. Develop the two-digit numbers: the numbers 20–99 first, then the teen numbers 11–19.
4. Build up the next multiple: 10 tens are 1 hundred and develop the three-digit numbers, including internal zeros and renaming to see that, for example, 406 has 40 tens.
5. Extend place value to each new multiple in turn: 10 hundreds are 1 thousand, 10 thousands are 1 ten thousand, 10 ten thousands are 1 hundred thousand and so on, linking the new place to the earlier numbers, including the many forms in which internal zeros and renaming occur.
6. Build in the additional place value aspects: numbers are grouped in hundreds, tens and ones of thousands, millions, billions and so on, as numbers are extended beyond six digits.
7. Extend the ways of writing numbers to include exponents, standard form and scientific notation.

One-digit numbers: the basis for all numbers

Children acquire their initial number understanding prior to beginning school, gaining early number words and building up informal methods for adding and subtracting numbers based on counting. Indeed, many children begin school with an ability to recite the numbers to ten, twenty, one hundred or beyond. However, this ability often does not extend to an understanding of all of these numbers—all that many children can do is reel off numbers in a fixed order, just as they would chant the words of a song. They would not necessarily correctly identify them with a group of objects or with a written symbol indicating the number of objects in the group.

Numbers may also be repeated or omitted from their number sequence altogether—for example, young children often leave out fifteen when they count beyond ten. It is also common for children to say the number names in a fixed but non-standard order. A child of three or four, for instance, might 'count' one, two, three, four, nine, eleven, fifteen, eight. In this case, the last number said does indeed name how many objects there are, but the sequence of words used would not allow other groups of objects to be counted accurately. In fact, it is usual for the standard number sequence to be built up in this way; gradually the stem of correctly ordered numbers is extended but the idiosyncratic tail of numbers will stay fixed over a short term until it undergoes modification to become more like the standard counting sequence (Fuson, 1988).

Thus, children's initial counting tends to be simply a repetition of a sequence of number names, largely learnt in the home, from other children or when playing games. These early number concepts and understandings are then extended and broadened during the first years of school, largely, but not solely, as the result of teaching specifically directed at mathematics. The first task is to provide full meaning for the number words, matching them to objects, and building from this understanding to the number symbols used to represent them.

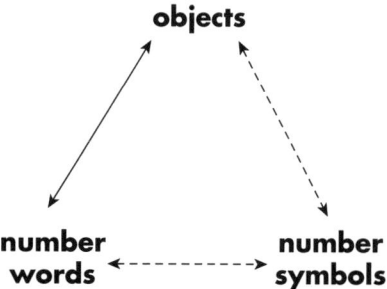

Many different forms of number symbols evolved throughout the world to match local needs, mostly based on a system of ten reflecting the number of fingers on two hands. In some systems, such as those used by the ancient Egyptians or Romans, there were symbols for ones, tens, hundreds and so on. Eventually, a more efficient system incorporating only ten symbols (called *digits* after the Latin word for finger) and using a positional notation called *place value* to denote numbers, small and large came into being. This system of number words and symbols is commonly called the Hindu-Arabic number system, reflecting its origin in the early Hindi numbers in India around the seventh century, although these in turn owe much to the earlier numbers of the Chinese, and their transmission to Europe via the Arab culture with the spread of Islam.

It was the acceptance of zero as a number that really marked the development of a number system as such rather than simply a systematic method for recording numbers. The last of the fundamental numbers to be developed, the concept of zero took shape in Hindu India between the fifth and eighth centuries (Kaplan, 2000, p. 42) as a means of representing 'none of something' to parallel the way '3' represents 'three of something'. This origin is revered in literature, as well as scientific circles, as being one of the greatest inventions of humankind:

> 'Can you teach him to bat properly?' Nakul pointed at his twin. 'He was out for zero again.'
>
> 'Zero?' the sadhu laughed. 'Well that is nothing to be ashamed of. The English game of cricket would never have taken shape with the Indian zero . . . our ancestors were the first to conceive of zero. Before that mathematicians, from the Arabs to the Chinese, left a blank space in their calculation; it took Indians to realise that even *nothing* can be something. Zero . . . embodies the unchanging reality of nothingness.'
>
> 'But zero's still zero', Nakul said.
>
> The sadhu roared with laughter. 'Not quite,' he said. The Indian zero is no empty shell . . . It is empty of numerical value. But it is full of non-empirical possibilities. It is nothing and everything . . .'.
>
> Source: *The Great Indian Novel*, Tharoor 1984, p. 169
> © Shashi Tharoor. Reproduced by permission.

It was soon adopted by Arab mathematicians who marvelled at the idea of writing 'something for nothing' and named it cipher (symbol) which eventually came to us as the number name zero. Just as this development took some time to occur in mathematics and needed even more time for its usefulness and importance to be widely appreciated, so individual children often find difficulties in coming to terms with the notion of a number that signifies none of something.

Naming numbers

The first numbers are quite arbitrary in nature, as evidenced by the fact that modern Hindi and Arabic numbers have a different appearance from those we use, and the symbols and words for the numbers 0–9 have to be attached to groups containing these amounts by a process of association. Language plays a crucial role in this process by providing the link between objects and symbols and giving meaning to the numbers. While any objects could be used to develop this understanding, in order for children to move beyond counting to a representation of numbers in their own right, it has been found that the use of patterns which show how one number builds on and links to other numbers is most helpful. In particular, the use of a *ten frame* on which counters such as Unifix can be placed in a doubles-based pattern has proved to be most effective, building in a meaningful notion of zero and laying the basis for the concept of ten:

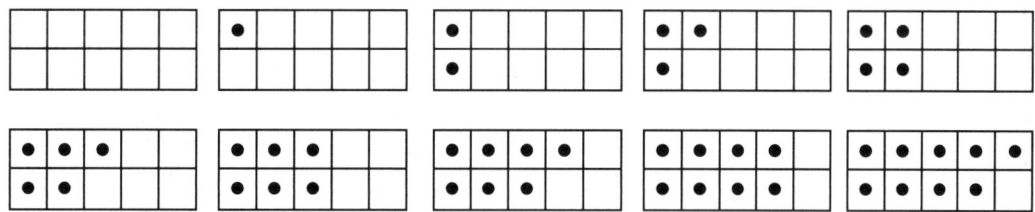

When children have learnt the oral number names and related them to the ten frame patterns, the number symbols can be introduced. At first these need only be matched to the objects and oral number names, but as soon as practicable an ability to write the number symbols and words needs to be developed. Care needs to be taken in introducing efficient ways of forming the symbols in a similar manner to the introduction of the letters of the alphabet as a precursor to meaningful writing. Number symbols with similar appearance such as 6 and 9 or 2 and 5 are frequently confused and children sometimes have their own idiosyncratic assignment of names to the number symbols since there is nothing to indicate how a particular symbol relates to the number of objects it represents:

0 1 2 3 4 5 6 7 8 9

While the number symbols are difficult enough for children, the written number words seem to be even more arbitrary and involve particularly difficult letter and sound combinations:

zero one two three four five six seven eight nine

For example, although one has a 'w' sound, there is no letter w in one, while in two the 'w' is not sounded. Nonetheless, a complete conception of the one-digit numbers requires an ability to write these number words and they should be built up as soon as possible, using carefully developed teaching materials.

◆ Activities with ten frames and counters such as Unifix and cards, dice or spinners:
  ◇ Each child can be given 10 Unifix and a ten frame. Cards with the symbols and words are placed in the centre of the table, at first only numbers 1–4 might be used. Later zero

and the remaining numbers 5–9 can be introduced. Each child in turn picks up a card and puts Unifix onto their ten frame to match. A game can be made of this activity where the first child to get five correct is the winner.

◇ A board game can be used with the ten frames on a spinner and the words and symbols on the board. At each turn, a child moves his or her marker to the next space that corresponds to the number on the spinner. (A computer version of this game is available on *Win with Maths!* NZCER 2002.)

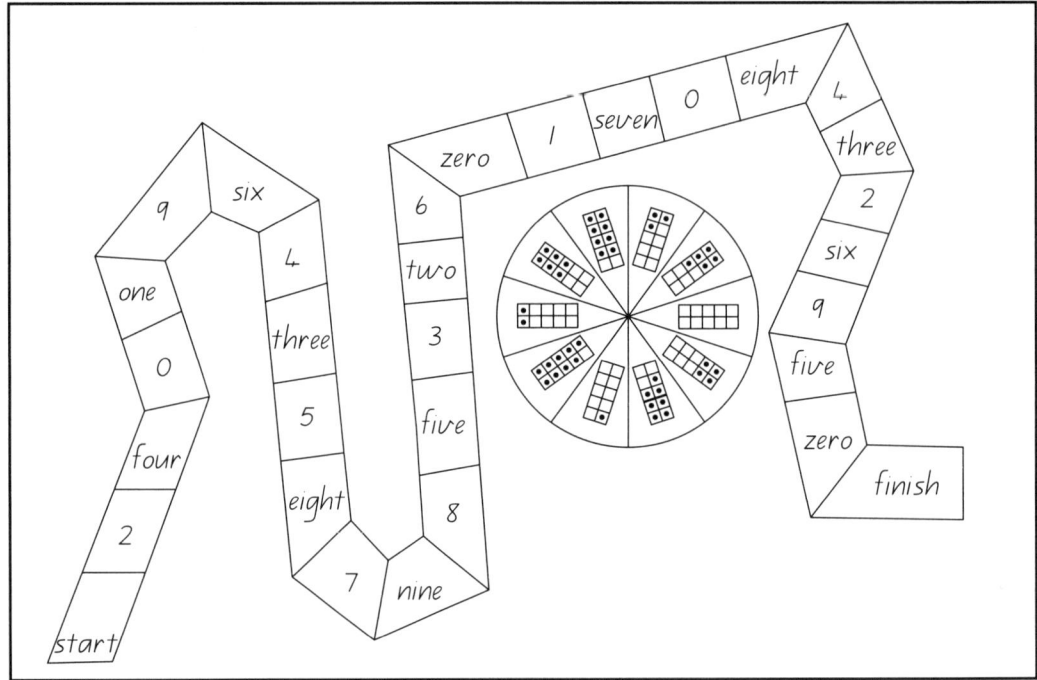

◇ Similar games can be made without a spinner, where the ten frames and number words are on the board and a ten-sided dice is rolled to indicate the number symbol 0–9.
◇ A chart showing the ten frames, number symbols and number words can be provided for children to check that they are making the correct match. It should be removed, or downplayed when children have formed the connections in their minds.

◆ 'Mix and match' cards for each of the numbers 0–9 such as:

There are many commercial forms of this activity available.

◆ Sets of cards corresponding to each of the numbers 0–10 can be used to play games such as 'fish', 'snap', 'concentration' and so on.

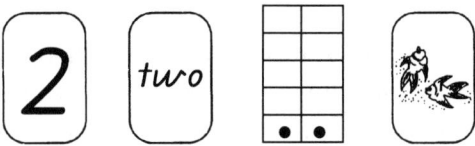

Ordinary playing cards may also be used.

◆ 'Bingo'—children place a counter on the number word or number of objects displayed on the bingo board corresponding to the number symbol that is shown on the card.

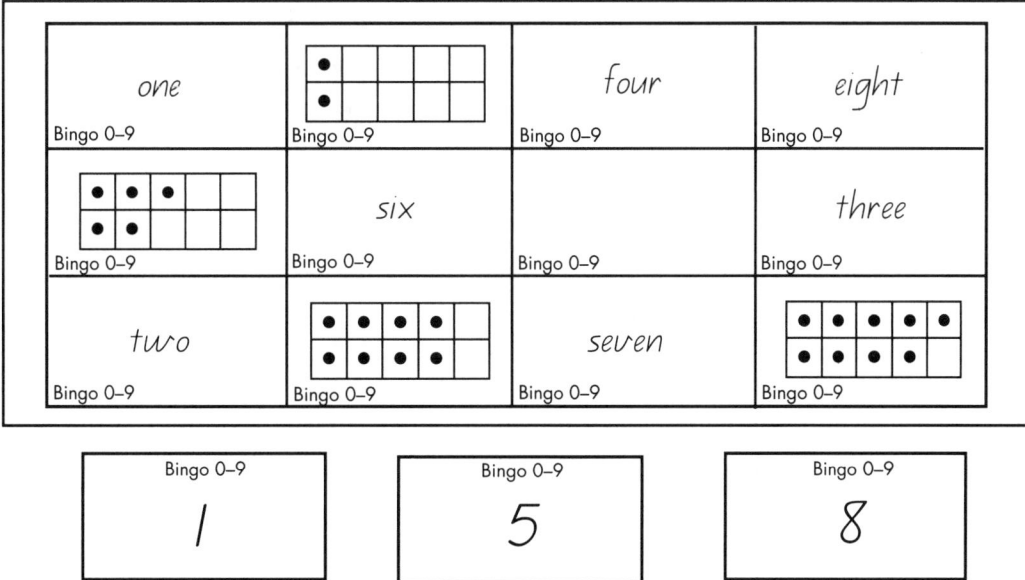

◆ Many board games that children play in and out of school can be adapted to develop number understanding rather than used after this has been established. Most games use the commonly available six-sided dice with dot patterns that have no relationship to one another, so that children are often led to count the dots to determine a move. There seems little reason to restrict the numbers used to only 1–6 when children need to come to terms with all the one-digit numbers, zero in particular. These can be replaced by dice or spinners which have the number symbols, number words or the dots displayed in the ten frame patterns in the first instance and to include all digits 0–9 as soon as possible. Ten-sided dice with the digits 0–9 are readily purchased but there seems to be no special reason to use dice with 8, 12 or 20 sides, except for variety and interest.

Similarly, rather than play proceeding by simply counting spaces to correspond with the number shown on the dice or spinner, it is preferable to make a move to a word or symbol that matches the number of objects. In this way, the links among these three representations come to be the fundamental understanding of early numbers so that children are less likely to depend on counting as a basis.

Counting

When meaning for each of the one-digit numbers is established, some time needs to be spent reviewing and reinforcing them, including zero, to build up the usual counting sequence. This should involve the use of materials on a ten frame, not just the number words or symbols alone, so that a mental representation is formed in which questions such as 'What number comes just after . . .?' prompt the inclusion of 1 more object and the consequent changing of associated number word and symbol.

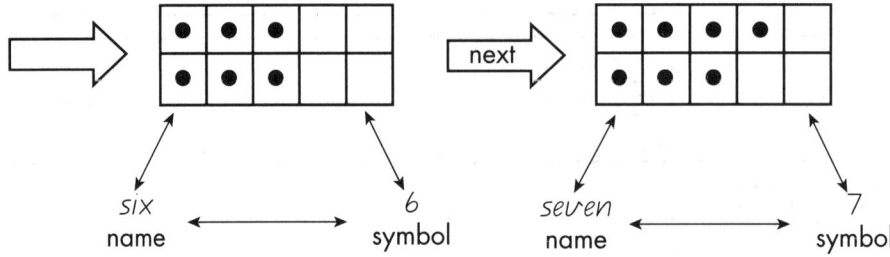

In time, this process will become automatic, and only the number words or symbols will be needed, but most children will need a good deal of practice involving objects first. Activities such as using a calculator to practise matching of oral names and number symbols, counting forwards and backwards by ones, finding the number that is 1 more or 1 less, finding the number between, and finding missing numbers are all necessary to consolidate the number sequence. *Counting*, then, is a complex process that involves coordinating several different ways of thinking:

1 Stable, ordered sequence of number words—number names have a fixed order. Young children realise this when they learn to recite the number names, irrespective of whether the order they use is the standard one.
2 One–one match of number words and object—when counting a collection of objects each item is tagged with one and only one number name.
3 The last number name said tells how many—this is the *cardinal* aspect of a number.

Children also need to accept that objects may be counted in any order. When we count a collection of objects we order them as we assign each number name. However, it does not matter which order we select; the total number will be the same. Nonetheless, even after children have acquired meaning for each number, teachers may observe them omitting some items when counting a group of objects or assigning two or more number names to a single object. This is particularly prevalent when the objects to be counted are scattered randomly rather than placed in a straight line and difficulty in counting objects placed in a circular arrangement is still common among older children.

Early-mathematical thinking

In the past, research based on the work of Jean Piaget (1896–1980) suggested that the teaching of number may be ineffective or even detrimental unless preceded by the development of conservation of number and the abilities to sort, classify and order objects. This traditional view of these logical operations as prerequisites for number led to them being called pre-number skills. Conservation of number meant an ability to determine that the number of objects in a collection remained constant when their relative positions were altered:

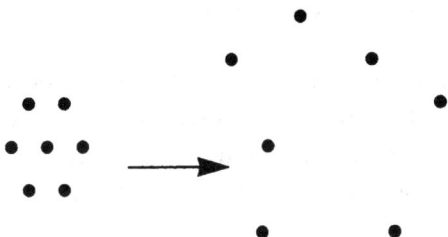

However, many children actually determine this by counting, so that conservation can hardly be a prerequisite to counting. A more meaningful definition of *conservation* is the ability to determine how many objects regardless of their spatial arrangement.

Mathematics education research has suggested that complex counting activities form a more appropriate basis for introducing number understanding and processes to young children. They also need to be provided with opportunities to solve challenging problems as part of their regular classroom experiences. In the early years, these problems usually demand that children perceive particular features of objects or events, classify them according to familiar criteria, or make perceptual comparisons and ascribe some order based on these comparisons. Other problems require the detection of patterns. This investigation of patterns and relationships helps children extend their understanding of number and develop rules for operating on numbers.

The main value of the traditional *pre-number* skills is as logical operations underpinning more complex skills involved in problem solving. They still have a place in mathematics in the early school years alongside the development of matching and counting, but they are more meaningfully described as *early-mathematical thinking*. Specifically, they are:

- Conservation of number—the ability to determine how many objects regardless of their spatial arrangement.
- Classifying—grouping according to specified criteria such as colour, shape, size, texture, thickness or number.
- Comparing—establishing a relationship between two objects on the basis of some specific attribute such as height, mass, thickness, texture, number. Comparing is the forerunner to ordering and measuring.
- Ordering—ordering builds on comparing. It involves an arrangement of objects, groups of objects or numbers that reflects some rule.
- Patterning—the ability to recognise patterns is basic to mathematical insight. By using real objects, pictures and drawings children can be made aware of patterns and can practise seeing, describing, extending and completing or repeating given patterns.

Young children usually develop the ability to *subitise*, that is, to determine the number of objects in a small group through sight alone. As their experience with number increases, they are able to instantly recognise

●
● as two counters

● ●
●   as three counters

without counting. This ability needs to be fostered and extended to include up to five objects, the maximum number most adults can perceive without counting, or it will be displaced by counting as means of obtaining the correct number in all situations. An ability to subitise is important when counting larger numbers of items, which can then be seen in terms of smaller groups:

● ● ●   can be seen as 3 and 2
● ●   or 2 and 2 and 1

● ● ● ● ●   can be counted
● ● ● ● ●   2, 4, 6, 8, 10

Later, it will be necessary to develop *counting on and counting back*; that is, to start counting from a given number either forwards or backwards rather than counting all numbers:

4, 5, 6, 7 or **8**, 7, 6

This is not at all straightforward as it requires the coordination of several abilities, in particular the ability to consider number words as objects to be counted. For instance, counting on three more from 4 requires that counting one elicits the response 5, counting two elicits the response 6, while counting three elicits the response 7. For many children, counting backwards is even more complex, so that they need to experience many activities requiring the next number or they will simply count all the objects they see. Activities include:

◆ Begin counting from one; then stop and point to a child who gives the next number in the sequence.
◆ Make cards to show each of the numbers 0–9 in symbol, word and ten frame form. Place a mix of the representations starting from zero in ascending order, then turn all of the cards face down. Turn the first 3 cards right way up and ask what the next card must be. Check by turning the card over. Repeat for 3 cards in different positions and gradually reduce this to showing 2 cards or 1 card. Use the cards in descending order from 9 in the same way.
◆ Roll a six-sided dice (digits or words). Children make that number on a ten-frame with Unifix or counters of one colour. Then place 1 more counter of second colour and say how many. Build up to counting on 2 more and 3 more.

- Place six marbles in a tin and write a large '6' on the outside. Drop another marble into the tin. Ask how many marbles are in the container now. Repeat for other initial numbers and add 1, 2 or 3 marbles having the children count on as they see and hear the marble fall.

- Use two dice, one with numbers such as 4, 5, 6, 7, 8 and 9 written on it, the other with 1, 2, or 3 dots in ten frame patterns. Roll the dice and ask the children how many there are altogether. Use these dice when playing board games.
- A calculator can be used for counting on: enter a number, then press + 1; when = is pressed the next counting number appears. Ask children to close their eyes and count on as they press the = key, opening them to see if the number they counted to is the same as the one they keyed in. To count back, press − 1 to begin.

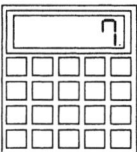

Comparing, sequencing and ordering

One further number process essential for using numbers is the ability to determine which has more. Many children attempt to do this by a counting strategy, counting each group to see which has more which in time can be related to their ability to subitise to see numbers in manageable groups. Others will use a one-to-one matching basis, in effect lining up one object against another, either literally with materials or in some form of picture in their head. Neither of these approaches provide a basis for working with larger numbers nor are they very efficient. Instead, it is preferable to base comparison on the number concept directly, using a visualisation where ten frame patterns are matched to number words and symbols. To achieve this, it is necessary for children to have practice where they compare:

1	Objects with objects:	which has more?	••/••	or	••/•••
2	Objects with symbols:	which has more?	•••/•••	or	3
3	Symbols with symbols:	which has more?	5	or	8

It is sufficient for a child to say which is larger or underline or circle the larger number and the mathematical symbol > is not needed or appropriate at this time. Once children can successfully compare two numbers, activities where they are called on to find the greatest, middle or least number of three numbers are needed. In due course, this can be extended to comparison of several numbers, leading to placing them in ascending or descending order.

Activities that focus on ordering several numbers are made more achievable when the numbers are written on cards that can be moved about as they are compared pair by pair.

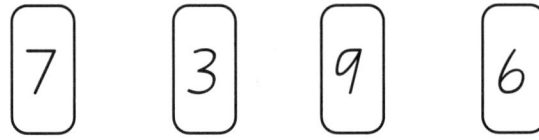

Arrange in order from least to greatest

Card games also provide a very good means of extending an ability to compare numbers to one of sequencing.

BEFORE AND AFTER: A GAME FOR 2–4 PLAYERS

1 Use a pack of playing cards from which the picture cards have been removed or a set of 40 cards showing the numbers 0–9 in symbol, word, ten frame and picture form as on p. 85.
2 Deal out 5 cards to each player.
3 Place the remaining cards face down on the table and then turn the top card face up and place it alongside the pile.
4 Play begins with the first player putting out a card that comes either one before or one after the card that is face up.
5 If a player is unable to put out a card, he or she picks up a card from the pile of cards that is face down.
6 The first player to dispose of his or her cards is the winner.

SIXES: A GAME FOR 3–4 PLAYERS

1 Use a pack of playing cards from which the picture cards have been removed or a set of 40 cards showing the numbers 0–9 in symbol, word, ten frame and picture form as on p. 85.
2 Deal out all of the cards).
3 Play begins with the first player putting out a 6 of any suit (or one of the four representations of 6 if using the set of 40 cards.
4 At each turn thereafter, a player can put out a card which comes immediately after or before and is of the same suit (or representation) as a card already in play or else puts out 6 of another suit (or representation).
5 If this is not possible, play moves to the next player.
6 The cards are put out one under or one on top of the other so that only the numbers are left showing (as in the game of *Patience*).
7 The first player to dispose of his or her cards is the winner.

Ordinal numbers

As well as being used for ordering numbers and objects, the ordinal numbers are essential in helping to construct new number names such as thirty rather than threety, for teen numbers such as fifteen, and all fraction names, thirds, fourths, fifths and so on. Yet, many children lack a proper understanding of ordinal numbers when they begin school. They might know the meaning of first but tend to associate it with last as an assessment of the relative value of the position rather than with the other ordinal numbers second, third, fourth and so on, as an expression of relative position. Since most ordinal numbers are simply given by adding 'th' to the corresponding number name, it is sensible to begin with the most regular ordinal numbers, ninth, eighth, seventh, sixth and fourth, before considering the irregular fifth, third, second and first. These can be shown with rows of pictures where a particular place is coloured:

What place is the grey pelican in?

What is the place 3 birds after the grey pelican?

Further experiences and discussion to come to terms with this new form and use of number can be provided by a matching game based on bingo. An everyday context that is readily available is the spelling of words; they have a clearly defined order of letters and can be chosen to fit the abilities and needs of particular learners. The cards used carry instructions such as 'find the sixth letter in ordinal' and the card is then placed on top of the appropriate letter on the bingo board (in this case the letter a). A computer version of this game is available on *Win with Maths!* (NZCER, 2002).

d	u	c	k
p	o	n	d
f	a	r	m

What is the fourth letter in bundle?

This card would be placed on top of the letter 'd' on the board

The concept of ten

The aspects of early number ideas which require particular attention are the idea of zero and the notion of ten as a unit composed of 10 ones. Ten frames have been specifically developed to provide each of these ideas with meaning and make them seem a natural part of the number system. The part/whole notion whereby 1 ten is 10 ones requires more than simply accepting the value of a new number. While using counters on a ten frame, children can see the 10 ones completely filling the frame to make 1 ten, and internalise this relationship which is so crucial to the meaningful development of larger numbers.

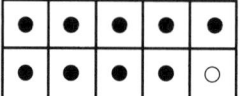

10 ones make 1 ten

While the written symbol 10 seems obvious to those who already know numbers, for young children it is largely seen as just a new way of writing a symbol similar to 9 for nine. Yet in reality, it uses place value and is composed of a 1 in the tens place and a 0 in the ones place. For this reason, the oral and written number names should be given most emphasis at this point. The number symbol 10 will need to be recognised and written, but its full meaning will not be available until the two-digit numbers have been met.

Children's early number concepts are developed as children match objects or pictures of objects with the corresponding number names and number symbols for the one-digit numbers and ten. These should be introduced so that knowledge of one number is used to extend an understanding to the next number with the additional notions of comparison, ordinal numbers and counting on and back built in once the sequence of numbers is established:

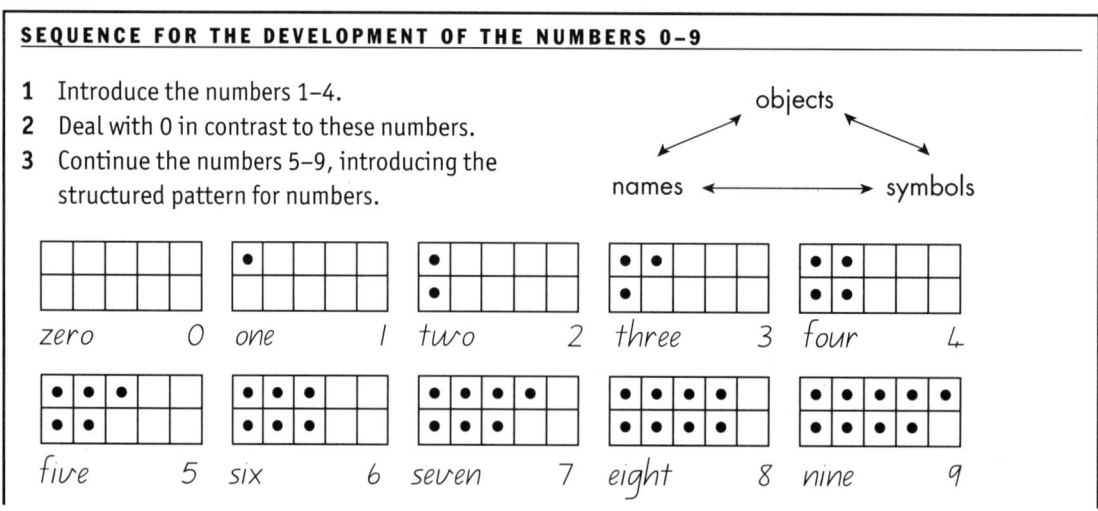

Stress the discrimination of one number from another.
4 Sequence the numbers 0–9.
5 Compare numbers: 1 more, 1 less.
6 Introduce the ordinal numbers: regular cases of ninth, eighth, seventh, sixth, fourth before irregular cases of fifth, third, second, first.
7 Treat counting on and counting back.
8 Introduce the number ten.

•••••
•••••
ten   10

Likely difficulties

Difficulties with one-digit numbers occur when children have not linked number names and symbols to the objects they represent. This shows up in their counting, writing of the digits and in confusion as to the meaning of zero.

- The digits 7, 2, 3 and 5 may be reversed and written as 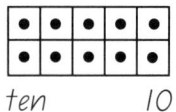.
- A child may confuse the digits 6 and 9 or 2 and 5.
- Zero may be viewed as nothing rather than a number representing none of something.
- Counting may be confused—non-standard number sequence, objects and names not matched, the final count may be 1 more or 1 less when an object is omitted or counted more than once.
- Fingers, tapping of fingers or a ruler may be used to *count all* rather than *count on*.
- When attempting to *count back*, a child may count forward then back 1 to find each number in turn.

Investigate and discuss

1 Counting is such an automatic activity for adults that it is often difficult to comprehend the difficulties that young children have. The activities that follow provide situations where the various skills underlying counting can be examined. Spread out a collection of up to ten counters in front of you. Count them.
   a What number did you get?
   b At what point did you know the number telling how many?
   c How could an error be made? What skills does accurate counting require?
   d Repeat with different patterns or arrangements of the counters. Try to find patterns that help to see how many more quickly than by counting each counter individually.
   e Does counting by twos or threes help in counting quickly? Why?

2 Use your understanding of counting skills to analyse each of the following errors and identify the underlying causes.

Child's response	Misconception or difficulty
There are 6 counters on a desk. A child counts from one to ten before her finger reaches the last counter.	
When asked to place 4 counters in a container, a child places the fourth counter in the container.	
If the same number of counters are put in a pile, a child thinks there are fewer than if they are spread out on the desk.	
When counters are placed randomly or arranged in a circular pattern, a child becomes confused while counting and has to begin again.	

3 Investigate the use of commercial materials and games that can be used to build up early-mathematical thinking—comparing, ordering, sorting and patterning. For instance, *Attribute Blocks* can be used to make, describe and repeat patterns:

■ ■ ● ● ▲ ▲ ■ ■ ● . . .

These can involve changes in size, thickness and colour as well as shape and can be spread over a page as well as simply in a linear pattern. Later, number words or symbols and the letters of the alphabet can be used for making or continuing patterns.

Logic People, Three Bear Family and other sorting shapes, everyday objects and toys found in pre-schools and classrooms also readily allow activities for sorting, classifying and patterning.

Two-digit numbers: establishing a system of place value

Numbers 20–99

When the one-digit numbers and the concept of ten have been established, children are ready to extend their knowledge to two-digit numbers. For numbers greater than ten there are no new number symbols. Instead, the new numbers are generated from those that have already been learned. Two-digit numbers have traditionally been introduced on the basis of counting with eleven introduced as the next number after ten, followed by twelve and so on. However, being able to count to a given number and recognise its symbol does not mean that a child knows the number. It is more important that larger numbers are thought of in terms of place value; 78 needs to be seen as 7 tens 8 ones, for example, rather than simply the number that comes after 77. Materials that show tens and ones and highlight the nature of ten as a new unit (1 ten is 10 ones) are essential in this development.

Ten frames are best used to introduce this thinking for two-digit numbers as they provide a visual model for tens and ones that links readily to the earlier understanding of one-digit numbers. Building from this knowledge to the multiples of ten is the most natural first step in this development as it is no more difficult to show 2 tens, 4 tens, 7 tens or 9 tens than it is to show 2, 4, 7, or 9 objects. Activities and games that use ten frames and Unifix or other counters are very helpful in making this extension.

MAKE TENS

Materials required:
- playing board with 9 ten frames for each player
- dice (ten-sided 0–9 is preferable)
- Unifix or counters
- egg timer

Method of play: Each player in turn rolls the dice and places the number of Unifix shown on the upper face onto one of the ten frames. When there are enough Unifix to make ten the player begins to fill the next ten frame. Play proceeds until the egg timer runs out. The player with the most tens is the winner.

When children can readily make tens, they are ready to learn their formal number names. Although many children will be familiar with these names from counting activities, few will have an understanding of their place value meaning. Recording them will not make sense at this stage and should be left until later, when the recording of tens and ones is introduced. At this point, the major focus must be on building meaning by relating the number of tens to the sound of the number names and thus giving a means of visualising the number of tens when the number name is heard or seen. It is important to sequence this introduction from the regular language that builds on the number words saying how many tens to the irregular tens where such cues do not exist.

    9 tens is 9t   8 tens is 8t   7 tens is 7t   6 tens is 6t

Forty is also regular (4t for 4 tens), but instead of 5t we say fif t, and use thir t rather than 3t. These less regular numbers, based on the ordinal number names rather than the counting numbers, should be left until later. The least regular, twenty, which builds on the number word two rather than the reading of number symbols, is best dealt with last of all. Some teachers highlight the correspondence between 'two' and 'twin' and 'twenty' to give meaning to the initial sound for each word, but it is realising that we say twen t for <u>two</u> <u>ten</u>s that is crucial to understanding this difficult number name.

SEQUENCE TO INTRODUCE THE MULTIPLES OF TEN

1 Begin with the regular numbers: 9t, 8t, 7t, 6t, 4t
2 Continue with those based on the ordinal numbers: fif t, thir t
3 Finish with the irregular twen t based on blending the number word <u>two</u> with <u>ten</u>.

Since the names for the multiples of ten are based on those for the underlying one-digit numbers, it follows that the most appropriate counting to encourage for forming the tens is 'one, two, three, four, five, six tens; sixty' rather than 'ten, twenty, thirty, forty, fifty, sixty'. Not only does this latter counting have little practical significance, it forces the children to learn a new set of number names as if they were quite independent of the earlier learning of the numbers to ten, so denying the importance of the place value system altogether.

Following this, numbers with both tens and ones can be formed. Again, beginning this extension using ten frames is very helpful as children can readily see and name the number of tens from the filled ten frames and simply say the number of ones in the partly filled ten frame. For example, when there are 4 ten frames filled and one with 6 objects on it, children can name the tens and ones, and put these together to give the full number name:

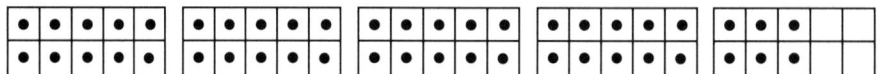

4 tens 6 ones or 4 t 6

The tens name based on the number of tens is simply followed by the name for the number of ones. When talking about these numbers, especially when asking children to show them, it is helpful to refer to a number as 6 tens 8 ones, rather than 6 tens and 8 ones, since the number name will be '6 t 8' not '6 t and 8'.

The 'Make tens' game can now be played so that when the egg timer runs out, each group can say the actual number shown. In this way, an understanding of two-digit numbers in terms of tens and ones is built in from the outset. However, in view of the difficulties that are often experienced with the teen numbers, the introduction to two-digit numbers should focus on numbers with two or more tens. Teen numbers, which have only one ten and are read quite differently, are best introduced later and integrated into an existing understanding of place value.

When the idea of forming objects into tens is secure and the naming of two-digit numbers has been established, Bundling sticks can be used in a similar manner to consolidate and extend this thinking:

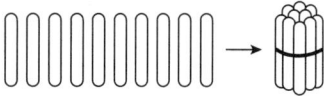

10 ones make 1 ten

At first this should focus on making and naming tens, but can soon move to using the number of tens and ones to provide the number names. A bundling game similar to 'Make tens' can be used with these new materials. Use of the same board will provide a visual reminder to bundle ten sticks before placing them on the spaces on the board.

BUNDLE A TEN

Materials required:
- playing board with 9 spaces for each player
- dice (ten-sided 0–9 is preferable)
- Bundling sticks (ice-cream sticks or coffee stirrers with small rubber bands)
- egg timer

Method of play: Each player in turn rolls the dice and collects the number of sticks shown on the upper face of the dice. When there are enough sticks to make a bundle of ten the player forms the bundle and places it on one of the 9 spaces. Play proceeds until the egg timer runs out. The player with the most tens or largest number is the winner.

Using Bundling sticks and objects on ten frames readily links the use of materials to the number names and lays the basis for introducing the number symbols. This requires writing digits 0–9 in particular positions, so it is crucial that the children be guided to recognise the tens and ones places. At this early age, they are often unable to distinguish left from right, so a place value chart with an arrow to indicate the ones place is used to structure the placing of the tens and ones.

Teaching Primary Mathematics

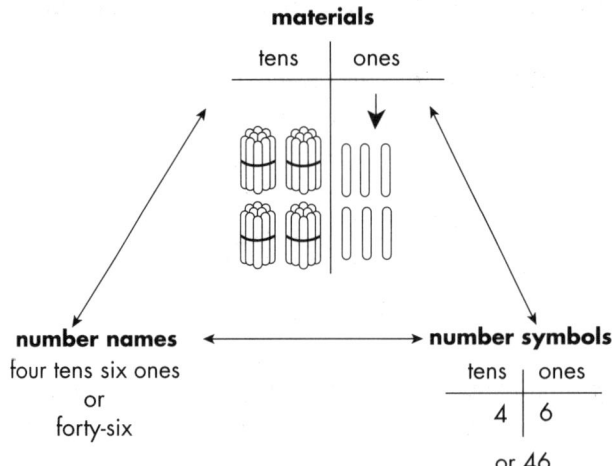

Activities and games with Bundling sticks on a place value chart build up an ability to relate the number names to the tens and ones places, consolidating the naming of two-digit numbers and laying the foundation for recording them.

TENS AND ONES GAME

Materials required:
- Bundling sticks
- dice (ten-sided 0–9 is preferable)
- place value chart

tens	ones
	↓

Method of play: Each player in turn rolls the dice and places the number of ones shown on the upper face in the ones column. When there are enough ones to make a ten the player forms a bundle of ten and places it in the tens place. Play proceeds until a nominated two-digit number such as 75 is reached or exceeded.

Note that:
- At this stage in the development of two-digit numbers, it is best to start the game by asking the children to first place tens and ones on their place value chart to show a given number greater than twenty on the chart so as to avoid consideration of the teen numbers.
- Play is best carried out in groups where each player in turn rolls the dice and puts out the Bundling sticks with the groups competing to be first to reach the nominated number. Later, when the idea of the game is grasped, the children can play against each other within the groups.
- At each roll of the dice, questions can be asked to bring out the number of tens and ones and the corresponding number names.
- The 'Tens and ones' game also lays the foundation for the renaming needed later in the algorithms for larger numbers. The development of this thinking can be brought out through

activities where 5 tens 13 ones are shown and children are asked 'How many is that altogether?'.

◆ Later, this game can be used as a basis for developing comparing and ordering. When one group has won, a representative of each group can take the number they have to the front of the class and place themselves in ascending or descending order. The children can be asked to do this without talking so that they need to look at the tens, then the ones, to sort out the correct order.

Using a place value chart to structure the placing of the tens and ones when numbers are made and named leads in a natural way to writing how many tens or ones there are in each place. If the place value chart is laminated or covered with contact it will be possible to write the number names directly beneath the materials:

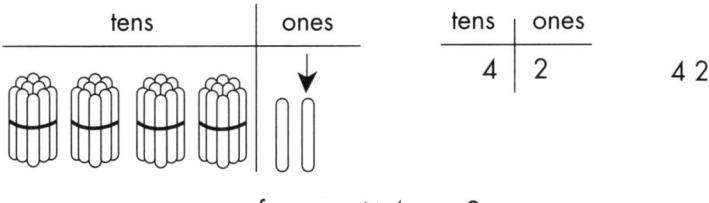

forty-two is 4 tens 2 ones

However, writing the symbols for the number with 4 tens is not so obvious. If the materials were shown on a place value chart, it is reasonable to assume that only the 4 in the tens place would be recorded.

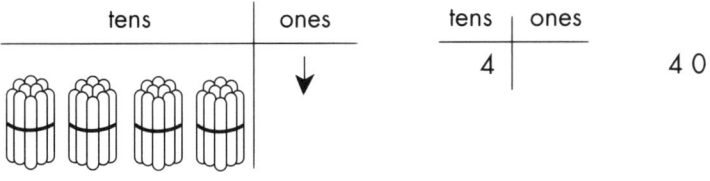

forty is 4 tens

What is needed is a realisation that a digit is written to show what is in both the tens and the ones place, leading to the writing of 40 for the number with 4 tens. This will be much more reasonable and significant if the recording of numbers with tens and no ones follows a good deal of experience with numbers that have tens and ones, for this builds up the expectation that numbers with tens require two digits when recording.

Without using materials that show tens and ones and placing these onto a place value chart, many children think of larger numbers in a similar manner to the numbers to ten, using the sound of the number names and counting as the basis of their understanding. When place value is introduced, it is seen as an added aspect of the numbers rather than the fundamental basis for them. Difficulties in writing numbers such as 40 often follow because there is no sound to hear that might guide them to write a zero in the ones place. Zero is actually recorded in the ones place to show that the number has no ones and an awareness of this can only be built in after there has been a lot of experience in recording 1–9 ones.

The teen numbers

This sequence for developing the two-digit numbers leaves the teen numbers until after the numbers 20–99 have been developed. While the teen numbers are written and represented with materials in exactly the same way as the other two-digit numbers, there is a conflict between the reading of the teen number words and the order of the symbols used in writing them. For example:

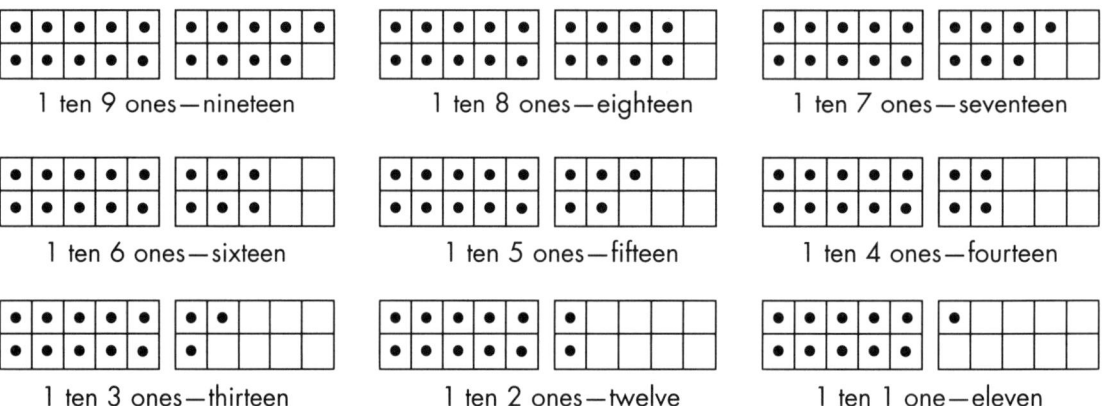

The way the teen numbers are said is unfortunately different from the way all other two-digit numbers are read. Usually the ones digit is said first and the word 'teen' is used for 1 ten. If the pattern of the other two-digit numbers had been followed, a number such as 19 would be said 'one-t-nine' rather than the back-to-front 'nineteen'. Even worse, the teen numbers eleven and twelve give no indication of their tens and ones structure at all. This does not really cause any difficulty with 11 as it is not possible to write it wrongly although many children initially name it 'eleventeen' as they try to make it conform to the pattern of the other teens, but twelve is frequently misunderstood because it is so different from the other teen numbers.

These irregularities with the teen numbers create uncertainty for many children, leading to reversals when the symbols are written. This is particularly the case when they are introduced immediately after the numbers to ten based on counting rather than place value. If the next stage in children's number learning then focuses on exploring the numbers to twenty they usually develop an understanding of the numbers solely on the basis of these associations. This is the main reason for the confusion exhibited between number names such as 'sixteen' and 'sixty' and also lies behind the writing of 60 instead of 16. Indeed, in attempting to come to terms with the structure of the teen numbers, some children even generalise the pattern they find to other numbers, reversing the tens and ones places and writing '64' for forty-six.

Yet by the time the other two-digit numbers have been learned, place value ideas have been fully developed and children are readily able to cope with the use of materials to represent teen numbers and record symbols to match them. In other words, they have already built up a good basis for the teen numbers:

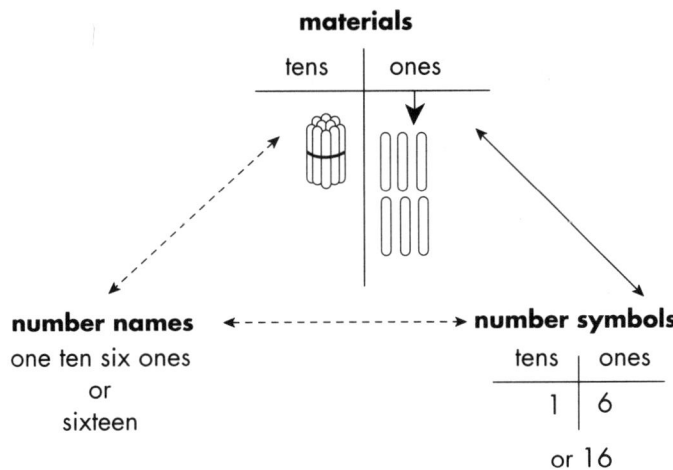

All that remains to be done is to build in the usual number name using place value understanding and to draw attention to the different way the numbers are read, proceeding from the most regular to the most irregular. Many children would be able to count the total number of sticks to obtain sixteen and eighteen but others would need to be introduced to the number names. In either case, it would then be necessary to highlight the way the number is said, contrasting it with the other two-digit numbers which are read from left to right. This is best done by indicating how the teen numbers are read from right to left, pointing to the ones digit as it is said and highlighting that 'teen' is said to indicate the 1 ten while pointing to the 1 in the tens place.

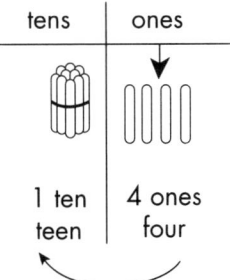

It is also useful to draw out the 'teeeen' sound to give an emphasis to the meaning of teen to signify 1 ten and to provide a means of distinguishing it from the 't' sound which means tens. In this way, children will be far less likely to confuse numbers such as fourteen and forty.

These names are best developed from the most regular 19, 18, 17, 16 and 14, which build on the number names 9, 8, 7, 6 and 4, before 15 and 13, which relate to the ordinal names fifth and third similarly to the development of the multiples of ten. Additional difficulties occur with the number twelve. While it is linked to the number two by having 2 ones, this relationship is not apparent in the way the name is said. Rather, it is seen in the way two and twelve are written. On the other hand, twelve is written in a similar manner to the number word twenty. It is also said similarly and has received a lot of attention when the other two-digit numbers were developed. A consequence of this is that many children will be led to write the 2 first for twelve, reversing the digits to give 21, because of the distracting initial sound. It is another teen number reversal but it has its roots in an understanding that has been helpful until now. With the additional difficulty that no 'teen' is said for this number, it is hardly surprising that twelve is almost the final two-digit number that children come to terms with. Eleven and twelve are both problematic in this way and are best left until last and established by building on 1 or 2 more than ten with ten frames or Bundling sticks on a place value chart to focus on the place value notions they also exhibit.

SEQUENCE TO INTRODUCE THE TEEN NUMBERS

1 Begin with those based on the number names: 19, 18, 17, 16, 14
2 Continue with those based on the ordinal numbers: 15, 13
3 Finish with those with no teen to indicate 1 ten: 12, 11

The 'Tens and ones' game can now be used to consolidate an ability to make, name and understand all two-digit numbers, including the teens by starting from an empty place value chart. Cards showing the various aspects of two-digit number understanding can also be made to use in games such as concentration, fish or in conjunction with board and bingo games.

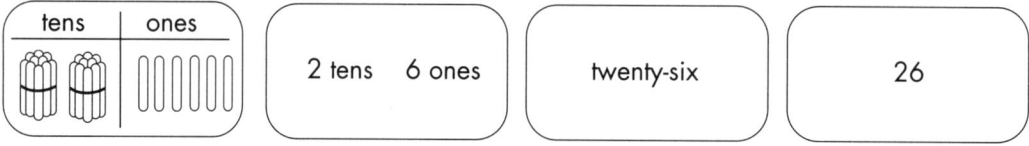

Place value and the Base 10 structure of the number system provide the foundation for the computational algorithms as well as for the numbers ten and greater. An increased awareness of the sources of children's difficulties with computation together with the broader exposure to the numbers 1–10 children receive before they reach Year 1 have suggested the earlier treatment of place value and larger numbers. In particular, children need to be encouraged to think in terms of tens and ones as soon as they try to come to terms with two-digit numbers. This understanding needs to build on initial number ideas, focusing on the language used to form the bridge between the materials and the symbolic recording and introducing those aspects where the materials, language and symbols are in accord before those that are in conflict.

Number processes

When this meaning for the two-digit numbers has been acquired, further processes such as comparing, sequencing and counting on and back can be built up. Problem solving, estimation and the development of computation demand an ability to sequence and compare numbers. Whereas the single-digit numbers are ordered or compared by simply knowing the number sequence, two-digit numbers demand a more systematic analysis. Once again the key is place value, for an efficient comparison strategy requires that the tens digits be compared first; if they are the same it is then necessary to compare the ones digits. The development of this process should also begin with materials and link through language to the number symbols only when the children understand what is needed.

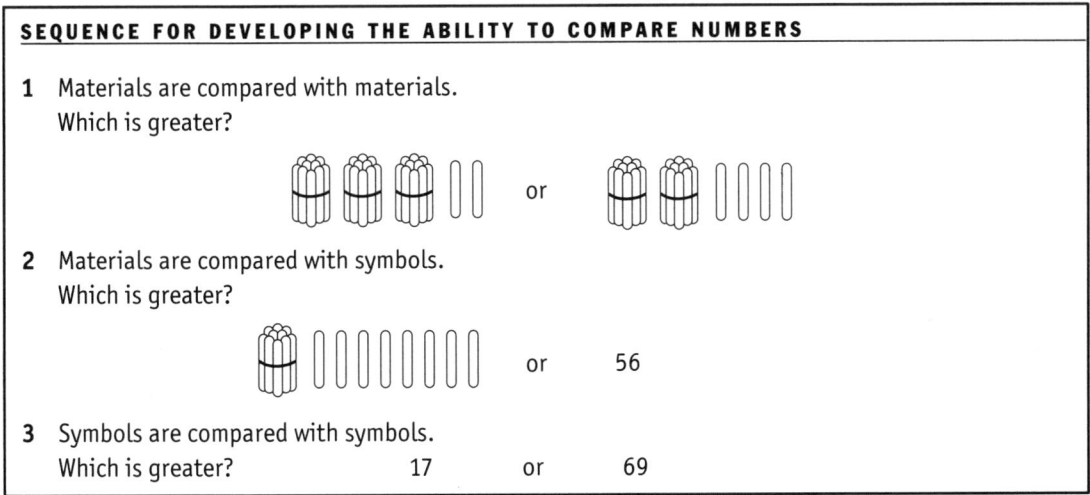

A full understanding of the comparison process based on place value ideas needs to be fully built up before the > symbol is introduced. The use of comparison and the ability to order and sequence numbers that flows from it in everyday situations do not involve the comparison symbol. Instead the emphasis is on the interpretation and assessment of the situations themselves and this ability in turn will later allow the correct use of the symbol. Consequently, it is not usually introduced until four-digit comparison has been met, and at that stage it is sufficient to introduce only one symbol, which is simply read from both left and right:

> 36 > 29 shows that 36 is greater than 29 and that 29 is less than 36
> Later, this can also be written in both ways: 36 > 29 and 29 < 36

The open end of the symbol indicates which number is greater while the symbol points to the number that is lesser.

The ability to count on and back in both tens and ones is also best built up using materials before it is treated as an oral or written activity. It is the creation of mental images in terms of tens and ones that gives meaning to the processes of counting two-digit numbers.

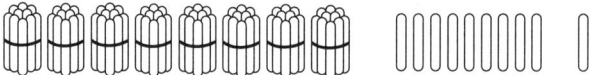

90 comes after 89 because 8 tens 9 ones and 1 one makes 9 tens 0 ones

When the ability to compare numbers at the symbolic level is grasped, one useful device for further practice that also leads into sequencing and consolidating counting is the 0–99 board. This shows very readily the tens and ones basis for the two-digit numbers since across the rows the numbers increase by ones and the rows themselves increase by tens. Note that the number zero is included on the board and that the rows begin with a multiple of ten and no ones and end with a multiple of ten and 9 ones. It also highlights the building to ten that is occurring as counting proceeds and enables children to see the counting that is involved in counting on or back by tens from a number such as 53. This arrangement of the numbers means that the three-digit number 100 has no place; it is not the conclusion of counting past ninety but in fact is the first of the next unit in the numeration system.

⟵——— Increase by ones ———⟶

0	1	2	3	4	5	6	7	8	9
10	11	12	13	14	15	16	17	18	19
20	21	22	23	24	25	26	27	28	29
30	31	32	33	34	35	36	37	38	39
40	41	42	43	44	45	46	47	48	49
50	51	52	53	54	55	56	57	58	59
60	61	62	63	64	65	66	67	68	69
70	71	72	73	74	75	76	77	78	79
80	81	82	83	84	85	86	87	88	89
90	91	92	93	94	95	96	97	98	99

⟵——— Increase by tens ———⟶

SEQUENCE FOR THE DEVELOPMENT OF THE TWO-DIGIT NUMBERS

1. Build up the multiples of ten; materials and language only.
2. Deal with the numbers 20–99 in terms of tens and ones, focusing on the relationships between the materials, language and recording. Record tens and ones before tens and no ones.
3. Relate the previous steps to the teen numbers 11–19; materials and recording first, then focus on the irregular language.
4. Compare the numbers: 1 more, 1 less; 10 more, 10 less.
5. Sequence the numbers: before, after; ordinal names.
6. Consolidate counting: count on and count back by ones and by tens.

When there is a good understanding of the two-digit numbers, further materials can be introduced to consolidate this base of understanding and prepare for the development of larger numbers. Small wooden cubes are used to represent ones, with ten represented by a piece equivalent to 10 of these cubes.

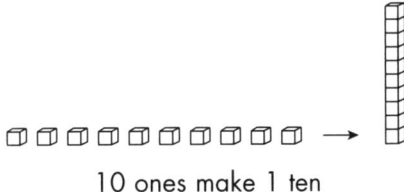

10 ones make 1 ten

These materials have been used in mathematics since the latter part of the nineteenth century (Chapter 1, p. 13) and came into particular prominence during the 1960s through the work of Zoltan Dienes who used a collection of them to provide a concrete basis for numbers in other bases. He called these materials Multi-base Arithmetic Blocks, which was soon

tens	ones	tens	ones
4	6	1	6

shortened to MABs. However, the materials now used in schools to show numbers only reflect Base 10, and cannot show any other base. For this reason, they should be called Base 10 blocks or materials, not MABs. They are most readily introduced through the 'Tens and ones' game, calling on children's understanding of ten to exchange 10 of the ones for 1 ten as they build to the required number. Two-digit numbers can then be represented on a place value chart in the same way as with the Bundling sticks:

Likely difficulties

Difficulties with two-digit numbers occur most frequently with teen numbers which are read differently from the other numbers. The unusual tens names—twenty, thirty and fifty—may also cause difficulties and some children may confuse the similar sounding names, twenty and twelve, thirteen and thirty, forty and fourteen and so on.

- A child might show 3 tens 4 ones for 43 or 5 tens 1 one for fifteen.
- A child who writes a teen number such as 17 correctly may overgeneralise and write 'seventy-four' in the same way as 47. Conversely, a child who writes the numbers 20–99 correctly may write the teen numbers in reverse, writing 61 for sixteen.
- 21 or 20 might be written for 'twelve' because the initial sound suggests 'twenty'.
- When comparing numbers, a child might think that 'seventeen' is larger than 'forty-six' because 'seven' is larger than 'four' and these are the number words that seem to be said first.
- Counting on and back by ones and tens might be confused. For example, a child might count '58, 48, 38, 28' then stop, not realising that '18' and '8' should follow.

Investigate and discuss

1. Why would it be more appropriate to count tens '1, 2, 3, 4, tens, forty' rather than 'ten, twenty, thirty, forty'? How would you encourage a child to count out tens and ones to show forty-seven?

2. Counting activities involving tens and ones should focus on counting on and back, by tens (23, 33, 43, 53, 63, . . .) or by ones (53, 52, 51, 50, 49, 48, . . .). How might you use the two-digit number board to assist children in these types of activities? Would there be a role for materials in this development?

3. A child was asked to show fifteen using Bundling sticks. She carefully put out 1 bundle of ten, hesitated, then put out another 4 bundles of ten and stopped to think. After a while, she put 1 one to the left of the 5 bundles.
Asked how she knew to do this, she explained 'These numbers are a bit tricky. They have a one in front'.
   a What understanding of numbers does the child have?
   b What difficulty might she have?
   c What might be the source of this difficulty?
   d What could you do to lead the child to see the error?
   e How would you help the child to overcome this difficulty? (Consider the role of materials and language in your response.)

Three-digit numbers: consolidating place value and establishing renaming

The development of the two-digit numbers provides the foundation for numeration ideas, establishing the fundamental importance of place value and the need to proceed in an order and manner that allows children to construct new number meanings from those they have already learned. In many ways, the three-digit numbers are easier to learn than the earlier numbers as the language is much more straightforward and the concept of building a new unit from 10 of an earlier unit has already been established. Base 10 materials can be used to review the way in which 10 ones make 1 ten and introduce the new number name, 10 tens make 1 hundred. Since these materials are more abstract than the Bundling sticks used to build up two-digit numbers, where the 10 ones actually became the 1 ten, activities and games that focus on exchanging 10 ones for 1 ten and 10 tens for 1 hundred are helpful in building this understanding.

MAKE ONE HUNDRED

Materials required:
- playing board with 10 ones, 10 tens and 1 hundred for each player
- dice (ten-sided 0–9 is preferable)
- Base 10 materials—15 ones, 10 tens and 1 hundred per playing board

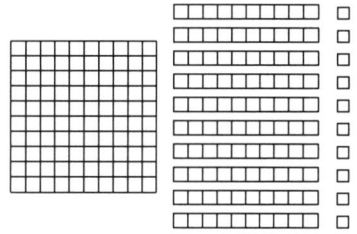

Method of play: Each player in turn rolls the dice and places the number of ones onto the small squares. When all the small squares are filled, the 10 ones are moved onto the first tens space. When the player needs more ones, the 10 ones can be exchanged for 1 ten piece, which is placed onto the tens space. Play proceeds until a player has filled all the tens spaces and realises they can be exchanged for the 1 hundred piece. The first player to place the hundred piece on his or her board is the winner.

Naming hundreds is then very easy. Unlike the two-digit numbers where a new name had to be derived from the tens (4 tens is 4t), the hundreds place is said explicitly—4 hundreds is simply said four hundred. This new place can then be integrated with the existing two-digit numbers using Base 10 materials on a place value chart to build up the language for three-digit numbers in general:

hundreds	tens	ones
3 hundreds	6 tens	4 ones

3 hundred and 6 t 4

The only new learning required is that the word 'and' is usually said between the hundreds and tens places. To assist with this, it is helpful to ask children to show 5 hundreds *and* 8 tens 2 ones on their place value chart rather than 5 hundreds and 8 tens and 2 ones or 5 hundreds 8 tens 2 ones. Activities and games with Base 10 materials on a place value chart build up an ability to relate the number names to the hundreds, tens and ones places, consolidating the naming of three-digit numbers and laying the foundation for recording them.

HUNDREDS, TENS AND ONES GAME

Materials required:
- Base 10 materials
- dice (ten-sided 0–9 is preferable)
- place value chart

hundreds	tens	ones
		↓

Method of play: Each player in turn rolls the dice and places the number of ones shown on the upper face in the ones place. When there are enough ones to make a ten the player exchanges them for 1 ten and places it in the tens place. When there are enough tens to make a hundred the player exchanges them for 1 hundred and places it in the hundreds place. Play proceeds until a nominated three-digit number such as 362 is reached or exceeded.

Play is best carried out in groups where each player in turn rolls the dice and puts out the materials, with the groups competing to be first to reach the nominated number. Later, using a different coloured dice for the tens and for the ones, so that both tens and ones are put out at each move, enables the players to build to the given number more quickly and lays an informal foundation for the renaming that will be developed later. This game can also be used to develop comparison and ordering. When one group has won, each group can say their number and use their hundreds, then tens, then ones materials that they have to place themselves in ascending or descending order.

The recording for these numbers can then be introduced by writing a digit in each place to reflect the number of hundreds, tens or ones, leading to a full understanding of three-digit numbers in terms of materials, language and symbols:

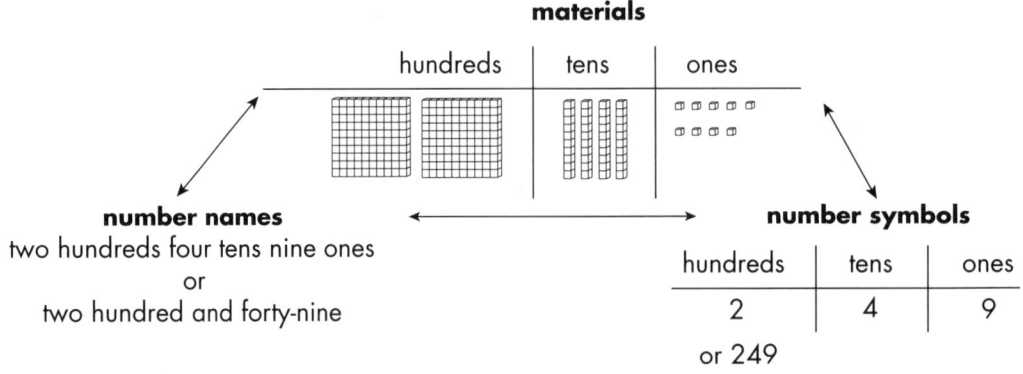

It is again helpful to introduce the most regular three-digit numbers first, those with a digit from 1–9 in each place such as 264, before moving to those that involve teen numbers such as 417. Making them on a place value chart with Base 10 materials and reading them while touching the materials in each place in turn builds an understanding of the different ways the numbers are read:

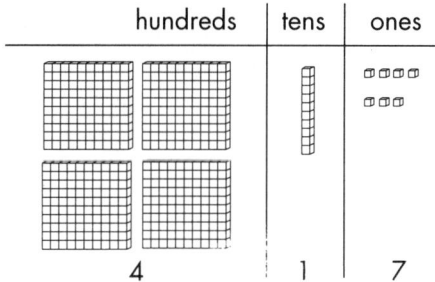

two hundred and sixty-four   four hundred and seventeen

This will help diminish any continuing difficulties with the teen numbers such as confusing 417 with 471 or 315 with 350. Using materials in this way will also provide meaning for numbers with internal zeros, such as 209, 508, 207, . . ., 901, 400, which are best left until last so that children appreciate that a zero is written in the tens place when only the hundreds and ones names are said:

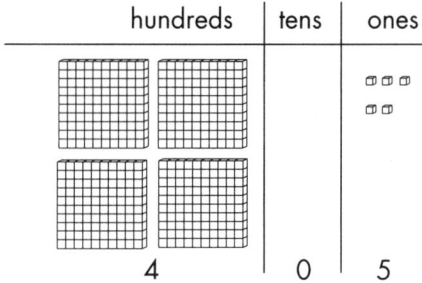
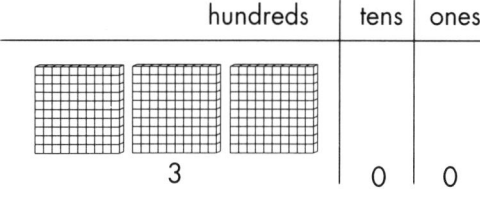

four hundred and five   three hundred

In this way, confusion in reading 406 as forty-six or writing 4006 for four hundred and six can be replaced by an understanding of the significance of zero in indicating when a number has no tens.

Renaming numbers

One further aspect of numeration is introduced with the three-digit numbers. So far, the emphasis has been on the number of hundreds, tens or ones in each place, for this place value understanding of numbers is crucial to using and interpreting them. However, it is not the only way that numbers need to be understood. Representing them by using Base 10 materials shows that they can be named in different ways by exchanging 1 hundred for 10 tens or 1 ten for 10 ones. For example, while the number 348 has 3 in the hundreds place, 4 in the tens place and 8 in the ones place, it can be *renamed* as 34 tens 8 ones or 348 ones. These alternative ways of naming numbers are important for comparison and rounding, counting on and back, and, later, in the algorithms for subtracting and dividing larger numbers.

	hundreds	tens	ones	
then renamed as	▦▦▦	▤▤▤▤	∷∷	3 hundreds 4 tens 8 ones
or	▦▦	▤▤▤▤▤▤▤▤▤▤▤▤▤▤	∷∷	2 hundreds 14 tens 8 ones
or	▦	▤×24	∷∷	1 hundreds 24 tens 8 ones
or		▤×34	∷∷	34 tens 8 ones
or		▤×33	∷×18	33 tens 18 ones
or			∷×348	348 ones

348 can be renamed as 2 hundreds 14 tens 8 ones or 1 hundred 24 tens 8 ones or 34 tens 8 ones or 348 ones

When an understanding of how a number can be renamed in different ways has been built up using materials, a number expander can be introduced as a way of showing this with the symbols alone:

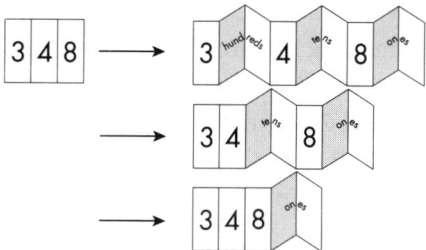

It is essential to introduce and make the distinction between the number of tens or ones in each place and the number of tens or ones that the number has altogether

The processes of comparing, ordering and sequencing also need to be integrated into this understanding of three-digit numbers. They use place value and renaming to build on the thinking for two-digit numbers, again starting with materials only, then linking through language to use the symbols on their own.

SEQUENCE FOR DEVELOPING THE ABILITY TO COMPARE NUMBERS

1 Materials are compared with materials.
  Which is greater?

2 Materials are compared with symbols.
  Which is greater?

3 Symbols are compared with symbols.
  Which is greater?   247   or   325

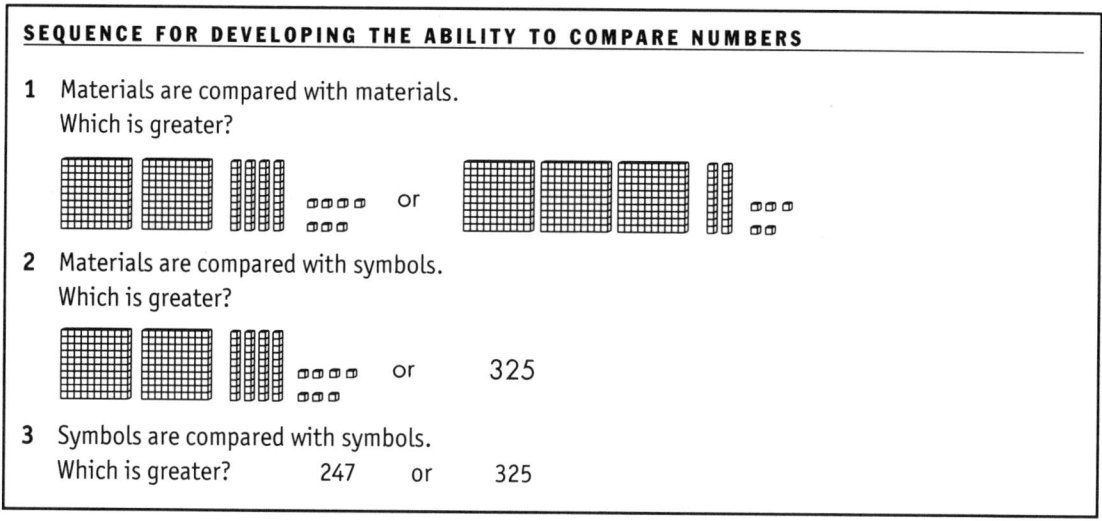

By this stage in the development of numeration, it may be reasonable to introduce the > symbol to indicate which is larger, although it is not essential and may be delayed until the thousands are met. Children's difficulties with comparison are more often problems in knowing how to use the greater than or less than symbol than in determining which number is larger. When symbols are introduced, it is preferable to emphasise the use of only one symbol that is simply read from both left and right:

427 > 356   427 is greater than 356 and 356 is less than 427
Later, this can be written both ways: 427 > 356 and 356 < 427

Counting on and back also needs to be extended so that children can count by hundreds, by tens and by ones beginning with a number such as 245. While this is mostly straightforward, difficulties are often experienced when counting by tens across the hundreds, such as counting backwards: 235, 225, 215, 205, 195, . . . Base 10 materials can be used to show what is occurring by making the number 235 then removing 1 ten at a time, and the understanding developed with a number expander is particularly helpful with this more complex thinking. Since 245, 235, 225, 215 and 205 have 24 tens, 23 tens, 22 tens, 21 tens and 20 tens, the next number must have 19 tens. Counting on and back with tens can be shown to simply build on the earlier two-digit counting:

245, 235, 225, 215, 205, 195, . . . or, **245, 255, 265, 275, 285, 295, 305,** . . .

A calculator can also be used to count back in this way by entering −10 then seeing the number of tens change each time the = key is pressed. Dealing with counting when the three-digit numbers are well understood not only provides a meaning for these difficult tasks, it also consolidates the place value on which larger numbers depend.

SEQUENCE FOR DEVELOPING THE THREE-DIGIT NUMBERS

1 Build up the new unit, 10 tens make 1 hundred, and extend the use of materials and language to hundreds, tens and ones.
2 Record three-digit numbers. First introduce the regular numbers such as 692 which have a non-zero digit in each place, then the irregular numbers such as 459 and 815 where the reading is not so straightforward and leave those with zeros until last. Numbers such as 508 have little match between the reading and writing while 300 needs to be seen in terms of place value where the 3 is written in the hundreds place with the other places having zero to show that there are no tens and no ones.
3 Rename three-digit numbers to highlight the distinction between the number of tens in the tens place as opposed to the number of tens the number has altogether. It is particularly important to bring out the understanding that a number such as 804 has 80 tens.
4 Compare, order and sequence three-digit numbers using materials and a number expander to provide meaning.
5 Count on and back by ones, tens and hundreds to consolidate understanding of three-digit numbers, place value in particular.

Likely difficulties

Difficulties with three-digit numbers occur most frequently with the occurrence of zero in the ones or tens places, when a number contains a teen number or when renaming is called for. Other children may not be secure with the concept of place value established with the two-digit numbers. If they simply see the tens digit comes first in a number and the ones digit last, they cannot have any basis to deal meaningfully with the three-digit numbers. The number names for the new hundreds place are unlikely to cause difficulties, but the unusual tens names—twenty, thirty and fifty—may still cause difficulties within the new numbers.

- A child might read 568 as '5 hundred and sixty and 8', overgeneralising the use of the word 'and'.
- A number such as 401 may be read as 'forty one' because of the way it appears, especially when it is read from a calculator.
- When a number such as 'six hundred and seventeen' is read out, a child may write 671 or 670, writing digits in the order in which they are heard.
- When a number such as 'six hundred and eight' is read out, a child may write 6008, simply writing the numbers they hear.
- When asked how many tens in a number such as 932 or 502, a child might answer '3' and '0', focusing on the tens place instead of the number of tens shown by renaming, '93' and '50'.
- Counting on and back by tens might be hesitant and become confused when changing from one hundred to the next, for example a child might count '346, 336, 326, 316, 300, 200, 100'.

Investigate and discuss

1 a Use Base 10 materials to show 3 hundreds, 4 tens, 8 ones on a place value chart. Cover the tens and ones and say the number name for what you can see (three hundred). Then cover the hundreds and say the number name for the tens and ones (forty-eight). Join the two parts using 'and' to say the full number name (three hundred and forty-eight), touching the materials in each place as you say that number name. Repeat for 2 hundreds and 7 tens, 3 ones; 5 hundreds and 7 tens, 1 ones; 6 hundred and 5 tens, 2 ones.

b Now put out 3 hundreds and 1 ten, 8 ones. Read this number (three hundred and eighteen) while touching the materials in each place as you say the number name. What do you notice about the order of reading when a teen number is involved? Repeat for 317, 316, 314 and 313. Then try 312 and 311. What additional difficulties occur with these last numbers?

c Make the number with 3 hundreds, no tens, 6 ones and read the number while touching the materials in the hundreds and ones places. What difficulties might these numbers cause for children? How would you explain the way these numbers are read?

d How is the number with 3 hundreds said? Would the reading of numbers like this cause any difficulties?

e In what order would you introduce the three-digit number names?

2 Show 7 hundreds and 2 ones on a place value chart. Then read the number out loud and write it in symbols.

a Why might a young child write 7002 when he or she hears this number?

b Why might a child read 'seventy-two' when 702 is written?
c How would you lead a child to see that these ways of writing and reading this number are incorrect?
d A child reads the number 700 correctly as '7 hundred'. When you ask how she knows this, she says 'the 7 tells me seven and the 2 zeros tell me hundred'. How would you respond to this child?
e In which order would you introduce the writing of these numbers?
400   493   451   450   416   407   412

3 A child was given the following question:
There are ten coloured pencils in a packet. If you had two packets and five extra pencils, how many pencils would you have?
When asked to write the number and read it out loud, the child wrote 205 and said 'twenty-five'.
  a What understanding of numbers does the child have?
  b What difficulty might he have?
  c What might be the source of this difficulty?
  d What could you do to lead the child to see the error?
  e Which steps in the teaching sequence for the two-digit numbers has the child apparently failed to grasp?
  f How would you help the child to overcome this difficulty? (Consider the role of materials and language in your response.)

4 Children's difficulties with ordering numbers usually result more from keeping track of what they are doing rather than extending the comparison processes. Writing the numbers onto cards that can then be shuffled into the correct order before the numbers are written in order is often helpful in moving the children towards the full thinking process.
  a Write these numbers onto cards and use them to order them from largest to smallest:
    400   510   390   410   472   623   403   501
    Comment on how you were able to go about this task.
  b What further understanding is needed to move from comparing two numbers to ordering three or more numbers?
  c Cards with the digits 0–9 can also be used to consolidate the skills of comparing and ordering. Children can be asked to make the smallest or largest three-digit number from three cards such as

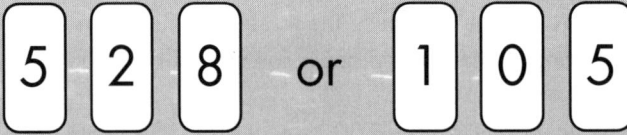

Thousands—four, five and six-digit numbers: consolidating renaming and extending place value

Four-digit numbers

Extending numeration understanding to include the four-digit numbers is readily achieved using Base 10 materials to build the new unit, 10 hundreds make 1 thousand. Naming the thousand multiples again causes no difficulties and the full four-digit numbers can be developed by matching the language and symbols to the Base 10 material shown on a place value chart:

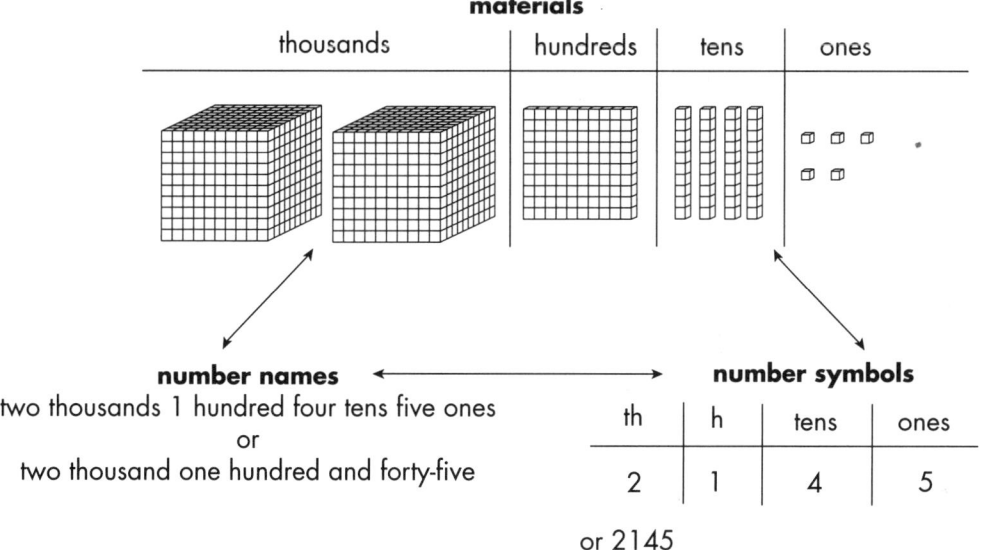

Teen numbers and internal zeros should be left until after the regular numbers have been built up, and the distinction between the number of tens or ones in each place and the number of tens or ones that the number has altogether also needs to be built into the larger numbers. This renaming can be introduced using materials on a place value chart and is readily shown using a number expander:

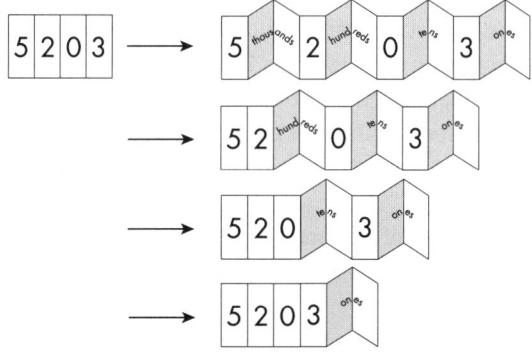

A further recording pattern is also introduced with the thousands. It is usual to leave a space between the digit in the thousands place and the digit in the hundreds place to facilitate reading these numbers. This convention is optional with four-digit numbers which are read place by place in the same way as the earlier numbers, but will be necessary for larger numbers when each place is no longer said explicitly. Consequently it is best not to introduce leaving a space until the five- and six-digit numbers are met to provide a reason for what would otherwise appear to be an arbitrary convention. After this, all numbers with thousands, including the four-digit numbers, can be recorded with a space between the thousands and hundreds places. As larger numbers are met, the pattern for writing numbers with a gap before each group of three digits will become even more important as a second place value system is developed.

Comparing, ordering, sequencing and counting also need to be integrated into this understanding of four-digit numbers. Comparison of numbers needs to be carried out place by place, looking at the thousands, then the hundreds, then the tens and finally the ones in order to determine which number is the lesser or greater. Children also need to be able to count by thousands, by hundreds, by tens and by ones beginning with a number such as 6827. Counting by tens and hundreds across the thousands is again something that many children have found difficult, and the thinking developed with the number expander can show how this builds on earlier three-digit counting:

6827, 6817, 6807, 6797, ... or 6827, 6927, 7027, ...

Again, a calculator can be used to show counting on and back by entering a four-digit number and −10, −100 or −1000 and seeing the number of tens, hundreds or thousands change each time the = key is pressed. Note that on most calculators the four-digit numbers do not show a space, except when a number with only 1 thousand such as 1629 is entered and 1 629 is shown. Care also needs to be taken when numbers such as 4316, which looks like 431 6, and 7123, which looks like 71 23, are entered.

SEQUENCE OF DEVELOPMENT FOR THE FOUR-DIGIT NUMBERS

1. Build up the new unit, 10 hundreds make 1 thousand, and extend the use of materials and language to thousands, hundreds, tens and ones.
2. Record four-digit numbers. First introduce the regular numbers such as 8723 which have a non-zero digit in each place, then the irregular numbers such as 5137 or 3417 where the reading is not so straightforward and leave those with zeros, such as 2460, 8306, 7014, 9002 and 1008 until last.
3. Rename four-digit numbers to highlight the distinction between the number of hundreds or tens in each place as opposed to the number of hundreds or tens the number has altogether. It is particularly important to bring out the understanding that a number such as 4005 has 40 hundreds and 5 ones or 400 tens 5 ones.
4. Compare, order and sequence four-digit numbers.
5. Count on and back by ones, tens, hundreds and thousands to consolidate understanding, place value in particular.

Five- and six-digit numbers

It is not feasible to continue to develop and use Base 10 materials beyond the thousands place. Rather, children need to rely on a sound understanding of the pattern that 10 of a place make one of the next place, and use this to build each new number name in turn. In this way, they can visualise the meaning of larger numbers and their relative sizes. The naming system that is followed for the next place should be sufficiently helpful in itself; just as 10 hundreds make 1 thousand, 10 thousands form the next place which is simply called the ten-thousands place. In turn, 10 of the ten-thousands form 1 hundred thousand to give the next remaining thousands place.

However, the five- and six-digit numbers introduce a new way of reading numbers. For a number such as 6542, each place is said: '6 thousand, 5 hundred, and 4 t(ens), 2'. But for five-digit numbers such as 86542, while the 8 is in the ten-thousands place, the number is not read '8 ten thousand, 6 thousand, 5 hundred, and 4t, 2'. Instead, there is a new way of reading larger numbers: '86 thousand, 5 hundred, and 4t, 2'. For this reason, it is customary to introduce the convention of leaving a space between the thousands digit and the hundreds digit at this point: 86 542. This facilitates the reading of the numbers, allows each of the thousands places to be readily identified and can be used to assist the children in learning to read larger numbers that use this new pattern. It is helpful to suggest that the space is left in the number so that when we read it we remember to put the word 'thousand' at that point.

$$86\underset{\text{thousand}}{\wedge}542$$

Since there is little confusion in reading four-digit numbers it seems better, from a teaching point of view, to omit mention of the space between the thousands and hundreds digit when they are first met and to stress its use when the five- and six-digit numbers are introduced, pointing out that the same convention applies to all numbers with thousands. Thus, six thousand five hundred and forty-two can be written either as 6542 or 6 542, while a number such as 786 542 is read '786 thousand, 5 hundred and forty-two'.

```
    786          542
   h t ones    h t ones
     of
   thousands
```

This new way of reading numbers reflects an extension of place value where large numbers are also grouped in hundreds, tens and ones of thousands. A second place value system, based on hundreds, tens and ones of ones, then hundreds, tens and ones of thousands, has been established that lays the basis for the development of all large numbers. However, care needs to be taken that a child can still identify the place values as well as read these larger numbers, for example to see that 8 is in the ten-thousands place while 6 is in the thousands place for 86 542 and not to say that the number has 86 in the thousands place.

The five- and six-digit numbers are developed in a similar way to the four-digit numbers, linking the language to the underlying places, renaming numbers to see the distinction between the digit in a place such as the ten-thousands and the number of ten-thousands it has. There will need to be a continuing focus on the aspects of these numbers brought out by the use of a number expander and the zeros in the internal places will cause some difficulties, but in general the patterns and thinking established for the four-digit numbers will follow through to the other thousands.

> **SEQUENCE OF DEVELOPMENT FOR THE THOUSANDS**
>
> 1 Build up the new units, 10 thousands make 1 ten-thousand and 10 ten-thousands make 1 hundred-thousand, and integrate this language with the earlier ways of naming and writing numbers.
> 2 Record these numbers, introducing the convention that a space is left between the thousands and hundreds digit to facilitate the reading of larger numbers.
> 3 Rename numbers to highlight the distinction between the digit in each place as opposed to the number of ten-thousands, thousands, hundreds or tens the number has altogether. It is particularly important to bring out this understanding for numbers that include zeros.
> 4 Compare, order, sequence and count on and back to consolidate numeration understanding, place value in particular.

Likely difficulties

Difficulties with thousands also occur with internal zeros, teen numbers or when renaming is called for. The new way of reading numbers may also cause difficulties if children do not realise the way place value has been extended for the five- and six-digit numbers.

- A child may read a number such as 947 285 as '94 thousand 7 thousand 2 hundred and eighty and 5', especially when the number is seen without a gap as on a calculator.
- A number such as 6007 may be read as '6 hundred and seven' because of the way it appears.
- When a number such as 'seventy thousand and three' is read out, a child may write 70 000 3, simply writing the numbers they hear.
- When asked how many ten-thousands in a number such as 907 651 2, a child might answer '0', focusing on the ten-thousands place instead of the number of ten-thousands shown by renaming, '90'.
- Counting on and back by thousands, hundreds or tens might be hesitant and become confused when counting across a thousand or hundred.

Investigate and discuss

1. Use Base 10 materials to show 2 thousands and 7 ones on a place value chart. How would you read this number? How would you write it? What would you do to ensure that children understand the reading and writing of these numbers? Now put 2 thousands 1 ten 4 ones on the place value chart. What possible difficulties do these numbers introduce?

2. A child reads the number 9000 correctly as '9 thousand'. When you ask how they know this, the response is 'The "9" tells me nine and the three zeros "000" tells me thousand'. What really says that the number is read as 9 thousand? What difficulties might they have in writing nine thousand and two?

3. Enter the digits 1, 5, 6, 7 onto a calculator and read the number shown out loud. Repeat for 5, 1, 6, 7; 6, 5, 1, 7; and 7, 6, 5, 1. Would any of these numbers prove difficult for young children to read? Why is it important that children (and adults) learn to read numbers onto and from a calculator rather than simply spell them out digit by digit? What difficulties might a child experience in reading the number that results when 2, 0, 0, 1 is entered onto a calculator?

4. In which order would you introduce the writing of these numbers? Why?
   9051   9681   9300   9304   9012   9009

5. Make a number expander for four-digit numbers. Cover the blank rectangles with transparent tape so that numbers can be written on the expander with a water-based pen to allow re-use. Then fold the expander concertina fashion:

   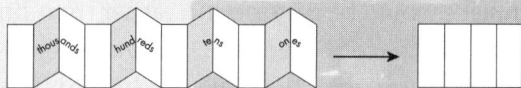

   a. With the number expander completely folded, write the digits 5, 4, 1, 3 in the rectangles to display 5413. What number name is behind the 5? What number name is behind the 4? What number name is behind the 1? What number name is behind the 3?

   b. Open the expander out to show how many hundreds there are altogether (54), how many tens there are altogether (541) and how many ones there are altogether (5413).

   c. Repeat for 3506, 7013 and 2001.

6. a. What difficulties would you anticipate with the reading and writing of five-digit numbers?

   b. How could you use a number expander to help in reading these numbers?

   c. How would you assist children in coming to terms with internal zeros in the five-digit numbers?

   d. How would you use the number expander to bring out all of the ways in which these numbers can be renamed?

7. Why might children find the reading of a six-digit number such as 729 451 easier than a five-digit number such as 35 607?

Rounding

Situations in everyday life often demand estimations and approximate calculations, requiring a process of rounding to the nearest ten, hundred, thousand or larger place. This further aspect of numeration is usually introduced after the three-digit and four-digit numbers are secure, building first onto existing two-digit thinking before being extended to larger numbers. However, for many children and adults rounding is a procedure that seems to have been arbitrarily imposed and devoid of understanding. Yet a meaningful approach to rounding can be readily established and become a process used for all numbers no matter how large or small. The key to this is to relate it to place value and renaming, building an intuitive conception of 'closer to' into a process that seems to be reasonable and unambiguous.

> 58 has 5 tens; since there are 8 ones it is closer to 6 tens or 60
> 732 has 73 tens; but since there are only 2 ones it rounds to 73 tens or 730

Traditionally, the technique that has been given was to look at the digit that is immediately to the right of the place to which the number is to be rounded. If that digit was 5 or greater the number was said to 'round up'; if it was less than 5 the number was said to 'round down'. Occasionally, numbers were even said to 'round off'. Thus if 6378 was to be rounded to the nearest ten, the digit in the ones place was considered and 6378 was 'rounded up' or 'rounded off' to 6380. Several aspects of this traditional technique have contributed to the confusion and doubt about its use and validity. First, if the number is to be rounded to a particular place, surely consideration of that place would be paramount, rather than simply looking to see the size of the digit in the next smaller place? Given that the rounding associated with numbers greater than or less than 5 seems reasonable, the issue of why 5 rounds 'up' and not 'down' is largely avoided and simply given as a 'rule' that should be followed. With this lack of reasons, many children then think that if numbers can round 'up' or 'down' it is simply a matter of choice and decide to round 8762 'up' to 8770. Even if this is not done all of the time, they are likely to move from 'up' to 'down' according to circumstance or in response to different questioning from a teacher.

If children are to understand and apply rounding, this uncertainty needs to be replaced by a conception of rounding that is unambiguous and easily applied. The first step is to change the language used to describe rounding. Since there is only one correct answer, it is more appropriate to talk about 'rounding to the nearest' and very confusing to talk about rounding 'up', 'down' or 'off'. Each of these implies a degree of arbitrariness or at least choice, yet only one result is considered to be mathematically valid.

The second aspect that needs to be altered is to change the uncertainty when the digit '5' occurs in the place that is being rounded. Largely a justification for the 'closeness' has relied on the use of a number line model in which a number containing the digit 5 always appears in the middle of the two numbers to which it might round. Arbitration is then needed to decide which is to be chosen. While this seems reasonable to those who know how to round numbers, it is not so convincing to those who are learning about rounding for the first time. There is no structure to the number line, so the points chosen on it to 'show' the thinking seem just as arbitrary as the rule that it is supposed to lead into. In contrast, the 0–99 number board has proven useful in revealing place value structure and will also provide an insightful model to introduce rounding for two-digit numbers and provide a way of thinking that can be readily extended to other places. As a result, the procedure of looking only at the digit in an adjacent place can be replaced by considering the number of tens, hundreds or whatever the number has to see what it might round to, then determining which is closer. This same thinking can later be used for all numbers no matter how large and include decimal fractions.

Rounding two-digit numbers

Two-digit numbers can be rounded to the nearest ten. This is readily shown with a 0–99 board:

0	1	2	3	4	5	6	7	8	9
10	11	12	13	14	15	16	17	18	19
20	21	22	23	24	25	26	27	28	29
30	31	32	33	34	35	36	37	38	39
40	41	42	43	44	45	46	47	48	49
50	51	52	53	54	55	56	57	58	59
60	61	62	63	64	65	66	67	68	69
70	**71**	**72**	**73**	**74**	**75**	**76**	**77**	**78**	**79**
80	81	82	83	84	85	86	87	88	89
90	91	92	93	94	95	96	97	98	99

the tens the number has
7 tens

the next ten
8 tens

A two-digit number board highlights the procedure

Consider the row which contains all of the numbers with 7 tens. The first numbers, 70, 71, 72, 73 and 74, are close to seven tens and so round to seventy. The numbers at the end of the row, 79, 78, 77, and 76 are close to 8 tens and so round to eighty. Five of the numbers (70, 71, 72, 73 and 74) round to the tens they have, 7 tens or seventy; four of the other numbers (76, 77, 78 and 79) round to the next ten, 8 tens or eighty. It is only the remaining number, 75, which requires any further analysis. There are ten numbers in each row. Five numbers round to the tens they have and four of the others clearly round to the next ten. The symmetry inherent in all mathematics suggests that all of the other five numbers should round to the next ten. In other words, 75 should also round to the next ten, 8 tens or eighty, which is in fact the way rounding has always been.

It follows that there is no doubt about numbers with a 5 in the ones place; they are not 'in the middle' but simply join the numbers which round to the next ten. Whenever we consider the numbers in a given row, half will round to the ten that they have, the other half will round to the next ten. This is indicated by the heavy line through the middle of the 0–99 number board and shows that half of all two-digit numbers round to the ten they have, while the other half round to the next ten. This explanation has proven to be a very effective means of introducing the concept and process of rounding to children who have used the 0–99 number board to consolidate their understanding of numbers in terms of tens and ones.

However, rather than maintaining a focus on which row in the two-digit number board a given number lies, it is more appropriate to ask directly how many tens the number has. This, of course, is equivalent to identifying the row, but moves from a direct reliance on the board to a basis in the tens and ones pattern on which it is constructed. For instance, to round the number 47 to the nearest ten, note that it has 4 tens so can only round to 4 tens or to 5 tens. Since 47 is closer to 5 tens than it is to 4 tens, 47 rounds to 5 tens or 50. This process for rounding two-digit numbers to the nearest ten involves three steps.

ROUNDING TO TENS

1 Determine how many tens the number has.
2 Which tens could the number round to (the tens it has or the next ten)?
3 Which ten is closer?

Activities and games that call on children to round a number to the nearest ten can be used to build up this thinking. Providing a 0–99 board for reference allows a child to check to see whether his or her answer is correct; when he or she stops looking at the board for confirmation, the way of thinking has been internalised.

ROUNDING

Materials required:
- playing board with one spinner to show 0–9 tens and one spinner to show 0–9 ones
- 3–6 bingo strips each showing six numbers selected from 10, 20, 30, . . ., 90
- 6 coloured markers for each bingo strip

Method of play: Each player spins the spinners and rounds the number indicated to the nearest ten. The 0–99 board can be used as a guide. If the rounded number corresponds to a number on the player's bingo strip a marker is placed on that number. The winner is the first player to cover his or her bingo strip.

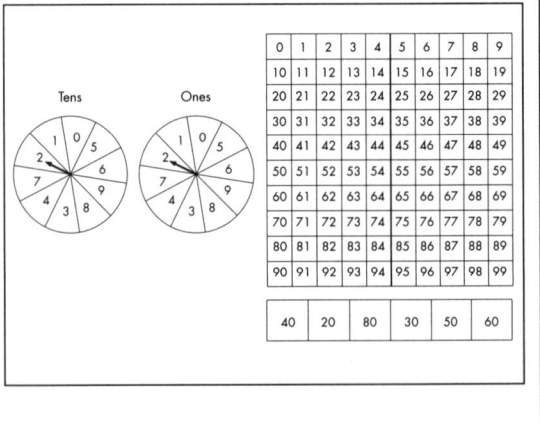

Rounding three-digit numbers

Three-digit numbers can be rounded to the nearest hundred or to the nearest ten. Unfortunately, because of a reliance on a rule that looks not to the place being rounded but to whether a digit is greater than, less than or equal to 5, many children generalise from their rounding of two-digit numbers to round larger numbers only to the leading place. For example, when asked to round 428 to the nearest ten they answer 400 rather than 430. A three-digit number board can be used to build an understanding that there are two possibilities and then can be integrated with the use of a number expander for hundreds, tens and ones to extend the rounding process to the nearest ten or nearest hundred.

400	401	402	403	404	405	406	407	408	409
410	411	412	413	414	415	416	417	418	419
420	421	422	423	424	425	426	427	428	429
430	431	432	433	434	435	436	437	438	439
440	441	442	443	444	445	446	447	448	449
450	451	452	453	454	455	456	457	458	459
460	461	462	463	464	465	466	467	468	469
470	471	472	473	474	475	476	477	478	479
480	481	482	483	484	485	486	487	488	489
490	491	492	493	494	495	496	497	498	499

ROUND 428 TO THE NEAREST TEN

1	How many tens does the number have?	42
2	Which tens could the number round to?	42 tens or 43 tens
3	Which tens is closer?	43 tens

428 rounds to 43 tens or 430

ROUND 428 TO THE NEAREST HUNDRED

1	How many hundreds does the number have?	4
2	Which hundreds could the number round to?	4 hundreds or 5 hundreds
3	Which hundreds is closer?	4 hundreds

428 rounds to 4 hundreds or 400

Rounding four-digit and larger numbers

This thinking pattern can be extended to round four-digit numbers to the nearest thousand, hundred or ten, using a number expander to highlight how many hundreds or tens the number has. Thus, 5267 rounded to the nearest thousand is 5000, rounded to the nearest hundred is 5300, and rounded to the nearest ten is 5270. Rounding can then be integrated into the numeration understanding for each new place as it is built up, simply becoming another fundamental way of thinking about numbers along with comparing, ordering and sequencing.

This way of thinking is also the key to the way money is rounded to the nearest five cents. When 1 cent and 2 cent coins were abolished at the end of 1990, at first notices appeared based on a misunderstood rule:

NOTICE TO CUSTOMERS

From 1 October 1990 the Commonwealth Government will cease issuing 1 and 2 cent coins.

As these coins become unavailable, the guidelines for rounding CASH REGISTER TOTALS will be as follows:

1 & 2 cents rounded <u>DOWN</u> to nearest 10

3 & 4 cents rounded <u>UP</u> to nearest 5

6 & 7 cents rounded <u>DOWN</u> to nearest 5

8 & 9 cents rounded <u>UP</u> to nearest 10

When notification was required on each bill, common (and mathematical) sense prevailed:

'In line with the government's decision to abolish the 1 and 2 cent coins, your account is round to the nearest 5 cents'

The same way of thinking that allowed ready rounding to the nearest ten also applies to rounding to 5 cents. Taught this way, after the concept of rounding has been introduced, rounding money to the nearest 5 cents also becomes a meaningful and manageable task.

Likely difficulties

Difficulties with rounding usually occur because of a focus on a rule which asks whether a digit is greater than, less than or equal to 5 rather than on the likely result when the number is rounded to the nearest ten, hundred or whatever. Consequently, a child may focus on rounding only to the leading place or be confused about whether a number with 5 in a place rounds to the next number of that place or the one that the number has.
- When asked to round 567 to the nearest ten, a child might answer 600 rather than 570.
- A child may round 4572 to 4500 rather than 4600.

> • When asked to round a larger number such as 637 412 to the nearest ten thousand, a child may focus on rounding to ten in the thousands and answer 640 412, correctly rounding 637 to the nearest ten, and then including the digits in the other places in the result rather than rounding to 640 thousands.

Investigate and discuss

> **1** Often children will only consider rounding a three-digit number to the nearest hundred, a four-digit number to the nearest thousand and so on. For instance, asked to round 576 to the nearest ten a child might answer '600', or asked to round 8318 to the nearest hundred or ten a child might respond '8000'.
> **a** Why do you think such a pattern might have eventuated?
> **b** How would you help a child to see that these responses are inappropriate and then lead him or her to understand how to obtain the correct answer? Consider the uses that might be made of a three-digit number board and a number expander.

Large numbers: extending the system for naming numbers

A second place value system

While it is difficult to conceive of or appreciate the size of large numbers, their frequent use in modern society means that children must develop an awareness of their meaning. Discussions of budget deficits, international debt, population explosions, scientific discoveries and astronomical distances involve numbers in the millions, billions, trillions and larger. Yet it is very difficult to visualise such large quantities. Often the only way we can make sense of them is by understanding their order of magnitude, based on the way in which they are written or spoken. Such an awareness depends on a meaningful extension of the place value patterns previously met. A new number structure is followed for larger numbers and this was introduced with the thousands. After the new thousands name had been introduced there were no completely new number names, instead they were simply called the 'ten thousands' and 'hundred thousands'. This pattern of ones, tens and hundreds (of thousands) is continued for the naming of larger numbers.

The next name to introduce is million: 10 hundred thousand make 1 million. This name is derived from the Latin word which meant 1000 thousands. When the idea of millions has been grasped, the further aspect of place value showing the patterns of ones, thousands and millions can be introduced. The first three digits on the right show hundreds, tens and ones of ones, the next three digits show hundreds, tens and ones of thousands and the next three digits show hundreds, tens and ones of millions.

4 5 6	7 8 9	1 2 3
h t ones	h t ones	h t ones
of millions	of thousands	of (ones)

This structure is indicated by the space left between the third and fourth and sixth and seventh digits. In the past, this division was represented by a comma, thought to be an adaptation of the Romans' use of a cross in each third column of their abacus. The comma, like the cross used by the Romans, recognised and drew attention to this second level of place value structure in large numbers. International convention has now decreed, however, that a space (rather than a comma) be left between groups of three digits to the right and left of the decimal marker. One reason for this is that a comma is an acceptable alternative to a decimal point in many parts of the world. Number expanders can help children to come to terms with reading these large numbers and also show the total number of thousands, hundreds, tens or ones rather than just the digit in each place.

| 9 | 8 | 7 | 6 | 5 | 4 | 3 | 2 | 1 |

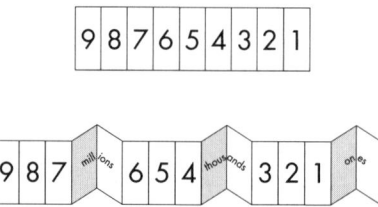

When the numeration system continues for even larger numbers, the next place to be created is the billions place but there is sometimes confusion as to how big one billion is. The pattern based on hundreds, tens and ones of each place has been the key to extending the numbers beyond the thousands place. This is recognised in the convention for writing numbers where every group of three digits is separated by a space and would make the billions the next place:

h t ones	h t ones	h t ones	h t ones
of billions	of millions	of thousands	of (ones)

This convention shows that 1 billion must be 1 thousand millions and the pattern is readily continued giving a new place value name for 10 hundred billions (or 1000 billions) as 1 trillion and so on.

The alternative interpretation sometimes put forward is that 1 billion is 1 million millions. This follows the earlier creation of the name million (1 thousand thousands) and suggests the places would be: millions, ten millions, hundred millions, thousand millions, hundred thousand millions, billions. Both generalisations of earlier patterns have a logical basis, but a little consideration shows that this alternative pattern gives increasingly large orders of magnitude between each new place value name unlike the uniform growth in the pattern based on hundreds, tens and ones of each unit. For this reason, the system which has now been adopted internationally is indeed the one based on the extended pattern of hundreds, tens and ones of each new unit. Perhaps the choice of the initial name, millions, was a poor one in retrospect, but the next name, billions, readily indicates that it is the second place after the thousands in the new naming system. After that the number names are: trillions, quadrillions, quintillions, sextillions, septillions and so on; the prefix is simply based on the Latin counting numbers for 3 (tri), 4 (quad), 5 (quin), 6 (sex), 7 (sept) and so on, giving the number of new places after the thousands.

It should be noted that in order to come to terms with any pattern at least three instances are needed. Children need to investigate numbers to millions or beyond to fully comprehend the way in which the place value pattern is extended to large numbers. Simply stopping with the hundred thousands and then 1 million leaves them uncertain as to how to deal with large numbers, especially when they are met on a calculator which usually does not show any spaces. One fruitful source of large numbers is the Internet, not only to look up uses and definitions, but also as a creator of a need for ever larger numbers. For instance, it was reported in *The Economist* of July 1999 that the Internet Assigned Numbers Authority had begun to hand out Internet addresses 128 binary digits long, rather than the traditional 32. The effect of this was to increase the number of available addresses from some 4 billion to a 'third of a duodecillion' (340 282 366 920 938 463 374 607 431 211 456). Interestingly, rather than round this amount to 340 undecillion, 1 third of the next place was used, probably reflecting uncertainty on the part of the authors of the article as to how the number after decillion should be named.

Likely difficulties

Difficulties with large numbers usually occur because of little understanding of the second place value system and the names used for each new unit. For example, many children erroneously state that the numbers after the billions are 'zillions', which came to us from comics and cartoons as a number that was exceedingly large, unable to be counted or determined.

- When asked to read 764 837 562 319 a child might begin 764 thousand 837 . . . no, 764 million 837 thousand 562 . . . I don't know.
- When a number with the digit 1 is shown on a calculator, for example 2346 1456, a space appears to be left before the 1, leading to attempts to read this as 2 thousand 346 thousand one thousand 456.
- Confusion as to whether 1 billion is 1 million millions or 1 thousand millions is still quite common.

Investigate and discuss

1. List as many examples as you can of the everyday use or occurrence of five-digit and six-digit numbers that children in primary school might encounter. Look through advertisements, news, business and science reports in the media for use of billions or trillions. Even larger numbers are frequently used in the reporting of DNA testing for crime and medical reports. Consider any impact or difficulties they might convey to both adults and children.

2. Although the place value of each digit after the thousands place is no longer explicitly said, there is a second place value pattern which helps to give meaning to large numbers. Read out loud the number 294 468 325. Each group of three digits is read as if it were hundreds, tens and ones.
   a. What additional words are used to give the value of each group of three digits?
   b. How do you know which word to add when you read the whole number?
   c. Make a number expander (as shown on p. 126) to help in introducing the reading of these numbers. What additional help is given by modifying the expander to show only the millions, thousand and ones names?

3. Write the number 4 billion 2 thousand and six. How do you know how to write a number with internal zeros? What difficulties would you anticipate that children might have in reading and interpreting a number such as 560 031 008? How could you use a number expander to come to terms with internal zeros in large numbers?

4. Another situation where children encounter large numbers is on a calculator. Enter the number 39 282 408 on a calculator. What is shown on the display? Would there be any difficulty in reading or understanding this number? What happens when 239 282 408 is entered? Why might a child read a number such as 700013 as 7 thousand and 13 when it appears on a calculator?

Exponential and scientific notation

When large numbers involving millions, billions and trillions are used for financial and population purposes, these are usually rounded amounts. It is also usual for these to be written as 19 million, 7 billion or 4 trillion rather than recorded in full as 19 000 000, 7 000 000 000 or 4 000 000 000 000. This grows readily from the names used in the second place value system, and in turn lays the basis for a further way in which they can be written succinctly using powers of ten and exponential notation. For example, 1 million is the product of $10 \times 10 \times 10 \times 10 \times 10 \times 10$, where 10 is a factor six times. Rather than always write this in full, a new notation is used to write 1 million as 10^6 where the small 6 written above the 10, called an *exponent*, indicates how many times 10 is a factor. Similarly, 1 billion is written 10^9 and 1 trillion is written 10^{12}. In turn, 19 million can be written 19×10^6, 7 billion is written 7×10^9 and 4 trillion is written 4×10^{12}. Not only does this allow large numbers to be expressed concisely, it also allows comparisons of magnitude to be made more readily and simplifies calculations involving them.

In order to develop this new way of recording large numbers a good understanding of multiplication is needed, including the idea that the numbers that are multiplied are called factors. For example, since 4 multiplied by 9 is 36, 4 and 9 are factors of 36; 36 is also the product of the factors 3 and 12 or 2 and 18. In the primary school this idea has frequently been introduced by the use of factor trees, where a number is broken down into the product of its prime factors. A prime number is a number which has no factors other than itself and 1—for example, 2, 3, 5, 7, 11, 13 and so on—so a prime factor is just a factor which is also a prime number. Thus 36 can be written as a product of the prime factors $2 \times 2 \times 3 \times 3$:

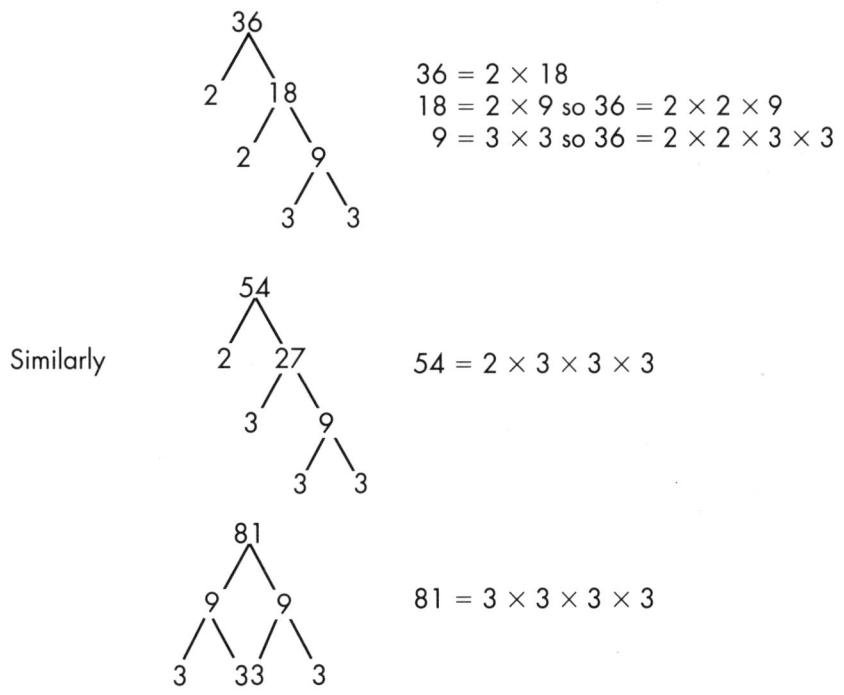

When the same factor occurs several times a more concise notation using exponents is used to record the product; for example, $3 \times 3 \times 3 \times 3$ is recorded 3^4. However, because many children confuse the use of exponents with multiplication, their introduction should not occur until large numbers are familiar, multiplication itself is very secure and children see the need and power of this new way of recording numbers. The first step to build up this new notation is to develop the idea of exponent as a shortened means of recording the result of multiplying the same factor several times. A calculator provides a very successful means for this.

For instance, children can be asked to use their calculator to find how many 2s need to be multiplied together to get a three-digit number. This result can be expressed orally as 2 multiplied by 2, multiplied by 2, multiplied by 2 and so on but it is hard to remember how many 2s are involved. Even recording this result as $2 \times 2 \times 2 \times 2 \times 2 \times 2 \times 2 = 128$ still requires a further step of counting the number of times 2 occurs. Asking other questions such as how many 2s are multiplied to get a four-digit number, five-digit number and so on soon suggests that some more immediate way of recording the number of times a factor is multiplied is needed. Recalling that the number is a factor in each of these multiplications suggests that the results can be more concisely stated by saying that 2 is a factor 7 times to get a three-digit number, 10 times to get a four-digit number, 14 times to get a five-digit number and so on. The notion of an exponent can then be introduced as a shortened recording to show how many times 2 is a factor:

product	number of times 2 is a factor	shortened recording
$2 \times 2 = 4$	2	2^2
$2 \times 2 \times 2 = 8$	3	2^3
$2 \times 2 \times 2 \times 2 = 16$	4	2^4
$2 \times 2 \times 2 \times 2 \times 2 = 32$	5	2^5
$2 \times 2 \times 2 \times 2 \times 2 \times 2 = 64$	6	2^6
$2 \times 2 \times 2 \times 2 \times 2 \times 2 \times 2 = 128$	7	2^7
$2 \times 2 \times 2 \times 2 \times 2 \times 2 \times 2 \times 2 \times 2 \times 2 = 1024$	10	2^{10}
$2 \times 2 \times 2 \times 2 \times 2 \times 2 \times 2 \times 2 \times 2 \times 2 \times 2 \times 2 \times 2 \times 2 = 16\,384$	14	2^{14}

The smaller numbers written above the 2 are called 'exponents'; they tell how many times 2 is a factor. When 128 is written as 2^7 it is called 'exponential notation' and read as '2 to the power 7' or '2 to the seventh power'.

Using exponential notation, since $36 = 2 \times 2 \times 3 \times 3$, 36 can be expressed as $2^2 \times 3^2$, $54 = 2 \times 3 \times 3 \times 3$ or 2×3^3 and $81 = 3 \times 3 \times 3 \times 3$ or 3^4. This notation is especially important in expressing products of ten, since the fundamental idea of place value is that each place is ten times as large as the preceding place:

$$50 = 5 \times 10$$
$$500 = 5 \times 10 \times 10 = 5 \times 10^2$$
$$5000 = 5 \times 10 \times 10 \times 10 = 5 \times 10^3$$
$$50\,000 = 5 \times 10 \times 10 \times 10 \times 10 = 5 \times 10^4$$
$$500\,000 = 5 \times 10 \times 10 \times 10 \times 10 \times 10 = 5 \times 10^5$$
$$5\,000\,000 = 5 \times 10 \times 10 \times 10 \times 10 \times 10 \times 10 = 5 \times 10^6$$

Numbers written in this way using a digit in the 1–9 range together with 10 as a factor several times (a power of ten), for example when 5 million is written as 5×10^6, are said to be written in *standard form*. Thus, 4 trillion is written as 4×10^{12} in standard form since 1 trillion is 1 000 000 000 000 which is $10 \times 10 \times 10 \times 10 \times 10 \times 10 \times 10 \times 10 \times 10 \times 10 \times 10 \times 10$ or 10^{12}. However, when 19 million is written as 19×10^6 this is not in standard form since 19 is not a digit in the 1–9 range.

At a later stage this standard form can be extended to a shortened notation to cater for all numbers, not just those expressed as multiples of ten. Since 19 000 000 is 19 × 1 000 000 and 19 is 1.9 × 10, we could also write 19 000 000 as 1.9 × 10 000 000 or 1.9×10^7. This more compact form of writing large numbers, where a number between 0 and 10 is multiplied by a power of ten, is preferred in scientific work and hence is called *scientific notation*. For example:

12 billion is 12×10^9—since 12 is 1.2 × 10, in scientific notation
12 billion is written 1.2×10^{10}
346 trillion is 346×10^{12}—since 346 is 3.46×10^2, in scientific notation
346 trillion is written 3.46×10^{14}
6.02×10^8 is $6.02 \times 10^2 \times 10^6$ or 602×10^6—6.02×10^8 is 602 million

Note that in many cases the standard form and scientific notation will coincide since the product already uses a digit in the 1–9 range.

While this way of writing numbers readily communicates their meaning and relative size, it can cause confusion when such large numbers are compared. Then it is more helpful first to write each number in terms of the same power of ten so that only the initial multiples need to be compared. For instance, in comparing 3×10^6 with 7×10^5 it is easy to get an incorrect answer by comparing 3 and 7. It has to be realised that 3×10^6 is the same as 30×10^5, so that 30, not 3, is compared with 7. Similarly, in comparing 3.4×10^5 with 7.6×10^4, it is first necessary to express them in terms of the same multiple of 10, 3.4×10^5 and 0.76×10^5, and then realise that since 3.4 is greater than 0.76, 3.4×10^5 must be greater than 7.6×10^4.

Naming and recording large numbers

thousand	1 000	10^3
million	1 000 000	10^6
billion	1 000 000 000	10^9
trillion	1 000 000 000 000	10^{12}
quadrillion	1 000 000 000 000 000	10^{15}
quintillion	1 000 000 000 000 000 000	10^{18}
sextillion	1 000 000 000 000 000 000 000	10^{21}
septillion	1 000 000 000 000 000 000 000 000	10^{24}
octillion	1 000 000 000 000 000 000 000 000 000	10^{27}
nonillion	1 000 000 000 000 000 000 000 000 000 000	10^{30}
decillion	1 000 000 000 000 000 000 000 000 000 000 000	10^{33}
undecillion	1 000 000 000 000 000 000 000 000 000 000 000 000	10^{36}
duodecillion	1 000 000 000 000 000 000 000 000 000 000 000 000 000	10^{39}
tredecillion	...	10^{42}
quattuordecillion	...	10^{45}
quindecillion	...	10^{48}
sexdecillion	...	10^{51}
septendecillion	...	10^{54}
octadecillion	...	10^{57}
novemdecillion	...	10^{60}
vigindecillion	...	10^{63}

Likely difficulties

Difficulties with numbers written using standard form or scientific notation usually occur because of a lack of understanding of the concept of factor and the use of exponents in writing the product of several of the same factor.

- When asked to name the number 3^4, a child might answer 12 confusing the fact that 3 is a factor 4 times with the common use of 'times' to indicate multiplication, rather than $3 \times 3 \times 3 \times 3$ or 81.
- A child who works out that 10^6 is 1 million by simply writing six zeros after the 1 may then think that 100^6 is 100 000 000 or 100 million instead of $100 \times 100 \times 100 \times 100 \times 100 \times 100$ which is 10^{12} or 1 trillion. Similarly, he or she might write 20×10^6 as 2 000 000 by simply writing 0 until six zeros occur.
- When comparing numbers in scientific form, a child may simply compare the decimal fractions in front of the powers of ten even when they are not the same power, or be unable to rename the number in terms of a smaller power of ten, for example to see that 24×10^8 is the same as 2.4×10^9.

Investigate and discuss

1 a One very helpful way to introduce exponents is to use the multiplication constant on a calculator. For instance, pressing $3 \times =$ gives $3 \times 3 = 9$, $3 \times \times =$ gives $3 \times 3 \times 3 = 27$ and so on. Use your calculator to find how many 3s you need to multiply together to get a four-digit number. How many 4s do you need to multiply together to get a five-digit number? How many 5s do you need to multiply together to get a six-digit number?

b What number does 2^{10} represent? What is 5^8, 8^8, 9^8? What happens when you try to use your calculator to work out 8^9 and 9^9? What numbers would they be?

2 The number 8^6 is often said to mean 8 multiplied by itself 6 times.

a Literally, this would be $\mathbf{8} \times 8 \times 8 \times 8 \times 8 \times 8 \times 8$
so why would it be interpreted as $8 \times 8 \times 8 \times 8 \times 8 \times 8$?
Is there any similar ambiguity with the language '8 is a factor 6 times'?

b Why might a child introduced to the number 8^6 as 8 multiplied by itself 6 times conclude that 8^6 is 48?

3 Express these numbers in scientific notation:
      3256   34 567   123 456   67 001

4 How would you introduce comparison of numbers involving exponents to children?

Section II: Fraction ideas

FRACTION ideas refer to the different mathematical ways of dealing with parts of things. In today's world, their use largely focuses on decimal fractions through which measurement is expressed, and in per cents which are a convenient way to discuss proportion and change in matters ranging from finance to sporting achievements. Despite their name, common fractions are no longer used very often, except to name particular amounts, and any computations involving them are usually converted to a decimal calculation. Nonetheless, an ability to name fractions in general is consistent with the ideas of common fractions. Similarly, the concept of renaming as equivalent fractions required for common fractions applies equally to decimal fractions and per cents. Renaming also underpins the ability to express a given fraction amount in each of these equivalent written forms. Further, once the new ways of naming and renaming fractions have been established, computation, comparison and rounding processes are largely simple extensions of those used for whole numbers.

Numeration, then, is the key to understanding and using fraction ideas in the same way as it underpins all work with whole numbers. However, while there is a need to understand and use decimal fractions, per cents and common fractions to deal with situations and problems in the real world, the extent to which these fraction ideas are dealt with in school far exceeds most practical needs. This is because they also provide a crucial basis for the development of further mathematics. Notions of proportionality are implicit in many fraction concepts and processes. Building these notions extends a child's capacity to generalise, leading to the development of important higher-level cognitive processes. It is these understandings, developed through fraction ideas and the associated computational processes, that provide the basis for mathematics in secondary school rather than the techniques *per se*. To reflect this, the teaching of fraction ideas has changed from the rote learning of techniques to understanding the way in which the fraction concepts and processes are extended from those already developed for whole numbers. The challenge though, as Smith (2002, p. 3) points out, is that 'no area of school mathematics is as mathematically rich, cognitively complicated, and difficult to teach as fractions, ratio and proportion'.

Whereas whole number understanding builds on children's everyday experiences with natural numbers, an understanding of fraction ideas is not so strongly rooted in the real world. Children tend to focus on the size of the parts, rather than their occurrence as part of a larger whole. Indeed, focusing on the relationship between parts of numbers requires a significant change in thinking, and has led to fractions being described as *artificial* numbers (Smith, 1925). Not every civilisation extended their conception of number to include fractions and their origins in many different mathematical systems only came together as a complete topic in recent times. Thus, the first ideas of fractions were given to us by the ancient Egyptians who talked about 1 part out of a whole (Gillings, 1982, pp. 126 and 127). If a loaf of bread was to be divided among 8 workers, the last worker received the eighth part of the loaf. The 7 workers left then divided the remaining loaf and the next worker received the seventh part, and so on.

the last worker received the eighth part   the next worker received the seventh part

This use of ordinal numbers to name the part given out eventually led to each of the 8 parts being named as eighths and is the origin of the fraction names now used. It was the ancient Hindu mathematicians who developed this idea further and began to talk of 3 eighths or 5 sevenths and established the basic methods of calculation with common fractions. The Romans focused on 12 parts for most of their measurements but introduced the notion of paying a tax at so many parts for each hundred (per centum) which gave rise to the per cents used for taxes, profit margins and so on used today. Decimal fractions are comparatively recent, although the ancient Babylonians used a place value system based on 60 rather than 10 to denote numbers smaller than one as well as large numbers. In ancient China, there were also essentially decimal fractions for small parts building on their Base 10 place value system as well as forms of common fractions, but these were not related to each other and existed as different forms of numbers used for different purposes (Needham, 1959). Decimal fractions as we know them did not arise until the system of place value for whole numbers had been accepted in the Arab world around the tenth century and in Europe after the fifteenth century. Interestingly, while much of our mathematics builds on ideas developed by Greek mathematicians, the ancient Greeks did not accept the idea of a fraction at all, arguing that since 1 was the source of all numbers being grouped to show tens, hundreds and so on it could not be broken into smaller parts. Instead, they focused on the ratio or proportion of one number to another to indicate parts smaller than one whole.

The language used to describe fractions is also problematic. A given number can have many names—1 half ($\frac{1}{2}$) is also 5 tenths (0.5) and 50% as well as 2 fourths, 3 sixths, 4 eighths and so on. Different words are also used when fractions are expressed in different ways—there are improper fractions and mixed numbers as well as common fractions that may have common, low or lowest denominators and be equivalent or irrational. Not only is this diverse legacy of fraction ideas difficult to meld into a mathematically cohesive whole, it suggests why children may find the extension of their whole number understandings to include parts of 1 one difficult and potentially confusing.

Consequently, an initial conception of a fraction as a part of something needs to be made more precise. The use of materials is as important in establishing fraction ideas as it is for whole numbers, although one of something is now broken into parts rather than built up to form larger numbers. Physical materials as such are not necessarily helpful. When smaller parts are formed from something representing 1 one, a child is just as likely to see several smaller ones as to realise these are equal parts of the original. Further, children acquiring fraction understanding need to experience this partitioning for themselves rather than be given materials already formed into parts. For these reasons, a model in which equal parts are completed and shaded is usually preferable to cutting something into equal pieces or to reconstructing 1 one from ready-made parts.

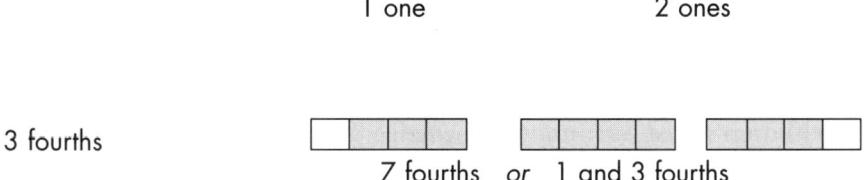

When 1 one is partitioned into equal-sized parts, this fraction is referred to as a *proper fraction*, since it denotes a number between 0 and 1. By extending this idea to fractions which are parts of more than 1 one, other fractions can be formed and seen as numbers in an extended number system rather than simply as a way of using already known whole numbers, separated by a point (3.46) or one on top of another ($\frac{3}{4}$). Thus an *improper fraction* may be a part of 2 ones (e.g. $\frac{7}{4}$), 3 ones (e.g. $\frac{8}{3}$), 4 ones and so on and these can also be expressed as a *mixed number*, that is, as a whole number together with a proper fraction (e.g. $1\frac{3}{4}$ or $2\frac{2}{3}$). Of course, each of these amounts could also be expressed as a decimal fraction or per cent, and the fractions can be seen as numbers that lie between any two whole numbers:

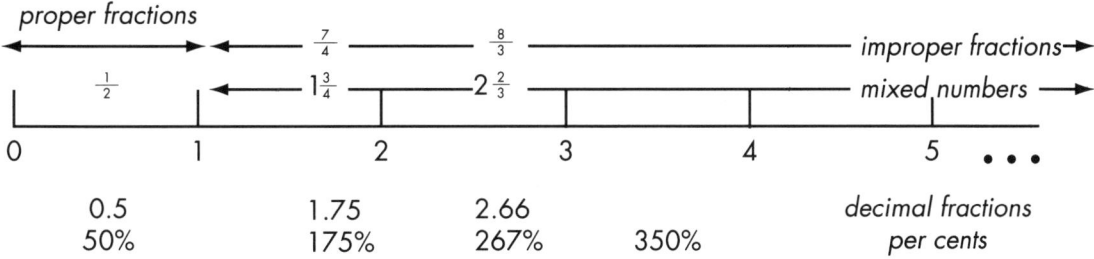

A number line shows that fractions are parts of 1 one, 2 ones, 3 ones, . . . and names them as proper and improper fractions, mixed numbers, decimal fractions and per cents

Just as 1 one was the base for all whole numbers (10 ones make 1 ten, 10 tens make 1 hundred and so on), partitioning 1 one in this way also shows that it is the basis for all fractions as well. The parts of 1 one give the initial fractions halves, thirds, fourths and so on, and in turn these generate all numbers by allowing for parts of 2 or more ones such as 1 and 3 fifths, 2.7, 350%. A rectangular region model has proven to be most suited to this initial introduction. Number names can then be related to parts of the whole and in due course, fraction symbols can be introduced and associated with the models and number names.

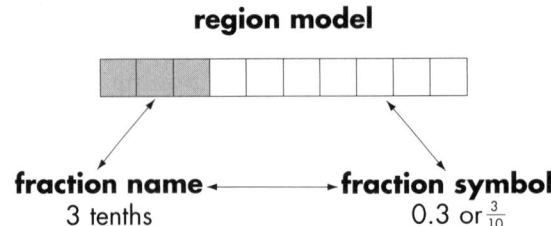

region model

fraction name ← → **fraction symbol**
3 tenths   0.3 or $\frac{3}{10}$

Although the region model and the language are the same for all forms of fraction, the symbols used to record them are quite different. Decimal fraction notation and numeration are based on the same place value ideas as whole numbers, providing a consistent link and basis for assimilation to children's existing number knowledge. In contrast, common fractions look very different as it is not possible to extend the place value to cope with all of the new fractional names. Furthermore, the use of decimal fractions in everyday situations is far more prevalent than common fractions, particularly with the widespread use of electronic calculating devices and metric measurement. Children consequently have a considerable exposure to decimal notation on which the decimal form of fraction recording can be built while the recording of common fractions is rarely seen. These considerations have led to decimal fraction notation being the first form of recording to be formally introduced to children in school, although in practice the introduction of common fraction symbols follows soon after, not least because the recording of some fractions immediately raises questions as to how the other fraction amounts might be written. Nonetheless, the only real difference between these two fraction forms is the method of recording: both grow out of the same real-world experiences, the part/whole model and a language that describes the number and size of the parts. Similarly, per cents can be linked to decimal fractions involving hundredths as another way of recording fraction amounts using a new symbol, %, to indicate a proportion for each hundred.

Decimal fractions are introduced by partitioning a square region representing 1 one into 10 equal parts and developing the language of tenths to describe these parts. Earlier numeration shows that 1 thousand is made up of 10 hundreds, 1 hundred is made up of 10 tens and 1 ten is made up of 10 ones so that the focus on 1 one being made up of 10 tenths follows naturally from the Base 10 structure of the number system. This in turn suggests the extension of place value to decimal fractions and leads to recording

ones	tenths	
6	3	as 6.3

The next place, hundredths, can be introduced by a similar partitioning process of a square region to show one hundred equal parts but the development of further places cannot be based on a physical model. Partitioning 1 one into one thousand equal and very small pieces is too difficult while a region model does not allow the generalisation to a third dimension in the way that thousands blocks built on hundreds blocks in the construction of whole number numeration. Instead, succeeding places need to be developed by a continuation of the Base 10 pattern, 1 hundredth is made up of 10 thousandths, 1 thousandth is made up of 10 ten-thousandths and so on.

Once the initial ideas have been understood for each new place, numeration for these decimal fractions can be built up in a manner paralleling that used for whole numbers. Comparing, ordering and rounding skills can be based on place value ideas and an ability to interpret decimal fractions in different ways can be promoted by the use of diagrams showing the parts to lead into the use of a decimal number expander.

6.3	is	6 and 3 tenths
	or	63 tenths
6.37	is	6 and 37 hundredths
	or	6 and 3 tenths 7 hundredths
	or	63 tenths 7 hundredths
	or	637 hundredths

The reading and writing of whole numbers has been based on place value so that, for example, 6341 is read as '6 thousand 3 hundred and 4t (for tens) 1' rather than spelled out digit by digit, 'six three four one'. When decimal fractions are read, it is essential that this too is based on place value so that meaning and direction are provided in the same way as for whole numbers. Building on this earlier explicit use of place value and the ideas developed through the partitioning process, it follows that 6.3 should be read as '6 and 3 tenths' rather than spelled out as 'six point three' or 'six decimal three'. In the same way 6.37 is read as '6 and 37 hundredths' rather than 'six point three seven' although there will also be a need for the interpretation '6 and 3 tenths 7 hundredths'.

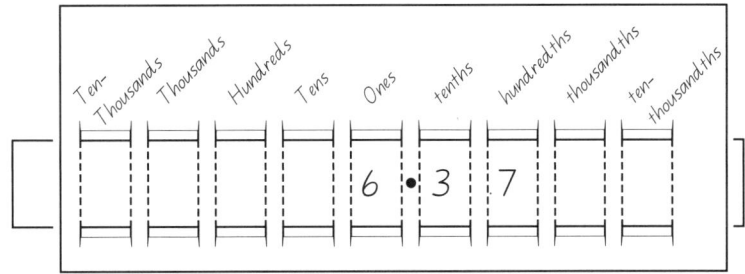

However, when there are more than three decimal places, saying the places explicitly no longer helps make the meaning clear and it seems reasonable to spell out a number such as 6.3745 as '6 decimal point 3, 7, 4, 5'. In fact, since a number with more than three decimal places is most likely to occur with the use of a calculator, spelling out each place is a useful procedure to adopt in keying in numbers or in writing out the result given for a calculation.

While the recorded form of common fractions differs from that of decimal fractions, the use of a region model and the language used to describe the ones and parts are identical for both. Careful introduction of this new recording is necessary to ensure that it is meaningful and that it relates to the steps used in constructing the model. The use of the language '3 parts out of 5 equal parts' is the key to establishing this understanding, so that there is a progression from the initial fraction concept to the recorded form for common fractions:

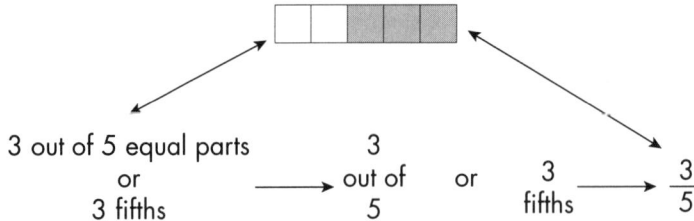

In this way, the top number tells how many parts (hence the formal name numerator from the Latin word *numare*—to number), the line symbolises 'out of', and the bottom number names the size of the parts (giving rise to the formal name denominator from the Latin words *de nomen*—name). From this development of the recording of common fractions, it follows that a meaningful shortened form of reading a common fraction such as $\frac{37}{143}$ would be '37 out of 143' rather than to spell it out as '37 over 143'. This parallels the consideration given to decimal fractions where reading was associated with the meaning given by place value. Here it is given by reference to the parts which constitute the fraction idea.

In addition to establishing this meaning for the fraction symbols, it is crucial for an understanding of the ways in which common fractions can be renamed to be addressed from the outset. In the first instance, this means that the fraction concept is extended so that $1\frac{3}{5}$ can be seen to be 8 of the fifths and written:

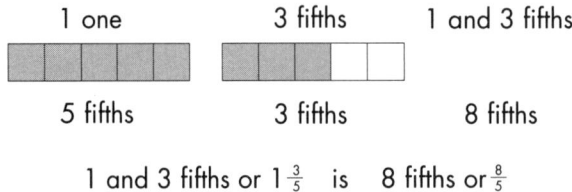

Later this renaming needs to be extended to include equivalence of fractions—$\frac{3}{5}$ is the same as $\frac{6}{10}$, $\frac{9}{15}$, $\frac{12}{20}$ and so on—and to the equivalence between common fraction and decimal fraction forms.

Another form of fraction, per cents, also needs to be developed and linked to decimal fractions and common fractions by the same use of the region model and language of parts of 1 one. In the past, per cents were often taught independently of fraction ideas and given little relationship to the decimal fraction ideas to which they most closely relate. Indeed, since the use of electronic calculators for computation is universal in business, science and industry, it is crucial for per cents to be seen in terms of decimal fractions to allow more efficient calculations. With the understanding that per cent is simply another way of expressing hundredths, the interpretation and computation of per cents then becomes a relatively straightforward application of decimal fraction skills. In particular, a realisation that one hundred per cent, 100%, is just another way of representing 1 one allows notions such as 250% (2 ones and 50% of another one) and 1 half of 1% or $\frac{1}{2}$% to become meaningful and devoid of confusion. Per cents are also essentially a form of proportion, representing so many for each hundred. In due course, this thinking needs to be extended to consider proportionality in general, leading to the concept of ratio that focuses on relative rather than absolute changes in situations involving growth.

Developing the fraction concept

Language and models

Just as a sound basis of numeration is essential for the development of computation for whole numbers and their application in problem solving, a full understanding of the fraction concept is a prerequisite for the use of decimal fractions, common fractions and per cents. Whole number numeration is established through the use of materials that are grouped to show the place value ideas that provide the basis of all numbers, however large. Fraction ideas also need to grow out of the use of materials that represent ones, although 1 one is now broken into parts and a fraction language needs to develop from this partitioning process. Activities that involve forming and shading parts on a region model are needed to make a child's intuitive understanding of a fraction as a 'part of something' more precise, building into the part/whole model on which initial fraction ideas depend.

SEQUENCE FOR THE DEVELOPMENT OF THE INITIAL FRACTION CONCEPT

1 Realisation that 1 one must be partitioned into equal-sized parts.
   Which of these show equal parts?

2 Naming the number of equal-sized parts to be considered.
   How many parts are shaded?

3 Naming the number of equal-sized parts in 1 one and thus naming their size.
   10 parts altogether—they are called tenths

This last step of forming the fraction name is best achieved by relating the total number of equal-sized parts to the corresponding ordinal names:

number of equal-sized parts	9 parts	8 parts	7 parts	...	4 parts	3 parts	2 parts
ordinal names	ninth	eighth	seventh	...	fourth	third	second
fraction names	ninth	eighth	seventh	...	fourth	third	half

Basing this development on the ordinal names means that there is only one new name to be learnt—half—and most children have some meaning for this even before entering school. At first, half may be taken to mean one of two parts, whether or not they are equal sized. This meaning grows out of the tendency of parents, and even teachers sometimes, to talk of the 'small half' when sharing things among children so that the older child gets a larger amount. By the time children reach pre-school or school, they are in situations where all children are about the same age and size and will no longer countenance unequal shares. So the understanding that parts must be equal dominates, even to the point that children at this stage will call three equal parts three halves. During this first year of school, children usually put both aspects together to develop a meaning that 1 half is one of two equal parts so that the only new name is usually well established by the time fraction ideas are formally introduced. Of course, there is still the difficulty that the plural of half is halves, whereas for all other fraction names an 's' is simply added to the name, for example sixth and sixths.

While the name for one of four equal-sized parts is often called a quarter, many children view 'a quarter' as 'half of a half' due no doubt to frequent experiences seeing fruit and sandwiches cut into quarters. This view is not incorrect, but it is not the meaning of equal parts of 1 one that is needed. Developing the name 'fourths' at first enables children to fit these fractions in with the general pattern by seeing this as 1 part out of four equal parts just as 1 sixth is one part out of six equal parts. In due course children need to be introduced to the alternative name, quarter. However, if it is used initially, this pattern may not be seen and children may be led to learn each fraction name independently. It is also very important that fractions are always talked about in terms of 1 part, such as 1 fifth or 1 tenth, rather than as a part, a fifth or a tenth. The use of 1 fifth relates directly to the way it will be recorded, $\frac{1}{5}$ not $\frac{a}{5}$, and enables it to fit in with the pattern for all of the other fractions, 2 fifths, 3 fifths, even 7 fifths, when improper fractions are met. In this way, fraction names can be seen to form a meaningful numeration system, adding to and building on that already established for whole numbers.

The rectangular region has proven to be the most effective means of establishing the fraction concept as it not only shows the partitioning and naming processes but also allows an informal development of the notion that as the number of parts increases their relative size decreases.

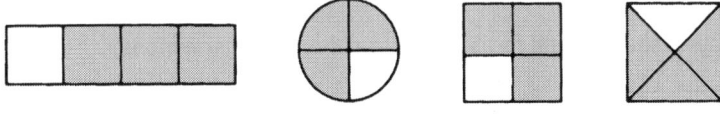

3 out of 4 equal parts		3 fourths
3 out of 5 equal parts		3 fifths
3 out of 6 equal parts		3 sixths
3 out of 7 equal parts		3 sevenths
3 out of 8 equal parts		3 eighths
3 out of 9 equal parts		3 ninths

It is important at first for the rectangles used to denote each fraction to be the same size in order to bring out this subtle relationship. However, to consolidate the fraction concept, once the rectangular model has been internalised, children may be asked to recognise or shade representations based on different regions.

Each of these regions shows 3 fourths

While it is necessary to use a variety of shapes to consolidate the fraction concept, using a rectangle to establish this meaning is most helpful. The use of a circular region to introduce the fraction concept is not recommended, despite the common occurrence of materials of this form in the primary school. This is because the size of the equal parts is really determined on the basis of the angle measured at the centre rather than the more intuitive measurement based on length used by the rectangular region. Children have not usually met angle measurements at the time the circular region is first used, so their familiarity with length often overrides the less obvious way of forming parts. It is not so difficult to choose and name already formed equal-sized parts on the less regular shapes, but errors such as these are frequently noted when children attempt their own partitioning.

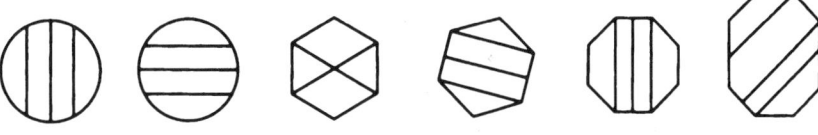

Children's misconceptions when drawing fourths

A motivating way of introducing and consolidating the fraction names is through the use of games that call on children to name and to match fractions (Booker, 2000). Cards which show parts shaded on rectangles and other shapes and both forms of fraction language such as '3 out of 8 equal parts' and '3 eighths' can be used for memory, fish and bingo games similar to those described for the teaching of one-digit numbers on p. 85. Other games that require players to find and colour given fractional amounts rather than attempt to draw equal parts are also invaluable in linking pictures of fractions to the corresponding fraction names:

COLOUR A FRACTION GAME

Materials required:
- 2 dice or spinners—one with digits in the range 0–9, the other with halves, thirds, fourths, fifths, sixths, sevenths, eighths, ninths, tenths and twelfths
- fraction playing sheet for each player

Method of play: Each player in turn rolls the dice or spins the spinner, locates the rectangle or shape that shows the number of parts indicated by the fraction name and then colours the number of parts shown on the dice. A particular row (fraction name) can only be coloured once. If it is not possible to colour a fraction on a particular turn, either because that row (fraction name) has already been coloured or because the particular fraction is not possible (e.g. 6 halves) the next player has a turn. Play proceeds until a player has filled in each row on his or her playing sheet.

1 At first, the playing sheet should focus on the use of rectangular regions.

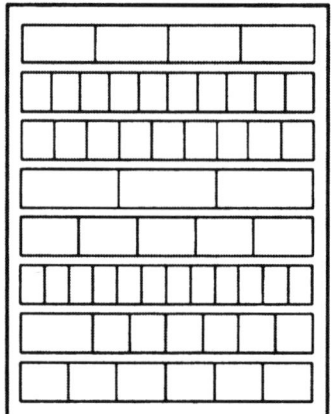

2 Later, different shapes should be used to consolidate the part/whole thinking.

3 Shapes like these where the parts are not equal, and hence fractions as such are not shown, can also be used to help focus attention on the need for all parts to be equal.

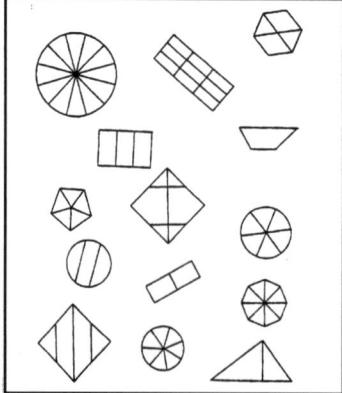

4 In order to lay the basis for renaming, playing sheets can also contain more than one of each rectangle or shape so that it is now possible to colour 6 halves, etc.

 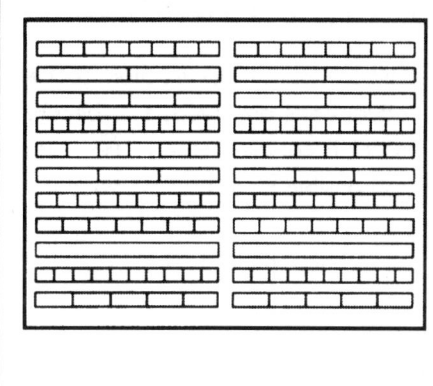

Although fractions are always concerned with part of something, this does not mean that they are solely conceived of as part of a region. They also refer to parts of a collection of discrete objects. In everyday use, this is commonplace as in 'half of the group' wanted to do one thing, or 'the motion was passed on a two-thirds majority'. When a basic conception of a fraction exists, it is possible to extend this meaning to collections, but its use may appear to be confusing and lead young children to misconceptions or misinterpretations. For instance, if a collection of 5 cars is shown, 2 of them black, many children will see 2 black cars, 3 other cars and name the fraction that is black 2 out of 3 or 2 thirds. In order to overcome this interpretation, it is necessary to extend the thinking used for the region model to the whole collection and then to the part of the collection that is to be considered.

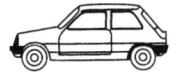 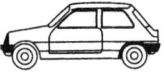   

How many cars in the collection?	5
What part of the collection is 1 car?	1 fifth
How many cars are black?	2
What fraction of the collection is black?	2 fifths

2 fifths of the cars are black

Practice with this way of thinking can be provided using cards of the form

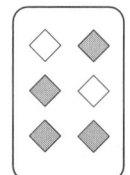 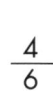 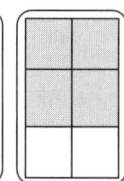

to play card and bingo games. Including the already known region model as well as the two language forms used to describe the fraction helps link the potentially confusing *fractions as collections* to the more intuitive initial representation.

Likely difficulties

Difficulties with the fraction concept usually occur because there is little understanding that equal parts are needed or because the process of naming the parts to provide the fraction name is not internalised. Use of a circular region as the mainstay of fraction teaching is often responsible for this, as the equality among the parts is not intuitively obvious to young children.

◆ A child might draw 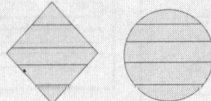 to show 4 fifths.

◆ When counting the parts to determine a fraction a child may use the counting numbers rather than relate them to the ordinal numbers to give the fraction names. For example, a region such as

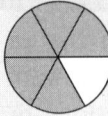 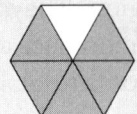 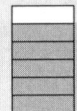

might be named 5 sixes.

◆ When fractions involving collections are shown

they might be named as 2 thirds rather than 2 fifths.

◆ The fraction may be associated with the objects within a collection, rather than the collection as a whole. For example, 1 third may be matched to 3 objects within the collection or applied to each object:

Circle one third of the fish

Circle one third of the fish

Investigate and discuss

1. It is relatively easy to use computer programs to draw fraction shapes. While a drawing program can create and shade a variety of accurate shapes, spreadsheet tools can also be used. Investigate how you can draw rectangular and circular regions using a program such as Excel.
2. How would you relate fraction ideas to length measurement? Consider the use of a ruler and the expression of distance travelled such as '2 thirds of the way home'.

Recording fractional amounts

Decimal fractions

The development of a systematic method for recording decimal fraction amounts is relatively recent. There have been a variety of means of recording common fractions from the so-called unit fractions of Egyptian times through competing Greek representations to the numbers and methods of calculation of the seventh century Hindu mathematicians that are essentially those in use today. While Chinese mathematicians used decimal fractions at least 3000 years ago, their recording of these, as with their whole numbers, uses additional symbols to state each place specifically. Even when Arab mathematicians first extended the place value system for whole numbers to include decimal fractions around the tenth century a form of recording was not clear cut or widespread. Thus, the extension of the place value notational system is generally credited to Simon Stevin of Bruges who in 1585 published a booklet, *De Thiende* [The Tenth], which extolled the use of decimal fractions over common fractions and gave methods for recording and computing with the new numbers. It is interesting to note that even Stevin's original notation required the explicit writing of each fractional place:

$$9\ 4\ 1\ ⓪\ 6\ ①\ 9\ ②\ 4\ ③$$

The ⓪ indicated the whole numbers, ① the first decimal place,
② the second decimal place and so on

It was some fifty years before the use of a decimal marker, either a full stop (decimal point) or comma, and implicit place value was adopted, so that this number is now written as either 941.694 or 941,694.

In both the Arabian and European situations, quite a large number of years needed to elapse from the first acceptance of the Base 10 system for whole numbers until it was sufficiently understood for the generalisation to a system catering for fractional amounts to occur. It is hardly surprising, then, that many young children do not readily accept decimal fractions for some time and are likely to make inappropriate generalisations from their whole number understanding to decimal fraction naming, reading, writing and processing. After all, if hundreds are larger than tens, why should tenths be larger than hundredths and thousandths? This difficulty is apparent both in naming the smaller places and in the region models with which they are represented. Even the writing of decimal fractions seems to go against the intuitive understanding gained with whole numbers when recording more digits in a number no longer necessarily means it is larger or even very large in itself.

Bundling sticks and Base 10 materials provided the basis for introducing the language and recording for whole numbers ten and larger. While this material could be adapted to allow tenths to be shown, it is not easy to allow the one in each case to be partitioned into ten equal pieces and certainly not into 100 parts. An alternative that is sometimes observed in schools is to rename the pieces so that, for example, the thousand block is considered to be one and the other blocks then show tenths, hundredths and thousandths. This tends to be bewildering for children who are not easily able to change their conception of the value of each block. Added to this is the confusion where now the hundreds block indicates 1 tenth and the tens block 1 hundredth! Consequently, it is not a practice to be recommended at all, as really what is occurring is that to 'see' this representation you first need to have the knowledge it is attempting to introduce.

There is also a problem with cutting up Bundling sticks or Base 10 materials. The essence of using materials for whole number numeration was to allow children to experience the grouping of 10 ones to form 1 ten, 10 tens to form 1 hundred and so on. Construction of a place value system for decimal fractions requires the experience of partitioning 1 one into 10 and 100 equal parts. It is really not feasible for children to actually cut the wooden pieces, let alone to do this accurately. At the same time, the children already have a good deal of knowledge and some well-established expectations about numbers on which this extension of the place value system can be built. Accordingly, it is not necessary that they use the actual materials that established numeration for whole numbers. Instead, it is sufficient that they build on the underlying principles used earlier.

For this reason, the use of a square region to represent 1 one is preferable. It allows children to construct, select and colour 10 equal parts, and later 100 equal parts, in a manner analogous to the use of the Base 10 hundreds, tens and ones. This use not only shows the place value pattern continuing, but highlights the decimal fraction parts in contrast to the whole number parts, an essential aspect of reading and interpreting written numbers with whole number and decimal places. However, it is not easy to determine where to draw the lines and it is potentially confusing to be measuring nine divisions to show ten equal parts. Accordingly, it is preferable to provide the beginning and end points of each line and to have children complete the partitioning without needing recourse to measuring techniques and the confusion they might bring. Shading the whole square also shows that 10 tenths is the same as 1 one, laying a foundation for the renaming that will be required later for rounding and computation.

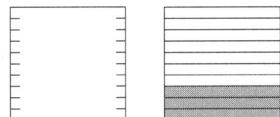

Children first need to construct the tenths for themselves before using squares where the lines are already drawn

When the recording of decimal fractions is first introduced, it is essential to start with both ones and tenths before moving to tens and no ones in much the same way as the recording of two-digit numbers began with tens and ones before considering tens and no ones. Squares are shaded to show both ones and tenths:

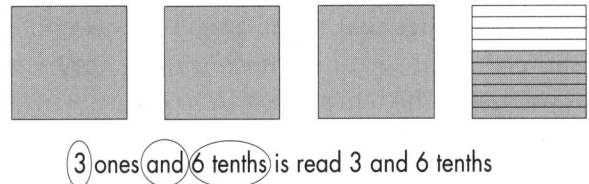

3 ones and 6 tenths is read 3 and 6 tenths

Decimal fractions can then be read in the same manner as whole numbers: 3 ones is simply read 'three', 6 tenths is read with its place value name in the same way as hundreds, thousands and so on. The connecting word 'and' is also said in the same way as 346 is read three hundred and forty-six, so that 3 ones and 6 tenths is read '3 and 6 tenths'. A place value chart can then be used to record the ones and tenths, extending the number system is to include decimal fractions. However, when a similar development was effected with the whole numbers, it was simply a matter of removing the place value chart to leave the usual recording once the children appreciated the place occupied by the ones and tens. If this was all that was done for the ones and tenths, there would be no way of knowing the value of the number that was shown.

ones	tenths
3	6

→ 3 6   Is this 36 or 3 and 6 tenths?

There needs to be some way of indicating whether a digit is in the ones or tenths place. The convention adopted has been to use a decimal marker next to the ones digit, a fullstop in many countries, but a comma in most of Europe, Africa and South America. It is helpful to have children enter one-digit numbers onto their calculator to see that a fullstop always sits alongside the digit, for example when five is entered, 5. appears. When larger numbers such as twenty-four or three hundred and seventeen are entered, the fullstop appears next to the ones digit, 24. and 317. The fullstop indicates that the whole numbers have ended; any digits after this must show fraction parts.

```
ones | tenths
  3  |   6      → 3.6
```

After the recording of numbers with ones and tenths is established first, recording a number with no ones follows quite readily.

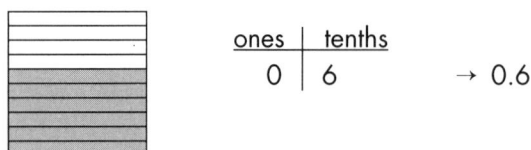

```
ones | tenths
  0  |   6      → 0.6
```

As well as being consistent with the pattern for numbers with some ones, having zero preceding the decimal marker ensures that the fullstop is not overlooked. When 6 tenths is written as 0.6, the zero shows there are no ones, the fullstop indicates that the whole numbers have ended, and the 6 is written in the tenths place. In contrast, when 6 tenths is incorrectly recorded as .6, it is often confused with 6.

This emphasis on place value in the formation of decimal fractions also extends to the way in which they are read. When 7.5 is read as '7 and 5 tenths' and 0.7 is read as '7 tenths', this allows meaning to be attached to the number by bringing the visual representation immediately to mind. On the other hand, saying 'seven point five' or 'zero point seven' is simply spelling out the digit in each place. Any understanding of the number that is being said depends on the child searching for what is meant without the benefit of any cues. In later computational work, comparing and rounding, this place value interpretation of decimal fractions will prove invaluable in allowing the ready extension of processes already established for whole numbers.

Once the naming of decimal fractions has been established, this understanding can be extended to include the way they are renamed so that a number with ones and tenths can be seen in terms of the number of tenths altogether. This can be readily seen with diagrams showing ones and tenths before leading into the use of a number expander in a similar manner to the way in which whole number understanding was extended.

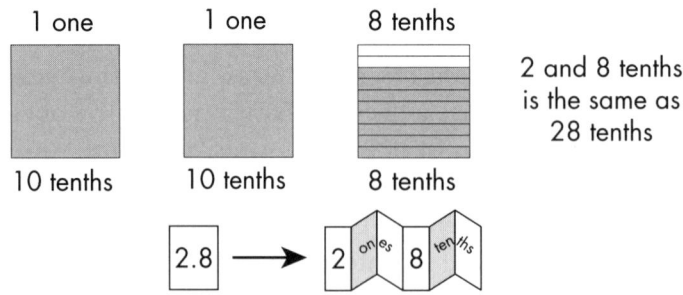

A decimal number expander also shows that 2.8 has 28 tenths

When the naming and renaming of decimal fractions is secure, it is usual to move immediately to addition and subtraction with them, using an understanding of decimal place value to build on the corresponding processes for whole numbers. Following that, the next decimal place can be introduced by partitioning 1 one to give hundredths. Again, children need to experience this partitioning process to see that one hundred smaller squares result:

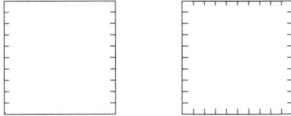

Children should first connect the horizontal lines to show tenths, then the vertical lines to show 100 parts altogether. 100 equal parts are hundredths

The recording of numbers with hundredths can then be introduced in a similar manner to the recording of ones and tenths, shading squares showing both ones and hundredths:

$$\begin{array}{c|c} \text{ones} & \text{hundredths} \\ 3 & 26 \end{array} \rightarrow 3.26$$

③ones (and) ㉖ hundredths is read 3 and 26 hundredths

When the recording for decimal fractions with hundredths is secure, renaming can be introduced to show the two different ways in which they can be read and interpreted. The first step builds on the diagram used to introduce hundredths to show that 10 hundredths complete 1 row and hence are the same as 1 tenth. Children need to readily see that 40 hundredths are 4 tenths and 7 tenths are the same as 70 hundredths as a basis for renaming numbers with hundredths in general. Diagrams showing ones and hundredths can then be interpreted in terms of ones, tenths and hundredths and lead into the use of a decimal number expander to show the different ways decimal fractions can be renamed:

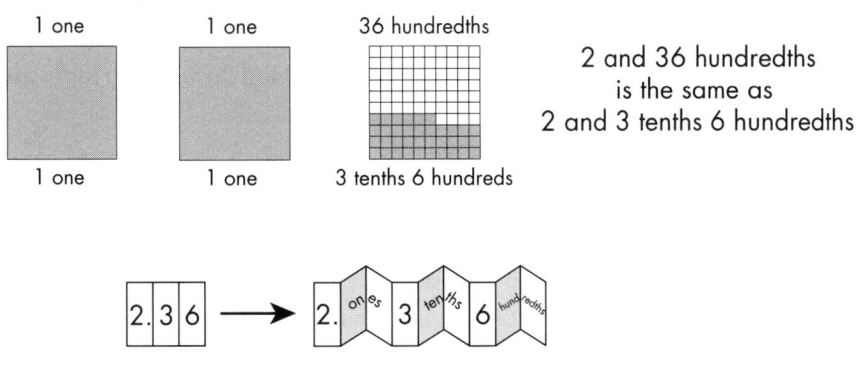

2 and 36 hundredths
is the same as
2 and 3 tenths 6 hundredths

It is not possible to partition the grid further to show thousandths. Instead, it is necessary to use an understanding of the place value patterns revealed so far to establish the thousandth and subsequent places. Writing a number such as 7849.52 shows the symmetry of the whole number and decimal places about the ones place:

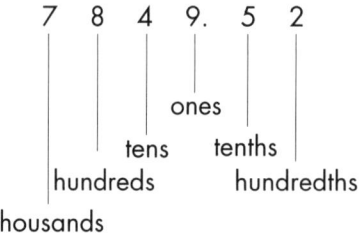

From the understanding that 1 thousand gives 10 hundreds, 1 hundred gives 10 tens, 1 ten gives 10 ones, 1 one gives 10 tenths and 1 tenth gives 10 hundredths, it follows that 1 hundredth gives 10 thousandths. The next place must be the thousandths place which fits with the symmetry of the place value pattern. Continuing this pattern, subsequent places will be the ten-thousandths, the hundred-thousandths and so on. Number expanders show that there are several ways in which numbers with larger numbers of decimal places can be read:

    49.523 is   49 and 523 thousandths
                 or   49 and 5 tenths 23 thousandths
                 or   49 and 52 hundredths 3 thousandths
                 or   49 and 5 tenths 2 hundredths 3 thousandths

However, when there are more than three decimal places, it is customary to spell out the number name rather than read it, as it is arguable whether a number read as 'nine and five thousand two hundred and thirty-eight ten-thousandths' is any more meaningful than saying '9 decimal point 5, 2, 3, 8'. Accordingly, in order to assist in the interpretation of these numbers, there is a further convention in their writing which leaves a space after every group of three digits from the decimal point, for example 9.523 824. This corresponds to the convention of leaving a space for every hundreds, tens and ones grouping of whole number digits, for example 9 523 824.

Although these conventions seem relatively straightforward, there are some subtle difficulties:

- When 9.523 is read as 'nine and five hundred and twenty-three thousandths' the '5' actually corresponds to 5 tenths, the '2' to 2 hundredths and the '3' to 3 thousandths in conflict with the five hundred and twenty-three that is said.
- When 9.523 824 is written, the grouping simply follows the pattern introduced with whole numbers. Whereas the whole numbers showed hundreds, tens and ones of each new place, with decimal fractions tenths, hundredths and thousandths are grouped, then ten-thousandths, hundred-thousandths and millionths and so on with no apparent pattern at all.

- There is also a conflict between the emphasis on reading decimal fractions for meaning developed here and the way in which most adults spell decimal fractions with no reference to place value. Children may find it difficult to understand what those adults are referring to and their own way of making sense of decimal fractions may seem unusual to many adults. However, the real difficulty is that very few people in our society really have any idea of the sense underpinning decimal fractions. Yet the world they now live in requires them to interpret and use decimal fractions in their work and everyday life. Perhaps children's language and understanding can be a means of building a firmer foundation of decimal fractions in society at large.

Likely difficulties

Difficulties with decimal fractions usually occur because place value is not used in naming the numbers, and they are spelled out as separate digit, for example 0.25 may be said 'zero point two five' rather than read 'twenty-five hundredths'. There may also be confusion about the new decimal places and the earlier whole number places. Often, when children meet decimal fractions a teacher finds that children do not have a sufficient understanding of the place value concept on which to base an extension to the fraction places, even though the children may have done quite well with whole numbers.

- There may be no understanding that the ones place is always recorded, for example writing .26 for 26 hundredths rather than 0.26.
- Reading and interpreting numbers with internal zeros such as 5.03 or 26.305 and numbers where teen numbers are read, such as 3.17 or 5.013, may cause similar difficulties to those experienced with whole numbers.
- Reading a number such as 9.764 as '9 and 7 hundred and sixty-four' may cause confusion as to where the decimal point occurs because the word 'and' which signifies it is also said within the number 764.
- Counting on or back may not incorporate place value. For instance, when counting on in tenths, a child may say '7 tenths, 8 tenths, 9 tenths, 10 tenths, 11 tenths' rather than rename to get '9 tenths, one, one and 1 tenth'.
- When asked which number is smaller, 4.7 or 4.19 a child may answer 4.7 reasoning that 7 is smaller than 19 because of the way he or she says or thinks about decimal fractions. By the same reasoning, 4.7 will be much smaller than a number such as 4.351.
- Children may also think that 7.4 is smaller than 4.19 or 4.327 by reasoning that the more digits a number has the larger it is—a procedure that produces correct answers for whole numbers.
- Rounding decimal fractions may be confused. When a number such as 4.327 is read as '4 and 3 hundred and twenty-seven thousandths' this may lead a child rounding it to the nearest hundredth to answer 4.3. Instead, the full place value needs to be considered, 4 and 3 tenths 2 hundredths 7 thousandths, to see that the answer should be 4.28.

Investigate and discuss

1. Write each of these in symbol form:
   a. four and three hundred and eighteen thousandths
   b. fifty and fifteen thousandths
   c. one and one hundredth
   d. eighty-seven decimal point nine, zero, six, eight, one
   e. ten decimal point zero, zero, zero, zero, three

   Comment on any difficulties that you experienced or might anticipate with young children.

2. a. Make a decimal number expander and use it to show how 42 hundredths can also be read as 4 tenths 2 hundredths.
   b. Write the two different interpretations for each of:
      i. 0.58
      ii. 0.36
      iii. 0.5
      iv. 0.14

3. For each of the numbers below, write the place occupied by the **bold** digit. How many thousandths are there in each of these decimal fractions?
   a. 5.789 **5**2
   b. 45.**1**26 407
   c. 13.**4**59 301
   d. 8.000 0**6**5
   e. 678.90**5**

4. Solve the following problem and consider how you might use such a problem to discuss decimal fraction meanings:

In the 100 m butterfly competition, the swimmers in each lane recorded these times in seconds:

Lane							
1	2	3	4	5	6	7	8
110.3	111.6	111.4	112.0	110.1	113.9	110.15	112.7

   a Which swimmer came first?
   b Which swimmer came last?
   c What was the difference in time between the first and last swimmers?
   d What was the difference in time between the first and second swimmers?
   e How much faster would the swimmer in lane 3 have had to swim to gain a place in the first 3?
   f List the times in order from slowest to fastest.
5 After learning decimal fractions for ones and tenths, a Year 4 child was then introduced to common fraction recording. Some time later when he was asked to write '7 tenths', he recorded 7.10.
   a What understanding of fractions does he have?
   b What difficulty might he have?
   c What might be the source of his difficulty?
   d How would you lead him to see the inappropriateness of this thinking?
   e How would you assist him to overcome his difficulty?

Per cents

In everyday life, fraction ideas are most commonly expressed and used in the form of per cents, especially in the way data is presented in government reports, financial markets and scientific areas such as health and agriculture. They occur frequently in newspaper advertisements, at discount sales, and in the media presenting information regarding the growth or decline of populations, trade figures, environmental issues or economic trends. While an understanding of the notion of per cent is needed by the end of primary school, a complete facility in its application is unlikely. On the other hand, the largely symbolic, rule-bound approach to the teaching of per cent in the past has meant that there was insufficient understanding to enable the interpretation and application of the concept and processes, even among otherwise numerate adults. For example, the *New Scientist* of April 2002 reported how many adults believed that 'If you pay for both gas and electricity by direct debit, you get 5 per cent off each, that makes 10 per cent off altogether' (p. 64).

   The term per cent means for each hundred deriving from the Latin words *per* (for) and *centum* (hundred). Thus a per cent is essentially another way of expressing fractions in terms of hundredths, and is most readily linked to an understanding of decimal fractions using the same form of square region to show hundredths:

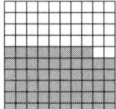

58 of the one hundred small squares are shaded

The diagram shows:

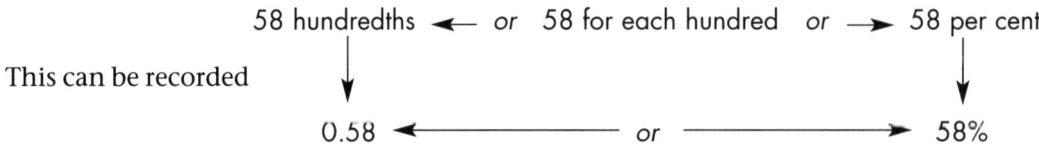

where the new symbol '%' is a mathematical way of writing 'per cent'. Like most other mathematical symbols, it was not used until the development of the printing press in the sixteenth century and is most likely derived from the use of $p\,c^o$ as an early abbreviation of *per cento*, the Italian expression for per cent. It is not a rearrangement of the symbols 100 used to denote one hundred as is often stated. Not only does the region model showing hundredths provide a meaningful conception of per cents, it also ensures that the equivalence of '58 hundredths', 0.58 and 58% is built in from the outset. Renaming between per cents and decimal fractions essentially builds on this common representation and language when 'hundredths' and 'for each hundred' are used interchangeably. However, this is not always straightforward, particularly when per cents also incorporate decimal fractions such as 17.5%. This needs to be seen as 17.5 hundredths or 175 thousandths, so that 0.175 is the same as 17.5%. Use of a diagram showing 17 small squares and 5 tenths or 1 half of another small square shaded and a decimal number expander can be used to support this thinking:

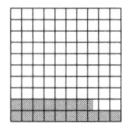

This thinking also is the key to coming to terms with per cents less than one which are often confused with decimal fractions, for example $\frac{1}{2}$% or 0.5% may be erroneously interpreted as 50% or 1 half or $\frac{1}{4}$% be taken as 25% or 1 fourth. Again, the use of the region model is critical in coming to terms with these potentially conflicting ideas:

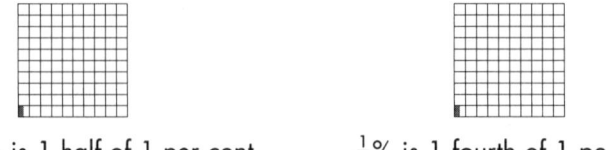

$\frac{1}{2}$% or 0.5% is 1 half of 1 per cent $\frac{1}{4}$% is 1 fourth of 1 per cent

A further difficulty occurs with the meaning for per cents greater than one hundred per cent which are frequently used to express large increases, such as the change in prices over a period of ten or twenty years. When a price increases by a factor of 3, for instance the price of milk increases from 50c to $1.50 per litre, this may be expressed as a 300% increase since three of the original price make up the new price. Many children and adults believe that it is only possible to have per cents up to 100% because of the emphasis on these initial ideas so have no sense of what this expression might mean. Yet, an understanding that per cents are just another representation of fraction ideas where 100% is 1 one allows the development of per cents larger than one hundred per cent, such as 250%, corresponding to the mixed numbers 2.5 or $2\frac{1}{2}$:

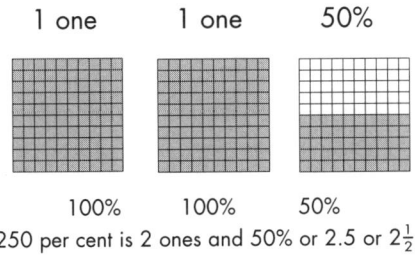

250 per cent is 2 ones and 50% or 2.5 or $2\frac{1}{2}$

When per cents are seen as another expression for hundredths and linked to decimal fractions, calculations involving them become quite straightforward. Unfortunately, the emphasis in teaching per cents in the past has largely linked them to common fraction ideas, resulting in a host of inefficient computational methods based on dividing or multiplying by 100. In contrast, most practical work with per cents will these days be done on a calculator, so the link to the decimal fractions that calculators exhibit is much more important. Seeing per cents in terms of decimal fraction equivalents allows the ready carrying out, reading and interpretation of calculations with them by entering and using decimal fractions directly. Eventually, however, this understanding needs to be extended to an ability to use the per cent key itself. It would be simplistic to treat this function as a simple keying procedure and in order to allow it to be used confidently there needs first to be a careful building up of a process for obtaining per cents of a given quantity through multiplication. A secure grounding of the notion of per cent in the general framework of fraction ideas with a particular emphasis on the region model and the language used for decimal fractions will ensure that this topic is developed and used surely and meaningfully. Nonetheless, there will still be a need to rename common fractions as per cents and show the equivalence of all three fraction forms.

Likely difficulties

Difficulties with per cents usually occur because of lack of a link to decimal fractions and a belief that they only use the whole numbers of 1–100. If per cents are developed simply as a computational procedure, children have no means of visualising their meaning and hence their significance.

- $2\frac{1}{2}$% may be confused with 250%.
- 17.25% may be interpreted as $\frac{17.25}{100}$ and hence will make no sense to a child, rather than related to the decimal fraction 0.1725.
- Most other difficulties occur within computation involving per cents, such as finding discounts, profits and increased prices due to inflation or changes in tax rates. When a conception of per cents is linked strongly to decimal fractions, these difficulties tend to be minimal.

Investigate and discuss

1. Collect examples of per cents in everyday use from newspapers and magazines. Use these to prepare an introduction to per cents for children in Year 5 of school.
2. Investigate how the per cent key on a calculator is used. Use the following problem to see how this function might be solved efficiently (how to enter the numbers, in which order to use them, can the memory keys be used, what effect might the % key have) and compare with another person who has solved the problem:

Salespeople at Superior Cars can purchase a new car at a discount of 15%. Sam wants to buy a small car which is usually sold for $12 500. How much will she have to pay? [*Hint*: Use a diagram to help organise the ways in which the problem might be answered.]

Common fractions

Although per cents build on the notion of decimal fractions, in practice common fractions are usually introduced soon after the development of decimal fractions with tenths, and per cents are left until hundredths are also built up. Largely this is because the fraction concept deals with parts in general, while the decimal fractions only concern the recording for a restricted number of fractions. Once the use of place value to record tenths has been established, the question arises as to how to record fractions that have more or less than ten equal parts. Some discussion of what might be required if a place value system was used soon leads to a realisation that this is no longer practical. For instance, if the ones place is followed by the tenths place, where would places for halves, thirds, fourths, fifths and so on be recorded? Not only would there need to be a large number of places between the ones and tenths digits, the same would occur later between the tenths and hundredths, between the hundredths and thousandths and so on. This would lead to uncertainty as to what a particular digit might represent as well as requiring large gaps or many zeros between digits. Talking about this difficulty with children prior to beginning common fraction notation raises the issue of how a meaningful recording might be established even before the unusual form, with one digit seemingly written on top of another, is introduced.

The meaning for a fraction such as 5 eighths has been established as '5 parts shaded out of 8 equal parts' and this language can be used to bridge from the region model to the common fraction symbols. The first step is to ensure that the meaning of 5 eighths as 5 out of 8 equal pieces is secure, then extend this thinking to the recording via a stylised form of writing these words that is later abbreviated to the mathematical symbols:

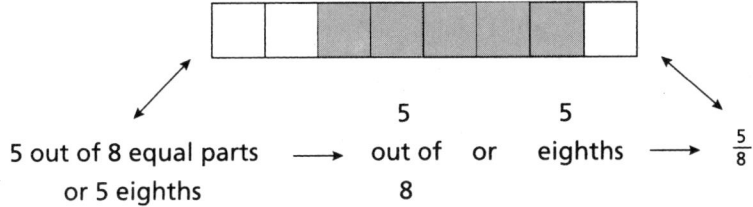

Children will need a good deal of practice to link the region model to the symbols and language. The 'Colour a fraction' games discussed on p. 142 can be adapted to help in this by modifying the dice or spinners to include symbols and having the players write symbols to match the numbers that they have shaded. Again, the rectangular region should be used first to build up this understanding, before different-shaped regions are introduced to consolidate the naming process. Since 9 ninths, 8 eighths, 6 sixths and so on form 1, they are simply recorded as 1 rather than the superfluous $\frac{9}{9}$, $\frac{8}{8}$ or $\frac{6}{6}$. Board games in which children match the language to the fraction shape or symbol are also helpful.

NAMING FRACTIONS GAME

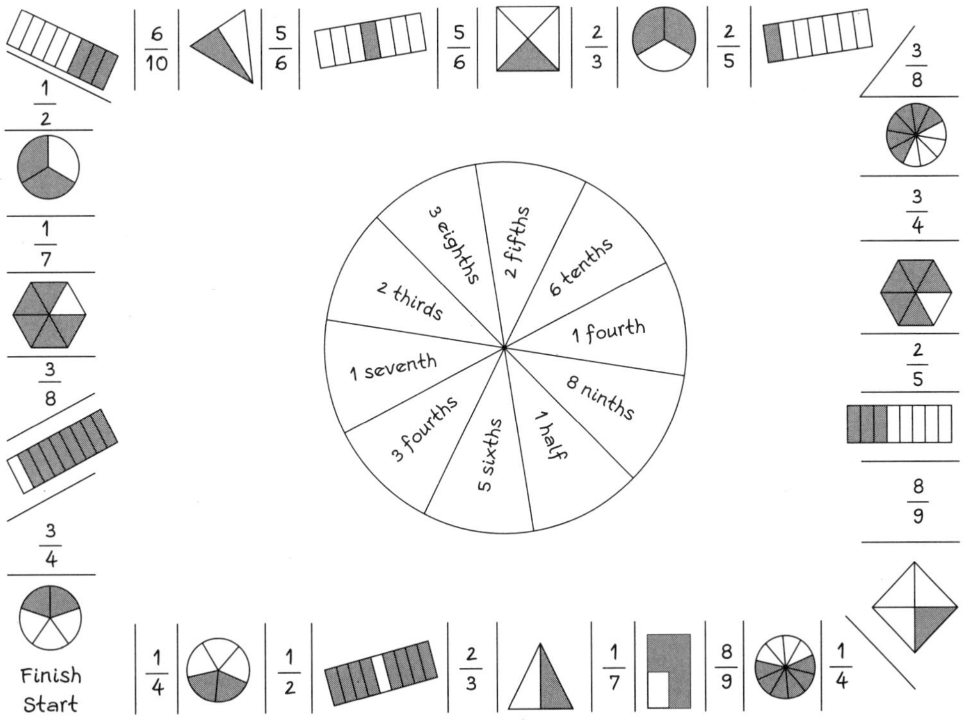

Materials required:
- playing board with fraction names and shaded regions
- spinner with fraction symbols to match the names and shaded regions on the playing board
- 6 coloured markers

Method of play: Each player in turn spins the spinner and moves his or her marker to the first shaded region or name that matches the fraction symbol indicated by the spinner. The first player to make three circuits of the board and pass the start/finish position is the winner.

1 At first, the playing board should use rectangular regions.

2 Later, different shapes should be used to consolidate the thinking.

It is also important to bring out the understanding that 1 (1 one) can be named in many ways to prepare for renaming common fractions in general:

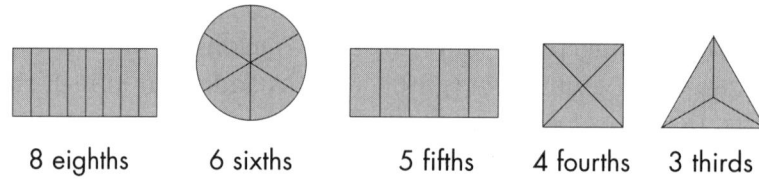

8 eighths   6 sixths   5 fifths   4 fourths   3 thirds

1 or 1 one can be named in many ways

Likely difficulties

Difficulties with recording common fractions usually occur because a link has not been made to the models used to introduce the concept. Often this results in children seeing common fractions simply as 'one number written on top of another', leading to difficulties with comparison, ordering and, later, computation.

- $\frac{7}{12}$ may be spelled out '7 over 12' instead of the more meaningful '7 out of 12' or '7 twelfths'.
- A child may write $\frac{5}{4}$ for 4 fifths.
- $\frac{3}{8}$ may be seen to be larger than $\frac{3}{5}$ because 8 is larger than 5.

Investigate and discuss

Many problems involving fractions can be solved readily by using a diagram based on the rectangular region used to establish the initial concept. For example:

Peter spent 3 fifths of his money and had $120.00 left. How much money did Peter originally have?

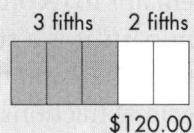

Since $120.00 is 2 fifths of his money, 1 fifth must be $60.00. So 5 fifths, all of his money, is $300.00.

Use diagrams to solve these fraction problems in a similar manner:
  a Peta spent 1 third of her money and then lost 1 half of what she had left. She then had only $20.00. How much money did she have to start with?
  b What is my number when
   i 5 ninths of my number is 35?
   ii 3 eighths of my number is 48?
   iii the product of 1 and 1 half and my number is 60?
  c Find my number:
   when 63 is added to 1 fifth of my number, the result is double my number
   when 13 is added to 2 and 1 half of my number, the result is triple my number
  d Would expressing these fractions in symbol form, for example $\frac{1}{3}$, $\frac{5}{9}$, $\frac{3}{8}$ or $1\frac{1}{2}$, make solving these problems easier or more difficult?

Renaming common fractions

Understanding that 1 one is 2 halves, 3 thirds, 4 fourths and so on is the first way in which common fractions are renamed. This can then be extended to renaming numbers with both ones and fraction parts (such as 3 and 4 fifths) as so many parts altogether (19 fifths) where there are more parts than are needed to make 1 one. Renaming of common fractions in this way is essential for the addition and subtraction algorithms, and also assists the comparison of different fraction amounts and in determining equivalence among common fractions, decimal fractions and per cents.

In the past, the technique for renaming a number such as $2\frac{3}{4}$ was to stress a rote rule based on multiplication: '4 twos are 8 and 3 is 11, 11 fourths altogether'. In contrast, a full understanding of what is occurring needs to be based on the region model to provide a more meaningful language-based process: '1 one is 4 fourths; 4 fourths and 4 fourths are 8 fourths, and 3 fourths are 11 fourths'.

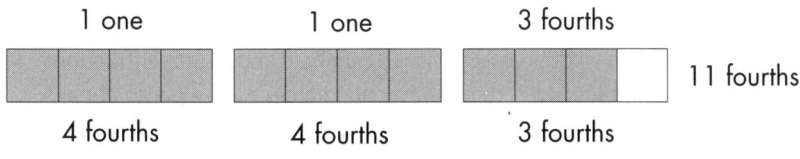

2 and 3 fourths is 11 fourths altogether

Once this thinking is secure, it only remains to record these two fraction amounts. For the conception in terms of ones and fraction parts it is a straightforward translation of what is read: '2 and 3 fourths' is $2\frac{3}{4}$. While recording the 2 and the $\frac{3}{4}$ is clear, some children include a full stop for the 'and' as they did with decimal fractions, $2.\frac{3}{4}$. A discussion of the convention of simply writing the ones and fraction part alongside each other in the same way that the hundreds digit and tens digit were written, for example 7 hundred and sixty-five is written 765, can help to overcome this tendency.

However, the recording of the alternative fraction form needs additional consideration because it seems to break with the initial part/whole concept. While the parts are undeniably fourths, what can 11 fourths mean? Certainly not the original notion, 11 out of 4 equal parts, but the meaning given in recording common fractions where 3 fourths was written $\frac{3}{\text{fourths}}$ to lead into recording $\frac{3}{4}$ can be readily extended to these new fractions made up of more than 1 one:

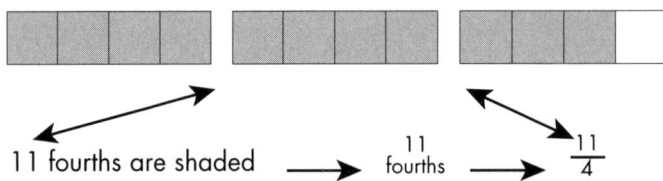

Thus, there are two ways of writing these fractions, $2\frac{3}{4}$ or $\frac{11}{4}$. A fraction such as $\frac{3}{4}$ which has parts of 1 one is called a *proper fraction*. All fractions of the form $\frac{11}{4}$ which have parts of more than 1 one are *improper fractions*, while those of the form $2\frac{3}{4}$ where there is a whole number together with a proper fraction are called *mixed numbers*. Being able to rename a mixed number as an improper fraction and vice versa is important in everyday applications of fractions, decimal fractions and per cents as well as common fractions (see p. 135 for the links among these various fractions when they are situated on a number line). Situations involving more than the number of parts required to give 1 one arise frequently and common fractions of this form are really no different from any of the earlier examples when understanding is tied to the region model.

Children need considerable practice in renaming between the two forms, using models and language to form the corresponding expressions to develop a meaningful process rather than a rote rule based on multiplying digits. A game similar to that used in naming fractions on p. 157 can be made by changing the board and using cards in place of the spinner.

RENAMING FRACTIONS GAME

Materials required:
- playing board with fractions written as both mixed numbers and improper fractions
- cards with region models shaded to match the fraction names on the playing board
- 6 coloured markers

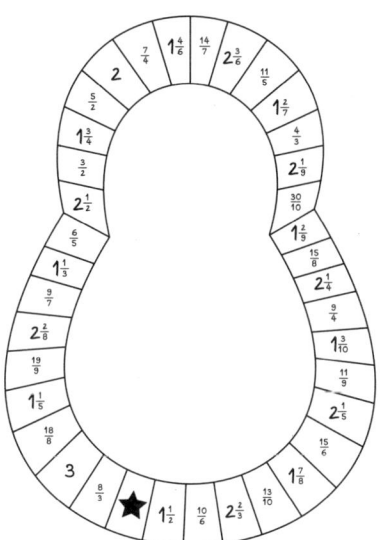

Method of play: Each player picks up a card, names the fraction indicated as both a mixed number and improper fraction and moves his or her marker to the first fraction symbol that matches the region model on the card. The first player to make two circuits of the board and pass the start/finish position is the winner.

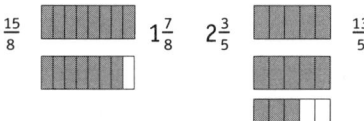

This understanding can now be used to compare and order common fractions. This is straightforward when common fractions have the same parts, simply a matter of determining how many of the part each number has and then comparing the resulting whole numbers. Common fractions with the same parts are called *like* common fractions, although they are sometimes referred to more formally as fractions with common denominators. Use of the word description 'like' provides a more immediate understanding so will be used in preference to use of the name denominator, and also suggests that common fractions with different parts are most helpfully called *unlike* common fractions. Thus, $\frac{7}{8}$, $\frac{15}{8}$ and $2\frac{1}{8}$ are like common fractions since they all involve eighths, while, $\frac{3}{5}$, $\frac{11}{4}$ and $2\frac{1}{2}$ are unlike common fractions. Comparing and ordering unlike common fractions is rarely as straightforward and reference needs to be made to the original rectangular regions so that the relative sizes of the parts can be assessed:

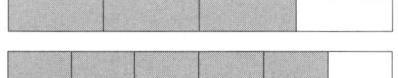

3 of the fourths are shaded, $\frac{3}{4}$

5 of the sixths are shaded, $\frac{5}{6}$

$\frac{3}{4}$ is less than $\frac{5}{6}$

Likely difficulties

Difficulties with renaming among improper fractions and mixed numbers usually arise from a lack of understanding that 1 can be written in many forms such 2 halves, 3 thirds, 4 fourths and so on. This often occurs because a child has no visual model of these fractions. In turn, such an inability to visualise these numbers leads to a reliance on computational rules to rename rather than a secure basis in the fraction concept.

- If the concept of improper fraction is not grasped, a child may find the idea of 7 fifths confusing, and record $\frac{5}{7}$ because of a belief that there cannot be more parts than make up 'a whole'.
- Counting on or back may not incorporate renaming. For instance, when counting on in fifths, a child may say '3 fifths, 4 fifths, 5 fifths, 6 fifths, 7 fifths' rather than rename to get '4 fifths, one, one and 1 fifth'.
- Renaming might be based on a multiplication rule that is not fully understood, for example $2\frac{5}{6}$ might be written as $\frac{17}{6}$ by reasoning '2 sixes are 12 and 5 is 17' rather than '1 one has 6 sixths, so 2 has 12 sixths and 5 more sixths is 17 sixths'. Such a multiplication rule can then be misapplied to multiply '5 twos are 10 and 6 is 16' by multiplying from the top.
- When improper fractions are written as mixed numbers, the renaming may occur in terms of 10 since that is what is done with whole numbers and decimal fractions. This can lead to $\frac{21}{6}$ being written as $2\frac{1}{6}$ or $\frac{7}{5}$ being left unchanged as there are only 7 parts not 10.

Investigate and discuss

1. Draw diagrams to show 14 thirds, 23 eighths, 33 twelfths and 17 fourths. Use these diagrams to show how they are named as improper fractions and mixed numbers.
2. A child answered four questions involving the renaming of common fractions:

    14 sixths is $1\frac{4}{6}$ 11 eighths is $1\frac{1}{8}$ $6\frac{4}{5}$ is $\frac{29}{5}$ $3\frac{5}{12}$ is $\frac{27}{12}$

    What thinking is being used to arrive at these answers?
3. Use diagrams to determine which fraction is larger:

    $\frac{3}{4}$ or $\frac{2}{3}$ $\frac{7}{12}$ or $\frac{5}{18}$ $\frac{9}{20}$ or $\frac{11}{24}$ $\frac{5}{8}$ or $\frac{3}{4}$

Renaming as equivalent fractions

Not only can common fractions take different forms through being renamed as mixed numbers and improper fractions, they can also be described using different fractional parts, for example 2 fourths or 5 tenths are also names for 1 half. When several fractions name the same amount, they are said to be *equivalent*, since they have the same value. This understanding can be developed through the use of the rectangular regions which established the initial fraction concept and proved helpful in comparing common fractions. In due course, the same model can be used to build a meaningful process for dealing with the fraction symbols directly. Renaming common fractions as equivalent fractions is essential for computation with fractions, so this will largely be left until the development of addition and subtraction in chapter 4, when a need for the somewhat complex thinking required is readily apparent. In time, this thinking will also need to extend to renaming among common fractions, decimal fractions and per cents, building on the relationship between the fraction concept and division.

Activities involving folding and shading paper rectangles have proven to be successful in introducing this idea:

1 Start with a rectangle that shows fifths. Colour 3 of the parts to show $\frac{3}{5}$.

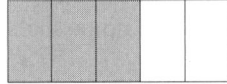

2 Fold the rectangle in two, lengthwise, then open it again.
Now there are 6 parts shaded and 10 parts altogether. The rectangle shows $\frac{6}{10}$.

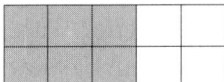

3 Fold the rectangle in two, lengthwise.
Fold it in two lengthwise again, then open it out.
Now there are 12 parts shaded and 20 parts altogether. The rectangle shows $\frac{12}{20}$.

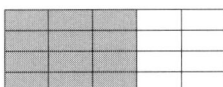

The amount of the rectangle that is shaded has not changed at all, so the fraction coloured has not changed.

$\frac{3}{5}$ or $\frac{6}{10}$ or $\frac{12}{20}$ are all names for this fraction

This idea can then be extended to include the other possibilities that paper folding cannot introduce by using a series of rectangular regions displayed in descending order. Including a full rectangle acts as a visual reminder that each rectangle represents 1 one partitioned into different parts:

Name the different common fractions that represent the same amount by aligning the vertical lines:

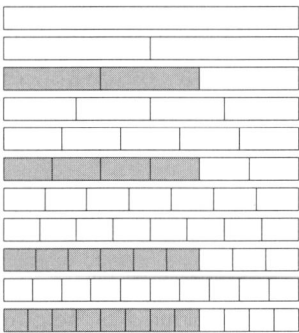

$\frac{2}{3}$ or $\frac{4}{6}$ or $\frac{6}{9}$ or $\frac{8}{12}$ all name the same fraction

It is not advisable to dwell on the paper-folding activities for too long in case children focus on the doubling pattern and develop a restricted conception of equivalence which will lead them astray in later computation. For instance, if they only rename by doubling, then they would rename $\frac{2}{3}$ as $\frac{4}{6}$ and $\frac{8}{12}$ but omit $\frac{6}{9}$. The focus needs to be on the general situation of renaming the initial fraction in terms of a fraction with 2 times as many parts, 3 times as many parts, 4 times as many parts and so on that can be drawn from the rectangular diagrams. This thinking can also be fostered by drawing several rectangles that show a particular fraction such as $\frac{3}{4}$, then drawing lines across each rectangle to show eighths, twelfths and sixteenths:

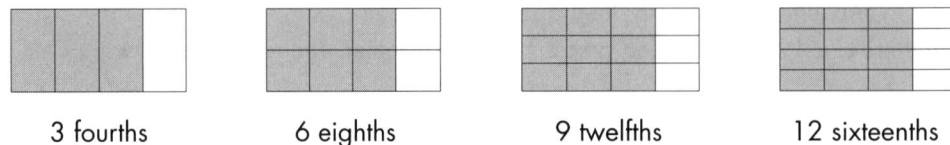

3 fourths   6 eighths   9 twelfths   12 sixteenths

The amount of the rectangle that is shaded has not changed at all, so the fraction coloured has not changed

$\frac{3}{4}$ or $\frac{6}{8}$ or $\frac{9}{12}$ and $\frac{12}{16}$ are all names for this fraction

Some children find the idea of renaming a fraction amount with different names difficult because of an underlying belief that each number should have a unique name. The value of using the concept of renaming throughout numeration is that it provides a basis for thinking of numbers in more than one way, and this is particularly important for work with fraction ideas. Nonetheless, it is best to leave any further development of the concept of equivalence until a need for it can be generated. Since its major use is to allow the addition and subtraction of unlike common fractions, it is only at this point that it makes sense to establish a formal renaming procedure where the initial fraction is multiplied successively by 2, 3, 4, 5 and so on:

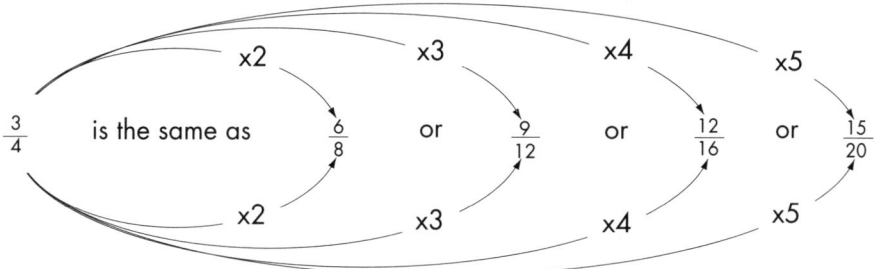

The development of this process in a meaningful way is discussed in chapter 4.

In everyday and problem situations, there is also a need for renaming among common fractions, decimal fractions and per cents. Renaming decimal fractions and per cents as common fractions is largely a matter of reading them using the language of hundredths and then writing in the required form. For example, 0.75 is 75 hundredths which can also be written as $\frac{75}{100}$. Similarly, 45% means 45 for each hundred or 45 hundredths, so is written as $\frac{45}{100}$. However, renaming a common fraction as a decimal fraction or per cent is not always so easy. Eventually division can be used in this renaming, but establishing and justifying its use is not easy for it requires a matching of the fraction concept, where a fraction can be seen to represent equal shares of the whole, and the division process, which involves sharing. This allows equivalent fractions to be found by dividing the numerator by the denominator:

Rename $\frac{3}{4}$ as a decimal fraction:

$\frac{3}{4}$ is $3 \div 4$

$$\begin{array}{r} 0.75 \\ 4\overline{)3.0} \\ \underline{2\ 8} \\ 20 \\ \underline{20} \end{array}$$

Rename $\frac{3}{8}$ as a per cent

$\frac{3}{8}$ is $3 \div 8$

$$\begin{array}{r} 0.375 \\ 8\overline{)3.0} \\ \underline{2\ 4} \\ 60 \\ \underline{56} \\ 40 \\ \underline{40} \end{array}$$

This process needs to be built up through the use of real situations and models in the first place and its development is best left until late in primary school when all of the underlying concepts and meanings are secure. A fuller treatment of this is given at the conclusion of the division section in chapter 4.

This approach to developing the recording of common fractions should ensure a smooth building up of the full fraction concept. However, it must be remembered that the basis for this notation is very different from the place value used for all other numbers. There is likely to be more difficulty in building competence with common fractions than the apparently straightforward sequence of development would suggest. In particular, the process of renaming so that the one fractional quantity will have a variety of names is quite unlike the situation for whole numbers where renaming is always in terms of place value, so that each number has essentially only one name. Accordingly, these equivalent fraction ideas often take a long time to become established and in turn are more often the source of difficulties that arise in computation than the algorithms themselves.

Likely difficulties

Difficulties with equivalent fractions also are largely sourced in a reliance on computational rules to rename rather than a basis in the fraction model and the process that develops from it.

- Often children only use doubling to find equivalent fractions. For example, they might rename $\frac{3}{4}$ as $\frac{6}{8}, \frac{12}{16}$ and $\frac{24}{32}$ but omit $\frac{9}{12}, \frac{15}{20}, \frac{18}{24}$ and $\frac{21}{28}$.
- Most other difficulties with renaming as equivalent fractions emerge in the computational contexts where they are used. For instance, in order to add or subtract, one or both of the common fractions involved need to be renamed as like common fractions. Some children may simply rename each fraction independently of the parts involved, for example to change $\frac{3}{4}$ to $\frac{3}{12}$ and $\frac{2}{3}$ to $\frac{2}{12}$ so that each number contains twelfths.

Investigate and discuss

Many problems involving proportions are solved by utilising the same thinking used for equivalent fractions. For example:

Sandra and Sandy are walking together. Sandy takes two steps for every three steps that Sandra takes. How many steps will Sandra take while Sandy takes 60 steps?

Sandy's steps	2	4	6	8	10	. . .	40
Sandra's steps	3	6	9	12	15	. . .	60

Sandra will take 60 steps while Sandy takes 40 steps.

This answer could be determined by completing the entire table, or by seeing patterns such as if the proportion is 6 steps for every 4 of Sandy's steps, this could be multiplied by 10 to get 60 for every 40 steps more directly, or when there are 10 steps for every 15 of Sandy's steps, multiplying each part by 4 would also give the answer more directly.

1 Use a table to solve these fraction problems in a similar manner.
   a In the cafeteria courtyard, 6 students sit at each table. If 90 students need a place at a table, how many tables are needed?

b Dog food was on special at the supermarket at two tins for the price of one. If one tin normally costs 89c and Simon buys 48 tins to feed his dog over the winter, how much did he pay?

c Simone discards 5 out of every 8 pictures she takes on her digital camera. If she deletes 60 pictures, how many did she take?

2 Using a table to show proportions also helps solve problems involving *per cents*. Use a table to solve these per cent problems in a similar manner to the fraction problems above.

a In a survey, 9 out of twenty people said that they disliked the amount of football shown on television. Assuming this survey is accurate, what per cent of people do not like the amount of football shown on television?

b A real estate agent sold a house for $320 000. She received a commission of 4% on the selling price. How much did she receive from the sale?

c During a drought, the amount of water in the town reservoir fell by 54%. If the reservoir held 18 000 000 kL of water before the drought, how much water was left?

3 How could you use a calculator to rename among common fractions, decimal fractions and per cents?

Ratios

There is one other mathematical idea that relates to fraction concepts and that is the notion of ratio. Sometimes it can be taken as another expression of a fraction, but at other times its meaning is more clearly related to multiplication. Because many of the ways of operating with fractions are essentially additive, this can cause confusion when ratios are used and interpreted. On the other hand, the way in which common fractions and per cents are renamed to provide equivalent fractions (developed in more detail in chapter 4) can be extended to a means of finding, simplifying and comparing ratios. It is important to understand that the relationship of ratios to fractions is not direct: sometimes it will be helpful, sometimes it may be confusing.

Ratios are commonly used in everyday situations ranging from collecting and exchanging swap cards, for example 'two of mine for one of yours', through buying lollies in quantities such as 3 for 10c, to discussing car performance in terms of litres per 100 kilometres or accelerating to reach high speeds within 10 seconds. They are applied routinely in many contexts, often without mentioning the word ratio at all. But without a careful introduction of the concept, language and meaning for ratio, a proper interpretation of situations involving their use is unlikely to occur.

The concept of ratio first needs to be carefully established in the context of comparison. Initially when things are compared, it is sufficient to know which is longer, larger or heavier. Later, the question of how much longer, larger or heavier arises and the difference is measured, using subtraction. But in many situations it is not the absolute difference but the relative difference that is important—how many times longer, larger or heavier one is than the other. In this way, one quantity might be seen as a multiple of another—a rhinoceros is 9 times as heavy as a lion—or one amount might be increased or decreased by a factor of the other—a lion is only 2 thirds as tall as a rhinoceros. The integration of ratio with multiplication facilitates the later extension of early ratio ideas to cover comparison of parts with parts to give rise to fractional ratios, for example a lion can move at 1.6 times the speed of a zebra, but a lion is only 1 sixth of the height of a giraffe. In order to allow a meaningful development of ratio ideas, it is important that children realise that ratio has its origins and uses in comparisons of this type, based on multiplicative ideas.

Comparison situations lead naturally to the use of tables (Streefland, 1985) in which values can be listed so that properties can be discovered and made conscious, giving rise to a meaningful use of the formal mathematical language and symbols for ratio, as in these examples.

1 To make drinks for the sports day the committee purchased fruit juice concentrate which was diluted in the ratio 1 to 4. How many litres were needed to fill the 40 litre drink cooler?

Litres of concentrate	1	2	3	4	5	6	7	8	9	10
Litres of water	4	8								
Litres of fruit drink	5									

2 Flavoured milk was made by adding flavouring to milk in the ratio 2 to 15. How many litres of flavoured milk could 3 L of flavouring make?

3 A fruit punch was made using ginger ale and pineapple juice in the ratio 5 to 3 and pieces of fresh fruit. If 64 litres of punch were made, how much pineapple juice and how much ginger ale was used?

4 Mathematically, we write 1:4 for the ratio of 1 to 4; the ratio of concentrate to water in the fruit drink is 1:4. How many litres of milk would be needed to fill the 40 litre drink cooler if the ratio of flavouring to milk was 1:7?

5 If 48 litres of fruit punch were made and the ratio of pineapple juice to ginger ale was 3:5, how much pineapple juice would be needed?

Such tables of values also provide a picture of proportion as an equivalence relation, allowing problems to be solved at a more intuitive level before formal procedures are built up. Seeing the values expressed in the same proportions highlights the similarities with the techniques for finding equivalent fractions and expressing fractions in simplest form. This allows the transition to a known way of thinking to express ratios simply before a purely multiplicative way of thinking is called upon. Since the use of tables also applies to every ratio situation, it allows the concept to be detached from its context, increasing the possibility of applying ratio to a variety of real applications such as those used in forming mixtures, distributing items proportionately and extending drawings or diagrams.

1 Proceeds from the sports day were used to purchase new equipment. The children were delighted when they found that the Education Department offered a subsidy of $2 for each $3 that was raised in this way. The drinks stall raised $372. How much subsidy should they receive? What would be the total amount of money raised for new equipment through their stall?

Using a table of values would take a long time!

Money raised	3	6	9	12	15	...	372
Subsidy	2	4	6	8	...	...	

These ratios, 3:2, 6:4, 9:6, ..., are equivalent, just like equivalent fractions.

3	6	9	12	15	...	372
2	4	6	8		...	?

The numbers in each row are multiples of the first number.
372 = 3 × 124, the corresponding number will be 2 × 124
The subsidy must be $248 and the total amount of money contributed by the stall is $520.

2 If the Education Department had given only $62, what would the rate of subsidy be then?

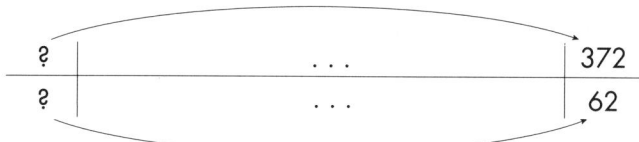

The numbers in each row would still be multiples of the first number. Now we have to find a number that is a factor of each number to express the ratio in simpler form.

372 = 2 × 186 or 3 × 124 or 4 × 93 or 6 × 62 or 12 × 31 = 2 × 6 × 31
62 = 2 × 31

2 and 31 are both factors of each number, so the ratio can be simplified to 6:1.
In this case the Education Department would only offer a subsidy of $1 for each $6 raised.

This visual integration of ratio with multiplication and equivalence of fractions also provides a basis for the development of an algebraic approach as problems become more complex beyond the primary years.

In the past, the teaching of ratio has been characterised by a very narrow approach, largely aimed at developing skills in reading and writing symbolic statements so as to lead into exercises with purely numerical ratios. There was little building up of the ratio concept, real applications were lacking, and the subject was taught in isolation from the mathematical ideas on which it rests. The difficulties and lack of interest that children often exhibited stemmed from such an approach. This was also exacerbated by too abrupt an introduction occurring too late for it to be firmly connected to ideas such as equivalence. Different properties that are mathematically the same were then not at all equivalent for these learners, in turn restricting the ways in which ratio ideas could be interpreted and applied. In particular, children often found it difficult to visualise ratios and lacked any meaningful schema for the numerical processing of ratio problems.

This brief discussion of ratio suggests how it can be introduced through meaningful activities, rather than a formal set of definitions. It also shows how ratios both relate to fraction ideas yet are also separate from them. If ratios are introduced in this way, children can construct their own intuitive understanding and move on to use it confidently and competently in the many real life applications that require ratios.

Investigate and discuss

1 Use tables to solve the problems in this section.
2 Solve these ratio problems in a similar manner:
   a A restaurant used 7 pieces of cutlery and 4 pieces of crockery to set each place. If it used 203 pieces of cutlery to set the tables last night, how many pieces of crockery were used?
   b If 6 boys fill 6 exercise books in 6 weeks and 4 girls fill 4 exercise books in 4 weeks, how many exercise books will a class of 12 boys and 12 girls fill in 12 weeks?
   c A team of 5 painters took 5 weeks and 2 days to paint a block of units. How long would it take a team of 4 painters to paint an identical block of units if each painter worked at the same rate? [A week has 6 working days.]

Patterning—moving to algebra

As we explore numbers, spatial figures and other relationships, and observe patterns and structures in them, we continually need to develop concepts and language in order to understand and describe them. Students need considerable facility with the observation and expression of generality prior to the introduction of algebraic symbolism—they must first be able to 'see' a pattern, [then] having identified it, they should develop ways of describing the pattern. Mathematics brings to the study of patterns an efficient and powerful notation for representing generality and variability, and for reducing complexity—algebra (The *National Statement on Mathematics for Australian Schools*, 1991).

While an ability to use and interpret numbers in a variety of ways is crucial for everyday uses of mathematics, it is a capacity to search for patterns and make generalisations that is central to mathematical thinking. Not only does the investigation of patterns and relationships build children's understanding of large and small numbers, but their 'descriptions and representations of patterns and their ability to express generalisations becomes more fluent'. The National Statement on Mathematics envisages that 'the investigation of patterns and relationships in number should be a focus of number work, and that children should identify, describe and continue patterns and create their own, using ideas that can be expressed with materials before moving to symbolic representations such as the figurate numbers and number sequences'.

PATTERNS IN NUMBERS

1 Use counters to continue this square number pattern:

1st 2nd 3rd 4th 5th 6th 7th

```
                ••••
        •••     ••••
••      •••     ••••
•       ••      •••     ••••
```

We have the pattern

1 : 1 + 3 : 1 + 3 + 5 : 1 + 3 + 5 + 7 : ...

if we consider the number of counters added to the previous square to make the next square. Each square number is the sum of odd numbers.

What is the sum of the first 4 odd numbers? The first 7 odd numbers? The first 100 odd numbers?

If we write S for square number, we can write the 4th square number as:

$S_4 = 1 + 3 + 5 + 7 = 16 = 4^2$ What is S_7? S_{12}? S_{100}?

How could you use this square number pattern to find

$$11 + 13 + 15 + 17 + 19 + 21 + 23 + 25?$$

How would you find the sum of even numbers?

Can you express the sum of the first 6 even numbers in terms of S_6?

2 Use counters to continue this triangular number pattern:

1st 2nd 3rd 4th 5th 6th 7th

> We have the pattern
>
> 1 1 + 2 1 + 2 + 3 1 + 2 + 3 + 4 . . .
>
> if we consider the number of counters added to the previous triangle to make the next triangle.
>
> If we write T for triangular number, we can write the 4th triangular number as: $T_4 = 1 + 2 + 3 + 4 = 10$.
>
> What is T_6? T_8? T_{25}?

An analysis of number sequences provides a good vehicle for this because the background knowledge is very familiar, making it easier for children to notice patterns and generalise from them. However, the expression of these observations in a verbal rule is often difficult, clumsy or long-winded. The inadequacy or imprecision of everyday language in turn lays the basis for the use of letters to signify numbers and number relations in a similar way to the use of letters as labels in geometry.

This first use of letters in arithmetic occurs as a natural abbreviation of the use of words and phrases and readily lays a foundation for their more extensive use to describe number patterns and the relationships between the numbers in these patterns. For number sequences, imprecise verbal rules give way to expressions in which each term is described in relation to the term number (the second, third, fourth, . . term in a sequence) and eventually these verbal rules can be replaced by the use of the letter n in an expression of the nth term of a sequence. This is analogous to the use of letters in measurement formulas for areas and volumes and allows a concise formulation of a relationship that is usually cumbersome to express in words.

1 Use counters to make S_4. Rearrange them to make 2 triangular numbers.

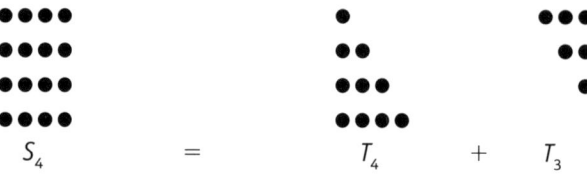

S_4 = T_4 + T_3

Does this relationship hold for all square numbers?
Can you write a general expression for the relationship?

2 We can call S_4 the 4th term in the sequence of square numbers, T_4 the 4th term in the sequence of triangular numbers; in general, we say that S_n and T_n are the nth terms in these sequences.

Find T_n when $n = 5$ $n = 10$ $n = 100$

If we examine the triangular numbers again using counters, we can see a general method:

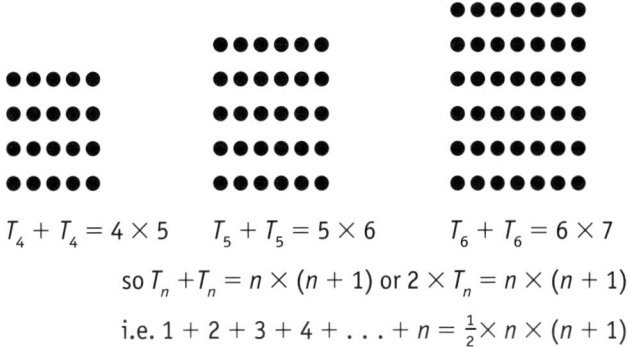

$T_4 + T_4 = 4 \times 5 \quad T_5 + T_5 = 5 \times 6 \quad T_6 + T_6 = 6 \times 7$

so $T_n + T_n = n \times (n+1)$ or $2 \times T_n = n \times (n+1)$

i.e. $1 + 2 + 3 + 4 + \ldots + n = \frac{1}{2} \times n \times (n+1)$

This now allows us to readily find the sum of the counting numbers.

This use of letters as substitutes for numbers or number relationships forms the first step towards the development of algebra, but still as a generalised arithmetic where a definite value is represented. Initially, the focus needs to be on building up an awareness of the power of using letters to assist in solving more complex number problems. Later, this development can be continued to further investigations of number topics, calling on computational as well as numeration understandings, where the usefulness of generalised arithmetic can be extended to the use of letters to find and verify number patterns.

1 Investigate the manner in which the mathematician Karl Gauss (1777–1855) made his discovery of the result

$$1 + 2 + 3 + 4 + \ldots + n = \frac{1}{2} \times n \times (n+1)$$

while only an 11-year-old school boy!

2 Use this result to solve the following problem:
To raise money for new pool equipment the swimming club held a swimathon. The 27 juniors decided that the first of them would swim 20 lengths, the next 22 lengths, the next 24 lengths and so on. If they were to be paid 10c for each length swum, how much money did they raise altogether? How much money did the last swimmer raise?

More specific algebraic skills can then be developed, although that is outside the scope of this book, but it is important that students arrive at that point with a prior realisation of the need for this way of thinking. Algebra will then be seen as a tool that extends earlier knowledge and understanding of number in a natural way to solve problems efficiently and that allows generalisations to proceed readily as a natural extension of earlier ways of thinking.

Investigate and discuss

1. Find other number patterns or sequences that could be used to introduce algebraic notation and reasoning using letters in place of general, unknown numbers.
2. Investigate the Tower of Hanoi puzzle. How might you use an investigation of the number of moves required for 3, 4, 5, 6, . . . n moves to consolidate algebraic recording and reasoning?
3. Take a sheet of paper. Fold it once, open it up and record the number of regions. Fold it again and record the results in a table:

Number of folds	1	2	3	4	5	6	7	8
Maximum number of regions								

Can you find a rule for n folds?

4. Draw polygons and put in all of their diagonals. Count the diagonals for each polygon and record the results in a table.

Number of sides of polygon	1	2	3	4	5	6	7	8	9	10
Number of diagonals										

Can you find a rule for a polygon with n sides?

CHAPTER FOUR
Computation and estimation for whole numbers, decimal fractions, common fractions and per cents

A major goal [of number and operations] is the development of computational fluency . . . having efficient, accurate and generalisable methods (algorithms) for computing that are based on well-understood properties and number relationships. Some of these methods are performed mentally, and others are carried out using pencil and paper to facilitate the recording of thinking. Students should come to view algorithms as tools for solving problems rather than as the goal of mathematics study. As students become fluent and flexible in computing, they should also develop computational estimation strategies for situations that call for an estimate and as a tool for judging the reasonableness of solutions.

Principles and Standards for School Mathematics 2000, p. 144

Teaching computation

COMPUTATION is concerned with the processes needed to add, subtract, multiply and divide numbers. These computational processes are important in their own right for the various tasks of everyday life, but they also underpin broader mathematical techniques, for instance finding averages and other statistical measures, determining length, area or volume, or constructing geometrical shapes. Advanced topics such as ratio and algebra rely on an ability to generalise the thinking built up with these fundamental operations, while most problem solving involves some form of computation, often used in a different manner to the way it was learned. To reflect the widespread use of technology in society, computation must include processes for estimation and approximation as well as exact calculations, an ability to calculate mentally, and a full adaptive use of calculators. Meaningful and well-understood *concepts* for each operation are essential in developing this full range of abilities. So too is ready access to *basic facts* involving combinations of the numbers 0–9 and a facility with the *algorithms* (or step by step processes) for calculations with larger numbers, decimal fractions and common fractions.

Ability with computation is also crucial in developing children's confidence in using mathematics. Those who understand and can carry out calculations readily are more likely to try new ideas and adapt their knowledge to unfamiliar situations. Conversely, children who are unable to complete the required computation easily and accurately soon develop a belief that they are poor at mathematics. This belief can then become self-fulfilling as they avoid taking part in situations where their lack of knowledge will be displayed, so that new ideas are not fully developed and applications are soon beyond their reach. In particular, there is often an incomplete grasp of the concepts underlying each operation, leading to attempts to master the algorithms by rote (that is, without understanding). Consequently, difficulties seen with computation usually involve confusion with the meaning of an operation or an inability to apply numeration ideas to direct the sequence of steps within a calculation. For instance, the use of renaming in one context may be inappropriately transferred to another, or numbers may simply be altered to allow a step to proceed without any consideration of the underlying place value meanings.

e.g.

$$\begin{array}{r} 72 \\ -49 \\ \hline 37 \end{array}$$

A focus on difference as the meaning for subtraction

$$\begin{array}{r} \overset{1}{6}2 \\ \times\ 7 \\ \hline 494 \end{array}$$

The additional ten is added prior to multiplying

$$\begin{array}{r} 5\ 6 \\ 5\overline{)2530} \end{array}$$

25 and 30 are divided by 5 without consideration of place value

In order to develop a numerate society, number work should reflect the balance of number techniques used in everyday life and work (*A National Statement on Mathematics for Australian Schools*, 1991). However, it is also crucial to consider the needs of the children who are to build up these understandings and abilities and use them throughout their adult lives. Estimation and approximation might be the main uses of mathematics outside of the exact calculations routinely done by calculators and computers, but an ability to calculate accurately and automatically is needed by children prior to the use of techniques which give answers that are 'near enough'. Indeed, moving from answers that are only slightly out but considered therefore to be wrong to situations where a similar answer is close enough is very difficult for young children. For instance, it is wrong to say that 7 and 4 is 10, but 100 is a good estimate for 37 and 64.

Being able to calculate accurately with an exact method that is understood is usually a prerequisite to building up the confidence to make an estimate. Only when a child realises that he or she can get an appropriate answer by using a meaningful process will it be possible to suspend that exact calculation and consider using a technique that will be close enough for the situation at hand. Of course, understanding and being confident with a given form of computation is also essential to knowing the context of the computation and thus to realising the degree of accuracy that is required. But the thinking underpinning the formal algorithms also gives rise to many of the techniques needed for estimation.

For example, when division is seen as a sharing-based process in which the largest place is necessarily shared first, it is not difficult to extend this thinking to an approximation technique:

$39\overline{)8397}$ can be viewed as sharing 840 tens among 40, leading to
$39\overline{)8397}$ is about 210

On the other hand, when many students
are asked to estimate

$$\begin{array}{r} 49 \\ \times\ 58 \\ \hline 392 \\ 2450 \\ \hline 2842 \end{array}$$

they multiply the answer in full

then round to the nearest hundred   2 8 4 2   to obtain an 'estimate' of 2800

While this is reasonably close to the answer, it is hardly an estimate, which is better given by considering each of the numbers being multiplied, and then carrying out a simple mental calculation:

49	rounds to	5 tens		5 tens		49
× 58	rounds to	× 6 tens	which is	× 6 tens	so that	× 58
				30 hundreds or 3000		is about 3000

Further consideration is also required of the differences between the uses made of mathematics in the daily life of an adult and the needs in the school life of a child. While mental techniques, whether in organising calculator or computer use, making estimates or carrying out exact calculations, might be most frequent for an adult, a child has yet to develop the mental pictures or structures on which these actions can be based. One aspect of computation not often used by an adult is crucial to developing such an understanding. Materials need to be used first to build up the conceptual base on which computation depends before developing a means of recording what is done. Not only does this recording allow the transfer of the intermediate steps from the child's limited memory to the extended written record, but the written steps can then be reviewed and reflected on to assist in the sense-making process. When an understanding of what is involved has been constructed, then estimation, approximation and mental operations can be built up and calculators can be used in efficient and effective ways.

After all, neither a calculator nor a mental operation—nor even the use of materials—leaves a record of all that was needed to complete the calculation. Pencil and paper calculations will continue to have a very important role in building up computational ability, though many of the demands of the past whereby individuals needed to develop automated skills have been replaced by automatic calculating machines. It is their role in developing understanding that can be applied to problem contexts and in assisting to construct a range of computational processes that is most important. In this way, children can develop a capacity to analyse and solve problems inside and outside the classroom, based on an ability to:

- determine which operations are needed (and in which order)
- select a means of carrying out the operations (materials, pencil and paper, mental or calculator)
- complete the calculations readily and accurately (using basic facts and numeration ideas)
- make sense of any answers that are obtained in the context of the problem situation

Learning computation

Children need to be led to actively construct their own meanings for computation and to develop meaningful mental and written processes in a similar manner to the way in which numeration was developed. Materials such as ten frames, Bundling sticks, Base 10, place value charts and number expanders that reflect the structure of the number system will again be crucial in developing computational abilities. They allow children to visualise the nature and effect of the operations and build a language that will give control over the thinking processes that are developed as well as link to the place value and renaming ideas on which they are based.

However, learning computation is more than just building an ability to calculate answers when called on. A full understanding of the concepts associated with each operation is essential in allowing these operations to be used in solving problems and applying mathematics in other areas of learning and work. Automatic, accurate access to the addition and multiplication basic facts is not only important in itself, but also needs to form the basis for dealing with subtraction and division situations. Combined with numeration understanding, basic facts provide the means of verifying and making sense of any answers that are obtained. The algorithms need to be seen as interconnected knowledge so ways of thinking can be used consistently for all numbers and provide a basis for generalisations to situations in algebra and other higher level mathematics. This will also allow operations to be used interchangeably when this is more efficient. For example, estimation with addition might be better carried out using multiplication, while calculations involved in determining area or volume, or when dealing with data, may require a mix of operations combined in different ways.

In the past, seemingly different methods of calculations were often taught in isolation from one another and some aspects either were not taught or were given insufficient emphasis. As a consequence, many children gained an incomplete computational ability, restricted to only one form of the algorithms for whole numbers. Anything to do with fractions or per cents was considered to be too difficult for them. Mental calculations were attempted using methods more suited to pencil and paper computation, and a calculator was simply used to process steps paralleling the written forms rather than as a tool to implement new ways of thinking. In contrast, this chapter outlines how a background of understanding as well as proficiency can be developed for all four operations, highlighting the role of materials and language in developing meaning, and ensuring that there is a consistent development between methods of calculation for the different forms of numbers, consistency from one operation to another, and a basis for the organisation and control of mental operations that underpin estimation, approximation and calculator use.

Concepts for the operations

The numeration ideas discussed in the preceding chapter are one essential prerequisite to the development of computational abilities. Developing a full set of meanings for the operations is another. This involves building the concepts for the operations in all their forms, using materials, stories and situations familiar to the child. For example, in constructing the concept of division, stories involving sharing are used at first:

There were 15 bananas.
Simon shared them among 3 of his friends.
How many did each friend get?

3 sharing

5 each
○ ○ ○ ○ ○
○ ○ ○ ○ ○
○ ○ ○ ○ ○

5
$3\overline{)15}$

Each friend gets 5 bananas

Later, this conception can extend to the use of the more formal language *divided by* and to situations involving repeated subtraction rather than sharing.

Language is crucial in establishing these meanings, yet a full understanding is only present when a child can match a given story with the materials used to show the concept and the symbols used to record it; provide a story and use of materials for a given written expression; and provide a story and symbolic expression for a given concrete situation. This ability directly corresponds to the relationship between materials, language and symbols stressed in the development of numeration, with the role of a particular language for each concept, such as share for division, crucial in making these links:

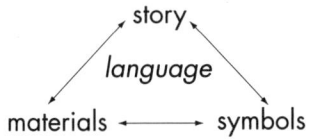

Basic facts

Basic facts are those fundamental number combinations on which all other computations depend. Essentially, these only involve facts for addition and multiplication using the numbers 0–9, since subtraction facts are derived from known addition facts and division uses the nearest multiplication fact and remainder. Addition and multiplication with 10 have already been built up in the development of numeration and will only need practice to bring to the point of automatic response rather than be learnt anew. For example, knowing the teen numbers shows that '4 and ten' or 'ten and 4' is fourteen while the tens are known because of the way the numbers are said: 4 tens is 4t. Multiplication involving 11, 12 and so on should be left as part of number sense rather than singled out as facts to be learned. Since 12 is 10 and 2, multiplying by 12 will be the same as multiplying by ten and adding 2 more. With this thinking, multiplication with 12 can be readily carried out with numbers such as 9, 15, 23 or 57 rather than be restricted only to numbers up to twelve as was the case when multiplication 'tables' built to twelve.

Showing children how to build on existing knowledge in this way, rather than start learning all basic fact combinations from scratch, is important in developing mathematical thinking. It not only gives rise to the strategies that will allow ready learning of the basic facts, for example that the twos facts for multiplication are really just a re-statement of the addition doubles, but also fosters the thinking that will be needed in problem solving. Learning basic facts should not be seen as a dreary but necessary step on the way to competence with computation, but as a pivotal stage in learning how to think mathematically. When children can not only provide answers to basic facts but also supply reasons for them, confidence in their mathematical knowledge is increased.

Having automatic recall of the basic facts associated with an operation is essential to building the corresponding algorithms for larger numbers. Children who do not have immediate access to these facts must resort to guessing or counting to determine answers to the many intermediate steps involved. It is hardly surprising that such children are unable to complete more complex calculations accurately. Diverging from the systematic processes to find these answers virtually ensures that some part will be displaced, even supposing the correct answers are obtained. In contrast, when basic facts are developed to the point of automatic recall, a child can attend fully to processing the steps in the algorithm, using the place value and renaming on which they depend.

The *Numeracy Benchmarks* (Curriculum Corporation, 2000) acknowledged the importance of basic facts by including them in the minimum sets of achievements for all children. By Year 3, this should incorporate 'basic addition facts to 10 + 10, the matching subtraction facts and extensions of these facts' to higher decades. By Year 5, 'multiplication facts to 10 × 10 and extensions of these facts' to multiples of ten would also be expected of all children. The use of thinking strategies such as count on, using doubles and make to ten are specifically mentioned for addition (Curriculum Corporation, 2000, p. 13), while multiplication facts are to be built up by strategies including using doubles and related facts (p. 33).

Learning these facts involves putting into long-term memory a large number of pieces of information so that at any time a specific item can be recalled readily. Since there are one hundred facts for each of addition, subtraction and multiplication and many more links from division to the nearest multiplication fact, this is a formidable task. Often this has been attempted principally on a basis of rote drill and practice. Children were introduced to the facts to be learnt in an orderly manner, and these were then mastered by activities that largely consisted of systematic repetition. While a child might have first met an addition fact like '3 and 4 through some enacted situation, the symbols were quickly introduced and associated with the response '7'. These two expressions were then presented together often enough until the stimulus '3 and 4' automatically evoked the response '7'. This approach has its origins in the work of E. L. Thorndike (*The Psychology of Arithmetic*, 1922) who analysed arithmetic into its constituent components or 'bonds' in the early part of the twentieth century. The task of the teacher was to carefully form the necessary bonds and combinations of them that would enable children to perform a full range of computations and to solve problems from everyday life. For example, sports, games, shopping for toys and buying lunch are common activities for children in which basic number combinations arise and require answers.

But simply having instant recall of number facts is not enough. Children also need to know why particular answers are generated so that they can develop a basis for estimation and mental computation processes. Estimation in any situation is dependent on a reasonably well developed familiarity with what is involved. We are able to estimate the time taken to travel between two points in our own environment, but in another city or town, even one considerably smaller than our own, we no longer have the same degree of accuracy. Similarly, a child's ability to estimate the size of an answer to a multiplication or division problem depends on a facility with multiplication facts. With the pervasive use of calculator-based devices, approximate calculations based on estimation to check the reasonableness of results have become a critical aspect of everyday life.

While practice is necessary in learning basic facts, care must be taken that it is not used prematurely. In the first instance, children must understand the concept of the operation being used. The second point to bear in mind is that researchers (Brownell 1935, Rathmell 1978, Thornton 1990, Isaacs and Carroll 1999) have consistently found that practice is much more effective when children have access to efficient strategies for generating the facts. Indeed, practice of itself does little to create new ways of determining answers, it merely speeds up those that the child already has. Consequently, it is important to help children develop efficient, non-counting strategies for each operation before practice is begun.

Such thinking strategies in the first instance grow out of the concept for each operation or from earlier knowledge built up in learning the one-digit and two-digit numbers. Others result from the exploration of number patterns and properties. In contrast to the enormity of learning more than 400 separate number combinations, children need a few powerful, but simple, thinking strategies. The order in which they are developed is also important, for the task in learning mathematics is not to learn a large number of things in isolation, but to construct new knowledge from that already possessed. Similarly, since the easiest order in which to think of a sequence of ideas after they have been formed is not necessarily the easiest order in which to learn them, a natural order may be better gained from studying the learner than from analysing the mathematics itself. For instance, rather than learn the multiplication facts in the order ones, twos, . . ., nines, children more readily learn the twos and fives first, and later relate the ones facts to a secure understanding of multiplication.

When practice is undertaken, various researchers (Thorndike 1922, Brownell 1935, Rathmell 1978, Thornton 1990, Isaacs and Carroll 1999) have provided important considerations to keep in mind. Each cluster of facts needs to be practised to the point of overlearning before proceeding to the next one. This means that answers to all the facts generated by a specific strategy can be recalled immediately without recourse to the strategy itself unless this is called for. Short, frequent periods of practice have proven to be more effective than longer time periods set aside only occasionally. For example, five minutes of oral practice activities every day is not only extremely productive but causes little or no distraction to the regular mathematics lesson. It should also be remembered that many facts, especially the easier ones, receive large amounts of additional practice in the process of developing further mathematics. More practice should thus be allocated to those facts that are giving, or are likely to give, most difficulty. As the thinking strategies are introduced, three phases may be discerned in the child's learning:

1 a way of thinking is internalised
2 there is increasing accuracy in the application of the strategy
3 responses are increasingly automatic

Motivating practice activities and instructional games should also be combined with the development of appropriate thinking strategies to move the child optimally through this development.

Algorithms for computation with larger numbers

Addition is the first operation children meet in everyday life and in school. Once the concept and basic facts have been learned, materials can be used to introduce addition with larger numbers, leading to a written algorithm in which all of the steps are recorded. Place value and renaming are simply combined with basic fact knowledge to provide a meaningful process that proceeds place by place; for example:

CHAPTER 4 • computation and estimation

```
   1
   4 5
 + 2 7
   7 2
```

Start with the ones
5 ones and 7 ones is 12 ones
12 ones are 2 ones 1 ten

- this requires **place value**
- this requires **basic facts**
- this requires **renaming**

Add the tens
5 tens and 2 tens is 7 tens
45 and 27 is 72

- this requires **place value**
- this requires **basic facts**

Addition with larger numbers uses the same steps for each place in turn:

```
   1 1
   8 3 7
 + 5 9 5
   1 4 3 2
```

Add the ones	7 ones and 5 ones are 12 ones
Rename	12 ones is 1 ten 2 ones
Add the tens	4 tens and 9 tens are 13 tens
Rename	13 tens is 1 hundred 3 tens.
Add the hundreds	9 hundreds and 5 hundreds are 14 hundreds;

837 and 595 is 1432

Later, the same thinking allows addition of decimal fractions and common fractions:

```
  1 1 1                    1
  3 4.6 7                  5 3/5
+ 6 7.3 7                + 7 4/5
  1 0 2.0 4                13 2/5
```

For all forms of number, this addition process consists of a cycle of three fundamental steps:

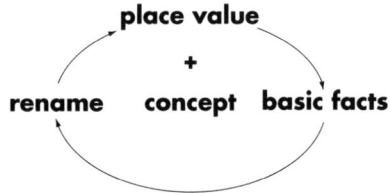

183

Further, both place value and renaming are fundamental to numeration, adding weight to the observation that the difficulties many children seem to experience with computation actually reflect insufficient understanding of numeration. It also reinforces the notion that teaching of the algorithm for larger numbers cannot begin until these numeration processes and a good understanding of the numbers involved are available. This means not only the numbers being added, but also the likely answers, so that addition with two-digit numbers actually requires a good understanding of three-digit numbers. Although it appears that the only other need for the addition algorithm is ready access to the basic facts, this of course builds on knowing the addition concept, so understanding the concept is also essential in developing meaning for each step.

Subtraction is usually introduced following the learning of addition, building from the concept and basic facts through the use of materials to the written process in a similar manner. The cycle of steps required for subtraction is:

$$\begin{array}{r} 7\ 12\ 11 \\ \cancel{8}\ \cancel{3}\ \cancel{1} \\ -\ 4\ 5\ 3 \\ \hline 3\ 7\ 8 \end{array}$$

Subtract the ones	1 one, can you take away 3 ones?	◆ requires **place value**
	No: rename 1 ten for 10 ones.	◆ requires **renaming**
	11 ones take away 3 ones is 8 ones	◆ requires a **basic fact**
Subtract the tens	2 tens, can you take away 5 tens?	◆ requires **place value**
	No: rename 1 hundred for 10 tens	◆ requires **renaming**
	12 tens take away 5 tens is 7 tens	◆ requires a **basic fact**
Subtract the hundreds	7 hundreds, can you take away 4 hundreds?	
	Yes 3 hundreds	
	831 takeaway 453 is 378	

For larger numbers, decimal fractions and common fractions, the process uses the same form of thinking. Thus, subtraction involves a cycle of three steps similar to those used in addition except that the order in which they are to be used is reversed:

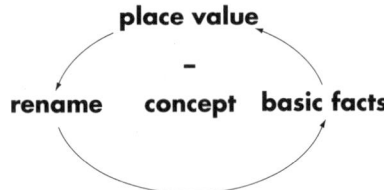

Multiplication introduces a different form of thinking from that used for addition and subtraction. The digits are no longer dealt with place by place; not only do ones multiply with ones, but ones also multiply with tens. Nonetheless, the algorithm demands a similar cycle of steps and understanding:

$$\begin{array}{r} 3 \\ 47 \\ \times\ 5 \\ \hline 235 \end{array}$$

Multiply the ones
5 sevens are 35
Rename 5 ones 3 tens
Multiply the tens
5 by 4 tens are 20 tens
Rename 20 tens and 3 tens are 23 tens

◆ this requires **place value**
◆ this requires a **basic fact**
◆ this requires **renaming**
◆ this requires **place value**
◆ this requires a **basic fact**
◆ this requires **renaming**

In fact, the pattern of thinking and understanding is the same as that required for addition:

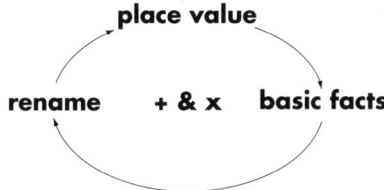

Multiplication with larger numbers simply involves a repetition of these steps:

$$\begin{array}{r} 43 \\ 47 \\ \times\ 65 \\ \hline 235 \\ 2820 \\ \hline 3055 \end{array}$$

Multiply the ones   5 sevens are 35; 3 tens 5 ones
Rename   5 ones 3 tens
Multiply ones by tens   5 by 4 tens are 20 tens
Rename   and 3 tens are 23 tens
Multiply tens by ones   6 tens by 7 ones are 42 tens;
Rename   zero ones, 2 tens, 4 hundreds
Multiply tens by tens   6 tens by 4 tens are 24 hundreds
Rename   and 4 hundreds are 28 hundreds
Add   235 and 2820 is 3055

Division has often been considered the most difficult of the operations to teach, with a significant proportion of children seemingly unable to learn the algorithm for larger numbers. This has usually been the consequence of an inappropriate approach to division or because the understandings on which it depends were not sufficiently developed. Yet, when the written division algorithm is analysed in the same manner as addition, subtraction and multiplication, it can be seen to depend on a similar cycle of steps, given meaning and direction by the underlying sharing concept, place value and renaming:

```
      9 2r7
   9)835
      81
      25
      18
       7
```

Share the hundreds first
Can you share 8 hundreds among 9?   ◆ requires **place value**
No: rename and share the 83 tens   ◆ requires **renaming**
Can you share 83 tens among 9?   ◆ requires a **basic fact**
Yes 9 tens each
81 tens are shared out, 2 tens remain
Can you share 2 tens among 9?   ◆ requires **place value**
No: rename and share the 25 ones   ◆ requires **renaming**
Can you share 25 ones among 9?   ◆ requires a **basic fact**
Yes 2 ones each
18 ones are shared out and 7 **remain**

The thinking and cycle of steps followed in division is the same as subtraction:

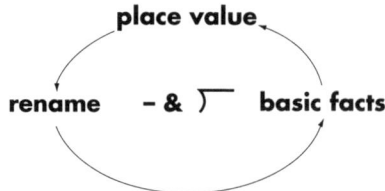

This analysis of the understanding needed to develop the written algorithms shows that there are fundamentally only two forms of thinking involved. It really is not asking too much that all children master these and develop an ability to calculate meaningfully and confidently. Since place value and renaming are the fundamental concerns of numeration, it also highlights why understanding numbers is the key to developing the operations with whole numbers, decimal fractions and common fractions. Indeed, when children experience difficulties with computation, it is as likely to be as a result of an inadequate grasp of the numbers that are being worked with as it is of the operation itself.

However, while the teaching and learning of written algorithms to provide exact answers will always be important, in many real life situations an approximate answer is sufficient. Indeed, with the widespread use of calculators, a feeling for the reasonableness of results is as crucial as an ability to calculate using pencil and paper. Estimation is an important part of this process of checking whether a particular result makes sense. On the other hand, although written computational processes do not of themselves contribute a great deal to skills with estimation and mental computation, they are vital in building up an understanding of each operation and in developing an awareness of how answers might be used and interpreted. Unfortunately, though, many people try to carry out calculations mentally in exactly the same way as the written algorithm, usually because they have not been introduced to alternative strategies. The strain of holding the very many partial results from the different steps and the renaming that is involved invariably becomes too much to manage without access to pencil and paper and any development of mental skills is curtailed from the outset.

In fact, the techniques that most readily allow mental calculations break with the manner of processing the algorithms, in particular drawing on a left to right order of proceeding. A mix of operations different from that used in a particular algorithm may also allow answers to be obtained more readily. For instance, when multiplying 67 by 29, it is easier to think of this as 67 by 30 and then to subtract 67 from the result.

Learning computation with an emphasis on exact answers can even be considered to present an obstacle to the learning of estimation. Formerly, responses that were only slightly different from the correct result were judged to be as wrong as those that were far from the answer. Now children are to be expected to put such experiences to one side and to find answers which are 'near enough' and even to sort out for themselves how near 'enough' is. This thinking demands a significant cognitive shift on the part of children and many find it difficult to adapt.

Nonetheless, estimation and mental computation are also underpinned by earlier number understanding. Important numeration skills, such as place value, renaming, rounding and the understanding shown by a number expander, are essential. There is also a need for secure basic fact knowledge across all of the operations and a capacity to extend this thinking to tens, hundreds and thousands. In particular, an ability to build across the higher decades in addition and subtraction is essential so that, for example, knowing 5 and 9 is 14 can be used to see that 5 and 19 is 24, 5 and 29 is 34, 35 and 9 is 44, and so on. A sound understanding of the two fundamental models for the operations, part/part/whole, which gives meaning to addition and subtraction, and the use of arrays, which allows a ready conceptualisation of multiplication and division, is also crucial in constructing strategies to manage numbers mentally. If mental strategies for dealing with numbers are to be built up, these fundamental understandings must be available. Even though the need for highly developed computational ability with the written algorithms may have diminished due to modern technology, the use of materials, language and recording in building them up will always be essential in providing the background of meaning on which to base estimation and mental computation skills. Consequently this sense of number and the range of particular mental strategies for dealing with numbers and calculations that can be based on them are now considered to be two of the most important outcomes for learning and teaching mathematics in the primary school.

When an estimation strategy is to be used, it must be mentally manageable as well as grounded in secure number sense. There are a great many ways in which estimates can be arrived at but it makes most sense to develop those that apply to almost all situations first and then to encourage children to construct their own variations to suit particular situations once the general process for estimation has been internalised. For this reason, it is suggested that computational estimation should not be begun until the written computations have been established for all operations and that a facility then be built up for addition, subtraction, multiplication and division in turn. At the same time, it is important for estimation to be viewed as a process to be applied across all operations rather than as a series of isolated techniques to be mastered separately. In particular, while many estimation techniques involve rounding, this is not necessarily the easiest to grasp for it imposes a significant demand on memory as numbers are rounded and calculations made with these new numbers. For example:

Estimate how much money this is:	$35.78	rounds to	$36
	51.19		51
	39.65		40
	23.87		24
	+ 18.05		+ 18
		about	$169

This is easy enough to do with pencil and paper, but mentally it requires that $35.78 be rounded (to $36) and remembered, $51.19 be rounded (to $51) and added to $36 (to give $87). This new answer needs to be remembered, $39.65 rounded to $40 and added to $87 to give $127 and so on.

A rounding-based process does not provide an insight into the need for estimation as it may seem just as easy simply to carry out the exact calculation. Thus, when estimation is introduced it is important to consider both the ease of the process and its capacity to build up an appreciation of the value of estimation as a further means of computation. An initial technique for estimation may consider the leading digits only, adding them to obtain a first estimate and then adjusting this by inspection of the other digits which make up the numbers:

Look at the tens first:	$35.78	$35.78	
	51.19	51.19	
	39.65	39.65	Look at the rest of each
	23.87	23.87	number; close to 30 more
	+ 18.05	+ 18.05	
	more than $140	about $170	

Such a front-end strategy is much easier for children to manage as a first step in building estimation. It also lays the basis for techniques that are developed according to ease of use rather than simply building on the exact computational processes. When a familiarity with looking to the largest place first is built up, it also seems much more reasonable to introduce the rounding on which other estimation relies. This in turn lays the foundation for mental computations where it is usually easier to begin with the largest place and then adjust the answer to account for the remaining places in turn. Keeping something close to the final answer in mind from the outset is easier than building up a series of steps which only look like the final answer after a lot of reasoning. Indeed, trying to calculate mentally by following through all of the steps in the written process is usually very difficult to manage as well as being intrinsically inefficient. Yet many adults report that the only strategy they have access to is to imagine a notepad in their head on which the calculations can be done. Unless alternative strategies are developed, using pencil and paper processes directly or in some imagined form will be the only alternatives to guessing. There needs to be a clear development from the estimation processes that provide answers that are close enough to techniques that allow this adjustment process to give exact answers. To achieve this, a very sound basis of numeration, basic facts and the concepts for each operation is necessary to allow meaningful short cuts to arise and to help in combining the various steps and ways of thinking. Thus, estimation is often a very good way to build up an initial ability with calculating answers mentally and opens the way for other strategies, using a combination of operations rather than just a more efficient adaptation of the written form.

Computation can now be viewed as a process to be applied over all operations that takes a variety of forms depending on the circumstances for which it is required. Its development demands coherency within each operation and connections among all the forms for each operation. Once the meaning for a concept is established, basic facts can be built up and the particular algorithm can be developed relying on numeration and basic facts. A consistent use of materials, language and recording to build from the initial concepts, through the basic facts to the recorded forms for computation with large numbers, decimal fractions or common fractions is required. In time, the understanding of the operation generated by the materials and by the reflection on the process that the recording permits will allow corresponding facility with estimation and mental computation, allowing the mental management of a variety of strategies for exact, approximate and calculator processes. Consequently, the foundations for powerful and meaningful computation can be delineated.

THE FUNDAMENTAL UNDERSTANDINGS UNDERPINNING COMPUTATIONAL FACILITY

1. Numeration: an understanding of numbers and an ability to think of them in more than one way.
2. Concepts for the operations: an ability to recognise the operation symbols, an understanding of what the operations do and a capacity to write and interpret symbolic statements involving them.
3. Basic facts: automatic recall based on efficient non-counting strategies for obtaining them.

4 Consistent development from the initial concepts, through the development of the basic facts to the various forms of the algorithms. This consistency includes the manner in which materials are used, the language that is formulated to describe their use and the way in which these lead to symbolic recording.

5 Estimation: builds on the concepts for each operation using extended basic fact knowledge integrated with numeration processes. An ability to estimate is needed to provide approximate answers and to assess the reasonableness of calculations and their applications to problem situations.

6 Mental computation and calculator use: an understanding of the fundamental concepts for numeration and each operation, coupled with a good sense of number, allows the mental management of strategies for organising computation in innovative and personal ways.

Section I: Overview to addition

ADDITION is the first operation young children meet at school, so it serves to introduce the general notion of recording computational processes and the use of symbols to indicate the operation, as well as the construction of specific addition concepts and processes. While most children are able to solve simple addition problems soon after entering school, a careful building up of addition is required from this first conception to the written and mental forms, using materials to model these problems and generate a language to give meaning to addition statements. At the same time, the recorded algorithm for larger numbers, which is the focus of initial instruction, must be borne in mind and the steps organised to allow a smooth and consistent development from the addition concept through learning basic addition facts to the introduction of the algorithm by means of materials. For this reason, a vertical recording of addition is preferable from the outset and use is made of a place value chart and Bundling sticks or Base 10 materials to lay the foundation for recording the written algorithm. The language for addition has been carefully developed to match the way the materials are manipulated and conversely the needs of the recorded form have governed the way the material is set out, handled and described.

SEQUENCE FOR DEVELOPING ADDITION

1 Develop the addition concept: verbal action stories
                                     model with materials
                                     vertical recording + symbol

2 Build up the basic facts: develop thinking strategies
                                     ensure automatic responses

3 Thinking in tens—extend basic facts to tens

4   2 digit
    + 2 digit   renaming 10 ones as 1 ten

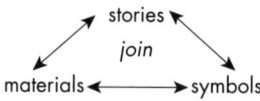

5	3 digit + 3 digit	renaming 10 ones as 1 ten renaming 10 tens as 1 hundred renaming both ones and tens
6	Addition of more than two numbers	
7	Addition with larger numbers	
8	Strategies for estimation, mental computation and calculator use	
9	Addition with decimal fractions:	tenths hundredths thousandths
10	Addition with common fractions:	like fractions need for equivalent fractions unlike fractions

The addition concept

By the time children commence school they have acquired a basic idea of addition from their everyday experiences and are able to find the answers to simple problems by counting. However, this does not usually extend to the written form, and they are unlikely to have an understanding of addition that will allow them to build abilities to handle larger numbers and more complex problem situations. In order to prepare for these higher level processes, children first need to build a full conception of the operation rather than simply focus on finding an answer. The role of materials and language in building this understanding is essential. Only when children are able to visualise and talk about addition can symbols be meaningfully introduced as a concise way of representing the operation.

Story situations presented as problem-solving tasks are used to introduce addition and focus attention on the underlying idea of joining. Materials such as counters or blocks can then be used to provide a picture of the whole addition process. It is helpful to build on the way that early numbers were represented on a ten frame, as this means that children can see each of the numbers involved by using the patterns with which they are familiar, rather than counting the objects one by one.

John had 6 apples   Jenny gave him 2 more apples   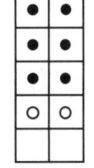 John has 8 apples altogether

The use of materials in this way enables the addition concept to grow naturally and meaningfully out of the numeration understanding that has already been built up and allows the parts that are joined to make a whole to be readily seen.

Initially, the language used for addition should focus on 'and' because it suggests joining and readily arises from the use of materials: 6 blue blocks *and* 2 red blocks; 8 blocks altogether. It is also most likely to be the word used in problem situations that require addition. Later, when the meaning for addition has been established, 'add' can be introduced as the mathematical word used for joining and other expressions that imply addition can be developed. For children, the word 'plus' has little or no relationship to addition or the operation of joining. If it is to be used, it should be left until much later and introduced as an alternative to words and expressions that are well known through their relationship to the use of materials and occurrence in addition problems.

A gradual introduction of symbols to match the use of materials and language used in making and representing addition situations will ensure meaning for addition statements. The addition symbol '+' can then be established as a mathematical way of writing 'and', indicating that the numbers are to be added.

'6 and 2 is 8' can be written as
$$\begin{array}{r} 6 \\ +\,2 \\ \hline 8 \end{array}$$

$$\begin{array}{r} 2 \\ +\,6 \\ \hline 8 \end{array}$$ can be read as '2 and 6 is 8'

Once this initial concept is secure, a complete understanding of addition requires an explicit awareness that parts are combined to form a whole. Problem situations that emphasise the notion of joining two distinct parts and reference to the use of different coloured counters for the two parts in the initial concept development are helpful in building up this understanding. Later, this understanding will be extended to joining more than two parts to form the whole for addition of several numbers. The part/part/whole notion also lays the foundation for subtraction and is crucial in building up an ability to distinguish between addition and subtraction situations, particularly in problem solving.

A full conception of addition is readily built up using the following steps:

SEQUENCE FOR DEVELOPING THE ADDITION CONCEPT

1 Introduce addition through verbal action stories that involve joining.
   There were 4 ducks on the pond.
   3 more ducks flew to the pond to join them.
   How many ducks are on the pond now?

2 Guide children in using materials to show these addition situations.
   Set out the objects vertically.
   Verbalise the joining process using 'and'.

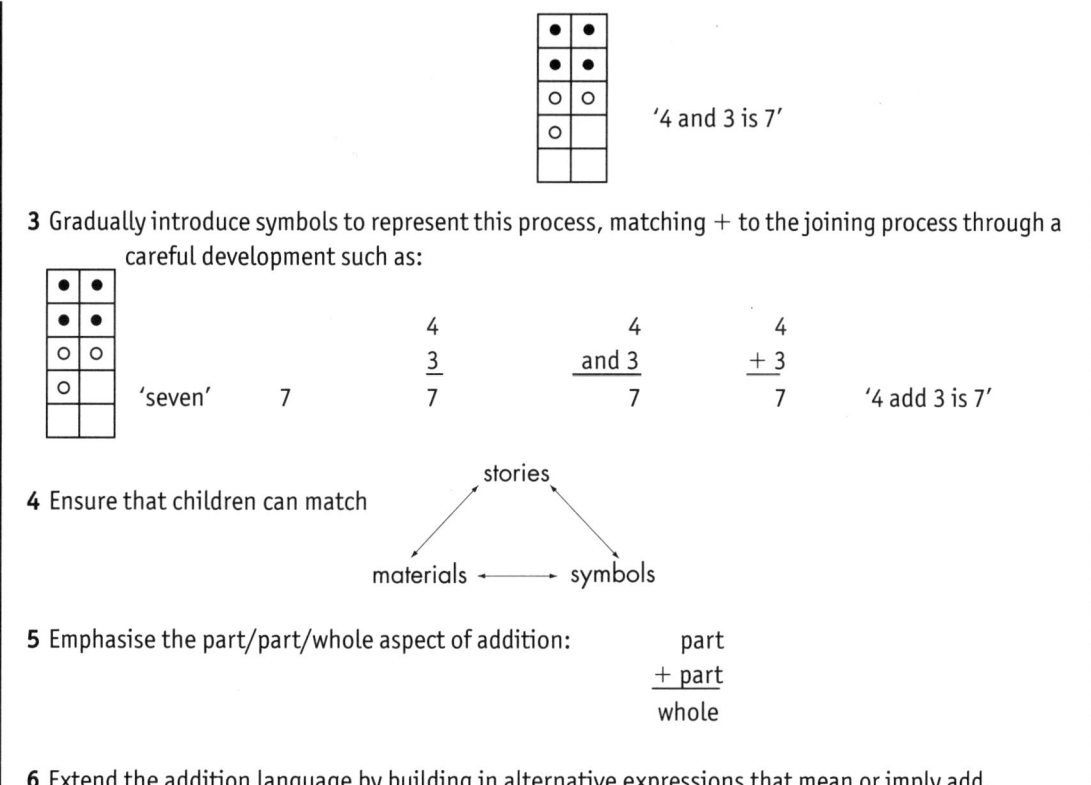

In order to facilitate the development of the written algorithm for addition, and later that for subtraction, addition statements should be read downwards from the outset. Children have grown used to reading in general proceeding down the page and will readily place their counters one group under the other when forming the different groups being added. This will be helped if pictures and situations involving addition involve vertical groupings and if expressions such as 'add up' are avoided—it is just as reasonable to ask children to simply add these numbers. When an understanding of addition as joining and the meaningful writing and reading of symbolic addition statements have been consolidated, the use of the word 'and' may be replaced by 'add', although it is quite acceptable to continue to use 'and', with its connotations of joining, at all times. In fact, the addition symbol has its origins in the Latin word for 'and', *et*, which was used in arithmetic until it became abbreviated to + as the first mathematical symbol; literally, + is 'and'. Further, in real life, the contexts in which addition problems arise involve lists of items linked by 'and' that need to be combined. Rarely is the word plus used in this way and its use even within purely mathematical situations should be delayed until addition is very secure. The words sum and total are more widely used and should feature at least as strongly as plus. Since these expressions do not indicate that the operation is addition they need to be carefully related to the addition concept so that they can be seen as another way of expressing informal statements such as and, join or put together.

At this point in the development of addition, horizontal recording such as $5 + 2 = \square$ or $3 + 6 = \square$ is not appropriate. An emphasis on left to right processing when addition is first recorded may lead to difficulties with the algorithm for larger numbers. In particular, if children process the written algorithm from left to right, simple examples may be answered correctly, but examples requiring renaming will be incorrect:

Working the algorithm from left to right

$$\begin{array}{r} 3\,4 \\ +\,4\,2 \\ \hline 7\,6 \end{array}$$ is correct, but $$\begin{array}{r} 3\,4 \\ +\,4\,8 \\ \hline 7\,12 \end{array}$$ is not correct

Not only does recording addition horizontally provides no link to the addition algorithm which is built up once the addition concept and basic facts have been acquired, it also requires the use of more symbols than a vertical form—the equals symbol (=) and a $\square$ symbol to indicate where the answer is to be written. In fact, the equals symbol is not really needed until much later when situations involving several operations are met. Recording addition vertically from the outset sets the scene for focusing on larger addition place by place, keeps the + symbol to one side of the numbers where it can be readily seen, and simply requires that the result be recorded under the line drawn beneath the numbers being added. Finally, the introduction of the = symbol actually involves the development of the equals concept, a concept now recognised to be very difficult for young children to come to terms with fully and one which is likely to be confused with the addition concept if introduced simultaneously with addition.

Likely difficulties

Difficulties with the concept usually occur because of a focus on the symbols and answers rather than the meaning associated with joining parts to give a whole. Consequently, a child may be unable to give a story to match a picture or symbols, may rush to give an answer rather than read the addition, and may not be able to show the situation using materials.

- When asked to read an example such as $\begin{array}{r}7\\+\,5\end{array}$ a child may not be able to read the expression with meaning, saying '7 plus 5' instead. This may show no understanding of addition other than as a command to provide an answer.
- A child may read addition upwards and say '5 and 7' which is likely to lead to difficulties with addition for larger numbers.
- When asked to use materials to show addition, the child may simply provide objects to match the answer rather than the parts involved in the story or written expression.
- When asked to provide an addition story to match a written expression or picture, a child may simply provide a story which matches the numbers but does not involve addition.

Investigate and discuss

1. Use counters or Unifix on a ten frame to show this problem: 'Jane has 3 toy trucks and 6 toy cars. How many vehicles does she have altogether?'
   a. What difficulties might the use of different types of vehicles cause?
   b. How would you draw out the part/part/whole notion in this example?
   c. Show 9 Unifix on the ten frame, using 4 in one colour and 5 in another colour. Make up a story to match the materials. Repeat for other combinations which make 9.
2. Further practice in relating addition statements to the use of materials can be provided by similar activities using counters:
   - Draw a centre-line across a piece of paper.
   - Take a small number of counters (10 or less), state the number you took and then drop them onto the paper so that some fall on either side of the line.
   - Describe the arrangement that results: e.g. 'there are 4 above the line, 5 below; 4 and 5 are 9'.
   a. Why do you think the activity referred to counters above and below the line rather than on the left and right?
   b. How could you use this activity to reinforce the part/part/whole idea?
3. The initial recording for addition uses a vertical rather than a horizontal form. Some reasons for this were summarised in the introduction. List these and offer any further reasons of your own either for or against this preference.
4. Horizontal recording will need to be introduced at some point after addition is well known. When several operations are involved there is really no alternative to a horizontal form and algebraic expressions are always written this way. Discuss any new understandings that will need to be established when horizontal recording for addition is introduced and used.

Learning the addition basic facts

Addition basic facts are those simple number combinations involving the numbers 0–9 that allow calculations with larger numbers to proceed such as:

$$\begin{array}{ccccccccc} 2 & 1 & 4 & 9 & 5 & 6 & 3 & 8 & 7 \\ +0 & +8 & +2 & +3 & +5 & +7 & +9 & +6 & +4 \end{array}$$

Having automatic answers to these facts enables a child to free up his or her mind to focus on managing the many steps and the renaming that the algorithms demands. Although combinations with ten are essential, they should already be known through the development of the teen numbers and simply need to be reviewed, rather than treated as a new set of facts to be learned.

Once children have come to terms with the addition concept and are able to write and interpret addition statements, they can begin to acquire the addition basic facts. Learning basic facts has always been an area that is fraught with difficulties for children, teachers and parents. The latter become agitated because children 'can never remember their facts' while children experience frustration in trying to do so. Many of these difficulties have occurred because of an inappropriate approach to the task, such as a focus on 'rote learning' based solely on repetition through drill and practice. In fact, since learning requires understanding and meaning, while rote means without meaning, 'rote learning' is an oxymoron—something is either learned and therefore has meaning, or it is rote and learning has not occurred. Neither is the traditional method of using 'tables' to introduce the facts conducive to learning. Rarely will facts be needed in the order promoted by tables. Instead, individual facts are what is required in everyday computation and it is more important to realise that knowing '3 and 4' also gives access to '4 and 3', '7 takeaway 3' and '7 takeaway 4'. In contrast, an alternative approach which teaches strategies for obtaining the facts before introducing practice enables basic facts to be learned in a way that makes the facts more meaningful as well as easier for children to learn and recall when needed.

The importance of teaching a few powerful, yet simple, *thinking strategies* to teach basic facts cannot be overemphasised. If efficient strategies are not taught, children will use or invent their own inefficient strategies, usually related to counting. However, it must be borne in mind that a strategy in itself is not sufficient. Before basic fact learning is commenced, it must be ensured that children have a full understanding of the concept for the operation. When the concept has been firmly established, the strategies need to be introduced one at a time, using materials and an associated language to develop understanding, and sequenced so as to assist the child to link new knowledge to existing knowledge. After the strategy for obtaining a given cluster of facts has been mastered, practice is required to obtain automatic recall of these facts and to ensure the basis for the next cluster of facts.

Answers to most addition facts are teen numbers. Many teachers have interpreted this to mean that they should first build up 'facts to ten', focusing on the size of the answer rather than the strategy that can be used to obtain them. Not only does this ignore the power of the various strategies and the advantages in linking together facts that are obtained by the same thinking, it has also led children simply to count on their *ten fingers* to obtain answers rather than use mental strategies. Inevitably, this habit then becomes the basis for obtaining all facts and is very hard to break.

A more significant interpretation of the occurrence of teen numbers answers is to be certain that these numbers are secure before the addition facts are formally introduced. Since the teen numbers are the last of the two-digit numbers to be acquired, this means that all of the numbers 0–99 should be known before the addition facts are met. Such an approach also ensures that children have had many experiences in counting on and forming ten ones to make 1 ten, thinking that underpins the way that many addition facts are best learned.

Three strategies have been found to be most effective in learning the addition facts—*count on from the larger number, using doubles* and *make to ten*. Within these strategies, the individual facts can be clustered into sets that draw on the same thinking, and in turn lead to further clusters that extend this thinking.

Count on from the larger number

Addition facts with 1, 2, 3 and 0 are best answered by counting on from the larger number, for example both *9 and 3* and *3 and 9* can be determined by counting **9** . . . 10, 11, **12**. Counting on was first introduced with numeration so there is little new in this thinking and counting on 1 more and 2 more in particular will probably be very well known from those earlier activities. Counting on 3 more requires some further thinking, while adding zero is developed last so that it is seen as a generalisation of counting 'no more'. These can be introduced by activities such as:

Dropping 1, 2, 3 or 0 marbles into a can containing a predetermined number of marbles. The children can see and hear the addition of more marbles and count on from the initial number to say how many.

Playing all board games with 2 dice, one of which has number words or symbols, the other 1, 2, 3 or no dots.

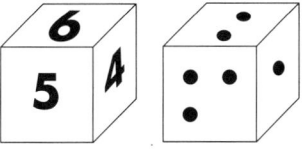

Count on 1 more: Facts like $\begin{array}{r}8\\+1\end{array}$ are answered by counting on the next number; 8 . . . **9**.

However, using the identical procedure for facts like $\begin{array}{r}1\\+8\end{array}$ counting 1 . . . 2, 3, 4, 5, 6, 7, 8, **9** is inefficient and difficult, so that children who attempt this usually count on their fingers to keep track of what is happening. Instead, they need to be led to first identify the larger number and then count on:

$$\begin{array}{r}7\\+\,1\\\hline 7\ldots\mathbf{8}\end{array}\qquad\begin{array}{r}1\\+\,8\\\hline 8\ldots\mathbf{9}\end{array}\qquad\begin{array}{r}6\\+\,1\\\hline 6\ldots\mathbf{7}\end{array}\qquad\begin{array}{r}1\\+\,5\\\hline 5\ldots\mathbf{6}\end{array}$$

Practice is needed with a mix of all of these facts to ensure children are able to use this thinking to answer readily any facts with 1 before moving to the next cluster of addition facts.

Count on 2 more: When any of the count on 1 more facts can be answered easily, the same thinking can be given to counting on 2 more. Again, the more straightforward facts like

$$\begin{array}{r}7\\+2\\\hline\end{array} \quad \text{or} \quad \begin{array}{r}5\\+2\\\hline\end{array} \quad \text{should be introduced first.}$$

Count on   7 . . . 8, **9**   5 . . . 6, **7**

The thinking strategy can then be extended to the other facts involving 2—start with the larger number, then count on:

$$\begin{array}{r}2\\+7\\\hline\end{array} \quad \text{or} \quad \begin{array}{r}2\\+5\\\hline\end{array}$$

7 . . . 8, **9**   5 . . . 6, **7**

Care needs to be taken that children do not inappropriately count on from the first number they see or hear, regardless of whether it is the smaller or larger number.

Count on 3 more: When any of the facts with 1 or 2 can be answered readily, the same thinking can be given to counting on 3 more:

$$\begin{array}{r}8\\+3\\\hline\end{array} \quad \text{and} \quad \begin{array}{r}3\\+8\\\hline\end{array} \quad \text{are both given by the same thinking.}$$

Count on 8 . . . 9, 10, **11**   8 . . . 9, 10, **11**

Count on zero more: Finally, when all of the facts with 1, 2 or 3 are known, facts involving addition with zero can be introduced as counting on 0.

6 marbles in the can	6	0 marbles in the can	0
no more are added	+ 0	6 more are added	+ 6
there must still be 6 marbles in the can	6	there must be 6 marbles in the can	6

Thus the addition *count ons* are learnt in four separate clusters, each learned to the point of automatic, accurate response for all facts in the cluster, before proceeding to the next cluster of facts. This one way of thinking—for facts with 1, 2, 3 or 0, count on from the larger number—generates 64 of the 100 addition facts. Almost all children can be expected to build up this way of thinking and become fluent with the count on facts early in their schooling. The confidence that this gives, along with an appreciation of ways of obtaining answers without simply counting will then provide the motivation and self-confidence to use thinking strategies to build the remaining addition facts.

Using doubles

Facts where the two numbers involved are about the same, such as '5 and 6', '5 and 7' or '6 and 7', are best built up by using the double based on the smaller number and then counting on 1 or 2 more.

Doubles: Many of the doubles facts have already been met in learning the numbers to ten. An emphasis on the ten frame as a way of structuring these numbers will have given a good picture of '1 and 1', 2 and 2', '3 and 3', '4 and 4' and '5 and 5'. For example, when children place 4 counters in one colour on a ten frame, and then 4 counters of another colour, they can see the two parts, '4 and 4', and also the result—8.

This thinking can be extended to the other doubles using two ten frames side by side. The first ten frame shows ten, the other forms the teen number that is the answer:

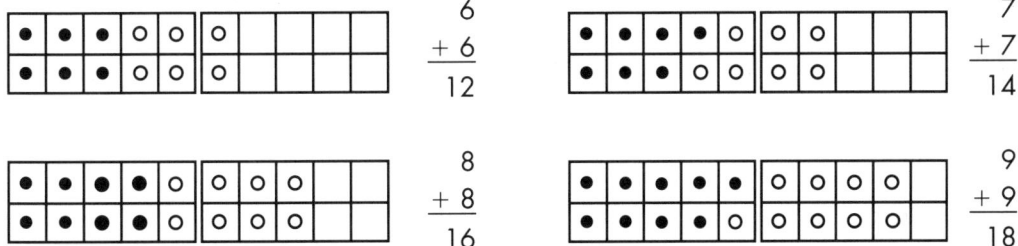

Doubles and 1 more: When any of the doubles facts can be answered easily, this thinking can be extended to those facts in which one number is 1 more than the other. Again, two ten frames provide a helpful way of structuring this thinking. The first number is put in a row across the top of the ten frames, the second number in a row underneath. Picking up 1 counter to leave a double shows the thinking used for these facts:

It is important to show this thinking with several facts one after the other to highlight the way a doubles fact is used, rather than to spend time on the same fact. For instance, have children make '8 and 9' to see how it is '8 and 8 and 1 more', then repeat for '8 and 7' to see it is '7 and 7 and 1 more', then '7 and 6', '6 and 5', '5 and 4' so that each fact is covered quickly. In this way, the process of removing one counter to make the double can be the focus of teaching, rather than individual answers. Note that the thinking is the same for both facts—identify the double, then add 1 more—building on the way the counting on strategy generated two facts.

Since the operation is addition, it really only makes sense to build up these facts by adding on to the known doubles. While it is possible for adults to use a form of adjustment by going to the next double and then subtracting 1, this is potentially very confusing for children, particularly as they are unlikely to have met subtraction in a formal sense at the time when they are learning addition facts. For this reason, the strategy that has been stressed here uses the thinking of adding 1 more or counting on 1 more from a known double, and the language that has been used to describe the strategy is doubles and 1, rather than plus 1, to be consistent with the meaning previously built up for addition.

Doubles and 2 more: When any of the doubles facts can be answered readily, two further pairs of using doubles facts can be built up in a similar manner using 2 more than a known double:

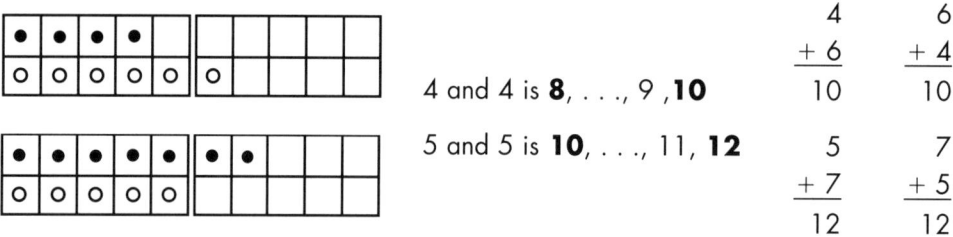

4 and 4 is **8**, . . ., 9 ,**10**

5 and 5 is **10**, . . ., 11, **12**

$$\begin{array}{r} 4 \\ +6 \\ \hline 10 \end{array} \qquad \begin{array}{r} 6 \\ +4 \\ \hline 10 \end{array}$$

$$\begin{array}{r} 5 \\ +7 \\ \hline 12 \end{array} \qquad \begin{array}{r} 7 \\ +5 \\ \hline 12 \end{array}$$

While there are other facts that could appear to be learned using doubles and 2 thinking, such as '7 and 9' or '8 and 6', in practice these have been found to be more readily achieved by reference to the next strategy, make to ten. The using doubles strategy gives all of the doubles and 1 facts and just these four doubles and 2 facts.

Make to ten

The remaining facts such as '9 and 5', '4 and 8' or '7 and 4' involve one number that is close to ten. These are best answered by first making to ten and then simply seeing the resulting teen number, for instance '9 and 5' is the same as '10 and 4'—14. This thinking has been integral to the development of numeration understanding so that once again there is likely to be a good deal of prior knowledge of these facts. The nines facts are built up first, because making to ten is easiest to see. Then the facts with eight can be developed in the same way, although care needs to be taken that children focus on using 2 to make ten, rather than over-generalise using 1 as with the nines facts. Leaving '7 and 4' until last means that most children almost immediately know to get the answer from '7 and 3 is 10, 1 more is 11'.

Nines: The nines facts are best built all at once, using counters with two ten frames, so that the thinking strategy dominates over the individual facts. Again, it is important to show this thinking with several facts one after the other to highlight the way making to ten is used, rather than to spend time on the same fact. For instance, have children make '9 and 7' to see how it is '10 and 6—16', then repeat for '9 and 6' to see it is '10 and 5—15', then '9 and 5' '9 and 4' so that each fact is covered quickly. In this way, the process of moving one counter to make ten can be the focus of teaching, rather than individual answers.

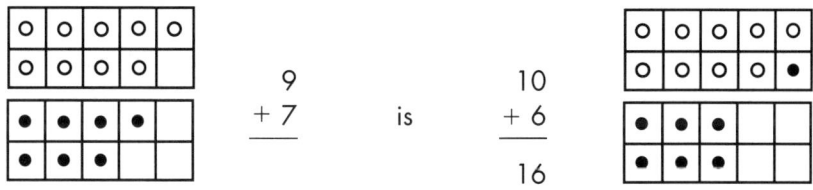

This activity should also be repeated starting with '7 and 9', then '6 and 9' and so on so both forms of the nines fact are developed.

Eights: When all of the nines facts are known, the eights facts can also be built up all at once, using the same thinking, except that 2 counters will need to be moved to make ten. Showing facts one after the other again highlights the way making to ten is used. For instance, have children make '8 and 6' to see how it is '10 and 4—14', then repeat for '8 and 5' to see it is '10 and 3—13', then '8 and 4' which is '10 and 2—12'. In this way, the process of moving two counters to make ten can be the focus and contrasted with the make to ten thinking used with the nines.

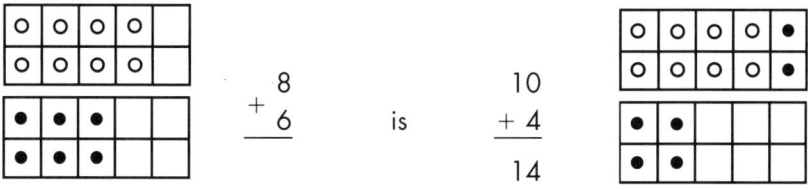

The activity also needs to be repeated starting with '6 and 8', then '5 and 8' and '4 and 8' to build both forms of the eights facts. Finally, the same thinking can be extended to the remaining fact, '7 and 4' or '4 and 7':

LEARNING THE ADDITION FACTS

All of the addition facts are readily learnt by the application of just three thinking strategies. Within each strategy, facts are organised in clusters that use identical thinking and need to be brought to the point of automatic, accurate answers to provide a secure basis on which to build the next set of related facts.

1 **Count on from the larger number.** Facts with 1 are built up first, then those with 2, 3, and finally those with 0 more; 64 of the 100 addition facts can be learned using this thinking.
2 **Using doubles.** Ten frames are used to visualise the thinking. The doubles facts are developed first to provide a base for the doubles and 1 more facts. Two further pairs of facts, '4 and 6', '6 and 4', '5 and 7' and '7 and 5' also use this thinking. Twenty of the 100 addition facts can be learned using this thinking.
3 **Make to ten.** Two ten frames provide a model for the thinking. Nines facts are built up first as they are easiest to see, then the eights facts use the same thinking and finally the last fact of all '7 and 4'; 16 of the 100 addition facts can be learned using this thinking.

Note that some facts could just as easily relate to more than one strategy. For instance, '9 and 3' can be accessed by making to ten or by counting on 3 more. The strategies used and the order in which the facts are developed here reflect experience in assisting children to learn basic facts readily. However, the most important point is that the children use a thinking strategy to obtain answers rather than count on their fingers or ruler. Once the facts have become automatic, any strategy should be pushed to the back of the mind and only accessed if there is a need to resurrect a fact or check that an answer is correct.

The facts learned through each strategy are shown on this number fact grid:

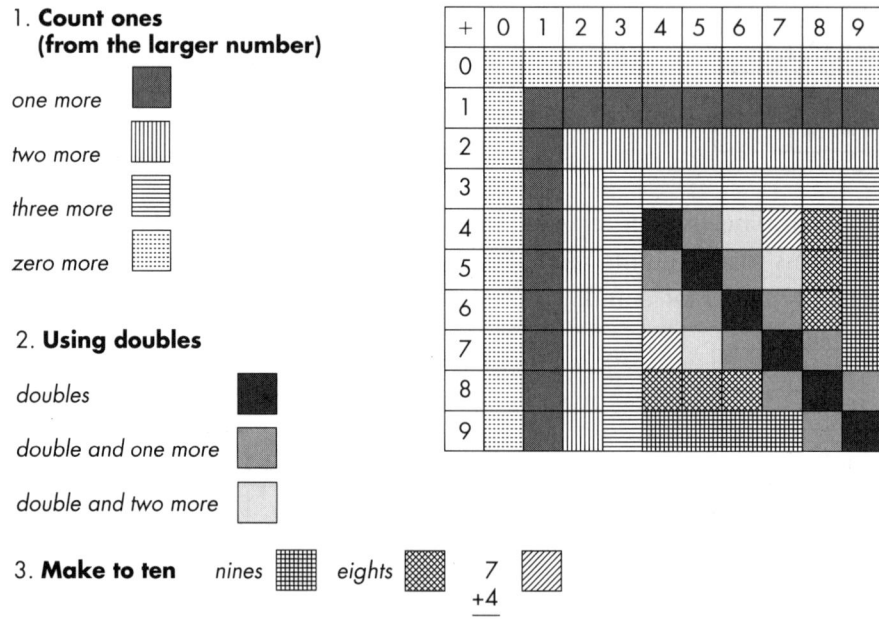

It must be noted, though, that this arrangement of the facts in a grid is simply to show a teacher how the three strategies build on each other and come together to build up access to all 100 addition facts. This is not intended as a teaching or learning device in itself; it serves to organise the presentation of clusters of facts in relation to the thinking strategies that most readily allow automatic responses.

Likely difficulties

Difficulties with the basic facts usually occur because a child lacks any strategies to obtain facts efficiently. Simply counting all numbers using fingers, dots on a page or tapping may be the only means the child has to try to obtain answers. Other children may base all facts on doubles, subtracting to adjust, but lose track of what they are trying to do. If there is confusion with another operation, this usually means that the concepts for both operations are not secure and will need to be built up appropriately.

- A child may guess or count on fingers to obtain answers rather than use more appropriate thinking. A fact such as '3 and 9' is often answered less readily than the matching fact '9 and 3'.
- When fingers are used to count all of each number, answers are often one too small when the last finger used for the first number is then used again in counting the second number; for example:

    seven and four: **10**;   six and eight: **13**;   five and seven: **11**

- Particular difficulties may occur with zero where a child thinks, for example, that 6 and 0 is 7, confusing zero with one, since to say that it is 6 looks like no addition has occurred.

Investigate and discuss

1. Collect 12 marbles and an empty container. Work with a partner to practise counting on. One person puts some marbles in the container and writes the number on the outside. The other person drops 1 more marble into the container, says how many marbles there are now and changes the number written on the container. The first person drops in 1 more marble and the process is repeated until all of the marbles are used.
   Repeat this activity dropping in 2 marbles and counting on from the number written on the side of the container. Then drop in and count on 3 marbles at a time.
2. Why should adding zero be the last of the *count on* facts to be learned?
3. Make two dice, one with the words or digits 1–6, the other with 1, 2 or 3 dots on each of two sides. Use these dice to play a simple board game such as *Snakes and Ladders*

> or one that you construct yourself. Discuss how this would foster counting on and how quickly a child might simply learn to answer the facts immediately.
>
> **4** Use two ten frames and Unifix in two different colours to show the *using doubles* facts as on p. 199.
>
> **5** Use two ten frames and Unifix in two different colours to show the *make to ten* facts as on p. 201

Developing the addition algorithm

Extending addition to deal with larger numbers is essentially just a matter of combining the addition concept and basic facts with numeration understanding. If the ones, tens and hundreds are considered separately, the addition of each part is simply a basic fact. The only other need is to rename 10 ones as 1 ten and include this in adding the tens, rename 10 tens as 1 hundred and include this in adding the hundreds and so on. This thinking is readily built up through the use of materials that, in turn, provide a visualisation of the entire process. Talking about the way the materials are combined can then lead to a meaningful language that will direct and give control over the addition algorithm. It is crucial that the order of proceeding is established from the outset: the ones are added first, then the tens, then the hundreds and so on. This is not always the way that young children would tend to proceed as it contradicts the way the numbers are written and (mostly) read. Thus, a means to build in the need to proceed from right to left is important within a teaching sequence focused on understanding the addition process.

It has been traditional to begin addition of larger numbers with examples that do not need any renaming, such as adding 62 and 24. Yet the way most two-digit numbers are read from left to right means that many children are then led to add the tens before the ones.

But addition with examples without renaming are often correct when the tens are added first:

e.g.
$$\begin{array}{r} 62 \\ + 24 \\ \hline 86 \end{array} \qquad \begin{array}{r} 92 \\ + 24 \\ \hline 116 \end{array}$$

Neither a teacher nor a child may realise that this will lead to difficulties if extended to examples which should involve renaming:

$$\begin{array}{r} 68 \\ + 24 \\ \hline 812 \end{array} \qquad \begin{array}{r} 98 \\ + 24 \\ \hline 1112 \end{array}$$

A further difficulty often seen in the teaching of addition is that two-digit numbers are added soon after the numbers themselves have been introduced. A child needs to be able to assess whether answers are reasonable. Addition of two-digit numbers needs to follow the development of three-digit numbers so that a child is able to be aware when answers are correct or incorrect.

Consequently, the most appropriate way to teach the addition algorithm is to involve renaming and right to left processing from the outset. Any addition without renaming can later be included among that with renaming to ensure the process is being thought about fully.

> **SEQUENCE FOR DEVELOPING THE ADDITION ALGORITHM**
>
> **1** Once the concept and basic facts have been learned, extend this thinking to deal with tens.
> **2** Addition with renaming is dealt with first to introduce the form of the algorithm, the need to begin with the ones, and the particular use of materials, language and recording.
> **3** Addition with no renaming is best included later among the examples with renaming as the thinking involved is identical, only there are no additional tens, hundreds or whatever to add.
> **4** This understanding is broadened to include the addition of several numbers.

Thinking in tens

If the addition facts have been introduced and written using a vertical recording, the only new thinking needed to extend

$$\begin{array}{r} 7 \\ +\,6 \end{array} \quad \text{to} \quad \begin{array}{r} 4\,7 \\ +\,3\,6 \end{array}$$

is an ability to deal with the tens. Consequently, the first need is to extend basic fact knowledge to what might be termed the 'tens facts' so as to encourage thinking in tens. Materials can be used to establish the connection that

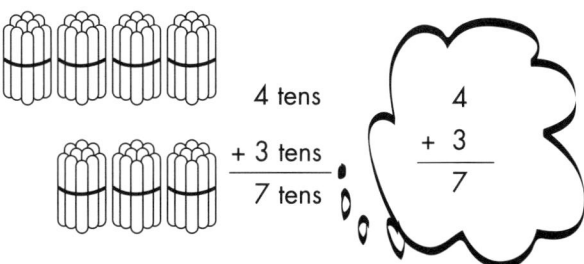

Once this link has been made, children need oral practice in the same way in which basic facts were built up after the thinking strategies were attained. This should in time include a mix of basic and tens facts since the processing of the algorithm will demand the flexibility of moving from one way of thinking to the other. Then, when children have a facility with these comparable to their fluency with the basic facts they are ready to combine these skills to add two-digit numbers.

Addition with two-digit numbers

At first this is simply a matter of having children make each number with materials, join them and read the result, although this needs to be coupled with the use of a place value chart to highlight the setting out of ones with ones and tens with tens. Bundling sticks are best for the initial development of the algorithm as the renaming process can be seen explicitly. It is also essential to direct the children to add the ones first then the tens, for they are likely to add the tens before the ones, both because of a conditioning to work from left to right and because of the attraction of the larger pieces used to represent tens. The inclusion of an arrow at the top of the ones place helps children to focus on where to start and provides a powerful non-verbal signal. For the renaming process that is needed with most examples, it is also helpful to ensure that the length of this arrow corresponds to the length of the Bundling stick or Base 10 ten. This will then leave a space at the top where any additional ten resulting from the renaming can be placed.

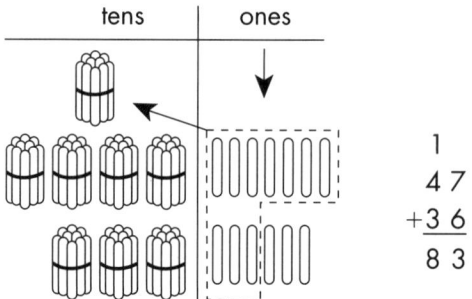

Consideration also needs to be given to the direction of reading and processing the two numbers being added. Traditionally, the approach has often been 'add up, check down'. Yet an almost automatic consequence of the arrow on the place value chart is that children add down from the outset. This of course is in accord with the way they are expected to complete all other written work (begin writing at the top of the page and proceed down the page). More importantly, subtraction demands a reading from top to bottom to match the sense of real life problems:

```
There were 65 apples on the tree              6 5           6 5
The children picked 24                  take away 2 4      − 2 4
There are 41 apples left on the tree          4 1           4 1
```

It is clearly sensible and helpful to introduce this directionality to the thinking as soon as written algorithms are introduced.

The recording for addition can then be phased in gradually, relating it to the use of materials and place value charts, and to the language used to describe them. A first step would be for the teacher to record the numbers at the board as the children make them, then the number of ones as they are joined, and finally the number of tens as these are joined. The second step would be for children to make numbers to match written examples, then join the ones and tens to form the answer and record this in the written algorithm. Only when the children are able to do this should they begin to complete written algorithms using symbols alone.

The major difficulty children encounter is remembering to join the additional 1 ten when completing the addition. Consequently, consideration needs to be given to a form of recording that will minimise this difficulty. Since the intention is to provide a consistent and smooth flow from the first introduction to the final recorded form, the recording should also build on the way children manipulate the materials.

When more than ten sticks result in the addition of the ones, children readily bundle 10 sticks to form 1 ten and realise that this additional ten should also be placed in the tens place. Since addition will not always result in an additional ten, it is advisable to place the 1 ten where it can be clearly seen and alternatively where the absence of an additional ten can be noted when there is no renaming. Further, a full-sized bundle of ten results, so that to be consistent with the development of recording as an expression of what the materials show, an additional ten should be recorded by writing a full-sized 1 in the tens place. This is also in accord with the instruction that children receive to write all numbers full sized.

It follows that the '1' must be recorded at the top of the tens place since there is not space to record it between the bottom number and the horizontal line. Later, the recording of subtraction also demands alteration to the top number, so this establishes early a process that will be useful at that time. It also means the additional bundle that results from grouping the materials should be placed in this position from the outset. The fact that the adoption of an arrow drawn in the ones place to highlight the starting position left a corresponding space at the top of the tens place was not at all coincidental! Children then move from placing the additional tens, when they result, at the top of the tens place to recording an additional 1 to show 1 ten above the tens place in the symbolic form.

The following table shows the language and recording for addition and their links to the use of materials:

Language	Materials	Recording
Show (record) the two numbers	tens | ones	
Add the ones 8 ones and 4 ones are 12 ones **Rename** 1 ten 2 ones Put (record) the 2 ones with the ones, the 1 ten with the tens		$\begin{array}{r} 1 \\ 68 \\ +24 \\ \hline 2 \end{array}$
Add the tens 7 tens and 2 tens are 9 tens	tens | ones	$\begin{array}{r} 1 \\ 68 \\ +24 \\ \hline 92 \end{array}$

68 and 24 is 92

Once the addition process is familiar, this thinking can be used to complete addition of two-digit numbers where answers in the hundreds result. Examples where there is no renaming can also be included as the process, language and recording to be followed is identical. Situations involving zero, either as a basic fact or in the answer, should also be left until later as these can cause difficulties for many children if they are met before the general thinking for the algorithm is secure.

$$\begin{array}{r} 1 \\ 86 \\ +57 \\ \hline 143 \end{array} \qquad \begin{array}{r} 47 \\ +32 \\ \hline 79 \end{array} \qquad \begin{array}{r} 1 \\ 36 \\ +24 \\ \hline 60 \end{array} \qquad \begin{array}{r} 68 \\ +20 \\ \hline 88 \end{array}$$

While Base 10 materials can provide a meaningful representation of the addition algorithm for two-digit numbers, it is preferable to use Bundling sticks first. This is partly because they are more familiar through being used to establish two-digit numeration initially, but also because they are better suited to introduce renaming. Only when the addition process is familiar should Base 10 materials be phased in, both to consolidate the thinking with two-digit numbers and to extend the process to adding three-digit numbers.

Addition with larger numbers

Extending addition to involve the hundreds place is relatively straightforward. Since addition examples where 10 tens were renamed to give 1 hundred have already been met, adding three-digit numbers is largely a matter of joining this to the hundreds digits in the numbers to be added. Base 10 materials should first be used to consolidate the addition process for two-digit numbers, then used to extend this thinking to addition with three-digit numbers. The first stage in this extension is simply to combine earlier three-digit addition with the renaming of 10 ones to form 1 ten. Then the new step of 10 tens make 1 hundred can be built in before addition involving renaming of both ones and tens is considered.

```
    1           1            1          1 1          1
   2 5        3 2 5        3 5 2       3 5 9       3 0 6
 + 4 9      + 4 4 9      + 4 9 4     + 4 9 5     + 8 5 4
   7 4        7 7 4        8 4 6       8 5 4     1 1 6 0
```

As with the earlier addition, situations involving zero are best left until the thinking is well established and situations will arise where the answer will be a four-digit number. The following table shows the language and recording for addition with three-digit numbers and their links to the use of Base 10 materials:

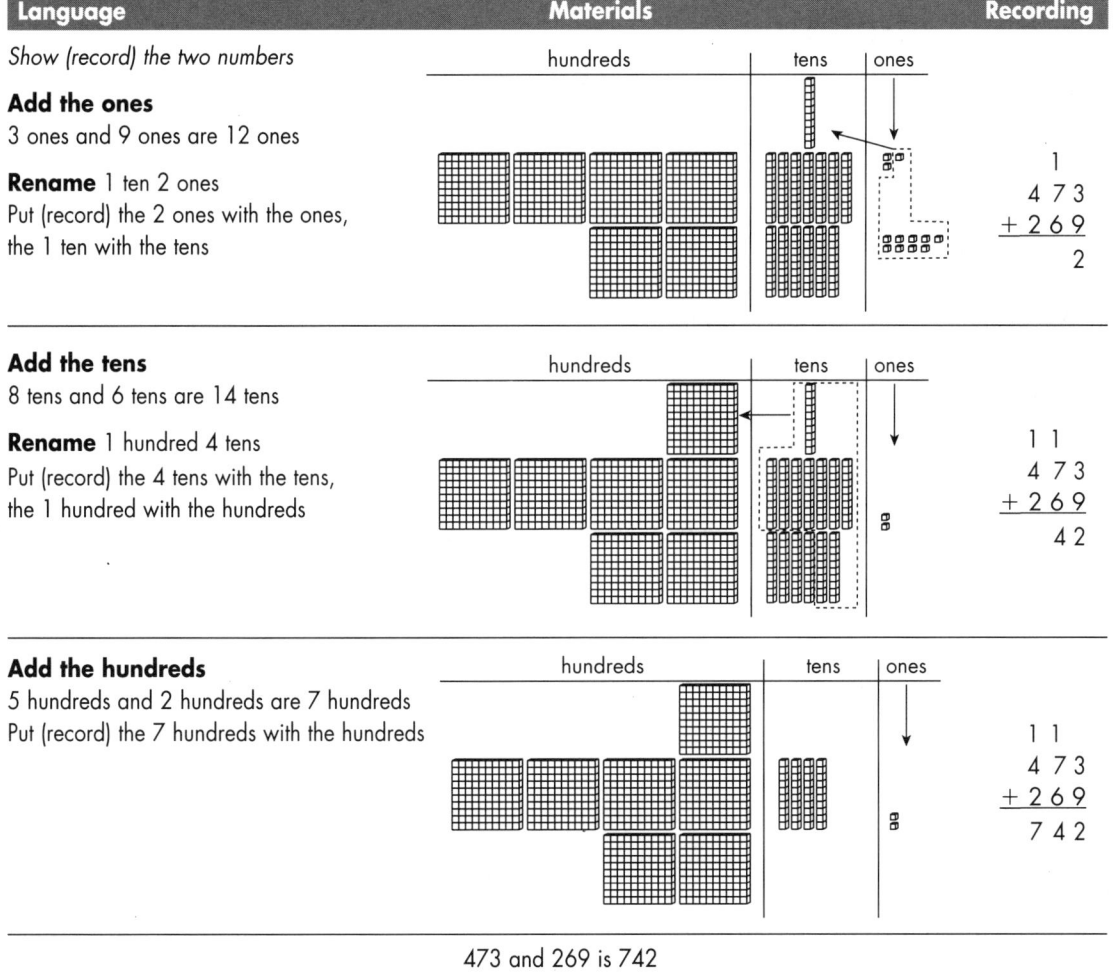

473 and 269 is 742

Addition with larger numbers is a straightforward continuation of the same processes used for two-digit and three-digit numbers; each place is considered in turn involving the simple application of a basic fact, although one or more renaming steps are often required. The understanding that 10 ones form 1 ten, 10 tens form 1 hundred, 10 hundreds form 1 thousand and so on is crucial. While materials can be used to give meaning and ensure a smooth transition to four-digit addition, their use should be minimal. Rather, language patterns and recording consistent with the earlier examples and the use of materials is the key to building facility with large numbers. The numbers resulting from renaming are recorded as a '1' written in full size in the appropriate place above the other tens, hundreds or thousands.

```
   1 1 1              1  1 1             1 1
   5 8 9 4            6 8 5 0 7          8 6 6 5 7 5
 + 7 5 3 8          + 2 9 3 9 5        + 2 0 3 8 6 3
  ─────────          ─────────         ─────────────
  1 3 4 3 2           9 7 9 0 2        1 0 7 0 4 3 8
```

Likely difficulties

Difficulties with the algorithm usually have their origins with inappropriate renaming, particular misconceptions with zero, or confusion with the recording of their working because of an inability to use place value. There may also be errors due to incorrect basic facts but these need to be addressed separately from the algorithm. Similarly, if there is confusion with the process used for another operation, this might mean that the concepts need revisiting or it may suggest that a rote procedure has been used rather than a meaningful process as built up here.

- *Place value:* Adds all the digits without regard to place value

```
        2 5
      + 3 2
      ─────
        1 2
```

When numbers are used from a problem, places may not be aligned

```
    3 8 2
  +   4 6
  ───────
    8 4 2
```

- *Renaming:* If addition without renaming is introduced first as a separate step, children often provide answers like

```
    4 6          3 5          6 4          7 8
  + 2 8        + 7 2        + 4 6        + 7 6
  ─────        ─────        ─────        ─────
   6 14        10 7         10 10        14 14
```

A renaming digit may be overlooked, especially when it is not recorded full size in the appropriate place:

```
  5 7          4 6 3
+ 3 5        + 5 7 8
  8 2          9 3 1
```

Teen number difficulties may occur when renaming:

```
        5
      3 8
    + 2 7
      10 1
```

- *Zeros:* These difficulties may relate to uncertainties with the concept of zero, with the concept for the operation or with renaming:

```
  7 0         5 4         5 4
+ 4 6       + 1 6       + 1 6
  11 0        6 10         6 1
```

- *Recording:* Numbers may be confused when the recording does not place digits in the correct places. Use of quad-ruled paper or a place value chart is helpful with children who misread their recording in this way.

Investigate and discuss

1 Use Base 10 materials and a place value chart to complete:

```
   68      393     408     369      37     463     786
 + 79    + 454   + 595   + 584    + 57   + 528   + 849
```

2 Record how these would be answered using the place value language outlined in this section. How could you use the materials to provide insight into the recording of each step?

3 Why might it be preferable to use Bundling sticks for the two-digit examples?

4 In which order would you introduce these addition examples to children?

Adding several numbers

While adding three single-digit numbers would appear to be a simpler task than adding two-digit and three-digit numbers with renaming, many children find it more difficult. This is partly because it has not been given sufficient teaching emphasis in recent times. However, difficulty is also caused by the lack of an obvious way to record the additional tens and the fact that often more than 1 ten results from the renaming. Adding more than two two-digit, three-digit or larger numbers introduces further difficulties in coping with the renaming and recording. In particular, there is a need to extend the basic facts to higher decade numbers so that this new thinking is available when required instead of calling for new facts to be worked out each time they are met. For example:

$$\begin{array}{c} 9 \\ +3 \\ \hline \text{since } 12 \end{array} \quad \text{then} \quad \begin{array}{c} 9 \\ +13 \\ \hline 22 \end{array} \text{(1 more ten)} \quad \begin{array}{c} 19 \\ +3 \\ \hline 22 \end{array} \text{(1 more ten)} \quad \begin{array}{c} 29 \\ +3 \\ \hline 32 \end{array} \text{(2 more tens)} \quad \begin{array}{c} 9 \\ +33 \\ \hline 42 \end{array} \text{(3 more tens) and so on}$$

Similarly

$$\text{since} \quad \begin{array}{c} 6 \text{ tens} \\ +8 \text{ tens} \\ \hline 14 \text{ tens} \end{array} \quad \text{then} \quad \begin{array}{c} 16 \text{ tens} \\ +8 \text{ tens} \\ \hline 24 \text{ tens} \end{array} \quad \text{(10 more tens)}$$

$$\text{since} \quad \begin{array}{c} 7 \text{ hundreds} \\ +9 \text{ hundreds} \\ \hline 16 \text{ hundreds} \end{array} \quad \text{then} \quad \begin{array}{c} 7 \text{ hundreds} \\ +19 \text{ hundreds} \\ \hline 26 \text{ hundreds} \end{array} \quad \text{(10 more hundreds)}$$

In this way, the memory demands introduced with addition of several numbers can be reduced and the process can be built securely onto the earlier ways of adding large numbers.

There may also be a conflict of method. Many children will continue to add down the places, while others add to ten wherever possible. The position taken here is that adding down is preferable in view of its consistency with earlier addition skills and that it helps to minimise the possibility of adding digits more than once in the process of making additional tens or whatever, and allows the development of a process that works identically for all situations. This last point is particularly important for those children likely to experience difficulties in mathematics.

Likely difficulties

- When children try to make to ten rather than continue to add down each place, digits may be omitted from the calculation, or used more than once.
- More than 1 ten, hundred and so on usually results from the renaming that occurs with each place. Some children may simply add 1 extra for each place by generalising inappropriately from the earlier addition involving only two numbers.
- Other children may simply record the total result for each place and not rename at all.

Investigate and discuss

1. Try adding these numbers:

   ```
      8     9     4     9     7
      5     5     7     7     9
      7     3     8     5     7
     +6     7     2     3     6
           +5    +6     8     3
                       +4     3
                             +4
   ```

   a. How did you add them? Downwards, adding each number in turn, or did you use pairs of numbers that sum to ten?
   b. Try both ways, then comment on their relative merits. Which is most consistent with previous addition? Which is least demanding on memory?
   c. What additional thinking is needed for this addition? What difficulties could children have with this? Can you suggest any form of recording that might assist such children?

2. Would there be any further difficulties in adding these numbers?

   ```
     68    82    38   816   465
     79    35    29   735   279
     32    58    81   362   667
     91    24    72   261   721
    +52   +71   +15  +582  +473
   ```

Estimation with addition

Coming to terms with the written addition algorithm is important as it provides a basis of understanding of the operation of addition and the ways in which it can be used. But in many real life situations, approximate rather than exact answers are required. Sometimes an estimate is needed to check the feasibility of completing a task, for instance to add the given lengths to see whether a particular piece of fabric or timber will allow for all of the smaller pieces required. At other times it may be simply to check that the amount asked for in a shop is about what it should be or to see whether a $50 or $100 note would be more appropriate to cover the bill.

For some people such estimations are easy to make and the processes they adopt reflect a well-internalised number sense. But for others the only technique is to try to treat any calculations as if they are to be written in their mind and the number of steps that need to be manipulated and remembered becomes too much. This may lead to a sense of powerlessness as they have to rely on the judgement of others, or it may lead to frustration when they are unable to carry out the tasks as quickly as the situation demands.

Given that 'good estimation and approximation skills enhance our ability to deal with everyday quantitative situations' (*National Statement on Mathematics in Australian Schools, 1991*) it is important for techniques allowing children to make computational estimations to be built in as soon as the children have an understanding of the concepts on which the techniques are based. The first of these is to see a need for estimation. Most estimation in everyday life involves measurements of some form and experiences with seeing how many, how far, how long and so on can be built into measurement activities from the outset as part of the process of constructing the concept of a particular measure. Later, approximate measures rather than exact measurements will be all that is needed, so children are able to see that something close enough is sufficient and the practical context provides a means of determining how close the measure needs to be. Experiences with estimation and approximation in measurement lay the basis for the whole estimation process and can in turn be used to build from the use of numbers in measuring to estimating with numbers alone.

A second foundation for estimation is a good grounding in number understanding and processes. Rounding will clearly play an important part and an understanding of place value and the ability to rename numbers is also essential. Similarly, the concept for addition and understanding of the processes built up with the use of materials and reflection on the recorded algorithm are fundamental in allowing estimation processes to be mentally managed.

While addition involving two numbers can be readily done by a process of rounding each number and then adding these rounded amounts, most often a need to estimate the result of adding will occur when there are several numbers to add. A process based on rounding would require a good deal of management and remembering of intermediate steps.

Estimate the total:
```
      683     Round 683 to 700
      527     Round 527 to 500, add to 700: 1200. Remember the total
      836     Round 836 to 800, add to the old total 1200: 2000. Remember the new total
      486     Round 486 to 500, add to the old total 2000: 2500. Remember the new total
    + 731     Round 731 to 700, add to the old total 2500: 3200
about 3200
```

A more manageable process is to use a front-end strategy of first considering the digits in the largest place common to the numbers being added. This provides a reasonable first estimate that can then be made even closer by considering the digits in the next place. For instance:

Estimate the total:
```
    683         first add the hundreds        6 ... ...
    527                                       5 ... ...
    836                                       8 ... ...
    486                                       4 ... ...
  + 731                                     + 7 ... ...
            the total is more than           30 hundreds
                                          then look at the tens        8
                                                                       2
                                                                       3
                                                                       8
                                                                     + 3
                                                            about 20 tens more
                                                            or 2 more hundreds
```
30 hundreds and 2 hundreds is 32 hundreds: the total is about 3200

When introducing this strategy at the board or on an overhead projector, it can be helpful to use a card to cover all but the largest place. Once the initial estimate has been obtained, slide the card over to show the digits in the next place to obtain a closer estimate. This should not involve adding; it is sufficient to make a quick overview of the digits to see how many more tens of that place could be formed as this will be all that is needed to adjust the estimate, as 10 of the second place represents 1 in the leading place. Use of the phrase 'more than' for the estimate obtained from the leading digits helps children to see that this will always be an underestimate and encourages them to look to the next or remaining digits to adjust to a closer estimate. Note that addition of the leading digits involves the higher decade facts similar to that required to add several digits. Leaving the development of estimation with addition until children are confident with exact addition not only allows them to accept approximations readily, it also provides the underlying numeration and extended basic fact knowledge on which estimation is most readily built.

Once children are familiar with the basic front-end strategy for estimation of addition, other techniques can be discussed and applied to particular situations. For instance, it is often possible to group numbers to form multiples of ten, one hundred or thousands and then add these, or, when the operation has been mastered, use multiplication to determine an estimate based on these groupings:

```
    837 kg ─┐ about 1000 kg           $15.35 ─┐ about $30.00
    378 kg ─┘                          11.85 ─┘ about $30.00
    612 kg ─┐ about 1000 kg             9.55 ─┐ about $30.00
    267 kg ─┘                          13.95 ─┘
            about 2000 kg altogether   21.10
                                       18.25
                                               about $90.00 altogether
```

When all of the numbers being added are about the same size, an estimate can be given by multiplying the number about which they all cluster by how many are being added:

      87c
      93c
      89c
      92c
      88c   Each number is about 90
+ 93c   6 nineties are 540
                About $5.50

Since multiplication is essentially repeated addition, there are many situations where it is possible to look for means of estimating addition using multiplication. Introducing these two commonplace examples is often enough to suggest to children the merits of multiplication in estimation and thereafter they tend to seek out particular shortcuts of their own. The most important task for a teacher in fostering estimation for addition is to encourage children to talk about the methods they use and so share both the particular techniques and the notion that estimation involves methods that are meaningful to and constructed by each person to suit their own ways of thinking and the particular needs of a given computational situation.

Adding numbers mentally

Just as there is often a need to obtain approximate answers rather than the precise ones given by pencil and paper methods or by calculators, so there is also a need for exact answers calculated quickly and on the spot without recourse to electronic aids or the written algorithms. This need is best satisfied by being able to 'work things out in one's head' but this is often easier said than done. In particular, it calls for different ways of thinking from those associated with the formal algorithms and for a good ability to organise and control the various thought processes entailed. Because they are less demanding on memory, efficient mental computation processes usually involve left to right processing similar to the thinking built up with strategies for estimation. In this way, the first calculation gets close to the answer and each subsequent step closes in on what is wanted so that there is no need to remember anything other than the result of each adjustment.

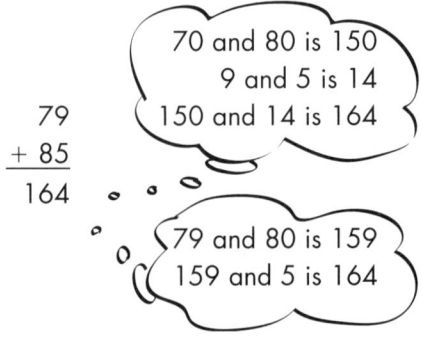

The development of the front-end strategy for estimation also prepares the way for this thinking and, in general, mental computation processes grow out of experiences with estimation far more than they build on the recorded algorithm.

There are many different ways in which answers can be determined mentally. Some will be personally constructed by individual children to suit particular sets of numbers, whereas others will use thinking strategies very similar to those used earlier in developing skills that underpin both exact calculations and estimations. For instance, extended basic facts are required for multiplication as well as addition so that developing the multiplication algorithm also builds up the need for and use of mental strategies for addition. For example:

```
     5
   7 8      7 eights are 56
 × 6 7      Rename: 6 ones 5 tens
   5 4 6    7 by 7 tens are 49 tens, and 5 tens is 54 tens
            and so on
```

Competency in mental computation demands a repertoire of methods adapted to suit particular numbers and situations. Children also need to see that mental arithmetic is a first resort in many situations where a calculation is required. A teacher's major role in this development is to encourage children to develop their own personal strategies, to explore and compare strategies used by others, and to choose those that fit their own strengths and inclinations as well as the problem contexts to which they might apply. Discussion about the merits, ease of use and degree of control a range of strategies allows should be a feature of the teaching of mental computation.

Using calculators in addition

The use of a calculator appears to be very straightforward to an adult who already has a well-developed understanding of numbers and a good deal of practical, everyday experience in the use of this knowledge. This has often led parents to deplore the use of calculators in school and to expect that children should learn about a calculator's possibilities in the same way as they did, that is, through a process of trial and error and self-teaching. Yet the use of a calculator is not straightforward. In the previous chapter, we saw that it is often no easy matter to enter numbers correctly or to read the numbers that are given. For addition, a difficulty often arises through the need to enter a great many numbers before an answer is obtained. All of this suggests that calculator use is in many ways another form of mental computation; the calculator only shows one step at a time and quickly erases all that went before. Unlike pencil and paper computation where a record is kept of everything that needs to be done, use of a calculator requires a mental plan to keep track of what is happening and to organise the various results that might be important along the way. A good understanding of the processes involved, the properties of the numbers being used and ways in which a calculator can be used to carry out the calculation more efficiently is necessary in the mental management of its use.

Unfortunately, because most people are self-taught, their use of a calculator often simply mimics the written processes they have learnt and fails to take advantage of either the new features calculators allow or their own knowledge of strategies gained from estimation and mental arithmetic techniques. This is most evident in the inability of many adults to use the memory features such as add to, subtract from and recall memory, especially the constant function that most calculators have. Since the memory of a calculator is rather limited (imagine if whenever you determined a new result you lost the previous one!) there is a need to use pencil and paper to record any intermediate result that might be useful in the calculation. Learning to use the non-writing hand to key in numbers and operations from the outset leaves the writing hand to record information simultaneously with its appearance on the calculator. Similarly, taking an overview to all that is required in the calculation before using a calculator will often show shortcuts even before data or operations are keyed in. A calculator is probably of most use when several operations are required, but this demands an understanding of the conventions that govern the order of operations and a sophisticated use of the calculator's memory to organise the sub-steps involved.

For addition alone, the use of a calculator is fairly simple. Once the concept is understood so that the order of keying numbers is apparent and the use of the addition symbol to do the addition and the equals symbol to obtain the result is known, children are ready to use a calculator. However, at first children may simply think that you key in all the numbers, then the + symbol to obtain the answers, much as was done in the first written forms. This may lead them simply to key in 4, 7, 2, 9, 6, + when carrying out the addition 4 + 7 + 2 + 9 + 6. Without an awareness of what addition is, a child who treats a calculator this way may be quite happy to accept the answer as 47296 since this is what the 'infallible' calculator shows. Instead, it is the reading and comprehension of what the computation is asking that is once again paramount in providing for meaningful use. A calculator can only perform as well as the understanding that children bring with them to its use and then apply in interpreting and assessing any results.

Use of a calculator for addition, then, needs to follow a good understanding of the concept and a proficiency with the basic facts so that an ability to assess the reasonableness of any answers may be given by use of these facts in conjunction with estimation. Calculator use can accompany the development of the written algorithm, but if begun too quickly might prevent some children from coming to terms with the steps involved and ways in which the concept and basic facts are integrated with numeration ideas to develop a coherent process. One way of introducing calculator use is by way of games in which basic facts, written computations or routine problems are answered at each turn, with one child using a calculator to check the accuracy of the answers. At the end of the game, the winner takes over the use of the calculator so that all players can practise the ideas behind the operation as well as gaining experience in using a calculator.

CHAPTER 4 • computation and estimation

A board game can be readily made by placing a game track on an attractive picture and making a set of cards with addition examples related to the level of ability of the players. When a player has correctly answered the addition on the card (checked by the child using the calculator), a dice can be rolled to nominate a particular place according to the type of number being used. For instance it might call for the thousands, hundreds, tens or ones place for larger whole numbers, or for the ones, tenths or hundredths place if decimal fractions are involved. The player can then move his or her marker to the next place that matches the digit in his or her answer.

For example, if the card shows $\begin{array}{r}784\\+\ 598\end{array}$ and the player correctly answers 1382, when the dice shows H for hundreds, the player moves his or her marker to the next 3 on the track.

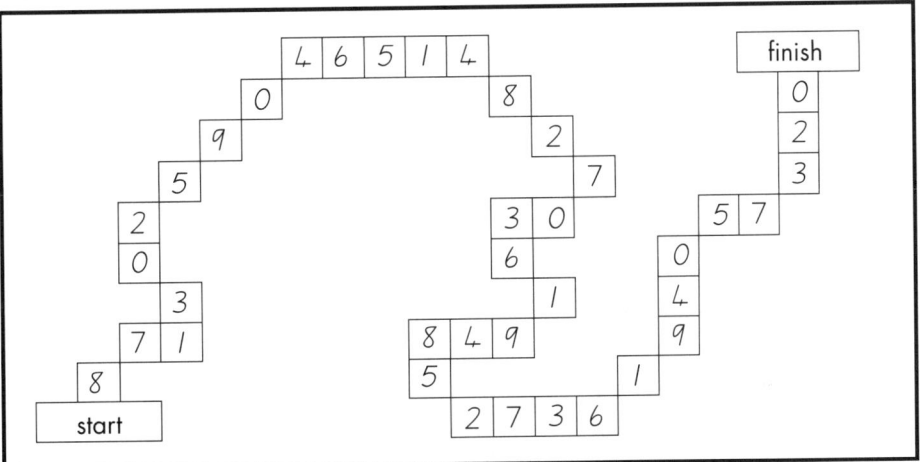

Calculators are also helpful in teaching addition by integrating their use with estimation, calling on children first to estimate then to compare the answer with that given by the calculator or to use estimation to assess the reasonableness of calculator answers, justifying the results to other members of the class.

Investigate and discuss

1. Make the game described above and play it with some children in middle primary school. Select addition with two-digit and three-digit numbers for the cards. Would it be possible to play this game at an early stage of development where the children use materials to answer the questions?
2. Another way of introducing addition with calculators is to give children problems to explore.

a Solve this problem and comment on its use in introducing the use of a calculator: Arrange the digits 1, 2, 3, 4, 5, 6 in these boxes to make the largest sum less than 1000 and the smallest sum more than 500.

```
  □□□              □□□
+ □□□           + □□□
```

b Write a number in the centre square. Find numbers to put in each of the corner squares so that they add to make the number in the centre:

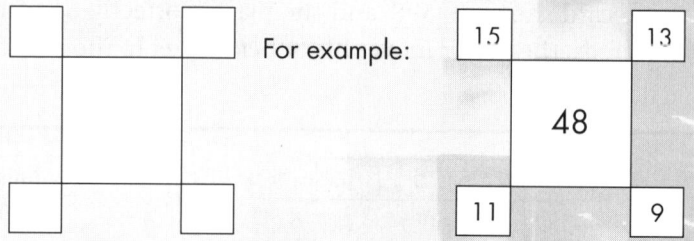

For example:

Repeat for other numbers.

c Change the rules to place numbers in each corner square so that each side adds to make the number in the centre.

3 A calculator game for two players, each with a calculator:
Both players enter the number 67 on the calculator at the start. Then each player in turn calls a number between 40 and 50, which both players add. The player who first calls the number that makes 600 appear on the display is the winner.

4 Find some other calculator games that you might use to introduce addition on a calculator.

Addition with decimal fractions and like common fractions

The key to adding whole numbers has been a good understanding of numeration, place value and renaming, along with facility with the basic facts. Addition for decimal fractions and common fractions relies on similar abilities together with a careful transition of the earlier thinking to the new types of numbers. In particular, for decimal fractions an ability to rename 10 tenths as 1 one, 10 hundredths as 1 tenth and so forth is almost the only new knowledge involved. Common fractions often require two forms of renaming—1 one needs to be renamed as 5 fifths, 6 sixths and so on and mixed numbers need to be renamed as improper fractions, for example 1 and 3 eighths is renamed as 11 eighths. A focus on a meaningful language for these fractions from the outset, together with the language developed for addition, allows a ready translation of whole number addition processes to each of these types of computation. The only new aspect is that the ones are no longer added first; addition begins with the place furthest to the right, be it tenths, hundredths or thousandths with decimal fractions or simply the fraction part for common fractions. Of course this has always been the case, for until fractions were considered the ones place always occupied the place furthest to the right in any number being added. The use of an arrow to indicate the starting place, in contrast only to stressing 'add the ones first', continues for fraction ideas, thus minimising possible confusion and avoiding a fixation on the ones place.

Addition with ones and tenths can begin when understanding for these decimal fractions, including renaming, has been established. Since materials cannot be used in the way they were for whole numbers, this means that a solid understanding of the representation for ones and tenths to provide a visualisation of the process is essential. This will allow a language based on ones and tenths to be developed and used to guide the recording in a similar manner to that for whole numbers.

Language	Recording
Add the tenths 7 tenths and 9 tenths are 16 tenths	1
Rename 16 tenths is 6 tenths 1 one	6.7
Record the 6 tenths with the tenths, the 1 one with the ones	+ 5.9
	6
Add the ones 7 ones and 5 ones are 12 ones	1
Record with the ones (place a full stop with the 2 to show it is in the ones place)	6.7
	+ 5.9
	12.6
6 and 7 tenths add 5 and 9 tenths is 12 and 6 tenths	

When hundredths have also been developed and children are comfortable with the thinking that 1 tenth is 10 hundredths, a similar extension can be made to adding numbers with ones, tenths and hundredths. Again, this process builds on a full understanding of the decimal fraction places, the relationships between them and the use of a meaningful language for addition based on the use of place value, renaming and basic facts. One difficulty that children might experience is in adding numbers with differing decimal places. In the past this was treated by asking children to 'line up the decimal points', introducing a new way of thinking to this form of addition. While this may help addition and, later, subtraction with decimal fractions, it is likely to confuse multiplication with decimal fractions where the decimal places do not 'line up'.

Rather than introduce this new rule, it is better to continue to treat addition in the same way as with whole numbers, aligning the places, ones with ones, tenths, with tenths, hundredths with hundredths and so on. A full stop is placed next to the digit in the ones place to indicate that the whole numbers have ended, and that this is the ones digit. In this way not only are all the place value positions aligned correctly, but the decimal points are in line as a consequence. The only difference is that what has been stressed is place value, which is after all one of the keys to all computation. Thus the extension of the addition algorithm to include all decimal fractions is a relatively straightforward process that can accompany the development of each new decimal place.

	1 1 1 1
Add the thousandths 7 thousandths and 5 thousandths is 12 thousandths	3 4 . 6 9 7
Rename 2 thousandths 1 hundredth	+ 5 7 . 7 6 5
Add the hundredths 10 hundredths and 6 hundredths is 16 hundredths	9 2 . 4 6 2
Rename 6 hundredths 1 tenth, and so on	

It should also be borne in mind that addition with decimal fractions fundamentally depends on numeration ideas rather than the size of the whole number portion of the numbers being added. To add 234.6 and 12 456.9 should be no harder than adding 4.6 and 6.9 when the addition algorithm for whole numbers is secure, which is of course a prerequisite for the extension of addition to fraction ideas.

Using a calculator to add decimal fractions is also very straightforward. The calculator automatically handles any differences in the number of decimal places the numbers have. All that is needed is an ability to read and key in the numbers and the operation symbols. Again, in order that children have the means to control and understand what the calculator can do so readily, it is important that they have some experience with addition using the written algorithm and are able to estimate the likely result before too much reliance is placed on the calculator alone. Situations involving the use of decimal fractions within the metric measurement system provide a good, real basis for this introduction.

After the addition of decimal fractions has been established and numeration for common fractions has been built up, the addition algorithm can be readily extended to common fractions, particularly if the first examples simply involve tenths. The thinking, language and recording is identical to that for the decimal fractions. Once again the use of materials after this point is not helpful; showing 3 eighths and 4 eighths is as likely to bring 7 sixteenths to mind as 7 eighths:

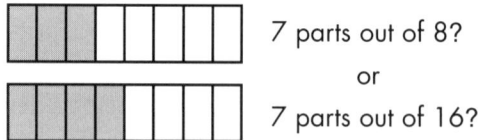

7 parts out of 8?
or
7 parts out of 16?

A full understanding of the common fraction concept and its recording is thus a necessary prerequisite for addition with common fractions. For this reason, it is usual to leave the development of the operations on the common fractions until some time after the introduction of the fraction ideas themselves.

Decimal fractions	Language	Common fractions
$\begin{array}{r}1\\5.7\\+3.6\\\hline 3\end{array}$	**Add the tenths** 7 tenths and 6 tenths is 13 tenths **Rename** 3 tenths 1 one	$\begin{array}{r}1\\5\frac{7}{10}\\+3\frac{6}{10}\\\hline \frac{3}{10}\end{array}$
$\begin{array}{r}1\\5.7\\+3.6\\\hline 9.3\end{array}$	**Add the ones** 6 ones and 3 ones are 9 ones	$\begin{array}{r}1\\5\frac{7}{10}\\+3\frac{6}{10}\\\hline 9\frac{3}{10}\end{array}$
	5 and 7 tenths add 3 and 6 tenths is 9 and 3 tenths	

Once the recording of addition with common fractions has been established for tenths, it is a very easy matter to extend this to all other fractions with the same denominator. As introduced earlier, such fractions will be called *like fractions*, not only because this readily highlights the main idea that only those fractions composed of the same type of parts can be added, but it also gives a much more friendly connotation to the process. At first, examples would involve only situations where there is no renaming to familiarise learners with both the recording and reading of this different-looking form of computation, but renaming can be begun immediately the underlying numeration knowledge is secure.

Language	Recording
Add the fractions 3 fifths and 4 fifths is 7 fifths **Rename** 2 fifths 1 one **Record** the 2 fifths with the fifths, the 1 one with the ones.	$\begin{array}{r}1\\4\frac{3}{5}\\+3\frac{4}{5}\\\hline \frac{2}{5}\end{array}$
Add the ones 5 ones and 3 ones is 8 ones	$\begin{array}{r}1\\4\frac{3}{5}\\+3\frac{4}{5}\\\hline 8\frac{2}{5}\end{array}$
4 and 3 fifths add 3 and 4 fifths is 8 and 2 fifths	

Adding like common fractions is a very simple extension of the addition algorithm, when recorded vertically, brought about by the use of a similar meaningful language in every case and dependent on a few fundamental numeration ideas. Adding unlike common fractions is not so simple. New understandings related to equivalent fractions are needed and confidence in one's own ability is required to have control of the process. For this reason, the extension of addition with common fractions to the general case involving different denominators is usually delayed until the end of primary schooling and is not discussed here until later in the chapter along with subtraction involving unlike fractions.

Likely difficulties

♦ If teachers treat addition without renaming as a separate introductory step, children often provide answers like:

```
   4 6        3 5        6 4        7 8
 + 2 8      + 7 2      + 4 6      + 7 6
  ─────      ─────      ─────      ─────
   6 14       10 7       10 10      14 14
```

What might have led them to believe that this is an appropriate way to add numbers?

Investigate and discuss

1 a Answer each of these examples using the recording and language outlined in this section:

$$\begin{array}{c} 5.6 \\ +4.8 \\ \hline \end{array} \quad \begin{array}{c} 9.23 \\ +6.79 \\ \hline \end{array} \quad \begin{array}{c} 7.2 \\ +3.89 \\ \hline \end{array} \quad \begin{array}{c} 13.06 \\ +28.14 \\ \hline \end{array} \quad \begin{array}{c} 5\frac{4}{5} \\ +4\frac{3}{5} \\ \hline \end{array} \quad \begin{array}{c} 9\frac{7}{8} \\ +6\frac{3}{8} \\ \hline \end{array} \quad \begin{array}{c} 7\frac{5}{12} \\ +3\frac{8}{12} \\ \hline \end{array}$$

b How does the cycle of steps discussed in the overview to this chapter:

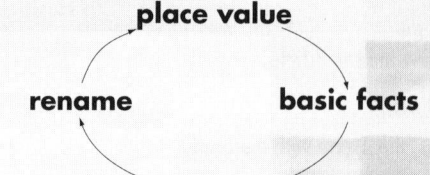

apply to the addition of these decimal fractions and common fractions?

c What new aspects of number understanding are required at each step in the addition cycle?

2 In order to use computation in applications and problem solving, a learner needs to be able to:
  ♦ decide what operation to perform
  ♦ select a means of carrying out the operation
  ♦ perform the operation
  ♦ make sense of the answer

What aspects of addition knowledge (concepts, basic facts, algorithms—written, mental, calculator and estimation) relate to each of these steps?

3. Solve the following problem and determine how you would assist a child to come to terms with what is required. Consider the general model for problem solving given in Chapter 2 as well as your answers to question 2 above and think about the ways in which a child might give an incorrect answer.

> At the zoo, the giraffes eat 263 bales of hay. The seals eat 392 buckets of fish. The elephants eat 426 bales of hay. How many bales of hay are eaten altogether?

4. How would you extend the mental computation and approximation strategies developed for whole numbers to decimal and common fractions? What new thinking is required in making an estimate for decimal places and for like common fractions? How might you deal with unlike common fractions?

5. Many people learnt to add common fractions in a horizontal fashion.

$$3\tfrac{7}{8} + 5\tfrac{5}{8} =$$

Find out how this was done and contrast it with the method introduced here. Is the same fundamental cycle of steps still being followed? Can you suggest any reasons for adopting a horizontal recording over a vertical one?

Section II: Overview to subtraction

SUBTRACTION is introduced once children have a good understanding of addition, using similar experiences with materials and problem situations to build a complete picture of the operation. Indeed, addition is the key to a meaningful learning of subtraction in that the concept is in many ways a different statement of the part/part/whole model (the whole and one part are known, the other part needs to be found). Most young children have some intuitive knowledge of subtraction based on their everyday experiences, but they tend to find answers to simple problems by counting back. In contrast, subtraction basic facts are best viewed as a reinterpretation of the addition facts (12 take away 5 is 7 because 7 and 5 is 12).

The recording of subtraction is also similar to addition, so that there is again a need for a careful sequencing of development from the initial introduction to the written and mental forms, using materials and language to provide meaning. A vertical recording allows a smooth transition to the processes for larger numbers and also provides a focus on the directionality implicit in subtraction; it is not simply the difference that is found but the result of having the top number, then taking away the bottom number. The use made of a place value chart with Bundling sticks or Base 10 materials shows this by asking only that the top number be shown while the bottom number is literally taken away. The thinking patterns established with this use of materials take on additional importance because of the way the original number changes as a result of renaming and because materials do not assist in situations with internal zeros.

> **SEQUENCE FOR DEVELOPING SUBTRACTION**
>
> 1. Develop the initial subtraction concept: verbal action stories
>   model with materials
>   vertical recording —symbol
>
>   Later extend to missing part and comparison
>
>    ```
>         story
>        /    \
>       /  takeaway
>      /        \
>    materials ←→ symbols
>    ```
>
> 2. Build up the basic facts: think of addition strategy
>   ensure automatic responses
>
> 3. Thinking in tens—extend basic facts to tens
>
>   2 digit
>   − 2 digit   without renaming. Later use the same thinking for
>
>   3 digit
>   − 3 digit
>
> 4. 2 digit
>   − 2 digit   renaming 1 ten as 10 ones
>
> 5. 3 digit
>   − 3 digit   renaming 1 ten as 10 ones
>   renaming 1 hundred as 10 tens
>   renaming both
>
> 6. 3 digit
>   − 3 digit   with internal zeros
>
> 7. Subtraction with larger numbers
>
> 8. Strategies for estimation, mental computation and calculator use
>
> 9. Subtraction with decimal fractions: tenths
>   hundredths
>   thousandths
>
> 10. Subtraction with common fractions: like fractions
>   need for equivalent fractions
>   unlike fractions

The subtraction concepts

Everyday experiences have usually given children some familiarity with subtraction, but they often see it as simply finding the difference between two numbers. Difficulties in carrying out computation or solving subtraction problems evident later are often due to this inadequate or incomplete understanding of subtraction. The first step to building a full understanding of subtraction focuses on the fundamental process of taking away an amount from a given total, but it is not the only notion that is needed. Problem situations often involve comparing two amounts or finding the amount required to make a particular quantity. Children need to experience all three types of subtraction, known as *takeaway*, *comparison* and *missing part*, before an appropriate basis for applying subtraction can be attained. The takeaway concept is developed first as it provides a firm basis on which the other two conceptions can be built and also establishes the thinking which underpins the algorithms for dealing with larger numbers.

Subtraction as takeaway

Subtraction is first introduced through story situations centred on takeaway. Materials can be used to provide a picture of the operation, although their use is not as straightforward as it was for addition. Since the initial number is shown while the other number is removed, it is not possible to use counters in two different colours to highlight what happens. Nonetheless, the use of the part/part/whole way of thinking should be focused on from the outset to show that in subtraction the whole is known, and one part is taken away to leave the other part.

Ten frames can be used to build up these initial ideas in a similar way to that for addition. The whole is placed on the ten frame and the other part readily removed using familiar number patterns so as to minimise counting:

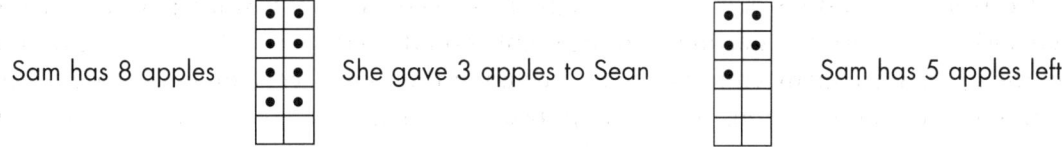

As even initial subtraction will involve numbers up to 20, two ten frames can be used for examples involving teen numbers, such as '14 takeaway 8'.

Most of the time, takeaway is the language used for subtraction because it readily arises from this use of materials and brings to mind the way in which the operation will be carried out. In due course, when subtraction is very familiar, subtract can be introduced as the mathematical word used for takeaway and other expressions that imply subtraction can be developed. For children, the word 'minus' has little or no relationship to subtraction or the operation of takeaway (after all, the operation is called subtraction not minusition). Many other forms of language for subtraction arise in real life situations such as 'how many left' and 'difference between'; minus can join these other forms at some later time when subtraction is secure and these expressions can be related to it through their occurrence in problem situations.

A gradual introduction of symbols as the final step in the sequence for developing the initial subtraction concept will ensure meaningful interpretation of subtraction statements. The symbol '−' can then be established as a mathematical way of writing 'takeaway', indicating that the numbers are to be subtracted.

$$\begin{array}{r} 8 \\ -3 \\ \hline 5 \end{array}$$

8 takeaway 3 is 5 can be written

$$\begin{array}{r} 14 \\ -8 \\ \hline 6 \end{array}$$ can be read as 14 takeaway 8 is 6

At this point in the development of subtraction, horizontal recording such as $8 - 3 = \square$ or $14 - 8 = \square$ is not appropriate. Use of the symbol $=$ would involve the equals concept and may lead to its confusion with the subtraction concept or to confusion between the two symbols $=$ and $-$. As well, horizontal recording requires the use of more symbols than the vertical format used in the development of the written subtraction algorithm. More importantly, the subtraction operation involves an initial amount from which something is removed. At the concept and basic fact stages, this initial amount is always larger, but for the algorithm there may not be enough ones, tens or whatever to allow straightforward taking away in each place. Consequently a common error with the algorithm is that the smaller number is subtracted from the larger regardless of the meaning involved. Such errors have been called reversal errors but they represent a much greater difficulty than simply misreading a written format, and a very conscious awareness of the directionality needed for subtraction must be built in from the outset so that children appreciate the need for renaming.

Horizontal recording does not highlight this need as readily since it would require verbalising the process as starting with the number on the left then taking away the number on the right. Since many children are not secure in their knowledge of left and right, they simply see the more obvious relationship 'take the smaller number away from the larger number'. In contrast, their understanding of top and bottom is established even before reaching school and can be used with a vertical recording to highlight the thinking 'start with the top number, takeaway the bottom number'. In this way identical recording and thinking is used for all subtraction so that a consistent and necessary way of thinking can be established from the outset.

Reading subtraction statements downwards also facilitates this understanding, and care needs to be taken with the language used for subtraction so that it is consistent with this thinking. For example, with subtraction such as 42 takeaway 17, saying '2 ones, can I takeaway 7 ones' reflects the subtraction process while reading subtraction in terms such as '2 from 7' breaks with the directionality of top to bottom. Indeed, this use is likely to lead to inappropriate subtraction of the smaller number from the larger as it seems to be unrelated to any meaningful conception of subtraction altogether. Similarly, use of the word minus, which has no obvious link to the subtraction process, also seems to lead many children simply to find the difference, rather than consider the directionality implicit in subtraction situations.

With these points in mind, the teaching sequence that has proven to be most effective in building up understanding of the subtraction process is:

SEQUENCE FOR DEVELOPING THE SUBTRACTION CONCEPTS

1 Introduce subtraction through verbal action stories that involve *takeaway*
There were 6 birds in the cage.
When the door was opened, 2 flew out.
How many birds are in the cage now?

2 Guide children in using materials to show the initial number, then remove or cover the amount that is taken away.
Set out the objects vertically.
Verbalise the subtraction process using takeaway.

3 Gradually introduce symbols to represent this process, matching − to the takeaway process through a careful development such as

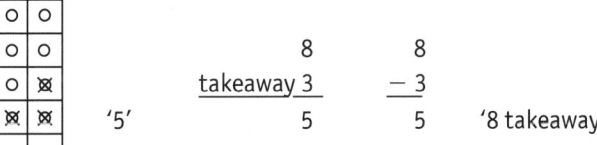

'8 takeaway 3 is 5'

4 Ensure that children can match

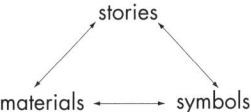

5 Emphasise the part/part/whole aspect of subtraction:

whole
− part
part

6 Extend the initial subtraction concept to include *missing part* and *comparison* as well as alternative expressions that mean or imply subtract.

Focusing on the part/part/whole model from the outset not only helps to establish the link between subtraction and addition, it also builds in a means of distinguishing between the two operations. When two parts are known and the whole has to be determined, the situation must involve addition. If the whole and one part are known then subtraction is needed to find the other part. Using this thinking allows children to determine for themselves which operation is required, rather than ask 'Is it add or takeaway?'.

Investigate and discuss

1 a Use counters or Unifix on two ten frames to demonstrate how you would expect a child to show this problem: Andrea caught 12 fish. Seven of them were too small so she put them back in the water. How many fish did she take home for dinner?

b Show 12 counters on the ten frames. Then remove 7. How would you relate this use of materials to the problem? (Consider the use of materials before and after the subtraction has been carried out.)

c Place 14 counters on the ten frames. Make up a subtraction story which you can show by removing 5 counters. Repeat for other pairs of numbers such as 13 and 8, 11 and 4 and so on.

d What advantages are there in using teen numbers when establishing meaning for the concept? (Consider what is needed for later examples like 75 subtract 25.)

2 Another way of overcoming the difficulty associated with the manner in which subtraction has before and after states is to use artwork to set up problems such as:

   a Write a subtraction story that can be associated with this picture.
   b What difficulties might children experience in seeing the initial number of objects in these situations?

3 Many teachers build up 'families of facts' as soon as subtraction is introduced. For example:

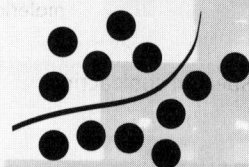

can be interpreted as

$$\begin{array}{cccc} 5 & 7 & 12 & 12 \\ +7 & +5 & -7 & -5 \\ \hline 12 & 12 & 5 & 7 \end{array}$$

When do you think such connections should be made? Justify your answer by reference to the strength of this knowledge and any likely interference across the operations.

Extending subtraction

While subtraction begins with a takeaway conception, problem situations occur that do not involve taking away and children need to be introduced to these other forms when the subtraction as takeaway is secure. For example, the word 'more' also suggests subtraction, either as *comparison* where the difference is needed, or when a *missing part* is required. Children need to be led to analyse these situations to identify the whole and the parts and then use subtraction to find what is needed:

Form of subtraction	Problem	Recording
Takeaway ◆ which is the whole ◆ which is the part?	There were 10 apples in the bowl James took 6 apples to give to his friend Jenny How many apples are left in the bowl? 4 apples are left in the bowl	10 − 6 ――― 4
Missing part ◆ which is the whole? ◆ which part is needed?	Jasmine has 7 apples She needs 12 apples to make a tart How many more apples does she need? Jasmine needs 5 more apples	12 − 7 ――― 5
Comparison Requires two steps: ◆ who has more? ◆ how many more?	James has 5 apples Jasmine has 9 apples How many more apples does Jasmine have than James?	9 − 5 ――― 4

Insufficient experiences with the range of these forms of subtraction or the introduction of the three subtraction types too soon or too closely together frequently lead children to misinterpret missing part and comparison situations as addition. Because they involve new and different language, focused on the use of the word *more*, it is necessary to treat their development in the same way as the initial takeaway concept was introduced, providing a variety of activities with problem situations and materials prior to any recording. Unfortunately, the use of materials has often been reserved solely for takeaway situations, leaving a superficial treatment of missing part and comparison, while the nature of the questioning used by the teacher in dealing with these examples does not always bring out their basis in subtraction. Since it is crucial that children integrate this new knowledge into their existing view of subtraction to build a unified conception of the operation, development of the other forms of subtraction needs to be dealt with explicitly soon after the basic takeaway idea has been learned. In this way, these other forms can be related to the initial concept but also differentiated from it.

Subtraction as missing part

Learning the subtraction basic facts largely depends on a secure understanding of missing part as this builds an ability to use the corresponding addition facts to determine answers. For this reason, subtraction as missing part is the first extension of subtraction to be developed and needs to be met prior to building the subtraction facts. Situations involving missing part usually ask the question 'How many more?'. If children do not fully understand what is meant by such a question they cannot even begin to solve the problem. Therefore it is essential that children experience a number of preliminary activities similar to the following.

missing part — do & draw

- Put 7 same-coloured counters on a ten frame. Now put more counters in another colour to fill the ten frame. How many more counters did you need?

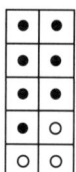

- Place 8 marbles in an egg carton. Now put more marbles to fill the egg carton. How many more marbles did you need?

- Put 9 big teddy counters in a row. Put out enough small teddy counters to make a row of 15. How many more teddy counters did you need?

When completing these activities children should be encouraged to focus attention on the parts that make up the whole rather than simply count to find the answer. Asking questions such as 'What do you have to find?', 'What do you already know?' is helpful in promoting the understanding that these problems also provide the total amount and one of the parts; the other part can be found as an application of subtraction.

In the past these examples were often recorded in an inappropriate form that caused children much distress as well as confusion. When these were written 3 + □ = 7 children were able to see two numbers connected by an addition symbol together with a place to put the answer, so were very surprised when their answer of 3 + 10 = 7 was rejected. The earlier meanings and interpretations of the symbols conflict with this new use to indicate a missing amount. The horizontal recording gave rise to the name 'missing addend' that is often used in place of the more meaningful missing part which draws attention to the way the part/part/whole thinking can provide an answer. It is now recognised that it is inappropriate to record subtraction in a form that incorporates an addition symbol. Rather, it is necessary to link the understanding of the situations and the manner in which the answers are obtained to the usual vertical form for subtraction. Great care needs to be exercised in this and no harm is done if the recording is delayed for some time.

Subtraction as comparison

When the takeaway and missing addend concepts are secure and the part/part/whole model can be readily used, subtraction as comparison can be introduced as an application of this knowledge. To this end, it is helpful to introduce the notion to children as a two-step process, building on their earlier knowledge of comparison in numeration:

1 Which number is greater/lesser? Use number understanding to compare.
2 How much greater/lesser? Use subtraction to determine the difference.

Again, children need to realise that subtraction is the required operation and the use of materials and language is crucial in helping children make this interpretation. Initially, this comparison can be made using counters to represent each quantity to make clear the answer to the first question, 'Which number is greater/lesser?'. While the answer to the second question, 'How much greater/ lesser?', could be found by pairing off the objects and counting the number left unpaired, children need to be led to use subtraction to determine this difference.

$$\begin{array}{r} 8 \\ -5 \\ \hline 3 \end{array}$$

comparisons

How many more are apples than pears?

Comparison problems often contain terminology children can find both foreign and misleading. If the meaning of comparison language is not grasped in the early years, children will experience difficulty in solving these subtraction problems later. Furthermore, teachers in the upper school usually assume that children are familiar with this language and hence overlook the fundamental source of the difficulty.

Likely difficulties

Difficulties with the concept usually occur because of a focus on the symbols and answers rather than the meaning associated with having a whole and taking away a part. Consequently, a child may be unable to give a story to match a picture or symbols, may rush to give an answer rather than read the addition, and may not be able to show the situation using materials. There may also be confusion between addition and subtraction.

- When asked to read an example such as $\begin{array}{r} 12 \\ -7 \\ \hline \end{array}$ a child may not be able to read the expression with meaning, reading upwards to say '7 from 12' or saying '12 minus 7'. This may show no understanding of subtraction other than as a command to provide an answer.
- A child may be 'answer focused' and simply give the answer '5' when asked to read the subtraction or give '12 takeaway 7 is 5' when asked to give a story to match the symbols.
- Children may interpret a problem involving missing part or comparison forms of subtraction as addition because of the word 'more' and the manner in which these problems ask for 'how many'.
- When asked to use materials to show subtraction, the child may simply provide objects to match the answer rather than the whole and parts involved in the story or written expression.
- When asked to provide an subtraction story to match a picture, a child may simply provide a story which describes the situation but does not involve subtraction.

Investigate and discuss

1. Use this *missing part* problem to outline the sequence you would follow in leading a child from using materials to recording the subtraction that is involved.

   Margaret needs to dig 14 holes to plant the shrubs she has bought. So far she has dug six of them. How many more holes does she need to dig? How would you justify the recording

   $$\begin{array}{r} 14 \\ -6 \\ \hline 8 \end{array}$$

2. Discuss any problems which might arise if this were to be recorded as $6 + \square = 14$.

Learning the subtraction basic facts

Subtraction basic facts are those simple number combinations using the numbers 0–9 needed for calculations with larger numbers. Since subtraction facts build on known addition facts, they include numbers to 18, not just to ten:

$$\begin{array}{ccccccccc} 3 & 6 & 10 & 12 & 7 & 14 & 16 & 17 & & 11 \\ -0 & -5 & -2 & -3 & -7 & -6 & -8 & -9 & \text{or} & -4 \end{array}$$

Ready access to these facts is critical for the smooth development of the algorithms for subtraction of larger numbers, including estimation and mental calculation. Knowing that they can readily access these basic facts is also crucial in building children's confidence in their ability to learn and use subtraction processes.

When children have developed the subtraction concepts, including an ability to write and interpret subtraction statements, they can begin to learn basic subtraction facts. As with addition, an approach that provides a strategy for obtaining facts before introducing practice enables them to be developed in a meaningful way as well as making them easier for children to acquire. In contrast, if efficient strategies are not taught, children may resort to counting on from the smaller number, using their fingers to keep track as the differences become larger. Alternatively, a method of counting back from the larger number may be used, again using fingers to keep track. Use of these procedures is often revealed through answers that are 1 too large when the counting includes the first number rather than only those that are taken away. For example:

10 takeaway 5 may be counted 5, 6, 7, 8, 9, 10 and the answer given as 6 rather than 5
14 takeaway 5 may be counted 14, 13, 12, 11, 10 to give as 10 rather than 9

If these inefficient ways of obtaining answers are to be avoided, there must first be a sound grasp of subtraction in terms of missing part and the part/part/whole model. The facts can then be introduced in clusters making use of meaningful language—'13 takeaway 9'; '9 and how many are 13?'—to ensure understanding and make the link to existing addition knowledge. Practice will be required to obtain automatic recall of the facts and children will also need experience in distinguishing between addition and subtraction.

Think of addition

Facts that subtract 1 are usually available as soon as subtraction is met as they simply ask for the number one less or immediately before a given number. Subtracting 2 may also be obtained in this way, but care must be taken that children do not then make an inappropriate generalisation to counting back as a means of obtaining other subtraction facts. Instead, once subtracting 1 and some facts involving subtracting 2 are known, they need to be related to the corresponding addition facts so that this link can be focused on explicitly to develop the one strategy for obtaining all subtraction facts.

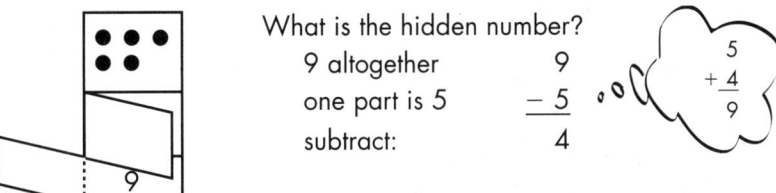

In order to extend this thinking to the other facts, it is helpful first to review each cluster of addition facts associated with a particular thinking strategy, *counting on 3 more, using doubles* and *make to ten*. The associated subtraction facts can then be built up using activities that call on children to select or recollect the corresponding addition fact to provide the answer.

What is the hidden number?
9 altogether
one part is 5
subtract:

$$\begin{array}{r}9\\-5\\\hline 4\end{array}$$

because $\begin{array}{r}5\\+4\\\hline 9\end{array}$

A number fact wheel provides one of the most effective ways of building this link and bringing the subtraction facts to the point of accurate, automatic response:

A basic fact wheel to develop the relationship between addition and subtraction facts is constructed by inserting a wheel with 10 equal divisions (to accommodate the digits 0–9 on the addition side) between a double-sided cover of the form indicated in the following illustration.

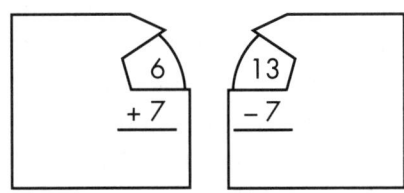

For this basic fact wheel based on adding and subtracting 7, first have one child nominate a number between 0 and 9, for example 6, then write this on the space on the wheel on the addition side. Since the answer to 6 and 7 is 13, this answer can be written on the space on the other side of the wheel. This subtraction side now reads 13 takeaway 7 and the answer 6 is found on the other side of the wheel. The addition fact has been used to generate the answer to the subtraction fact.

By repeating this for the remaining digits between 0 and 9 chosen in some random order, the activity brings out strongly the link between the addition and subtraction facts with 7. A similar activity can occur for each of the families of facts, using 1, 2, 3, 4, 5, 6, 7, 8 and 9. If the class is formed into groups to work with different facts, for example one group uses facts with 9, another facts with 6, another facts with 4, the general strategy can be emphasised over simply trying to memorise a particular set of facts.

Subtraction involving zero

Children often experience difficulties with zeros in subtraction. A similar use of materials to that used in introducing the thinking strategy for addition has proven beneficial in coping with subtraction of zero and subtraction of the same number to leave zero as the answer.

There were 8 marbles under the tin
0 have been removed
8 remain under the tin

$$\begin{array}{r} 8 \\ -0 \\ \hline 8 \end{array}$$

There were 7 marbles under the tin
7 have been removed
No marbles remain under the tin

$$\begin{array}{r} 7 \\ -7 \\ \hline 0 \end{array}$$

Automatic and accurate facility with basic addition and subtraction facts needs to be achieved by every child. Knowing the addition facts first is one key to this; another is to give continuing short periods of review of the way in which answers are obtained. Timed practice of basic facts reflecting the strategies by which they are learned, but also mixed among strategies and operations, should be an integral part of the mathematics program. Games and related activities are one aspect of this, but should be left until thinking strategies are available to obtain answers so as to avoid children guessing when they strive to win each turn. It may take some time for every child to be able to give accurate responses automatically, but further development of the algorithms for exact calculations, mental computation and estimation as well as practical applications demand this ability. Given appropriate thinking strategies, teaching and continuing practice, all children can be successful. On the other hand, if they are allowed simply to rely on counting or other inefficient means of finding answers, they will never feel compelled to look for more effective methods. In the same way, if a teacher rewards a correct answer regardless of the time taken or the method used to determine it, children will be unlikely to exchange inefficient ways of thinking for more efficient ones.

Learning any basic facts always involves two distinct, though related, steps:

1 to internalise effective thinking strategies
2 only then to use practice to obtain automatic responses with these facts

Too often teachers have been content with only the first step and children have suffered both in terms of a capacity to do further mathematics and in terms of a lack of self-confidence on which such further learning and applications depend so crucially.

Likely difficulties

Difficulties with the basic facts usually occur because a child lacks any strategies to obtain facts efficiently. Simply counting back using fingers, crossing out lines on a page or tapping may be the only means the child has to try to obtain answers. If there is confusion with addition, the concepts for both operations need to be revisited, particularly the part/part/whole notion which allows them to be discriminated.

- A child may guess or count on his or her fingers to obtain answers to all subtraction facts rather than use more appropriate thinking.
- When counting is used, answers are often one too large because the child counts back from the first number rather than only those that are taken away; for example:

    $$\begin{array}{cc} 12 & 9 \\ -5 & -7 \end{array}$$

    twelve takeaway five: **12**, 11, 10, 9, **8**   nine takeaway seven: **9**, 8, 7, 6, 5, 4, **3**

- Particular difficulties may occur with zero where a child thinks that any subtraction involving zero must result in zero, for example that 11 takeaway 0 is 0. There may also be confusion between zero and one, where 11 takeaway 0 is given as 10.
- A child may know how to work out answers rather than know facts to be able to give accurate answers automatically.

Investigate and discuss

1 It is often said that since addition and subtraction are inverse operations, they should be taught together.

   a At what point in the development of addition and subtraction would you introduce the subtraction concept?

   b When would you commence the learning of subtraction facts?

   c What is the role of 'families of facts' such as

   $$\begin{array}{cccc} 5 & 14 & 9 & 14 \\ +9 & -5 & +5 & -9 \\ \hline 14 & 9 & 14 & 5 \end{array}$$

2 Make sets of subtraction cards and number fact wheels as shown on p.235 and use them with children at about Year 2 or 3. Which facts would you use to initiate thinking that links needed subtraction facts to known addition facts?

3 How could you use counters and ten frames to extend the thinking involved in the initial concept to link subtraction facts to the corresponding addition facts?

Developing the subtraction algorithm

Just as for addition, subtraction with two-digit and larger numbers is essentially a matter of combining numeration understanding and basic fact knowledge. If the ones, tens and hundreds are considered separately, the subtraction of each part simply requires a basic fact, although renaming is often needed before this can be done. The ones are subtracted first, then the tens, following the order established with addition. However, the need to rename 1 ten as 10 ones is not considered initially to avoid confusing the renaming that occurs with addition with that used for subtraction. Subtraction of larger numbers is also introduced with materials, but their use is quite different from the way they have been used for addition. Only one number is made and the other number is simply removed place by place. Consequently, the first step in the development of the subtraction algorithm uses examples with no renaming to focus on the different thinking that is required. The new form of renaming needed with subtraction can then be introduced when the process is secure and addition is sufficiently well understood for it to be unlikely that children will confuse the two operations.

> **SEQUENCE FOR DEVELOPING THE SUBTRACTION ALGORITHM**
>
> **1** Once the concept and basic facts are secure, this thinking is extended to deal with tens.
> **2** Subtraction with no renaming is dealt with first to introduce the form of the algorithm and the particular use of materials, language and recording.
> **3** Subtraction with renaming is introduced later using materials to integrate this numeration understanding into the algorithm.
> **4** The process is broadened to include subtraction with internal zeros.

Subtraction without renaming

The subtraction facts have been introduced and written using a vertical recording, so the only new thinking needed to extend

$$\begin{array}{r} 9 \\ -6 \\ \hline \end{array} \quad \text{to} \quad \begin{array}{r} 89 \\ -56 \\ \hline \end{array} \quad \text{and} \quad \begin{array}{r} 139 \\ -56 \\ \hline \end{array}$$

is an ability to deal with the tens. Facility with the subtraction basic facts is readily extended to the 'tens facts' to enable *thinking in tens*, but practice similar to that given to build up basic facts will be needed. Later practice should be given to a mix of basic facts and tens facts since the processing of the algorithm will demand the flexibility of moving from one way of thinking to the other.

In contrast to addition, the use of materials for subtraction requires that only the top number is made; the bottom number is taken away. The answer is then obvious, for it is the number that is left, but the subtraction situation disappears when materials are removed. Children need to be guided in the use of materials to reflect this process and allow a meaningful recording to be developed. Bundling sticks are best used first because they most readily link to an understanding of two-digit numbers and will be needed when the renaming that occurs with most subtraction is developed. A place value chart with an arrow at the head of the ones place will ensure children subtract the ones first while the subtraction should be read from top to bottom in accord with the way the materials are manipulated.

tens	ones		
	↓	6 7	67;
🟫🟫🟫🟫🟫	‖ ‖‖‖‖‖	− 2 5	can you takeaway 25?

[handwritten note: a place value chart can be made with popsicle sticks]

The first introduction to subtracting two-digit numbers should be with materials only, so that the thinking behind the process is focused on rather than the symbolic recording that could lead to a mechanical interpretation of the subtraction algorithm. For this reason, the language 'can you takeaway?' is used from the outset so as to lay the foundation for subtraction with renaming where an answer no will suggest that 1 ten needs to be renamed as 10 ones, rather than inappropriately find the difference between the two digits. Care needs to be taken that children do not generalise from their familiarity with addition to make both numbers. Rather, they need to be led to make a given number on the chart, then to remove an amount representing the number that is taken away. One way to achieve a smooth introduction of this process is to ask the children to make the first number, then to follow quickly with 'Now takeaway . . . What do you takeaway first?'. This should ensure that the children focus on the ones place first as well as not leaving enough time for them to begin to put out the second number.

Language	Materials	Recording
Make the top number **Subtract the ones** 7 ones, can you takeaway 5 ones? **Yes** 2 ones	tens | ones	6 7 −2 5 2
Subtract the tens 6 tens can you takeaway 2 tens? **Yes** 4 tens	tens | ones	6 7 −2 5 4 2
	67 takeaway 25 is 42	

The recording for the algorithm should be introduced gradually, matching the symbols to the manipulation of the materials, but the example used needs to be written from the outset in order to indicate the two numbers involved when only one is made with the materials. At first, as the subtraction is carried out, the teacher should record the number of ones remaining, then the number of tens. When this is familiar, children can make numbers to match written examples, then takeaway the ones and tens to form the answer and record this in the written algorithm. Only then should children be asked to complete the algorithm using symbols alone.

When this thinking and recording is secure, Base 10 materials can be used to consolidate the process and extend it to subtraction of three-digit numbers without renaming. This only requires use of simple 'hundreds facts' and removal of Base 10 hundreds whereas moving to the subtraction of two-digit numbers with renaming at this stage would introduce a seemingly complicated process just when confidence with subtraction was being established. In contrast, the consideration of three-digit numbers without renaming is a straightforward step that helps build children's self-confidence by letting them easily cope with what seems to be quite large numbers.

Subtraction with renaming

Learning numeration built up the understanding that 1 ten is 10 ones to allow 74 to be renamed as 6 tens 14 ones. This thinking can now be combined with the initial algorithm to extend subtraction to two-digit numbers where renaming is needed before subtraction can proceed. The use of Bundling sticks on a place value chart is the key to extending the process and giving meaning and direction to the steps involved. They also serve to link the renaming required for subtraction to the earlier treatment of renaming in numeration. A language that describes the way materials are used follows on from that used earlier in introducing the subtraction algorithm and grows naturally out of unbundling 1 ten to give 10 single sticks. In turn, this language provides meaning and control over the recorded and mental algorithms that are the end-point of this development.

The major difficulty children encounter is in continuing to subtract the bottom number from the top number when at first sight it cannot be done. Consequently, consideration needs to be given to the language used and to a form of recording that will minimise these difficulties. If subtraction is read as takeaway for each place, an example like

renaming

$$\begin{array}{r} 7\,4 \\ -3\,9 \\ \hline \end{array}$$

will give rise to '4 takeaway 9'. When children see that this cannot be done, they may be tempted to resolve this impasse by reversing the situation to '9 takeaway 4' rather than to consider the need to rename 1 ten to supply enough ones to allow the subtraction to continue. Providing a cue to children that they need to consider whether subtraction can simply proceed or whether it will first be necessary to rename will minimise this difficulty. Using the question 'Can you takeaway?' at each step of the algorithm has proven a successful verbal cue. The first step in this subtraction is then '4 ones, can you takeaway 9 ones?'—the answer no induces a need to rename. At the next step, '6 tens, can you takeaway 3 tens?', the answer is simply yes and the subtraction of this step does not need renaming.

Since the intention is to provide a consistent and smooth flow from the introduction to the final algorithms, this consideration also needs to extend to the way children manipulate materials. When 1 ten is unbundled to give 10 single sticks, these must be placed in the ones place. If they are spread over the arrow on the place value chart, they can be readily combined with the number of ones in the original number by drawing on a familiarity with teen numbers rather than laboriously counting all of the sticks.

As there are now more than 10 ones in the ones place, to be consistent with the development of recording as an expression of what the materials show, this new number of ones should be written full sized in the ones place, while the reduced number of tens should be recorded full sized in the tens place. This is also in accord with previous expectations in numeration and addition that all numbers are written full sized. These changed numbers are recorded at the top of the ones and tens places in a similar way to the process developed earlier for addition to simply rename the top number in a manner that lets subtraction proceed—7 tens 4 ones is renamed as 6 tens 14 ones.

As children complete subtraction examples, their recording is usually written in pencil or ink of a different colour from that used to present the practice exercises. This draws attention to the way the original example has changed as the tens are renamed to allow the subtraction to proceed. Indeed, rarely is the number that is subtracted from actually the same as the original. This distinction is also helpful in initial teaching if one colour is used to record the required subtraction and another to record what is done in completing it, whether at the board or on the overhead projector. The table below shows the language and recording for two-digit subtraction with renaming and their links to the use of materials:

Language	Materials	Recording
Show the top number **Subtract the ones** 4 ones, can you takeaway 9 ones? **No: rename** 1 ten as 10 ones—6 tens 14 ones 14 ones, can you takeaway 9 ones? **Yes** 5 ones	tens | ones	6 14 7̶ 4̶ − 3 9 5
Subtract the tens 6 tens, can you takeaway 3 tens? **Yes** 3 tens	tens | ones	6 14 7̶ 4̶ − 3 9 3 5

74 takeaway 39 is 35

When subtraction with two-digit numbers has been established, Base 10 materials can be used to review the process and extend it to subtracting three-digit numbers. The first step in this process is simply to combine earlier three-digit subtraction with this renaming of 1 ten to give 10 ones. Then the new step of renaming 1 hundred to give 10 tens can be built in before subtraction involving renaming of both 1 ten and 1 hundred is considered.

```
   7 15        7 15         5 15       5 14 12
   8 5̶         6 8 5̶        6̶ 5 8       6̶ 5̶ 2̶
 −  4 6      − 2 4 6      − 2 6 4      − 2 6 4
   ────       ─────        ─────        ─────
   3 9        4 3 9        3 9 4        3 8 8
```

The following table shows the language and recording for subtraction with three-digit numbers and their links to the use of Base 10 materials:

Language	Materials	Recording
Show the top number **Subtract the ones** 2 ones, can you takeaway 4 ones? **No: rename** 1 ten as 10 ones—4 tens 12 ones 12 ones, can you takeaway 4 ones? **Yes** 8 ones	hundreds | tens | ones	4 12 6 5̶ 2̶ − 2 6 4 8

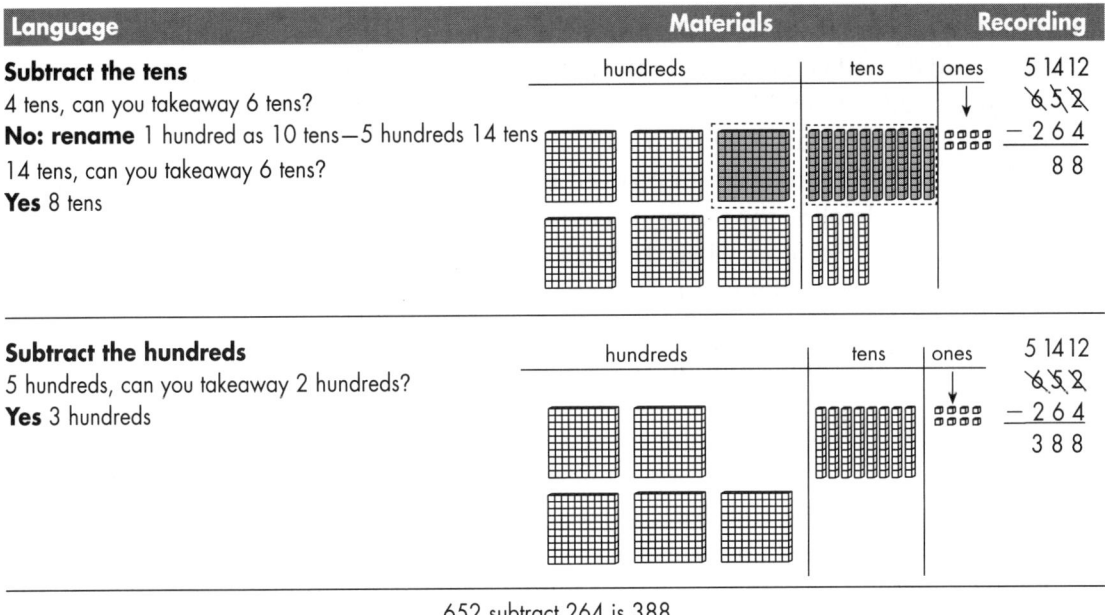

652 subtract 264 is 388

Subtraction with larger numbers is a straightforward continuation of the same processes used for two-digit and three-digit numbers; each place is considered as a separate entity involving the simple application of a basic fact, although a renaming step is often required. Materials can still be used to give meaning and ensure a smooth extension to four-digit subtraction, but their use should be minimal. Rather, the language and the recording, which is consistent with the use of materials, should be emphasised, building the understanding that 1 thousand is 10 hundreds onto the earlier 1 ten is 10 ones, 1 hundred is 10 tens.

There are no materials to use for numbers with five or more digits. Some children may encounter difficulty in recording the results of renaming neatly, but in practice most use this recording to complete one step and clear their head of that part of the calculation so as to be ready for the next step rather than use the recording directly to obtain the result for each step. The following recorded forms minimise this difficulty:

```
  5 111613          4 11 17 2 14         7 14 16   4 17
    6 2 7 3           5 2 7 3 4          8 5 6 4 5 7
  - 4 3 7 9         - 2 4 9 2 5        - 2 8 7 4 4 8
    1 8 9 4           2 7 8 0 9          5 6 9 0 0 9
```

The numbers resulting from the renaming are recorded in full size in the appropriate place above the other tens, hundreds, thousands and so on. Note that it is sufficient to write the 1 alongside the ten, hundred or whatever that has already been written above the original number to give the total number of tens, hundreds or thousands rather than to cross this out and write the numbers on another line above.

Subtraction with internal zeros

When the initial number has a zero in the tens place, subtraction of three-digit numbers can cause particular difficulties. Materials have proven invaluable in building up subtraction skills involving renaming for all other two-digit and three-digit numbers. Whenever there are insufficient ones for subtraction to proceed immediately, 1 ten is renamed to provide an additional 10 ones. In all of the examples considered so far there has always been 1 ten in the tens place to rename. However, when there is a zero in the tens place, materials confuse rather than help this further development. They suggest a two-step process in which 1 hundred is first renamed as 10 tens, and then 1 of these tens is renamed as 10 ones. This requires a renaming from left to right before any of the subtracting is begun. All the subtraction prior to this has used a single-step process, treating each place in turn from right to left. Clearly it would be preferable to use a single-step process consistent with the other examples in this case as well, subtracting ones, then tens, then hundreds, renaming 1 ten for 10 ones or 1 hundred for 10 tens immediately it is required.

In fact, a double renaming process is not necessary, for the language pattern that has been established earlier can be combined with children's knowledge of numeration to apply the same processing to the internal zero cases. A number expander can be used to draw attention to the total number of tens and so extend the same renaming of 1 ten for 10 ones used earlier.

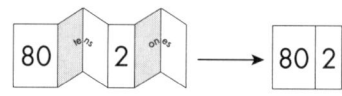

802 is 80 tens 2 ones

Language	Recording
Subtract the ones 2 ones, can you takeaway 7 ones? **No: rename** 1 ten as 10 ones 80 tens 2 ones is 79 tens 12 ones 12 ones can you takeaway 7 ones? **Yes** 5 ones	7 9 1 2 8̶ 0̶ 2̶ − 4 5 7 5
Subtract the tens 9 tens, can you takeaway 5 tens? **Yes** 4 tens	7 9 1 2 8̶ 0̶ 2̶ − 4 5 7 4 5
Subtract the hundreds 7 hundreds, can you takeaway 4 hundreds? **Yes** 3 hundreds	7 9 1 2 8̶ 0̶ 2̶ − 4 5 7 3 4 5
802 takeaway 457 is 345	

The process for subtracting four-digit and larger numbers is identical; when there are not enough ones, the total number of tens that the number has is considered. Since 6003 has 600 tens 3 ones, it can be renamed as 599 tens 13 ones.

$$\begin{array}{r} 5\,9\,9\,13 \\ \cancel{6\,0\,0\,3} \\ -\,3\,4\,1\,8 \\ \hline 2\,5\,8\,5 \end{array}$$

Again, a number expander may be helpful in showing how many tens, hundreds or whatever the number has, but it must be borne in mind that all that is required for the successful subtraction of numbers with internal zeros is a good understanding of renaming from numeration.

Likely difficulties

Difficulties with the algorithm usually have their origins with inappropriate renaming, particular misconceptions with zero, or confusion with the recording of their working because of an inability to use place value. There may also be errors due to incorrect basic facts but these need to be addressed separately from the algorithm. Similarly, if there is confusion with the procedure used for another operation, this might mean that the concepts need revisiting or it may suggest that a rote procedure has been used rather than a meaningful process as built up here.

◆ *Place value*: When numbers are used from a problem, places may not be aligned

$$\begin{array}{r} 7\,6\,5 \\ -\,5\,2 \\ \hline 2\,4\,5 \end{array}$$

◆ *Renaming*: Children may simply find the *difference* rather than rename

$$\begin{array}{r} 8\,2 \\ -\,3\,9 \\ \hline 5\,7 \end{array}$$

When an *internal zero* occurs, children may simply rename from the largest place each time:

$$\begin{array}{r} 5\,0\,2 \\ -\,2\,3\,8 \\ \hline 1\,7\,4 \end{array}$$

◆ *Zeros*: These difficulties may relate to uncertainties with the concept of zero, with the concept for the operation or with renaming

```
   60          78         512
 - 3 5       - 4 0       - 2 6 3
   3 5         3 7         3 6 9
```

- *Recording:* Numbers may be confused when the recording does not place digits in the correct places. Use of quad-ruled paper or a place value chart is helpful with children who misread their recording in this way.
- *Confusion with addition:* Children may add even when the examples are written with the subtraction symbol, either because they have recently been doing addition and simply continue the same process, or because they are unsure of subtraction and 'just do sums' that are easier for them.

Investigate and discuss

1. Answer each of these examples using the recording and language outlined in this section:

```
  63      824     702     420      58     961     815
- 58    - 369   - 453   - 268    - 25   - 654   - 532
```

   a How could you use Base 10 materials and a place value chart to provide insight into the recording of each step?
   b Why would materials not be helpful for the example with 702?
   c Why might it be preferable to use Bundling sticks for the two-digit examples?
   d In which order would you introduce these subtraction examples to children?

2. Answer each of these examples using the recording and language outlined in this section:

```
  7005      5017     6107     63 005     50 012
- 3689    - 3589   - 2838   - 39 416   - 37 380
```

What difficulties might children encounter in extending subtraction to these larger numbers?

Estimation with subtraction

Facility with the written subtraction algorithm is important in providing meaning for the operation, particularly the need to consider the initial amount and then remove a quantity from it. However, exact answers are not always needed and in many situations an approximation will suffice. An estimate is sufficient to check whether the change given in shopping is about what it should be or to see whether there will be money remaining to make another purchase. Subtraction involving time and measurements with length or mass will also more often need to be approximate rather than exact.

For many people, estimation is easy and the processes they use reflect a well-internalised understanding of subtraction and of the numbers involved. For others, the only technique is to try to treat any calculation in the same manner as the written process. The number of steps that need to be mentally manipulated and remembered then becomes too much. Powerlessness, frustration and a feeling of inadequacy often result in those people who are unable to carry out the seemingly simple tasks that others can do so readily.

The *National Statement on Mathematics for Australian Schools* (1991) stresses the need for estimation to be 'an ongoing part of children's study of numbers' with an emphasis on 'the development of a propensity to estimate'. Seeing a need for estimation is first built up through measurement experiences and extended to the computational context with addition. This needs to be continued with subtraction, involving real contexts as well as computation on its own. Particular strategies that permit computational estimations can be developed as soon as the concepts for subtraction and the meaning for the processes built up with the use of materials and reflection on the recorded process are secure. Rounding will again play an important part and an understanding of place value and renaming are also crucial.

While a front-end strategy has served to introduce the notion of computational estimation and provide a means of estimating the result of adding several numbers, there are difficulties in extending its use to subtraction. Often it would suggest an inappropriate estimate, sometimes significantly smaller, other times larger:

```
front-end  786        actually   768        front-end  602        actually   602
           - 405                 - 405                 - 292                 - 292
about       300                   363       about      400                   310
```

On the other hand, rounding provides a reasonable estimate for both situations:

```
    786   rounds to    800              602   rounds to    600
  - 405              - 400            - 292              - 300
          about       400                     about       300
```

Since there will only be two numbers to round and then subtract, this is easily managed mentally, so that all subtraction estimates can be made using a strategy based on rounding.

Nonetheless, there will be other ways of estimating subtraction and children are likely to seek out shortcuts of their own following their experiences in estimating with addition. For example, an estimate for subtraction is often determined by adding in a similar way to making change: 7689 − 6321 can be considered to be close to 7700 − 6300 and this can be determined by thinking 400 makes 6700, another 1000 to 7700, the difference is about 1400. Estimating will now involve both addition and subtraction and there will be particular methods related to seeing how a quantity in one part of the calculation will be cancelled by a subsequent subtraction, or a subtraction at one point will be cancelled by an addition elsewhere. It is important for a teacher to encourage children to talk about and share techniques such as these as well as develop strategies for single operations. A willingness to make approximate calculations requires methods that are meaningful to and constructed by each person to suit his or her own ways of thinking but exposure to the techniques others use will also help each person to adopt and adapt new ways of thinking.

Subtracting numbers mentally

While approximate answers may often be more important than precise ones given by the written algorithm or by a calculator, there may still be a need for exact answers calculated quickly and on the spot. This is best achieved by an easily managed mental process that relies on different ways of thinking from those associated with the written algorithm. Usually this will involve left to right processing similar to the way estimations are made. By focusing on the largest place first, the initial step will get close to the answer and each subsequent calculation then closes in on what is wanted with no need to remember anything other than the result of each adjustment. Experiences in giving or receiving change suggest how to subtract mentally, building up to the required number rather than simply subtracting place by place.

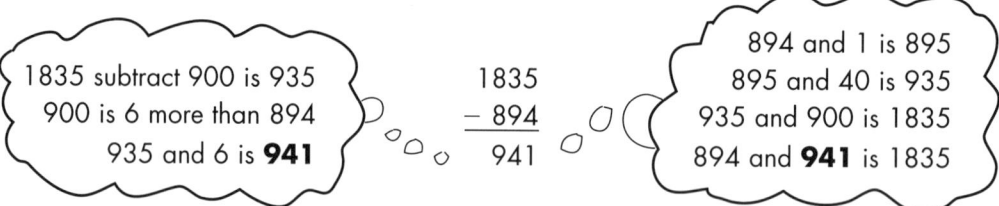

There will be many different ways in which answers can be determined mentally. Some will be personally constructed by individual children to suit particular sets of numbers whereas other children will use thinking strategies very similar to those used earlier in developing thinking that underpin both exact calculations and estimations. A range of methods adapted to suit particular numbers and situations will be needed for proficiency, but mental computation also needs to be viewed as a way of thinking that applies to all operations rather than a collection of separate strategies for each operation in isolation. Just as multiplication provides many of the techniques for addition, so subtraction can be used to make mental addition a simpler process. For instance, the nearest whole multiples can be added first then adjusted to take account of the additional numbers involved by subtraction:

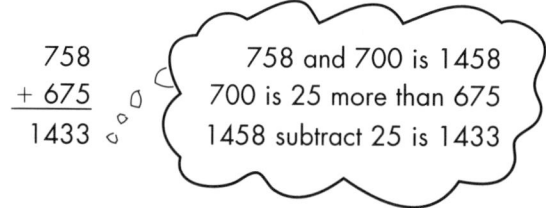

A similar use of subtraction and addition will also apply to multiplication:

$$\begin{array}{r} 94 \\ \times\ 58 \\ \hline 5452 \end{array}$$

94 by 60 is 5640
94 by 2 is 188
5640 subtract 200 is 5440
5440 and 12 is 5452

Mental arithmetic is the most used calculation in everyday situations. Children need to be confident in their ability to calculate mentally, to develop and use strategies that fit their strengths and to adopt and adapt those used by others. Discussing the merits, ease of use and degree of control a range of strategies allows is important in building up an inclination to choose to use mental computation whenever it is appropriate.

Using calculators in subtraction

Using a calculator is not as easy it may appear. Yet it can provide many benefits in freeing up thinking processes to focus on the problem being addressed, allow many answers to be obtained more quickly, and enable a series of calculations to be done much more efficiently. In many ways, use of a calculator is another form of mental computation; the calculator only shows one step at a time and quickly erases all that went before so that an overall mental picture of what is occurring is needed for the user to keep control of the steps and the answers that are given. A good understanding of the subtraction concepts, the properties of the numbers used and the manner in which a calculator can be utilised to carry out calculations more efficiently is essential. In particular, being able to use the various memory options—a calculator's built-in but limited memory, the constant functions and the use of pencil and paper to record important results that the calculator is quick to erase—is important to effective, ready use. Forming an overview of what is required in the calculation before using a calculator will often show shortcuts even before data or operations are keyed in. While a calculator may be of most use when several operations are required, it is also helpful when there are several subtractions involved as it can readily work from one result to the next in the string of computations. Mental and written methods usually require each subtraction to be considered on its own and be completed before the succeeding one can be begun.

The use of a calculator for subtraction is as straightforward as for addition. Once the concept is understood so that the order of keying numbers is apparent, use of the subtraction symbol to do the subtraction and use of the equals symbol to obtain the result are known, children are ready to use a calculator. Numbers and operation symbols are keyed in a similar manner to addition and calculator use can follow a good understanding of the various subtraction concepts. Proficiency with the basic facts is also essential to see whether answers make sense. Calculator use can accompany the development of the written algorithm, but if begun too quickly might inhibit some children coming to terms with the steps involved and ways in which the concept and basic facts are integrated with numeration ideas to develop a coherent process. The use of games as outlined in the addition section will again be a helpful way of introducing calculators as is their use alongside estimation. Children can be asked first to estimate an answer then compare it with that given by a calculator or to use estimation to assess the reasonableness of calculator answers, justifying the results to other members of the class.

When children are using calculators for subtraction, it is also likely that they will be involved with problems or calculations that require both addition and subtraction. A calculator can be used to give practice in focusing on the operation that is needed in a particular problem, where the task would be to identify the operation before doing any calculation, then use the calculator to obtain the answer using the chosen operation to see if the choice was correct. This may raise the possibility of what to do if the wrong operation is keyed in. Many people simply erase the calculation and begin again, but children need to come to terms with the idea that if a wrong operation is entered it can be overridden by pressing the appropriate operation key before any other numbers are entered. Similarly, a child may realise that he or she has added a given number rather than subtracted it, and needs to see that subtracting the number twice will get the calculation back to where it should be. Discussion of how these sorts of errors might arise and how they can be corrected will not only develop a good and competent facility with calculators but also assist in consolidating understanding of the concepts, allow children to distinguish between the operations, and promote their ability to apply operations to problem situations.

Investigate and discuss

1 There are many games that can be used to introduce subtraction with a calculator. Play this game with a partner using only one calculator.
   a Key in 100 on the calculator to begin. Then each player in turn presses – and any one number from 1 to 9 and =. The player who makes 0 appear in the display is the winner.
   b Make the game described on p. 219 and play it with some children in middle primary school. Use subtraction with two-digit and three-digit numbers for the cards. Would it be possible to play this game at an early stage of development where the children use materials to answer the questions?
   c Find some other calculator games that you might use to introduce subtraction on a calculator.

2 A calculator is also useful in allowing children to see the relationship between addition and subtraction.

a Complete the following exercises using a calculator:

```
   8003        .........      9 657          751
 - .........   - 6 782        .........      379
   4017        13 893       + 4 792        ......
                             20 537        + 826
                                            2815
```

b Comment on the operations required to complete each answer and the notions that might be introduced and reinforced with this type of question.

Subtraction with decimal fractions and common fractions

A thorough understanding of numeration, the renaming associated with the number expander in particular, and automatic access to the basic facts have been the key to subtracting whole numbers. Subtraction with decimal fractions and common fractions relies on a similar understanding together with a careful transition of the earlier skills to the new types of numbers. An ability to rename 1 one as 10 tenths, 1 tenth as 10 hundredths and so on is the only new thinking required for decimal fractions, while common fractions demand that 1 one be renamed as 4 fourths, 8 eighths and so on and that 2 and 4 fifths also be renamed as 14 fifths. A focus on a meaningful language for these fractions from the outset, together with the language developed for subtraction, allows a ready translation of the facility and understanding for whole number subtraction to each of these calculations. The only difference is that no longer are the ones subtracted first; subtraction begins with the place furthest to the right, be it tenths, hundredths or thousandths with decimal fractions, or simply the fraction part for common fractions. Use of an arrow to indicate which place to begin continues for fraction ideas, minimising possible confusion and avoiding a fixation on the ones place.

Subtraction with ones and tenths can begin alongside the development of addition for these decimal fractions. The language and recording used relates to the whole number algorithm and is similar to that established earlier for addition with ones and tenths. When hundredths have also been learned, particularly the understanding that 1 tenth is 10 hundredths, a similar extension can be made to subtracting numbers with ones, tenths and hundredths. Since materials cannot be used in the way they were for whole numbers, this means that a solid understanding of the representation for ones and tenths to provide a visualisation of the process is essential. This will allow a language based on ones and tenths to be developed and used to guide the recording in a similar manner to that for whole numbers:

Language	Recording
Subtract the hundredths 3 hundredths, can you takeaway 4 hundredths? **No: rename** 1 tenth as 10 hundredths 90 tenths 3 hundredths is 89 tenths 13 hundredths 13 hundredths, can you takeaway 4 hundredths? **Yes** 9 hundredths	8 9 13 9̶.̶0̶3̶ − 6.7 4 9
Subtract the tenths 9 tenths, can you takeaway 7 tenths? **Yes** 2 tenths	8 9 13 9̶.̶0̶3̶ − 6.7 4 2 9
Subtract the ones 8 ones, can you takeaway 6 ones? **Yes** 2 ones	8 9 13 9̶.̶0̶3̶ − 6.7 4 2.2 9

9 and 3 hundredths subtract 6 and 74 hundredths is 2 and 29 hundredths

Developing the subtraction algorithm to include further decimal fractions is a relatively straightforward process that can accompany the introduction of each new decimal place. Note that this depends on the numeration for decimal fractions, renaming in particular, and does not relate in any way to the size of the whole number portion of the numbers being subtracted. To subtract 578.9 from 3026.8 is no harder than subtracting 8.9 from 26.8 when whole number subtraction has been learnt, and this, of course, is a prerequisite for the extension of subtraction to fraction ideas.

Subtract the thousandths
4 thousandths, can you takeaway 8 thousandths?
No: rename 1 hundredth as 10 thousandths
50 hundredths 4 thousandths is 49 hundredths 14 thousandths
14 thousandths takeaway 8 thousandths is 6 thousandths

      615 14 9 14
      7̶6̶.̶5̶0̶4̶
    − 1 9 . 6 3 8
      5 6 . 8 6 6

Subtract the hundredths
9 hundredths takeaway 3 hundredths is 6 hundredths . . . and so on

Children may experience difficulty subtracting numbers with differing decimal places. Just as was suggested for addition, the key is to stress aligning the places, ones with ones, tenths with tenths, hundredths with hundredths and so on rather than talk of lining up the decimal point. This stress on place value in fact ensures that the decimal points are in line, but emphasises this in a meaningful way rather than through a rote rule.

Once subtraction with decimal fractions has been introduced at least as far as tenths and hundredths and numeration for common fractions is well understood, the subtraction algorithm can be extended to include common fractions. If the initial examples simply involve tenths, the thinking, language and recording can be seen to be identical, but just as for addition with common fractions the use of materials is no longer helpful. A very good internalisation of the common fraction concept and its recording is a necessary prerequisite for subtraction with common fractions. It is for this reason that the development of subtraction and addition with common fractions does not occur for some time after the introduction to the common fraction concept, recording and numeration ideas.

Decimal fractions	Language	Common fractions
$\begin{array}{r} 8.15 \\ \cancel{9}.\cancel{5} \\ -\,6.7 \\ \hline 8 \end{array}$	**Subtract the tenths** 5 tenths, can you takeaway 7 tenths? **No: rename** 1 one as 10 tenths 8 ones 15 tenths 15 tenths, can you takeaway 7 tenths **Yes** 8 tenths	$\begin{array}{r} 8\frac{15}{10} \\ \cancel{9}\cancel{\tfrac{5}{10}} \\ -\,6\tfrac{7}{10} \\ \hline \tfrac{8}{10} \end{array}$
$\begin{array}{r} 8.15 \\ \cancel{9}.\cancel{5} \\ -\,6.7 \\ \hline 2.8 \end{array}$	**Subtract the ones** 8 ones, can you takeaway 6 ones? **Yes** 2 ones	$\begin{array}{r} 8\frac{15}{10} \\ \cancel{9}\cancel{\tfrac{5}{10}} \\ -\,6\tfrac{7}{10} \\ \hline 2\tfrac{8}{10} \end{array}$
	9 and 5 tenths subtract 6 and 7 tenths is 2 and 8 tenths	

Once the recording of subtraction with common fractions has been established for tenths, it is a very easy matter to extend this to other like fractions. At first, examples would involve only situations where there is no renaming to familiarise learners with both the recording and reading of this different-looking form of computation, but renaming can be begun immediately the underlying numeration knowledge is secure.

Language	Recording
Subtract the fractions 3 eighths, can you takeaway 5 eighths? **No: rename** 1 one as 8 eighths, 5 and 11 eighths 13 eighths, can you takeaway 5 eighths? **Yes** 7 eighths	$\begin{array}{r} 5\frac{11}{8} \\ \cancel{6}\cancel{\tfrac{3}{8}} \\ -\,4\tfrac{5}{8} \\ \hline \tfrac{6}{8} \end{array}$
Subtract the ones 5 ones, can you takeaway 4 ones? **Yes** 1 one	$\begin{array}{r} 5\frac{11}{8} \\ \cancel{6}\cancel{\tfrac{3}{8}} \\ -\,4\tfrac{5}{8} \\ \hline 1\tfrac{6}{8} \end{array}$
6 and 3 eighths subtract 4 and 5 eighths is 1 and 6 eighths	

Subtracting like common fractions is a very simple extension of the subtraction algorithm when recorded vertically. The use of a similar meaningful language together with a few fundamental numeration ideas allows children to build up this thinking confidently and reduces the anxiety so often associated with subtraction of common fractions. However, extending subtraction to unlike common fractions (those with different denominators) requires new understandings related to equivalent fractions and is usually delayed until children feel secure with fraction ideas and with subtraction of simple common fractions late in primary school.

Subtraction and addition with unlike common fractions

Addition and subtraction of common fractions with different denominators is considerably more complex than the operations with like common fractions. For this reason, their development is usually delayed until late primary or even secondary school and it makes sense to teach them alongside each other to draw out the similarities in thinking that they involve. When the operations are attempted, the same problems arise as occurred with the comparison of unlike fractions. The parts are not the same portions of 1 one and so cannot interact directly. For comparison, it was possible to draw on the region model to see the relative sizes and this in turn led to the use of equivalence whereby common fractions were renamed to be described in terms of the same-sized part and compared on the basis of the number of these parts. Similarly, only when the parts are the same size can it make sense to combine or remove them. However, the direct use of fraction models for the operations can be confusing so use must be made of the concept of equivalence from the outset and a process for renaming common fractions is a necessary prerequisite for adding and subtracting unlike common fractions.

The basic notion of equivalence is introduced along with the concept of common fractions through the use of the region model and activities involving shading a series of rectangles representing a particular common fraction. In this way, a given fractional amount can be seen to have a variety of fraction names. However, any reason for doing this is not obvious to young children. They have been accustomed to each number having a unique name and do not readily see why it should be important to be able to find several names for the same quantity. For this reason it is best to leave any further development of the concept of equivalence until a need for it can be generated. Since its major use is to allow the addition and subtraction of unlike common fractions it is only at this point that it makes sense to establish a formal renaming process.

SEQUENCE FOR DEVELOPING ADDITION AND SUBTRACTION WITH COMMON FRACTIONS

1 Add and subtract like common fractions.
2 Establish a process for renaming common fractions as equivalent fractions.
3 Add and subtract unlike common fractions.

Addition and subtraction with like common fractions is relatively simple. There may be a need for renaming corresponding to that used for whole numbers, but basic fact knowledge allows these parts to be joined or taken away directly. Certain other fractions are almost as simple and suggest the way to proceed in general.

$$1\tfrac{2}{4}$$
$$3\tfrac{\cancel{1}\,2}{\cancel{2}\,4}$$
$$+\,4\tfrac{3}{4}$$
$$\overline{\,8\tfrac{1}{4}\,}$$

Can't add or subtract halves and fourths but 1 half is the same as 2 fourths

Changing to the same fraction makes adding or subtracting simple!

$$7\tfrac{5}{4}$$
$$8\tfrac{\cancel{1}}{4}$$
$$-\,5\tfrac{\cancel{1}\,2}{\cancel{2}\,4}$$
$$\overline{\,2\tfrac{3}{4}\,}$$

Examples involving halves and tenths or fifths and tenths tend to be familiar for children and can be used to build up the expectation that if common fractions can be renamed as like fractions, adding and subtracting them is simply a matter of applying the algorithm that has already been learnt. This understanding reveals two things. First, that the addition and subtraction of common fractions is really not difficult at all; it is only the methods used in the past that have caused any problems. Second, it highlights what really needs to be known. Common fractions with different denominators cannot be added or subtracted as such. They need to be brought to the same denominators and then dealt with as any like fractions. The fundamental thinking needed to add and subtract common fractions is an ability to *rename* them to like fractions.

Renaming common fractions as equivalent fractions

While the concept of equivalent fractions can be called upon to suggest the power of renaming for computation, of itself it will not be enough. Equivalence only suggests that fractions can have a variety of names without indicating how appropriate representations can be found. In order to allow addition and subtraction to proceed easily and smoothly, a process for renaming any common fraction is needed. The key to this is to use the patterns that can be revealed through subdividing a rectangular region to obtain successive names for the same shaded area:

Rename fractions:

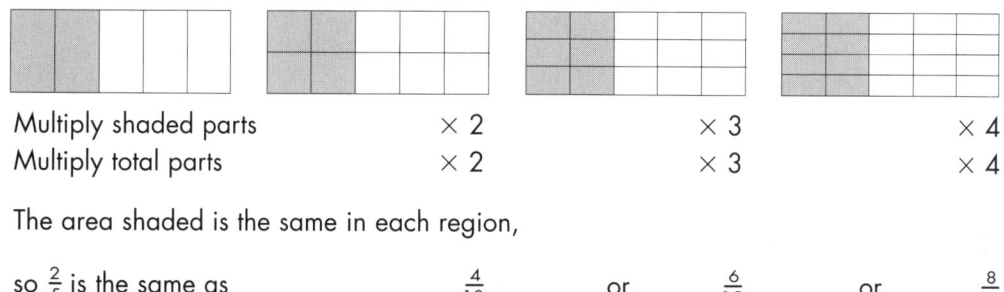

Multiply shaded parts   × 2   × 3   × 4
Multiply total parts   × 2   × 3   × 4

The area shaded is the same in each region,

so $\tfrac{2}{5}$ is the same as $\tfrac{4}{10}$ or $\tfrac{6}{15}$ or $\tfrac{8}{20}$

This can be summarised as:

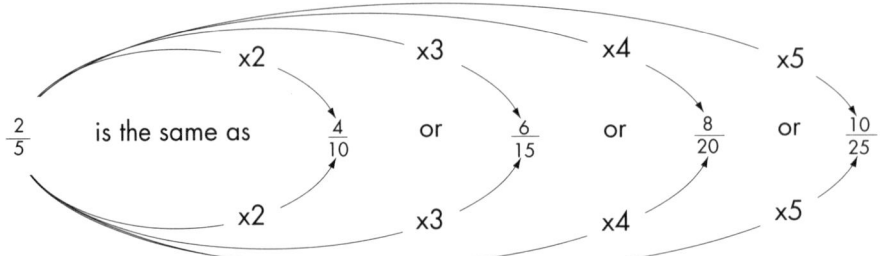

It is this pattern of multiplying the initial fraction by the factor by which the number of parts has increased that is the key to a formal process. If sufficient examples are given in a concrete form, this can be generalised to a method that generates equivalent fractions in turn; both parts of the initial fraction are multiplied successively by 2, 3, 4, 5 and so on. Finding the next equivalent fraction, rather than any fraction that names the same amount, is the key to allowing addition and subtraction of unlike fractions to proceed.

The use of a working space where the renaming of the unlike fractions can be systematically carried out is also helpful. Then, with a focus on renaming as the first step, all that is needed once the common fractions have been changed to equivalent forms is to use the earlier processes for like fractions:

Language	Recording	Working column
Subtract the fractions 2 thirds, can you takeaway 3 fourths? **No** They are not like fractions. **Rename** to be like fractions.	$12\frac{2}{3}\frac{8}{12}$ $-5\frac{3}{4}\frac{9}{12}$	$\frac{2}{3}=\frac{4}{6}=\frac{6}{9}=\frac{8}{12}$ $\frac{3}{4}=\frac{6}{8}=\frac{9}{12}=\frac{12}{16}$
8 twelfths, can you takeaway 9 twelfths?		
No: rename 1 one as 12 twelfths 20 twelfths, can you takeaway 9 twelfths? **Yes** 11 twelfths	$11\frac{20}{12}$ $12\frac{2}{3}\frac{8}{12}$ $-5\frac{3}{4}\frac{9}{12}$ $6\frac{11}{12}$	
Subtract the ones 11 ones, can you takeaway 5 ones? **Yes** 6 ones		
12 and 2 thirds subtract 5 and 3 fourths is 6 and 11 twelfths		

It is easiest for children simply to learn to write four or more equivalent fractions in turn, then to look for those that will allow both common fractions to be renamed as like fractions. The subtraction, or addition, is then simply a matter of applying the thinking and numeration understandings established earlier. However, there will sometimes be situations where only one of the common fractions needs to be renamed. These are in many ways more difficult than the case where both common fractions are renamed as it is easy to begin renaming the wrong one:

$$\begin{array}{r} 8\frac{5}{6}\overset{1}{} \\ + 9\frac{\cancel{1}}{3}\frac{2}{6} \\ \hline 18\frac{1}{6} \end{array}$$

Can you add 5 sixths and 1 third?
No. They are not like fractions
Rename 5 sixths to be the same: $\quad \frac{5}{6} = \frac{10}{12} = \frac{15}{18} = ...$
This will not give thirds

Rename 1 third to be the same:
5 sixths and 2 sixths is 7 sixths; 1 sixth, 1 one $\quad \frac{1}{3} = \frac{2}{6} = \frac{3}{9} = \frac{4}{12}$
9 ones and 9 ones is 18 ones

The thinking involved in completing these addition and subtraction examples not only highlights the need for a good understanding of equivalent fractions and the renaming process that allows them to be determined, but also suggests an appropriate sequence for building up computation and the process for renaming common fractions. Leaving the development of equivalence of common fractions until the impasse reached with addition and subtraction of unlike fractions is met provides a reason for renaming common fractions as well as delaying the introduction until children's mathematical understanding is more mature. In this way, not only do children obtain meaningful and usable algorithms for addition and subtraction of all common fractions, they also build up a complete and integrated understanding of fraction ideas.

Investigate and discuss

1 Answer each of these examples using the recording and language outlined in this section:

$$\begin{array}{r} 9.6 \\ -4.8 \end{array} \quad \begin{array}{r} 8.14 \\ -3.65 \end{array} \quad \begin{array}{r} 7.02 \\ -2.46 \end{array} \quad \begin{array}{r} 15.2 \\ -9.37 \end{array} \quad \begin{array}{r} 8\frac{3}{5} \\ -4\frac{4}{5} \end{array} \quad \begin{array}{r} 14\frac{3}{8} \\ -6\frac{7}{8} \end{array} \quad \begin{array}{r} 20\frac{5}{12} \\ -15\frac{8}{12} \end{array}$$

a How does the cycle of steps discussed in the overview to this chapter:

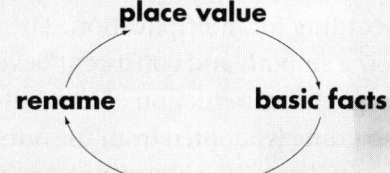

apply to the subtraction of these decimal fractions and common fractions?

b What new aspects of number understanding are required at each step in the subtraction cycle?

2. Solve the following problem and determine how you would assist a child to come terms with what is required.

   At Justine's birthday party, there were 57 party pies and 31 spring rolls on the table. During the party, 17 spring rolls and 37 party pies were eaten. How many party pies and spring rolls were left at the end of the party?

   Why might some children simply subtract the numbers in each sentence, while others add the numbers in each sentence and then subtract, and others simply add all of the numbers? Consider the general model for problem solving given in Chapter 2 as well as the place of knowledge of the addition and subtraction concepts and the part/part/whole model that both defines and distinguishes addition and subtraction.

3. Many people learnt to subtract common fractions in a horizontal fashion

   $$8\tfrac{2}{3} - 2\tfrac{3}{4} =$$

   a Find out how this was done and contrast it with the method introduced here.
   b Is the same fundamental cycle of steps still being followed?
   c Can you suggest any reasons for adopting a horizontal recording over a vertical one?

Section III: Overview to multiplication

MULTIPLICATION is more abstract than addition and subtraction, building on experiences with repeated addition rather than situations that children encounter in their daily lives. It can be related to real situations but care needs to be taken that it is not confused with addition. Groups of objects are used to show multiplication situations, but soon after the initial introduction of the concept the focus needs to be on arrays, where objects are formed into rows, to provide a representation for multiplication independently of addition. The use of arrays also lays the groundwork for learning basic facts, providing a link to earlier numeration and addition understandings to generate the thinking strategies that readily provide answers. Using materials to model problem situations with arrays also gives rise to a language that underpins a meaningful recording for multiplication. The steps leading to this written form need to be organised to allow a smooth and consistent development from the initial concept through learning basic facts to the introduction of the written algorithm for larger numbers. For this reason, a vertical recording is adopted from the outset and multiplication is read from bottom to top to match the way the written algorithm is performed. Since multiplication with larger numbers does not allow the use of materials to model what occurs to the same extent as addition and subtraction, the role of language is paramount and crucial. In particular, this language needs to draw on the underlying place value and renaming ideas to supply direction and meaning for the many steps involved.

SEQUENCE FOR DEVELOPING MULTIPLICATION				
1	Develop the multiplication concept:	repeated addition arrays to model these situations vertical recording, × symbol		
2	Build up the basic facts:	develop thinking strategies ensure automatic responses		
3	Thinking in tens	tens × ones tens	and	ones × tens tens
4	2 digit × 1 digit with renaming			
5	2 digit × tens using	ones × tens tens	and	tens × tens hundreds
6	2 digit × 2 digit combining the earlier steps			
7	Multiplication with larger numbers			
8	Strategies for estimation, mental computation and calculator use			
9	Multiplication with decimal fractions			

The multiplication concepts

Since multiplication initially builds on addition, rather than growing out of problem situations, its introduction should not occur before the addition concept is secure and numbers can be readily added. Activities involving materials in equal-sized groups will have been met well before the formal introduction of multiplication and these can provide a basis for the first conception of the operation in terms of *repeated addition*. This initial concept gives rise to a meaningful language for multiplication which can be used in the development of all forms of the concept, the basic facts and the algorithms for large numbers. For example:

Four books in each pile   4
3 piles   4
12 books altogether   + 4
                        3 fours are 12

Exposure to a range of similar examples leads naturally to this language in terms of the number of groups and the number in each group. Thus, 4 fives means 4 groups each with five objects, 5 threes means 5 groups of three objects, 2 eights is 2 groups of eight objects, and so on.

Unfortunately, making groups may also encourage counting to determine the total, yet repeatedly adding is not only impractical when larger numbers are involved, it also hinders the application of multiplication to problem situations. In particular, since both addition and multiplication problems often involve the expression 'how many altogether?' children may decide whether to add or multiply based on the size and frequency of the numbers rather than an analysis of the underlying situation. Not that grouping should be avoided, nor a repeated addition interpretation discouraged. In the early stages, this method is the only means children will have to find the total amount. However, a prolonged application of it may promote inefficient strategies for later work with multiplication facts:

9 fours?   4 and 4 is 8 and 4 is 12 and 4 is . . .

Further, a young child meeting multiplication solely through *grouping* activities is likely to view 4 threes differently to 3 fours, even though in each representation the total amount, 12, is the same.

4 threes   3 fours

When the multiplication facts are met, such thinking would require 100 facts to be learned, whereas seeing the equality of the two statements will halve the learning. The difficulty lies with the way the materials have been arranged. In contrast, if this is changed to rows of objects:

4 threes   3 fours

each arrangement can be seen to include the other; 3 fours and 4 threes are the same. Such an arrangement of objects in rows is called an array and the use of *arrays* is fundamental to developing multiplication as an operation in its own right independently of addition. As arrays not only facilitate the process of determining the product but also enable children to discriminate between multiplication and addition, they should take precedence over simply forming groups as soon practicable.

An array could of course be described in terms of columns instead of, or as well as, rows but column is neither a word nor a concept that children find easy. Rows have usually been part of life, schooling at least, for some years and it is much easier from the child's point of view to focus on just one way of describing what is being put out. The emphasis on rows will also be invaluable in the development of the multiplication algorithm, and in due course with division, particularly in making the link between multiplication and division that is fundamental to each step of the division algorithm.

As symbols are introduced to match these introductory activities, the emphasis will again be on a vertical recording from the outset. Largely this is to facilitate the movement to the written algorithm for larger numbers so attention will need to be given to this final form to guide the language and the way it is to be read. While addition and subtraction have been completed downwards, place by place, multiplication has traditionally been completed from bottom to top; for example:

```
                                        2 6
                                        5 7
Multiply by 9 ones                     × 4 9
  9 sevens are 63; 3 ones 6 tens       -----
  9 by 5 tens is 45 tens and 6 tens is 51 tens    5 1 3


                                        2 6
                                        5 7
                                       × 4 9
Multiply by 4 tens                     -----
  4 tens by 7 ones is 28 tens; no ones 8 tens 2 hundreds    5 1 3
  4 tens by 5 tens is 20 hundreds and 2 hundreds is 22 hundreds    2 2 8 0
Add                                                        -------
                                                            2 7 9 3
```

The reason for this difference is to draw attention to the different way the steps are completed—the ones multiply with ones, then multiply with tens; the tens multiply with ones as well as with tens. This crossing of places gave rise to the use of × for the multiplication symbol in order to highlight the different thinking that is needed. Reading and completing multiplication from bottom to top provides a further cue to the different way of processing. Thus, both the way the initial concept and basic facts are read should also be from the bottom to the top:

$$\begin{array}{c} 4 \\ \times\,3 \end{array} \text{ is 3 fours } \quad \text{and} \quad \text{4 threes is } \begin{array}{c} 3 \\ \times\,4 \end{array}$$

The language used to describe the way materials are used needs to be strongly linked to the fundamental ideas of multiplication so that the recording can represent a meaningful process, rather than a mechanical one. It is for these reasons that the commonly used language for multiplication 'times' is not advocated here. When there are 3 groups of four, this can of course be described as four, 3 times. But why this gives way to a reading '3 times four' is not at all obvious and is certainly not explicable to children at the time they first meet multiplication. Further, almost no one uses an expression '9 times 7' when completing an algorithm like that above; instead almost everyone says '9 sevens' as their first step. There seems to be little point in introducing a language at the beginning of multiplication learning if it is not to be used in later developments.

Nonetheless, times is a term that needs to be introduced at some stage so that children can understand and interpret multiplicative tasks that they might meet in other contexts, but even then the more formal language of multiply, product and factors and the informal language of real situations such as 'how many altogether?' will be of more importance. Similarly, the use of the expression 'lots of' is not encouraged as it has little connection with multiplication at all and, in fact, is more likely to be used to describe a situation in which there is to be no attempt to determine how many objects there are, such as 'there are lots of hundreds and thousands in the jar'. Indeed, children today rarely use the expression 'lots of' at all—they use 'heaps of'. It would sound as strange for adults to use heaps to describe multiplication as it does for children to use lots. Describing a multiplication situation in terms of lots or heaps may even suggest to children that it is too much to determine the answer at all.

SEQUENCE FOR THE DEVELOPMENT OF THE MULTIPLICATION CONCEPT

1 Introduce multiplication through activities that build on *repeated addition*.

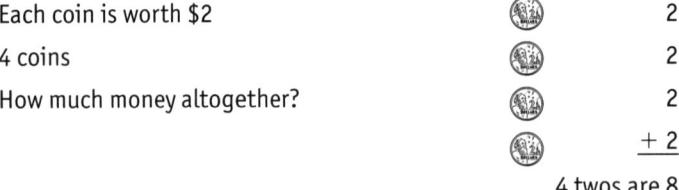

Each coin is worth $2
4 coins
How much money altogether?

$$\begin{array}{r} 2 \\ 2 \\ 2 \\ +\,2 \\ \hline \end{array}$$
4 twos are 8

Relate this initial repeated addition conception to *grouping* objects into equal-sized groups.

4 marbles in each bag
2 bags
How many marbles altogether?

2 fours are 8

2 Multiplication needs to focus on objects in *arrays* to establish it as an operation independent of addition.

4 objects in each row
3 rows
3 fours are 12

3 Gradually introduce symbols to represent this process, reading from bottom to top:

12 objects altogether   4 threes are 12

$$\begin{array}{r} 3 \\ \times\,4 \\ \hline 12 \end{array}$$

When the children are familiar with arrays, show how an array can be rotated to provide a matching fact:

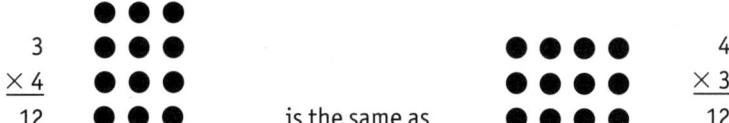

4 Extend the multiplication language to include multiplied by or by:

   3
  × 4
  ─────
  12   can be read as '4 multiplied by 3 is 12' or '4 by 3 is 12'

The use of 'times' is not helpful as it does not relate to multiplication in any meaningful way.

One fundamental difficulty that arises in the development of multiplication is that the concept is a generalisation from the earlier operation of addition rather than from the problem situations directly. In fact, any story that can be used to demonstrate multiplication can equally be said to show addition and it is the difficulty of this distinction that leads children often to confuse the two operations in problem solving. Consequently it is even more important that children quickly integrate the array conception into their fundamental notion of multiplication so that it can take on a meaning of its own, independent to that of addition, and that they learn how to relate this model to the problems and the way multiplication is symbolised. In order to build this facility, activities like the following can be given:

As facts are given, the children make the corresponding arrays. The important point to be gained is that the first number tells how many rows, the second the number in each row.

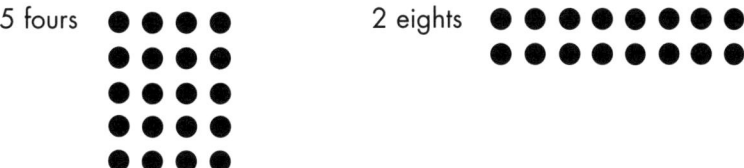

To make the activity more game-like and independent, two dice can be used, one with the digits 1–6, the other with the words, *twos, threes, fours, fives, sixes* and *eights*. One child from a group of 4–6 is selected to roll the dice. When the dice are rolled the other children make the array to match the roll of the dice. The first child to make the array rolls the dice for the next round.

THE ARRAYS GAME

Use of these dice with 1 cm grid paper makes a more competitive game. Children play in pairs, each with their own grid. Each player in turn rolls the dice and colours in the area on their grid indicated by the dice. The first player to colour all the squares on their grid wins.

  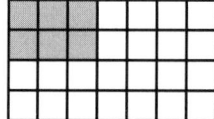

As play progresses, children will inevitably find there is not enough space to directly cover a particular throw and will be likely to suggest that '6 fours is the same as 2 fours and 4 fours'. In this way, the language is able to bring about an intuitive awareness of a very important principle of multiplication (formally, known as the distributive law).

At a later stage, players can also write the corresponding multiplication expression for each array on the coloured area.

  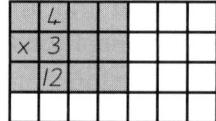

In addition to being able to form appropriate arrays unhesitatingly, a full ability with multiplication requires a good link between problem situations, the use of arrays and recording of the multiplication statements. In particular, this means that, given a multiplication statement, children should be able to:

◆ tell in their own words what it means
◆ make an array to illustrate it
◆ obtain and record the answer

It is also important that children develop an intuitive awareness of the fact that

$$\begin{array}{cc} 5 & 4 \\ \times\ 4 & \quad\text{and}\quad \times\ 5 \end{array}$$

give the same result. This is readily shown with an array once there is understanding of how 4 fives is to be represented. If the array is made on a piece of paper, it can then easily be rotated to show the alternative interpretation, 5 fours. This does, of course, depend on a firm awareness of the meaning of the expressions 4 fives and 5 fours. Hence the significance of the activities above in linking the directionality of multiplication to the language and symbolic statements used for it. This awareness is known formally as the 'commutative property' but this, like the earlier distributive property, is not language that would be used with children.

Once again there really is no need for a horizontal recording since there is no link whatsoever between initial statements recorded horizontally and the vertical form that will be required for the algorithm. With multiplication, this has even more significance in view of the alternative meanings possible in matching the use of symbols to the use of models. There has been a tendency to read $3 \times 2 = \square$ as '3 times 2'. Would this be the same as 3 twos or 2 threes? What would it correspond to when recorded vertically? Nonetheless, there will be a time when the horizontal recording of computations is needed. In the first instance this will occur when there are several computations in the one problem, especially if more than one operation is involved. This recording may also arise in the context of computation using a calculator. However, it is not reasonable to expect this new and different recording and new ways of thinking at this stage and it is a situation that really does not arise until the upper primary school, when it should be treated explicitly as something very different involving a rethinking of the way in which the operations and their recording have been conceived.

Likely difficulties

Difficulties with the concept usually occur because of a focus on the symbols and answers rather than the meaning associated with arrays. Consequently, a child may be unable to give a story to match a picture or symbols, may rush to give an answer rather than read the multiplication, and may not be able to show the situation using materials.

- When asked to read an example such as $\begin{array}{r}6\\ \times\ 4\end{array}$ a child may not be able to read the expression with meaning, saying '6 times 4' instead. This may show no understanding of multiplication other than as a command to provide an answer.
- A child may read multiplication downwards and say '6 fours' which is likely to lead to difficulties with multiplication for larger numbers when they may proceed place by place as for addition.
- When asked to use materials to show multiplication, a child may simply provide objects to match the symbols rather than show how multiplication gives the required solution to the story or written expression. For example, a child asked to show the example above put out materials ● ● ● ● ● ●   ●   ● ● ● ●   ●
and explained his result   'six   times   four   equals   24'
- A child may not be able to provide a story to match a written expression or picture, or may give a story which simply uses the numbers with another operation, even though their answer matches the meaning of the symbols or picture. For example, a child was asked to read this expression and provide a multiplication story:

    $\begin{array}{r}5\\ \times\ 3\end{array}$ Read as '5 times 3'
       Story 'There were 5 people and 3 wanted ice-creams'
       Answer '25 . . . I just know'

Investigate and discuss

1. Play the arrays game described above with a friend. Comment on the way in which it builds up a picture of multiplication in terms of arrays and links to a language for multiplication that then gives meaning to the vertical recording format.
2. Outline how you would build on this experience to develop an intuitive awareness of the commutative property, that 3 fours is the same as 4 threes, 5 sevens is the same as 7 fives and so on.
3. The array model has important applications in other areas of primary mathematics. List any that you can think of and discuss their relationship to multiplication.
4. Do you agree that use of the term 'times' should be phased out of primary mathematics? Discuss reasons for and against such a point of view. Consider a similar set of arguments for the expression 'lots of'.
5. How would you make children aware that the operation itself is multiplication when the word multiply is little used in the development so far?

Learning the multiplication basic facts

As with addition, an approach that provides strategies for obtaining the multiplication facts before introducing practice to obtain automatic recall enables them to be developed in a meaningful way as well as making them easier for children to learn. In contrast, when thinking strategies are not taught, many children use methods based on counting or repeated addition, or simply attempt to memorise by rote all 100 multiplication facts. This has even been encouraged by the tradition of learning multiplication facts by the mindless repetition of 'tables', as if the learners were devoid of any ability to think for themselves. Not only was this a complete denial of any earlier learning in numeration or addition, which in reality provides a very good basis for constructing knowledge of the multiplication facts, it was also terribly inefficient. Never was it mentioned that each fact was repeated twice in reciting the tables, nor was the link between these two forms brought out as a basis for learning. In contrast, an approach using a visual array model makes explicit use of earlier learning and attitudes to learning, for the power of thinking strategies is something that strikes children and appeals to their wish to understand and have control over their own mathematical thinking.

Many multiplication facts are already known as a consequence of knowing addition facts and earlier numeration knowledge and merely need to be reinterpreted and extended by means of the array model. For example, the twos facts are really just a restatement of the addition doubles:

```
     6      ● ● ● ● ● ●      6
    ×2      ● ● ● ● ● ●     +6
    ──                      ──
    12       2 sixes are    12
```

Other facts can then be built onto this knowledge, using the array as a visual model to show the thinking that is used:

```
    6      ● ● ● ● ● ●
  × 3      ● ● ● ● ● ●    2 sixes are 12
           ○ ○ ○ ○ ○ ○              and  6
                          3 sixes are 18
```

Once the array conception of multiplication is secure, multiplication facts can be introduced in an order that allows one set of facts to be constructed readily from those already known, using arrays to guide the thinking that is required. While the thinking for multiplication at all levels is more abstract than that needed for the other operations, this use of arrays provides a link to earlier knowledge and the simple language of '3 sixes' helps to ensure a smooth build up of known facts. Practice to obtain automatic recall will again be required once the strategy for obtaining a given cluster of facts has been understood.

The development of strategies that allow meaningful basic fact learning also contributes to the thinking required for effective problem solving. Solutions to many problems are assisted by an ability to interpret what is needed in terms of an array or area model. Such problems require the translation of familiar knowledge, building it into novel processes in a similar manner to the way in which the multiplication facts build on earlier numeration understanding or addition fact knowledge. Thus, rather than being a tedious aspect of initial learning in multiplication, building a capacity to provide automatic, accurate answers to multiplication facts is an important contributor to the development of mathematical thinking in general.

It is also the case that of all aspects of elementary mathematics that give children a sense of security about their own abilities, 'multiplication tables' feature highest. Children who do not know these facts have the feeling that they know very little about mathematics and this thinking can very quickly be self-fulfilling given the crucial part self-confidence plays in learning and applying a subject like mathematics. Even worse, many children who do not know some of the multiplication facts generalise to believe that they do not know multiplication at all and even that they 'are no good at mathematics'.

One way of showing children that they have already acquired many multiplication facts is to ask facts at random and black out the entries that are known in a table or grid that shows all of the facts:

A child can then see that many or even most facts are known and there are only a few to be learned. These can then be acquired within a reasonably short time by reference to the array that has given access to the clusters of facts to which they belong. Often, a child will know a fact such as '4 nines' but be uncertain about '9 fours', even saying that the nines facts are difficult. Use of an array to show how both facts are related is crucial in remedying this thinking and giving automatic access to related facts. Teaching which allows children to construct their own knowledge of the multiplication facts using arrays has proven to be readily accessible to young children and to be a key element in developing both their problem solving ability and their self-confidence.

Using arrays

If those facts that are most accessible to children are introduced first, then success and self-confidence can be built up from the outset. On the surface, it would seem easiest to begin with the ones facts, for example 7 ones are 7 and 1 seven is 7, but children do not at first readily match these results with their conception of multiplication. Instead, facts that directly relate to earlier knowledge of addition doubles provide a more secure foundation. In turn, the knowledge of these can be used to build clusters of facts involving 4 and 3. The less obvious ones and zeros facts are best left until there is a secure basis of known facts to which they can be related.

Twos facts

Multiplication facts with 2 are largely a restatement of the addition doubles. Children can be asked to show a row of 5 counters in one colour, then make another row of 5 counters in another colour beneath these. This shows 2 fives so that:

$$\begin{array}{c} 5 \\ +5 \\ \hline 10 \end{array} \qquad \begin{array}{c} \bullet\bullet\bullet\bullet\bullet \\ \circ\circ\circ\circ\circ \\ \text{2 fives are} \end{array} \qquad \begin{array}{c} 5 \\ \times 2 \\ \hline 10 \end{array}$$

The other doubles can be shown in the same way to allow children to see that their knowledge of addition simply needs to be reinterpreted using multiplication language to give many of the twos facts. Understanding that rotating the array keeps the result the same but shows a different fact—since 2 fives are 10, 5 twos are also 10—generates the remaining multiplication facts with two. The learning is a matter of linking multiplication to what is already known, rather than tediously learning a 'new' set of facts as if there was no prior knowledge.

Fours facts

Multiplication facts with 4 are built up by using 2 of the twos facts. Children can be asked to show 2 sixes with counters in one colour, then make another 2 sixes using counters in another colour beneath these. Each of these shows 2 sixes are 12, so altogether there are 4 sixes are 12 and 12, or 24:

```
                    ● ● ● ● ● ●
2 sixes are 12     ● ● ● ● ● ●          6
2 sixes are 12     ○ ○ ○ ○ ○ ○         × 4
         24        ○ ○ ○ ○ ○ ○         ──
                     4 sixes are        24
```

A similar use of arrays shows how the other fours facts—4 threes, 4 fours, 4 fives, 4 sevens, 4 eights and 4 nines—also build on known twos facts. When these are known, the related facts—3 fours, 5 fours, 6 fours, 7 fours, 8 fours and 9 fours—can be built up by rotating the arrays.

Threes facts

Multiplication facts with 3 also build on the twos facts. After the fours facts have been learned, children can readily use the same thinking to show 2 fives with counters in one colour, then another row of five counters in another colour so that 3 fives are 10 and 5, or 15:

```
                     ● ● ● ● ●
2 fives are 10      ● ● ● ● ●           5
       and  5      ○ ○ ○ ○ ○          × 3
          ──                            ──
          15        3 fives are         15
```

A similar use of arrays shows how the other threes facts—3 threes, 3 sixes, 3 sevens, 3 eights and 3 nines—also build on known twos facts. When these are known, the related facts—5 threes, 6 threes, 7 threes, 8 threes and 9 threes—can be built up by rotating the arrays.

Fives facts

The fives facts also relate to earlier mathematics, in particular learning to read a clock to 5 minute intervals. If there has been an emphasis on digital reading from the outset, children have learnt to associate the hands on a clock with the number of minutes after the hour:

The minute hand is on 3   The minute hand is on 7
3 fives are 15   7 fives are 35

This knowledge can then be used as a visual cue to relate the number of fives to the relative position on the clock face. Arrays can be used to consolidate this thinking and to extend the knowledge to the associated facts; for example:

In contrast, the statement often made that 'the fives are easy, they end in zero or five' is not really helpful as it is the number of tens that causes children difficulty and saying this does not mention what the number 'starts with'. Neither is counting '5, 10, 15, . . .' recommended as it ends up as another finger-counting exercise when children try to remember how many fives they have counted.

Sixes facts

Multiplication facts with 6 build on the fives facts in the same way that the threes facts built on the twos facts. After the fives facts have been learned, children can readily use this thinking to show 5 eights with counters in one colour, then another row of eight counters in another colour:

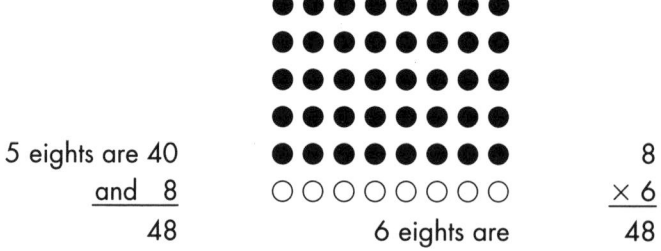

Sevens and eights facts

The only sevens and eights facts remaining to be learned are 8 eights, 7 sevens, 8 sevens and 7 eights. These have traditionally been viewed as the most difficult to learn, yet they can be readily generated from known facts using arrays in the same way as for the earlier facts. Children can be asked to show 4 eights with counters in one colour, then make another 4 eights with counters in another colour beneath these. This shows 8 eights so that:

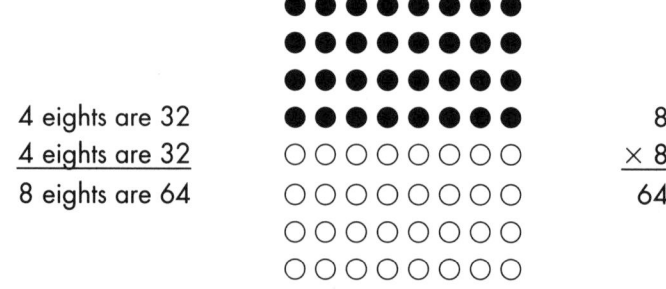

Children can then be asked to make 7 sevens and explore ways in which the answer can be generated:

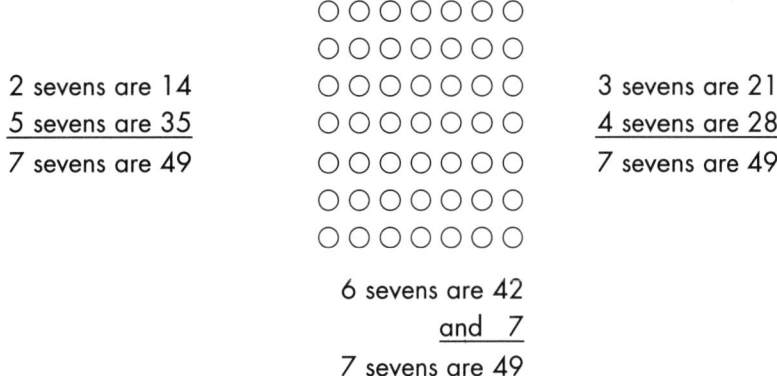

2 sevens are 14
5 sevens are 35
7 sevens are 49

3 sevens are 21
4 sevens are 28
7 sevens are 49

6 sevens are 42
and   7
7 sevens are 49

Adding seven to 7 sevens gives the last fact, 8 sevens, which has traditionally been among the most difficult for children to learn, yet, seen as an addition to a known fact, it is really quite easy to generate the correct answer:

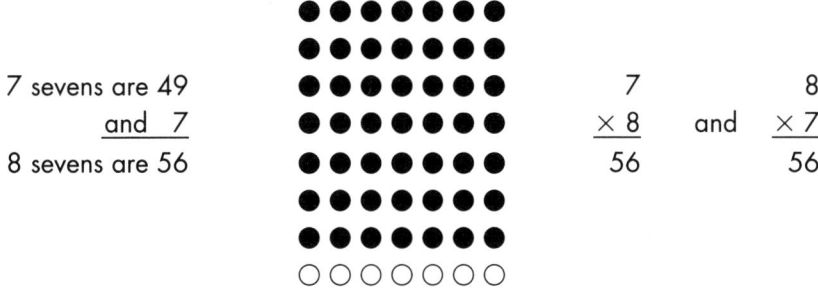

7 sevens are 49
and   7
8 sevens are 56

$$\begin{array}{r} 7 \\ \times\ 8 \\ \hline 56 \end{array}$$ and $$\begin{array}{r} 8 \\ \times\ 7 \\ \hline 56 \end{array}$$

Nines facts

By this stage in the development of the multiplication facts, many of the nines facts have already been met—2 nines and 9 twos, 3 nines and 9 threes, 4 nines and 9 fours, 5 nines and 9 fives. The remaining facts can be determined by using the digit pattern they exhibit to provide an effective thinking strategy for all of the multiplication facts with 9.

First, look at the pattern for the tens digits:

$$\begin{array}{r} 9 \\ \times\ 6 \\ \hline 5\ldots \end{array}$$ 9 is close to 10, so 6 nines must be close to 6 tens
there must be 5 tens

$$\begin{array}{r} 9 \\ \times\ 7 \\ \hline 6\ldots \end{array}$$ 9 is close to 10, so 7 nines must be close to 7 tens
there must be 6 tens

Then look at the pattern for the ones digits:

9	9	9	9	9	9	9	9
×2	×3	×4	×5	×6	×7	×8	×9
18	27	36	45	5...	6...	7...	8...
1 and 8 is 9	2 and 7 is 9	3 and 6 is 9	4 and 5 is 9		the digits sum to 9!		

Put the two patterns together:

```
    6    Answer is close to 6 tens, must have 5 tens
  × 9    5 and 4 are 9
   54    6 nines are 54 (and 9 sixes are 54)

    7    Answer is close to 7 tens, must have 6 tens
  × 9    6 and 3 are 9
   63    7 nines are 63 (and 9 sevens are 63)

    8    Answer is close to 8 tens, must have 7 tens
  × 9    7 and 2 are 9
   72    8 nines are 72 (and 9 eights are 72)

    9    Answer is close to 9 tens, must have 8 tens
  × 9    8 and 1 are 9
   81    9 nines are 81
```

The ones and zero facts

These are straightforward but tend to cause difficulties for children if introduced too soon. After all, why write anything when the answer is unchanged or is always zero? Any tendency of teachers to say that 'multiplication makes things bigger' will also create difficulties as multiplying by one leaves numbers the same while multiplying by zero gives zero, which for many children is as small as you can get. Indeed, zero has often been overlooked when the multiplication facts were introduced and children provide answers such as '6 zeros are 6' because of a belief that there ought to be some answer while zero is frequently viewed as nothing. However, it is comparatively straightforward to introduce multiplication with one and zero after the other facts have been learned as the meaning for multiplication, its language and recording will have been fully internalised:

```
    9                        1
  × 1   1 nine is 9        × 9   9 ones are 9 (obviously; it's what 9 ones is!)
```

When 1 is a factor, the product is always the other factor.

```
  9                          0
× 0   No nines are zero    × 9   9 zeros are 0 (there are no arrays with zero)
```

When 0 is a factor, the product is zero.

Using arrays in this way to provide a picture of what is involved enables them to be built up readily by most children soon after the concept has been understood. The only other thinking that is needed is an ability to add on to larger numbers mentally, particularly the extension of basic addition facts. For example, knowing 9 and 7 is 16 leads to 49 and 7 is 56, so that seeing that 8 sevens are 7 sevens and 7 then gives 8 sevens 56. This extended ability with basic facts is essential for adding several numbers and should be introduced some time before the multiplication facts are met. Although the strategies for learning multiplication facts involve a certain amount of new learning, it is based on earlier understanding central to addition. Further, this thinking also prepares for an essential aspect of the multiplication algorithm where the renaming step demands this type of addition. For example, when multiplying 9 by 48, the first step gives 72 while the second step asks that 36 tens and 7 tens be added. Thus, at the same time as providing an essential basis for multiplication for larger numbers, this thinking helps in establishing mental computation skills for both addition and multiplication.

In summary, learning the multiplication basic facts requires several distinct phases.

LEARNING THE MULTIPLICATION FACTS

All of the multiplication facts are readily built using the array concept for multiplication. Facts are learned in clusters that use identical thinking and need to be brought to the point of automatic, accurate answers to provide a secure basis on which to build the next set of related facts. Each cluster also includes the matching facts, for example linking 3 eights and 8 threes to the same thinking.

1. **Arrays** are used to introduce the thinking that most readily allows the learning of a particular cluster of facts.
2. This way of thinking is internalised before the **matching facts** are built up using the same thinking.
3. Some facts such as the *fives* and *nines* build on thinking well known from earlier learning of time and numeration. The ways of thinking for these facts should be established first, then the matching facts can be learned by reference to arrays or to the pattern that the arrays show.
4. Only after one cluster of facts has been learned to the point of automatic, accurate answers should the next set of facts be introduced. It is then necessary to ensure that the thinking for one cluster is not confused with the strategy for another and that there is an ability to provide immediate accurate answers for all facts.

The thinking used to develop the multiplication facts is shown on this number fact grid:

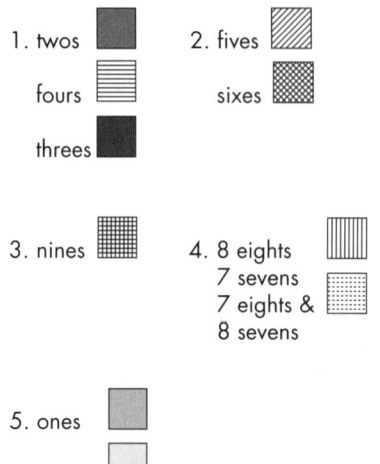

Likely difficulties

Difficulties with the basic facts usually occur because a child lacks any strategies to obtain answers efficiently. Often children endeavour simply to memorise all of the answers for multiplication facts, and then are unsure how to search through these numbers to find the appropriate result. As a consequence, children ask questions such as 'is 6 nines 56 or 54?' when they confuse the answers for 7 eights (56) and 6 nines (54) which are almost the same. Counting may also be used, using fingers, dots on a page or tapping. Other children may use doubling combined with addition or subtraction to adjust but lose track of what they are trying to do. If there is confusion with addition answers, this usually means that the multiplication concept is not secure and will need to be built up to show what multiplication is and provide a basis for building multiplication facts appropriately.

- A child may guess or count on his or her fingers to obtain answers rather than use more appropriate thinking. For example, a fact such as '6 eights' may be answered '8, 16, 24, 32, 40, 48', using fingers to keep track of the number of times 8 is added. Not only is this inefficient, it shows that the multiplication concept is focused on the very early repeated addition notion.
- Children who attempt to learn multiplication facts by reciting 'tables' are often unable to give responses automatically and mix the answers with the wrong facts when they try to speed up the process. For instance, an answer of 42 may be given for 6 eights when the children get out of step (6 sevens are 42; 6 eights are 48).
- Many children report that they 'don't know their sevens, eights and nines'. For example, they may be able to answer 5 nines are 45, but cannot give an answer for 9 fives. These children have not built up the thinking to link matching facts to the same array and way of thinking.

- Particular difficulties may occur with zero where a child thinks, for example, that 6 zeros are 6 or that 0 sevens are 7. Sometimes, children will not only treat zero as if it were 1, but add instead of multiplying to get 6 zeros are 7.
- They may also confuse other memorised answers for multiplication with those for addition and think that since 2 and 3 is 5, 2 threes are 5.

Investigate and discuss

1. Use different coloured pens to colour each cluster of facts in the number grid above (twos, fours, threes, fives, sixes, sevens and eights, nines). As you progress through the clusters of facts, what do you notice about the number of new facts to be mastered each time?
2. Why are the ones and zeros facts left until last in this development of the multiplication facts?
3. In contrast to the strategy approach, many children learn multiplication tables which they recite in sequence to obtain an answer, for example '4 ones are 4, 4 twos are 8, 4 threes are 12, 4 fours are 16, ...'. While this may assist them to obtain answers to multiplication facts, it is intrinsically inefficient. List any aspects of this approach to learning that render it less appropriate than the thinking strategy approach. Do you believe there is a place for this form of learning?

The multiplication algorithm

The multiplication algorithm is developed through a sequence paralleling that used to introduce the multiplication concept. Materials initially give meaning to the steps in the process and suggest a language pattern that will lead to the recording of the written form. However, their use is not practical for all of the renaming that occurs almost from the outset, and it is not helpful in building up skills for larger numbers. Rather, the numeration ideas behind the abstract number symbols are the key to the development of the algorithm. While a vertical recording similar to that used for addition and subtraction is also used for multiplication, the processing of the algorithm introduces a radically new step. Previously ones operated with ones, then tens with tens, and so on. For multiplication, ones are first multiplied by ones, which is simply basic fact knowledge, but then ones and tens are multiplied. Later, tens and ones are multiplied, then tens and tens. Instead of proceeding place by place, it is necessary to cross over places in completing the algorithm.

$$\begin{array}{r} 78 \\ \times\,49 \\ \hline 702 \\ 3120 \\ \hline 3822 \end{array}$$

This method of processing gave one of the first means of setting out the written algorithm as

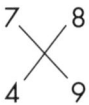

and led to the adoption of the symbol × to highlight this different process. The multiplication symbol itself can be a potent cue for children to remember to include all of the partial products when multiplying and help to overcome the common difficulty where children simply multiply ones with ones and record the result, then multiply tens with tens and record:

$$\begin{array}{r} 7\ 8 \\ \times\ 4\ 9 \\ \hline 28\ 72 \end{array}$$

In order to highlight this different way of operating with multiplication, the written algorithm is completed from bottom to top and, to be consistent, is also read from the bottom up. Renaming 10 ones to make 1 ten or 10 tens to make 1 hundred uses the same thinking as the addition algorithm so that its inclusion into the multiplication algorithm is not new. Instead, the focus from the outset needs to be on establishing the process for multiplying ones with ones then ones with tens in such a way as to provide meaning and to differentiate this from the way in which addition and subtraction are carried out.

SEQUENCE FOR DEVELOPING THE MULTIPLICATION ALGORITHM

1 Once the concept and basic facts have been learned, this thinking is extended to deal with tens.
2 Multiplication of one-digit by two-digit numbers includes renaming from the outset to introduce the form of the algorithm and the way materials, language and recording provide the meaning used to extend the algorithm to numbers where materials no longer provide insight.
3 Multiplication of tens with two-digit numbers is built on the thinking patterns that *ones by tens give tens* and *tens by tens give hundreds*.
4 The algorithm for multiplying two-digit by two-digit numbers combines multiplication of one-digit by two-digit numbers and tens by two-digit numbers.
5 This process and understanding is extended to multiplication with larger numbers, including estimation, mental computation and calculators.

Thinking in tens

Since multiplication has been introduced using a vertical format that has been read from bottom to top, the only new skill in extending

$$\begin{array}{r} 8 \\ \times\ 2 \\ \hline \end{array} \quad \text{to} \quad \begin{array}{r} 28 \\ \times\ 2 \\ \hline \end{array}$$

is an ability to deal with the tens. Consequently, the first need is to extend basic fact knowledge to the tens facts so as to encourage thinking in tens. This can grow out of the arrays conception of multiplication using materials to develop the understanding that ones multiplied by tens gives tens:

```
    8 tens            8 tens
  + 8 tens            × 2
   16 tens           16 tens    2 by 8 tens is 16 tens
```

Once this link has been made, children need oral practice in the same way that basic facts were built up after the thinking strategies were attained. In particular, drawing on the idea that since 2 by 3 tens is 6 tens, 3 tens by 2 is also 6 tens. A mix of basic and tens facts should also be included since the processing of the algorithm will demand the flexibility of moving from one way of thinking to the other. When children are as fluent with the tens facts as they are with the basic multiplication facts, they are ready to combine these to multiply one-digit by two-digit numbers with renaming.

Multiplication of one-digit by two-digit numbers

An initial reading of these multiplication examples as, for example, 2 thirty-sixes, is helpful in linking them to the use of the array model that allows the process to be developed in a similar manner to addition. Then it is simply a matter of having children make each number with materials on a place value chart, use the ones array to give the ones and rename, the tens array to give the tens, join any additional tens, and read the answer in terms of multiplication— 2 thirty-sixes are seventy-two or 2 by thirty-six is seventy-two. The inclusion of an arrow at the head of the ones place helps children to focus on where to start and in turn leaves a space at the top where the additional ten from renaming can be placed.

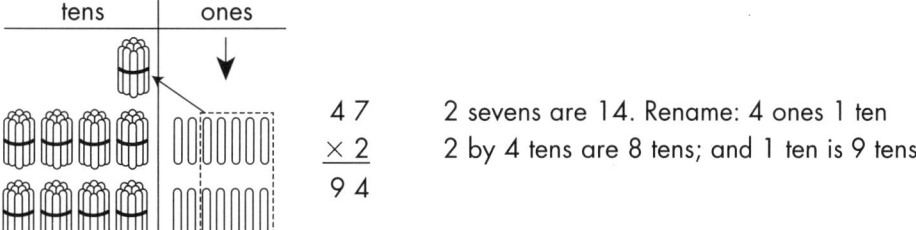

```
    4 7       2 sevens are 14. Rename: 4 ones 1 ten
    × 2       2 by 4 tens are 8 tens; and 1 ten is 9 tens
    9 4
```

The teaching of numeration and earlier experience with the addition algorithm has already established this ability to rename ones so renaming in multiplication is included from the outset. All that is needed is to merge the earlier addition process and knowledge of renaming with known basic facts and tens facts to provide a meaningful language and suggest a natural and appropriate recording. The first examples used should involve only the renaming of 10 ones to form one additional ten to maximise the similarity with addition and to give the children security with renaming in multiplication before the more general cases where several additional tens result. Since the use of materials is the same as for addition, it is natural that the additional ten also be placed at the top of the tens place and that the recording of this simply involve the writing of a full-sized 1 at the top of the tens place to indicate that there is one additional ten.

The recording for multiplication should be phased in gradually, with a teacher first recording at the board as children make the numbers and verbalise the process. The second step would be for children to make numbers to match written examples, then join the ones and tens to form the answer and record this in the written algorithm. When children are able to do this, they can begin to complete the written algorithms using symbols alone. Since there are only a few examples where there is no renaming, these can be introduced once the general process is understood, taking care that the language used continues to match the way the materials are joined, and focusing on the number of ones, 2 by 3 ones, and the number of tens, 2 by 4 tens.

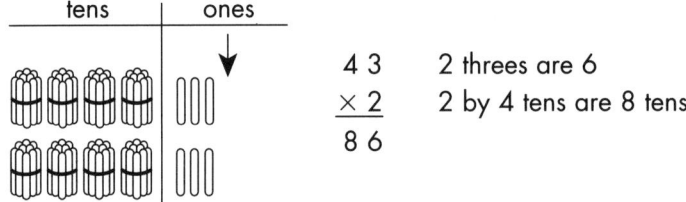

```
       tens | ones
                        4 3     2 threes are 6
                       × 2      2 by 4 tens are 8 tens
                        8 6
```

The following table shows the language and recording for multiplication and their links to the use of materials:

Language	Materials	Recording
2 forty-sevens **Multiply the ones** 2 sevens are 14 **Rename** 4 ones 1 ten Record the 4 ones with the ones, the 1 ten with the tens	tens | ones	1 4 7 × 2 4
Multiply the tens 2 by 4 tens are 8 tens **Rename** 8 tens and 1 ten is 9 tens Record the 9 tens with the tens	tens | ones	1 4 7 × 2 9 4
	2 forty-sevens are 94	

Note that the language used to describe multiplication reflects the place value basis for the algorithm. Ones are multiplied by ones, renamed and recorded, then ones are multiplied by tens; it is not helpful to talk of 'putting down the 2 and carrying the 1'.

The transition to the general case is assisted by using materials with some examples where 2 additional tens result. The new step of creating more than 1 additional ten can then be discussed and the recording introduced to match what the materials show. In this way, the tendency some children have to record a 1 in the tens place regardless of how many tens are regrouped can be minimised.

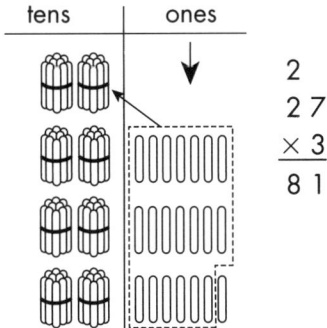

After this, materials are no longer helpful in illustrating the process as renaming ones to create an additional 5 tens, for example, would be so time consuming and demand so much manipulation that the multiplication being carried out would be overlooked altogether. Instead, reliance has to be placed on a language that grows out of and is consistent with the place value language and ideas. It will also be necessary to rename with large numbers, building on the skills established earlier with extended addition and when adding was used to develop the remaining multiplication facts. Nonetheless, this may cause some difficulties and care needs to be taken to ensure that children understand what is being done when the total number of tens is found. The expanded interpretation, which shows that a number such as 336 has 33 tens, is crucial in making this link to earlier renaming skills. The language for this final symbolic form is shown in the table.

Language	Recording
Multiply 6 fifty-sevens	4
Multiply the ones	5 7
6 sevens are 42	× 6
Rename 2 ones 4 tens	2
Multiply the tens	4
6 by 5 tens are 30 tens	5 7
Rename 30 tens and 4 tens are 34 tens	× 6
Record with the tens	3 4 2
6 fifty-sevens are 342	

At this point, it would be possible to extend multiplication to one-digit by three-digit or one-digit by four-digit numbers as there is really nothing new to be added to the process developed so far. Each place is considered as a separate entity involving the simple application of a basic fact although one or more renaming steps may be required. The understanding that 10 ones are 1 ten, 10 tens are 1 hundred, 10 hundreds are 1 thousand and so on is crucial. While materials can be used to give meaning and ensure a smooth extension to this large multiplication, their use should be minimal. Rather, the language and the setting out, which is consistent with the use of materials, should be emphasised. The following recorded forms are used:

$$\begin{array}{r} 53 \\ 675 \\ \times7 \\ \hline 4725 \end{array} \qquad \begin{array}{r} 215 \\ 7429 \\ \times6 \\ \hline 44574 \end{array}$$

However, in practice these forms are delayed until the children are secure with such large numbers, usually after the algorithm is extended to multiplication with two-digit numbers. But at that point it should be remembered that the extension is relatively straightforward and so need not receive much emphasis compared with the situations involving larger multipliers.

Likely difficulties

Apart from incomplete or confused basic fact knowledge, most difficulties with this early algorithm for multiplying one-digit and two-digit numbers is due to inappropriate generalisations from the earlier addition algorithm, where the renaming is either overlooked or carried out incorrectly.

- *Renaming:* A renaming digit may be overlooked, especially when it is not recorded full size in the appropriate place

$$\begin{array}{r} 39 \\ \times 6 \\ \hline 184 \end{array} \qquad \begin{array}{r} 312 \\ \times6 \\ \hline 1862 \end{array}$$

In addition, ones are added with ones and then all of the tens are combined:

$$\begin{array}{r} 1 \\ 57 \\ +36 \\ \hline 93 \end{array}$$ this may lead to $$\begin{array}{r} 4 \\ 57 \\ \times6 \\ \hline 362 \end{array}$$ where the tens are combined before multiplying.

It may also suggest that the tens need only be added: $$\begin{array}{r} 4 \\ 57 \\ \times6 \\ \hline 92 \end{array}$$

Investigate and discuss

1. Use Base 10 materials and a place value chart to complete:

29	37	24	16	28	48	51
×3	×2	×3	×4	×2	×3	×4

2. Record how these would be answered using the place value language outlined in this section.
   How could you use the materials to provide insight into the recording of each step?

3. How would you assist a child to see that this multiplication is not appropriate?

   ```
      5              2              4
     6 8           4 9            3 8
    × 7           × 3            × 5
    ———           ———            ———
    7 7 6         1 8 7          3 5 0
   ```

   How could you use a number expander to help the child focus on the number that is being multiplied?

4. What difficulties might result when multiplication by two-digit numbers is introduced if children or their teachers read these initial multiplication examples from top to bottom?

Multiplication with tens

It has been quite common in schools to move immediately from the one-digit by two-digit algorithm to the development of two-digit by two-digit multiplication. However, this substantial and abrupt move has been responsible for many of the difficulties experienced by children and their teachers in developing competency with the multiplication algorithm. Yet an analysis of multiplication by two-digit numbers shows that it is essentially multiplication by the ones digit followed by multiplication by the tens digit. The first part is simply the one-digit by two-digit multiplication developed so far while the second requires a careful building up of multiplication with tens.

Understanding that tens by ones give tens and tens by tens give hundreds is essential when multiplying by the tens digit if numeration knowledge is to be merged into the full multiplication algorithm and used to control and direct the process in the same way as for addition and subtraction. These reasons of control and understanding together with the large number of steps involved make it important for this new thinking to be developed before the full two-digit by two-digit algorithm is introduced and that the thinking underpinning this development be made more accessible and manageable to young children. At the same time, such a step by step development serves to delay the introduction of the full multiplication algorithm until the children are older and mathematically more mature, leading to an increased ability to complete and apply the algorithm meaningfully.

The use of materials to represent the multiplication of all except small numbers is impracticable. Instead, the thinking used in establishing the multiplication facts and the one-digit by two-digit algorithm needs to be extended to multiplication with tens. Since 3 by 4 tens is 12 tens, 4 tens by 3 must also be 12 tens. Knowing that ones by tens give tens is thus readily extended to tens by ones give tens. The next step, multiplying tens by tens, often proves to be more difficult. While most children understand that 10 tens are 1 hundred, generalising this to tens by tens give hundreds requires careful building up. Showing the pattern on a calculator has proven to be very helpful. For example, to multiply 6 tens by 8 tens, first 60 needs to be keyed in, then ×, then 80, then = is pressed. When 4800 results this is readily seen to be 48 hundreds, so that 6 tens by 8 tens is 48 hundreds.

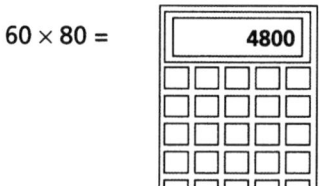

A teaching activity can be formed by providing each group of 4–6 children with one calculator. Ask one child to enter the number with 7 tens (70), the next child to press the multiplication symbol (×), the next child to enter 8 tens (80) and then the next child to press the = symbol (5600). The next child can provide the whole result 7 tens by 8 tens is 56 hundreds. The next child begins another round (e.g. 3 tens by 5 tens). As this activity is repeated, children become more familiar with the pattern and begin to generalise that tens by tens give hundreds. At this point, the child who presses the = symbol must state the result before pressing the = to check if his or her result is correct. Children can also work in pairs, one child nominating the numbers of tens to enter (e.g. 5 tens by 4 tens) and challenging the other to provide the result (20 hundreds).

However, even with this visual representation of the pattern, children may experience difficulties in incorporating the idea that tens by tens give hundreds into their existing thinking. After all, for the other operations met with up till now, tens have been added to or subtracted from tens to give tens. Why should that 'obvious' pattern change now? Yet this change in order of magnitude is fundamental to multiplication and to the real life situations to which it applies and understanding this pattern is the key to using and applying multiplication with whole numbers and decimal fractions. Later it will also allow the ready development of estimation skills. Multiplying 48 and 32 can be approximated by multiplying 5 tens and 3 tens. The estimate is then easily given as 15 hundreds or 1500 if the multiplication of tens by tens to give hundreds has been internalised. More importantly, at this point the incorporation of the thinking concerning multiplication with tens can be used to build up the second step in the final algorithm for two-digit numbers. The table below shows the language and recording that are used:

Language	Recording
Multiply 3 tens by 64	1
Multiply tens by ones	6 4
3 tens by 4 ones are 12 tens	× 3 tens
Rename 2 tens 1 hundred	2 tens
Record the 2 tens with the tens, the 1 hundred with the hundreds	
Multiply tens by tens	1
3 tens by 6 tens is 18 hundreds	6 4
Rename 18 hundreds and 1 hundred is 19 hundreds	× 3 tens
Record with the hundreds	1 9 2 tens
3 tens by 64 is 192 tens or 1920	

Later, when this thinking is secure, the recording may be related directly to place value and the fact that there are no ones recognised explicitly:

Language	Recording
Multiply 3 tens by 64	1
Multiply tens by ones	6 4
3 tens by 4 ones are 12 tens	× 3 tens
Rename 2 tens 1 hundred	2 0
Record the 0 with the ones, 2 tens with the tens, the 1 hundred with the hundreds	
Multiply tens by tens	1
3 tens by 6 tens is 18 hundreds	6 4
Rename 18 hundreds and 1 hundred is 19 hundreds	× 3 tens
Record with the hundreds	1 9 2 0
3 tens by 64 is 1920	

However, it should be stressed that a zero is recorded to show that there are no ones. This can only logically be stated after it is seen that the multiplication will give tens; it is not appropriate to talk about 'putting down a zero'. At best this might be interpreted as a generalisation that multiplying by the tens digit will never allow any ones, but it is just as likely to be seen as a mechanical means of avoiding writing the resulting digit in the wrong place. While this might give an appropriate response much of the time, it begs the question as to why the digit is written in the tens place and is likely to fall down when there is multiplication of the form 5 tens by 8 which gives 40 tens or 400.

Multiplication of two-digit by two-digit numbers

In order to develop the algorithm for multiplying two-digit numbers by two-digit numbers, it is now only a matter of combining the two parts developed earlier. This builds on the earlier idea that 6 fours is the same as 4 fours and 2 fours fostered by the arrays game and the particular language chosen to describe the initial multiplication concept. Multiplying 39 by 64 can be seen as the sum of 9 by 64 and 3 tens by 64:

```
    1 3              3                        1
    6 4              6 4                      6 4
  × 3 9            ×   9        and         ×  3 tens
    5 7 6            5 7 6                    1 9 2 0
  1 9 2 0
  2 4 9 6
```

The full language used is set out in the table:

Language	Recording
Multiply 39 by 64 **Multiply the ones** 9 fours are 36 **Rename** 6 ones 3 tens	3 6 4 × 3 9 6
9 by 6 tens are 54 tens **Rename** 54 tens and 3 tens are 57 tens	3 6 4 × 3 9 5 7 6
Multiply the tens 3 tens by 4 ones are 12 tens **Rename** 2 tens 1 hundred	1 3 6 4 × 3 9 5 7 6 2 0
3 tens by 6 tens are 18 hundreds **Rename** 18 hundreds and 1 hundred is 19 hundreds **Add**	1 3 6 4 × 3 9 5 7 6 1 9 2 0 2 4 9 6
39 by 64 is 2496	

Some children may experience difficulties adding the two partial products within the algorithm when there is no space at the top of the places to allow the recording of the renaming digits. In this case, transferring the numbers to a working space to complete the addition is helpful and enables a teacher to see that their difficulty is with the addition and not with multiplication as such. In time, they will come to add the digits and retain the renamed digits in their heads rather than needing to record them on paper.

Likely difficulties

Difficulties with the multiplication algorithm usually have their origins with inappropriate renaming, particular misconceptions with zero, or confusion with the recording of their working because of an inability to use place value. There may also be errors due to incorrect basic facts but these need to be addressed separately from the algorithm. Similarly, if there is confusion with the process used for another operation, this might mean that the concepts need revisiting or it may suggest that a rote procedure has been used rather than a meaningful process as built up here.

- *Place value*

```
Multiplies digits without regard to place value      4 8
                                                   × 5 7
                                                   3 3 6
No use is made of place value to see that          2 4 0
the answer is far too small                        5 7 6


Multiplies ones by ones, then tens by tens           4 8
without carrying out the steps which require       × 5 7
crossing places and simply records the results    20 56
```

- *Renaming:* Renaming digits may be overlooked when they are not recorded full size in the appropriate place

```
    4 8              7 3
  × 5 7            × 6 4
    2 8 6            2 9 2
  2 0 0 0          4 2 8 0
  2 2 8 6          4 5 7 2
```

```
                                                 5                5
Multiplies ones by ones, then renames before     4 8              4 8
multiplying tens by tens to get 45             × 5 7            × 5 7
or after multiplying tens by tens to get 25      4 5 6    or     2 5 6
```

No use is made of place value to see that the answers are far too small.

- *Zeros:* The answers are correct: thinking about the multiplication shows there must be zero ones or hundreds.

```
    49              49              49                        49
  × 60            × 60           × 500                      ×500
    0 0  There are zero ones   2 9 4 0      0 0  There are zero ones   2 4 5 0 0
  2 9 4 0  Multiply the tens                0 0 0  There are zero tens
  2 9 4 0                           2 4 5 0 0  Multiply the hundreds
                                    2 4 5 0 0
```

- *Recording:* Numbers may be confused when the recording does not place digits in the correct places. Use of quad-ruled paper or a place value chart is helpful with children who misread their recording in this way.
- *Adding to complete the multiplication algorithm:* Children may obtain incorrect answers through basic fact, renaming or place value difficulties with the addition step required to complete the multiplication. Since it is not the multiplication that is the underlying cause, separate review of addition will be needed.

Investigate and discuss

1 Answer these examples using the recording and language outlined in this section:

```
  57     89     64      8     90     48     90
× 94   × 50   × 5    × 40   × 20   × 9   × 50
```

In which order would you introduce these multiplication examples to children?

2 How could you use a number expander to help children bring together multiplication by ones and multiplication by tens into the full two-digit by two-digit algorithm?

Multiplication with larger numbers

Multiplication of larger numbers is essentially a continuation of the same process, thinking and language used for the two-digit by two-digit algorithm, although some confusion might result with the renaming unless the recording format is modified. This difficulty first arises with two-digit by three-digit multiplication when an additional line of recording is needed to keep track of the renaming and the question as to which of the digits in the hundreds is to be included:

```
      4
    1 3            When multiplying 6 tens by 3 tens,
    9 3 7          are 4 hundreds or 1 hundred to be added?
  ×   6 5
  4 6 8 5          6 tens by 3 tens are 18 hundreds,
      2 0          19 hundreds or 22 hundreds?
```

In order to overcome this uncertainty, as the additional tens, hundreds or thousands are added, the corresponding digit can be crossed out. In this way there is no confusion as to which of the additional hundreds need to be renamed and the process of writing the additional digits in their place is maintained by using one more line above. While this technique could have been used from the outset, there was no likelihood of confusion and hence no need to introduce a different method of renaming to that already developed for addition when the actual process is identical. On the other hand, discussion of the likely confusion as in the example above can lead children to see the need for this recording and allow them to control the large number of steps involved in these multiplications. As the numbers get larger, the same technique together with additional lines of recording for the regrouping of digits allows the multiplication to proceed smoothly:

Language	Recording
Multiply 379 by 482 **Multiply the ones** 9 twos are 18 **Rename** 8 ones 1 ten 9 by 8 tens are 72 tens **Rename** 72 tens and 1 ten is 73 tens. 9 by 4 hundreds are 36 hundreds **Rename** 36 hundreds and 7 hundreds is 43 hundreds	7̸ 1̸ 482 × 379 4338
Multiply the tens 7 tens by 2 ones are 14 tens **Rename** 4 tens 1 hundred 7 tens by 8 tens are 56 hundreds **Rename** 56 hundreds and 1 hundred is 57 hundreds 7 tens by 4 hundreds are 28 thousands **Rename** 28 thousands and 5 thousands is 33 thousands	5̸ 1̸ 7̸ 1̸ 482 × 379 4338 33740
Multiply the hundreds 3 hundreds by 2 ones are 6 hundreds 3 hundreds by 8 tens are 24 thousands **Rename** 4 thousands 2 ten thousands 3 hundreds by 4 hundreds are 12 ten thousands **Rename** 12 ten thousands and 2 ten thousands are 14 ten thousands **Add**	2̸ 5̸ 1̸ 7̸ 1̸ 482 × 379 4338 33740 144600 182678
	379 by 482 is 182 678

Likely difficulties

- If children do not record the renaming digits on separate lines and in the appropriate places, they may be confused as to which digit is to be used in each step.
- If the renaming digits are not crossed out as they are used, children may use a digit that has already been used in an earlier step or use both digits in that place and rename inappropriately.
- Children may try to follow a 'rule' of placing zeros rather than use place value explicitly for each step and be confused when multiplication of numbers such as 5 hundreds with 8 ones results in 40 hundreds, and record the 4 in the hundreds place rather than with the thousands.

Investigate and discuss

1 a Answer each of these examples using the recording and language outlined in this section:

$$\begin{array}{ccccc} 459 & 973 & 365 & 678 & 708 \\ \times\ 26 & \times\ 89 & \times\ 487 & \times\ 69 & \times\ 745 \end{array}$$

   b What difficulties might children encounter in extending multiplication to these larger numbers?

   c What need is there to develop a written algorithm for such large numbers? Consider any practical uses as well as building a realisation that multiplication can apply to numbers of all sizes.

2 Recent suggestions are that for larger than two-digit multipliers the use of a calculator is more appropriate than pencil and paper algorithms. What do you think? List reasons for and against this point of view to help make your decision.

Estimation with multiplication

Being able to readily complete the written multiplication algorithm based on a meaningful process is critical in providing understanding for multiplication and in building a capacity to use it in problem situations. But exact answers are not always needed and in many situations an approximation will be enough. An estimate is sufficient to see how much turf or carpet would be needed to cover a given area or how much soil or cement is needed to fill a space to a particular depth. Measurements involving area and volume are often approximate rather than exact and also involve the combination of smaller sections. This involves an estimate of each part before combining them in a fashion similar to the way the partial products are combined in the written algorithm. Thus, estimation for multiplication demands a familiarity with all aspects of the concept, basic facts and the written algorithm up to two-digits by two-digits if it is to be used flexibly and efficiently.

However, many people find estimation with multiplication difficult and tales where someone is left with a large quantity of quickly setting concrete or a large part of the garden or lawn without soil after an estimated quantity of material has been delivered are commonplace. Usually this is because the processes they employed did not build on a well-internalised understanding of multiplication or any real awareness of what the numbers that result would mean. A feeling of powerlessness and inadequacy usually results in the person having to rely on someone else's skills and there may even be a feeling that the 'expert' is taking advantage of the person's inability.

These examples show that good estimation and approximation skills enhance an ability to deal with everyday quantitative situations. This need for estimation in many everyday situations is first built up through measurement experiences and is extended to the computational context with addition and subtraction. Multiplication can continue to build up a propensity to use estimates and make approximate calculations through its use in a wide range of real contexts, and particular strategies that permit computational estimations can be built in soon after the multiplication concept in terms of arrays and the meaning for the recorded process is secure.

Strategies for estimating multiplication are based on rounding and an understanding of multiplying tens, hundreds and so on from the multiplication algorithm:

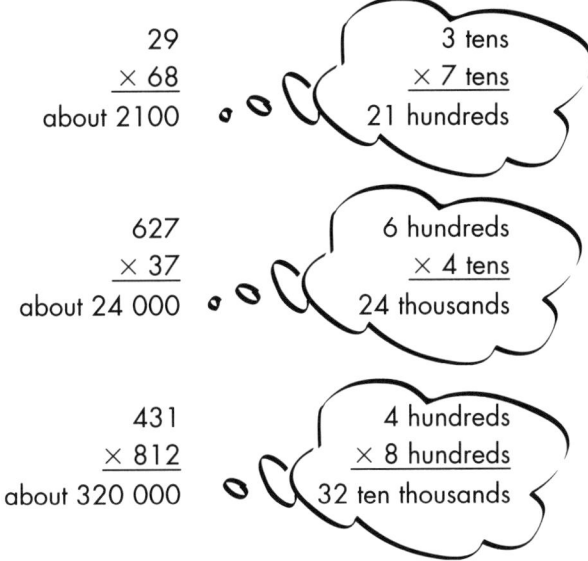

Unfortunately, with multiplication, estimates can be quite different from the actual outcome and children also need to develop a feeling for when their estimate is too large (as in the first example), or too small (as in the last example) or about right (as in the middle example). However, this understanding is itself part of the number sense that needs to be developed in all children and should grow out of an awareness of what the rounding process entails. It is another aspect of the need to assess the reasonableness of results that is so important in contemporary mathematics.

Since the development of skills and understandings with multiplication follow those for addition and subtraction, it is also reasonable to expect children to incorporate their understanding of all three operations into the way they will form estimates in various situations. Processes involving halving and doubling or using specific knowledge of multiplying by ten, 1 hundred or 1 thousand are also able to be grasped by young children. For example:

An important part of teaching about estimation is to allow time for children to investigate alternative ways of finding answers and to debate them with others so that they might be shared or adapted by others. It is also essential for children to build up an appreciation that there are many ways in which estimates and approximate calculations can be made and that some will give closer estimates than others. All are valuable and it is more a matter of deciding which methods are most suitable to a particular need or preferred way of thinking, and the degree of nearness that is warranted for each purpose.

Multiplying numbers mentally

Although approximate answers are often more important in multiplication than precise ones given by the written algorithm or a calculator, there will be a need for accurate answers calculated on the spot. This demands easily managed mental processes that utilise different ways of thinking from those associated with the written process. Particular ways of combining steps through the use of factors together with addition or subtraction can be used as well as left to right processing similar to the thinking built up with estimation. For instance, the nearest whole multiple can be obtained by breaking up the multiplier and using easier factors first. This first result can then be adjusted through subtraction if it is too large:

```
    72
  × 49
  3528
```

10 by 72 is 720
so 50 by 72 is 3600
subtract 72
49 by 72 is 3528

In some cases this may require several different mental computations in the same process, drawing on a range of thinking for both multiplication and subtraction or addition:

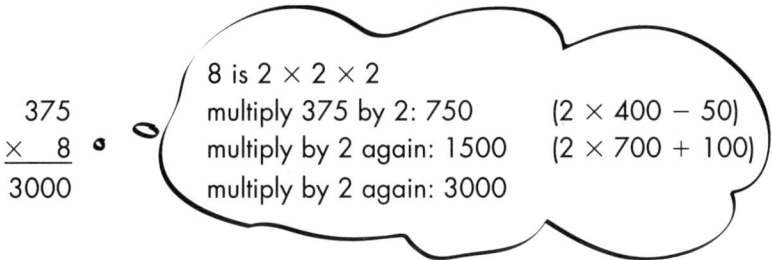

There are also more direct methods for certain factors that allow both multiplication and division to be performed efficiently.

```
    44      8 by 125 is 1000
  × 125     1000 by 44 is 44 000
   5500     divide by 8: 5500
```

Similar use can be made of the fact that 4 × 250 is 1000 or that 4 × 25 and 5 × 20 are 100. This thinking can extend to multiplying by 75 or 750 by considering multiplying first by 25 or 250 in this manner and then multiplying that result by 3. The possible ways of determining products mentally are boundless and call on a good understanding of factors and an ability to consider any of the four operations as a means of obtaining shortcuts that are more easily remembered on the way to the final result. Children need to be encouraged to construct their own methods and to explain and justify these to their classmates, teachers, parents and others in their daily lives. Discussing the merits and ease of use or control of a range of strategies is important in building up an inclination to choose to use mental computation in everyday situations. This will also assist in building up confidence in an ability to multiply mentally, using strategies that fit individual strengths as well as adopting or adapting those used by others. While this is likely to give rise to a range of methods adapted to suit particular numbers and situations, it is important that, through discussion, a teacher assists children to view mental computation as a way of thinking that applies to all operations as part of an expanding number sense rather than as a collection of separate strategies to be memorised.

A willingness to attempt to calculate mentally and to do so efficiently is most important and should lead in the longer term to skills in mental computation comparable to the automatic recall and use of number facts. By this time, children should be in a position to choose whether to use mental, pencil and paper or calculator processes to fit a given situation and have confidence in their ability to derive and interpret answers with them.

Investigate and discuss

1. Find as many ways as you can to obtain answers mentally for:

   $637 \quad\quad 6582 \quad\quad 78 \quad\quad\quad 125\overline{)7658}$
   $+586 \quad\quad -965 \quad\;\; \times 62$

   Which of these ways would you specifically introduce to children?

2. Can you find methods for multiplying and dividing by 5, 15, 20, 25, 55 and 75? How would you extend these to 50, 150, 200 and so on?
   Do you know any other shortcuts for finding answers in your head?

Using calculators in multiplication

The use of a calculator for multiplication can accompany the development of the written algorithm soon after understanding is available. However, if introduced too soon a calculator may prevent some children from coming to terms with the many steps involved and the manner in which the concept and basic facts are integrated with numeration ideas to develop a coherent process. But a calculator is also useful in highlighting these steps by showing how multiplying by tens will give tens:

$$\boxed{640}$$

4 tens × 16 is entered onto a calculator as 40 × 16 which gives 640 or 64 tens

and multiplying tens by tens gives hundreds:

$$\boxed{3200}$$

4 tens × 8 tens is entered onto a calculator as 40 × 80 which gives 3200 or 32 hundreds

Judicious use of a calculator also allows many answers to be obtained more quickly and often enables a series of calculations to be done much more efficiently. It also means that the user can focus on coming to terms with the problem being addressed rather than becoming bogged down in the many detailed sub-steps that multiplication requires. This overview of what is required will often show shortcuts even before data or operations are keyed in and suggest ways of solving problems that are different from those which might have been used with a pencil and paper or mental method. For example, consider this problem:

A book was opened and the reader noticed that the product of the two page numbers was 8971. What are the page numbers at which the reader had opened the book?

While this problem can be answered in many ways using guess and check techniques requiring many multiplications of likely numbers, an analysis of the problem and a very straightforward use of a calculator can supply a more insightful solution. Knowing that the pages must be consecutive, so that the page numbers differ by one, and that they must be reasonably large numbers allows an understanding that they are 'almost the same'. This means that 8971 is about the product of one of the numbers by itself; use of the square root key would suggest the answer.

Universally, children, teachers and other adults have then immediately stated that the pages are 94 and 95. An interpretation of the answer given by the calculator can give the answer to the problem without any further effort. But, the calculator was merely a tool in providing the solution. All of the thinking about the problem and its solution is provided by the thinking of the calculator user, thinking freed up by not having to focus on the actual computations that might be required.

Games similar to those outlined for addition and subtraction are helpful in introducing calculators as is their use alongside estimation where children are asked first to estimate an answer and then compare it with that given by a calculator, or to use estimation to assess the reasonableness of calculator answers, justifying the results to other members of the class. Discussion should also focus on ways in which use of a calculator might give rise to incorrect answers, for instance on the need to multiply like units of measurement, as well as on using the reciprocal nature of division and multiplication to correct when a given number has been multiplied rather than divided or divided rather than multiplied. Discussion of how these sorts of errors might arise and how they can be corrected will not only develop a good and competent facility with calculators but also assist in consolidating understanding of the concepts, allow children to distinguish between the operations, and promote their ability to apply operations to problem situations.

Investigate and discuss

1 How could you use the memory functions on a simple calculator to place the missing numbers on this table as efficiently as possible?

×	529	703	419
19			
54			
63			

2 A calculator is also useful in consolidating children's understanding of multiplication. Use what you know about multiplication to work out the missing digits. Then use a calculator to check your answers.

$$\begin{array}{cccc} ..4 & 4.. & 38.. & ..76 \\ \times\ 39 & \times\ 79 & \times\ 65 & \times\ ..8 \\ \hline 3666 & 3792 & 25025 & 50808 \end{array}$$

3 How would you use a calculator to multiply several numbers?
Try these multiplications: 349 × 568 × 973 × 459
23 × 4568 × 24 × 367
56 × 311 × 98 × 2005
Did you need to press '=' after each step?

Extending multiplication to fractions

Multiplication with decimal fractions

Traditionally, decimal fraction multiplication has simply involved multiplying numbers as if the decimal point were not there and then adjusting the answer at the conclusion, usually by a counting process. While this approach is basically sound, the mere counting of places has no meaning and is frequently misapplied by children. Yet the same form of language related to numeration that has governed whole number multiplication can also be used with multiplication of decimal fractions. At first, where only one factor involves a decimal fraction, explicit use can be made of the place value patterns to guide the recording of each step. Later, when both factors involve decimal fractions, this becomes too cumbersome and the more general process, multiply like whole numbers then place the decimal point (to show tenths, hundredths or whatever) will be needed. Since this will eventually be the way all decimal fraction multiplication will be conducted, it is necessary to build in and justify a pattern for placing the decimal point from the outset. The process will be the customary one, but the reason for placing the decimal point will be based on a meaningful numeration interpretation rather than a rote counting of places. The steps in building up such a facility can be summarised as:

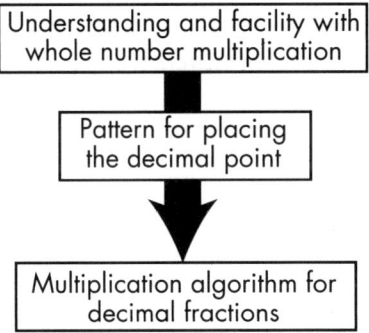

Multiplication of whole numbers and decimal fractions

Extending multiplication to situations involving decimal fractions and whole numbers is simply a matter of combining the meaning for decimal fractions with the concept, basic facts and processes for multiplying whole numbers. The first step is to build up the understanding that ones by tenths give tenths:

```
    0.6                  6 tenths                                    0.6
   × 7      means       × 7                                         × 7
                        42 tenths      42 tenths is 4.2      so     4.2
```

This thinking can then be applied to more general examples by multiplying and renaming each place in turn:

```
      4         7 by 4 and 6 tenths
    4.6         7 by 6 tenths is 42 tenths;
    × 7         Rename 2 tenths 4 ones
   32.2         7 fours are 28 and 4 is 32
```

Similar reasoning shows that ones by hundredths give hundredths and so on, allowing multiplication of numbers with more decimal places to proceed in the same way:

```
                  7 by 8 and 46 hundredths
     3 4          7 by 6 hundredths are 42 hundredths
     8.4 6        Rename 2 hundredths 4 tenths
     ×   7        7 by 4 tenths are 28 tenths and 4 tenths is 32 tenths
    5 9.2 2       Rename 2 tenths 3 ones
                  7 eights are 56 and 3 is 59
```

At first it is sufficient, and essential, for multiplication of whole numbers and decimal fractions to be carried out with a full numeration-based language to direct and control what is occurring. Later, these experiences can be built on to develop a more general process: when one factor has tenths, the answer will have tenths; when one factor has hundredths, the answer will have hundredths. Whole number multiplication can be used to determine how many tenths or hundredths there are while the decimal point need only be placed in the answer to show that these are tenths or hundredths:

```
    3                  3                      3 5              3 5
   5.6                56    tenths           6.47             647    hundredths
  × 6       ⟺       × 6  × ones           ×   8      ⟺     ×   8  × ones
  ────               ────                   ─────            ─────
  33.6               336   tenths           51.76            5176   hundredths
```

Multiply like whole numbers then place the decimal point

Similar reasoning allows this process to be extended to all situations involving the multiplication of whole numbers and decimal fractions. For example:

```
   7.45            745      hundredths
 ×   37    ⟺    ×  37    × ones
 ──────          ─────
 275.65          5215     hundredths
                 22350
                 ─────
                 27565    hundredths
```

On the other hand, the multiplication of decimal fractions by powers of ten is often treated separately and is subject to misleading rules. Some discussion of what is occurring with multiplication of decimal fractions by whole numbers shows that tens by tenths gives ones, tens by ones gives tens, tens by tens give hundreds and so on. This thinking allows the ready multiplication by powers of ten as shown on a number slide; multiplying by 10 means that the digits move one place to the left, as the number of tenths becomes the number of ones, the number of ones becomes the number of tens and so on.

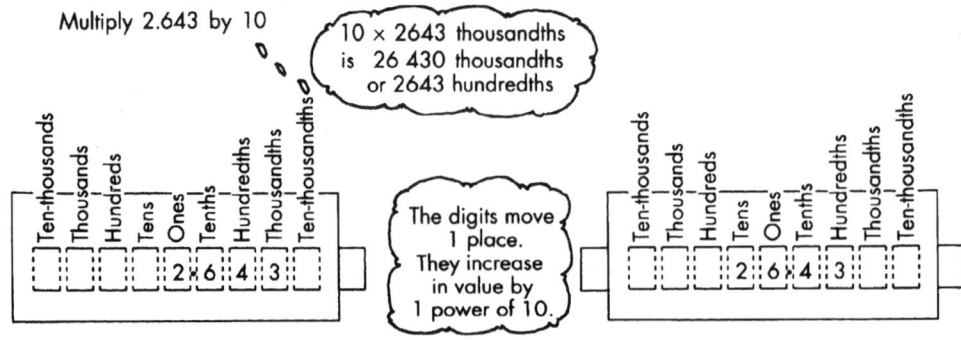

Multiply 2.643 by 10

10 × 2643 thousandths is 26 430 thousandths or 2643 hundredths

The digits move 1 place. They increase in value by 1 power of 10.

Note that it is not the decimal point that moves. The decimal point marks where the whole numbers end and the fractions begin. It cannot move. The digits move to the left as they increase in value by a power of ten and so occupy a place of greater value.

Multiplication of decimal fractions and decimal fractions

When both factors involve decimal fractions, further complications arise. Even a preliminary use of a full language to describe what is happening is likely to confuse. There are too many steps: the thinking involved, ones by tenths, tenths by ones and tenths by tenths, requires a lot of keeping track; and the recording necessarily breaks with the place value that in the past has assisted to control these many steps. For these reasons, rather than generalise the language used to begin multiplication with decimal fractions, it is the shortened process that is generalised, multiply like whole numbers then place the decimal point (to show tenths, hundredths or whatever). The understanding that ones by tenths gives tenths has already been developed and the children's earlier experience with whole number multiplication is likely to lead them to generalise from tens by tens give hundreds to tenths by tenths give hundredths. This pattern is readily shown with a calculator in a similar manner to that used to show that tens by tens give hundreds.

For example, to multiply 4 tenths by 9 tenths, first 0.4 needs to be keyed in, then ×, then 0.9, then = is pressed. When 0.36 results this is read as 36 hundredths, so that 0.4 tens by 0.9 tens is 36 hundredths.

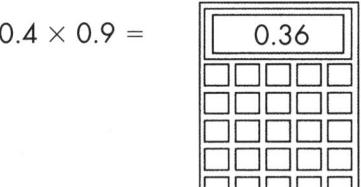

A teaching activity can be formed by providing each group of 4–6 children with one calculator. Ask one child to enter the number with 5 tenths (0.5), the next child to press the multiplication symbol (×), the next child to enter 3 tenths (0.3) and then the next child to press the = symbol (0.15). The next child can provide the whole result 5 tenths by 3 tenths is 15 hundredths. The next child begins another round (e.g. 7 tenths by 4 tenths). As this activity is repeated, children become more familiar with the pattern and begin to generalise that tenths by tenths give hundredths. At this point, the child who presses the = symbol must state the result before pressing the = to check if his or her result is correct. Children can also work in pairs, one child nominating the numbers of tenths to enter (e.g. 9 tenths by 6 tenths) and challenging the other to provide the result (54 hundredths).

In this way, a meaningful language-based pattern can be used for multiplying decimal fractions in contrast to the common rote procedure of counting decimal places.

$$
\begin{array}{r} 2 \\ 3.4 \\ \times\ 0.7 \\ \hline 2.38 \end{array}
\qquad
\begin{array}{r} 2 \\ 34 \\ \times\ \ 7 \\ \hline 238 \end{array}
\begin{array}{l} \\ \text{tenths} \\ \times\ \text{tenths} \\ \text{hundredths} \end{array}
$$

Multiply like whole numbers. Then adjust by placing the decimal point in the product to show hundredths.

Later, use can also be made of the calculator pattern to show that tenths by hundredths give thousandths, hundredths by hundredths give ten thousandths, and so on. It is only the process of placing the decimal point that is new. If children are unable to multiply the numbers as if they are whole numbers, then they should not even be attempting the multiplication of decimal fractions. Accordingly, once the patterns for multiplying decimal places are established, practice can focus on activities of the form:

$$
\begin{array}{r} 5.4 \\ \times\ 3.8 \\ \hline 2052 \end{array}
\quad
\begin{array}{r} 3.7 \\ \times\ 0.8 \\ \hline 296 \end{array}
\quad
\begin{array}{r} 58 \\ \times\ 6.2 \\ \hline 3596 \end{array}
\quad
\begin{array}{r} 9.3 \\ \times\ 4.6 \\ \hline 4278 \end{array}
\quad
\begin{array}{r} 7.5 \\ \times\ 9 \\ \hline 675 \end{array}
\quad
\begin{array}{r} 5.73 \\ \times\ 0.9 \\ \hline 5157 \end{array}
\quad
\begin{array}{r} 6.37 \\ \times\ 5.29 \\ \hline 336973 \end{array}
\quad
\begin{array}{r} 0.64 \\ \times\ 4.75 \\ \hline 30400 \end{array}
$$

Place the decimal point to show the correct answers

Children should also be encouraged to estimate to check that their answers are reasonable and some may develop this as an alternative means of placing the decimal point after multiplying the corresponding whole number.

Multiplication with common fractions

Multiplication of common fractions is an apparently simple process but fraught with difficulties associated with the simplification of the resulting products, the establishment of meaning for the process and the likelihood of it confusing the previously learned addition and subtraction algorithms. Indeed, this last point gives yet another argument for the adoption of the vertical format for addition and subtraction with common fractions. Once multiplication with common fractions has been met, children who have used a horizontal setting out for addition are likely to adopt the 'easier' method for all computation:

Multiplying $\frac{5}{6} \times \frac{4}{5} = \frac{20}{30}$ often leads to difficulties with addition of common fractions $\frac{5}{6} + \frac{4}{5} = \frac{9}{11}$

Incorrect generalisations of this form are much less likely to occur when addition and subtraction have always been recorded vertically as there is no obvious correspondence between the operations. Not only does a vertical recording facilitate the transfer of addition and subtraction skills from whole numbers and decimal fractions to common fractions, but it also builds in a defence against confusion with the multiplication process when this is introduced.

In reality, there are almost no practical uses of multiplication with common fractions. With the universal adoption of the metric system this need has diminished to the extent that it is no longer considered an appropriate topic for primary school. Indeed, the only aspect now considered at all relevant is the multiplication of whole numbers and common fractions. Examples of this form of multiplication arise when a recipe is increased. For instance, if 1 cup of concentrate is needed for 10 litres, 7 cups would be needed for 70 litres. These situations are analogous to the multiplication of ones by tenths used to introduce multiplication with decimal fractions so that the language and thinking used there can be extended to this form of common fraction multiplication:

$8 \times \frac{5}{6}$ 8×5 sixths are 40 sixths; 6 and 4 sixths: $6\frac{4}{6}$

$8 \times 3\frac{3}{5}$ 8×3 fifths are 24 fifths; 4 and 4 fifths $28\frac{4}{5}$

8 threes are 24 and 4 is 28

Everyday situations that involve finding a fraction of a given quantity also occur as examples of multiplication of a common fraction with a whole number. Using the understanding developed through the use of arrays, these calculations can be carried out in the manner developed above. However, consideration of the meaning of the fraction itself can also provide a soundly based method. For example, to find 4 fifths of something is only a matter of determining 1 fifth and multiplying this by 4:

$\frac{4}{5} \times 20$ $4 \times (1$ fifth of 20) or 4×4: $\frac{4}{5} \times 20$ is 16

or $\frac{4}{5} \times 20$ is the same as $20 \times \frac{4}{5}$: 20×4 fifths are 80 fifths or 16

Any multiplication of common fractions by common fractions is best left until needed in the context of further mathematics in secondary school. Indeed, it is helpful if the development of this multiplication is delayed for some time until children are totally secure with addition and subtraction and have a robust knowledge of the fraction concepts on which the process for multiplication can be easily and unambiguously built.

Multiplication with per cents

Multiplication occurs with per cents in the context of finding a per cent of a given quantity. Initially, this can be done by reference to equivalent fractions but in due course there is a need for a straightforward computational process. In the past, reference was made to common fraction multiplication and there are still many people who will even use a calculator to multiply whole numbers then divide by 100. Rote procedures such as these have little place in a contemporary mathematics program as they are likely to be inadequate when real life applications and problem solving arise. Rather, an emphasis on the relationship between per cents and decimal fractions together with the understanding of a per cent as 'so many for each hundred' can be combined with knowledge of decimal fraction multiplication to provide a meaningful process for both pencil and paper and calculator use.

300 children at The Glen School
40% of them travel to school by bus
How many children is that?
          120 children travel to school by bus

40% means 40 for each hundred
There are 3 hundreds: 3
                                × 40
                                120

While 40% can be considered as 40 for each hundred, the earlier work on the decimal fraction and per cent concepts has also shown that this is the same as 4 for each 10. Coupled with the understanding of how many tens a number has, this thinking provides another way to answer the problem (300 has 30 tens; 4 by 30 is 120) and extends the process to problems involving amounts less than 100:

60 children at The Glen School
play a musical instrument.
30% play the piano.
How many children is that?
          18 children play the piano

30% means 30 for each hundred
or 3 for each ten.
There are 6 tens: 6
                        × 3
                        18

The same form of thinking can also provide the development of a general computational process based on this understanding. Just as a per cent can be thought of as so many for each hundred or so many for each ten, it can also be extended to be so many tenths or hundredths for each one.

40% means	40 for each hundred	or 4 for each ten	or 0.4 for each one
40% of 300 is	3	30	300
	× 40	× 4	× 0.4
	120	120	120

30% means	30 for each hundred	or 3 for each ten	or 0.3 for each one
30% of 60 is		6	60
		× 3	× 0.3
		18	18

This process can then be extended to provide a general method for solving all per cent problems:

300 children at The Glen School   15% means   15 for each hundred
15% walk to school   or   0.15 for each one   300
How many children is that?   × 0.15

45 children walk to school   45

As this sequence of examples shows, the general process is: *to find a per cent of a number, change the per cent to a decimal fraction and multiply*. This process is a straightforward development from the underlying fraction conceptions and allows a meaningful process to be used at all times, whether with pencil and paper or calculator computations. Finding the per cent of a number is just an application of decimal fraction knowledge, once again emphasising that per cents are just another way of representing particular decimal fractions.

Investigate and discuss

1. Make a number slide as shown in the section on multiplying decimal fractions by whole numbers and investigate its use in multiplying whole numbers by 100, 1000, 10 000 and so on. How would you describe what is happening in a way that would help children to readily carry out this form of multiplication?

2. **a** Answer each of these examples using the recording, language and thinking outlined in this section:

        8.7   7.69   5.8   6.67   5.38
      × 9   × 8   × 9.7   × 8.3   × 9.74

   b How does the cycle of steps discussed in the overview to this chapter:

   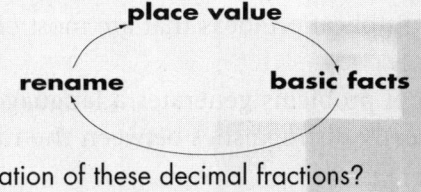

   apply to the multiplication of these decimal fractions?

c Would multiplication with per cents and common fractions also use this cycle of steps?

3 Solve the following problem and consider how you would assist a child to come to terms with what is required, building on the general model of problem solving given in Chapter 2.

Jane works at Floreat plant nursery. In the morning, she fertilised 46 pots and watered 34 rows of plants. During the afternoon she watered 52 rows of plants and fertilised 27 pots. If there are 68 plants in each row how many plants did she water?

4 How would you adapt the process for finding the per cent of a number to situations that involve per cent discounts or increases? Investigate how you might use a calculator for this. (There are several different ways this can be done—using multiplication by a decimal fraction, using two steps or one step to find an answer, and even by using the addition key.)

Section IV: Overview to division

DIVISION has often been viewed as the most difficult of the four operations to teach. Yet its basis in real world experiences, its strong reliance on what are now very secure numeration understandings and the fact that it is the last operation to be established should make it comparatively easy to learn. Many of the difficulties that children and their teachers have encountered have to do with the inappropriate language and rote procedures that have sometimes dominated its teaching. Variations on the meaningless ritual 'divide, multiply, subtract, bring down' are still being used and the initial conception of division is still sometimes related to a mysterious process of 'goes into'. The approach taken here instead focuses on a language to guide the division process that grows out of the sharing situations in which it is based and that relates strongly to the underlying numeration and multiplication ideas that govern the cycle of steps that is followed.

In contrast to the multiplication concept that is abstracted from addition, the division concept grows out of everyday experiences. Largely these are associated with sharing but there is a need to establish this in a more formalised way to lead into a meaningful interpretation of division and the way it is recorded. For this reason, the focus is on sharing objects to form arrays, rather than simply sharing them into equal-sized groups, so that a basis for the crucial link between division and multiplication can be formed from the outset. Later, the interpretation of division as repeated subtraction can also be built into the concept and related to sharing. However, repeated subtraction is more fundamental to certain problem situations than it is to the concept as such. It is the understanding of division as sharing and its link to already known multiplication ideas that are most crucial in developing division meaningfully.

The use of arrays to model problems generates a language that gives meaning to the recording of division and clearly distinguishes between the number being shared and the number it is shared among. This understanding will later be formalised as the number being divided and the number it is divided by, but it is the process of sharing to which these formal words refer that gives meaning to the process. One reason that children find division difficult is that there are no basic facts *per se*. An ability to use the nearest multiplication fact is required. Sharing materials in arrays links division to multiplication facts in a similar manner to the way in which subtraction facts were related to addition facts by means of the part/part/whole conception. It also suggests the use of the $\overline{)}$ symbol for division from the outset as this allows the ready identification of the number that is being shared in contrast to the number sharing. The alternative symbol, ÷, does not show this and children who meet this symbol first often believe that in division the larger number is divided by the smaller. As the division algorithm also uses a recording based on the use of $\overline{)}$, no use of ÷ need occur until the division concept is secure. For division with larger numbers, Bundling sticks or Base 10 materials can be shared into arrays in a similar manner to the way objects are shared for the concept, leading to a natural language that provides a meaningful process and gives direction to the recording of these steps.

One further complication arises with division. In contrast to the other operations where a basic fact associated with the operation gives the result for each step in the algorithm directly, in division the nearest multiplication fact is required to give a result as close as possible to the number that is being divided. This means that an additional level of understanding is required to determine each step and, in turn, some quantity will often be unable to be shared. When the division is completed, there are usually two parts to the answer: a number that has resulted from the sharing and a remainder that could not be shared. Consequently, the answer to a particular problem involving division may be the number that results, the number that remains or even a number that could be added to the remainder to allow further sharing. This source of difficulty in using and interpreting division highlights why the learning of a rote procedure is insufficient. Only a well-understood process will be able to be applied to the variety of problem situations that will arise.

THE SEQUENCE FOR DEVELOPING DIVISION

1 Develop the initial concept as sharing:
  verbal action stories
  use of arrays
  record using $\overline{)}$

2 Link division to the nearest multiplication fact

3 Division with renaming and remainders   1 digit$\overline{)}$2 digit then 1 digit$\overline{)}$3 digit

4 Extension to division of larger numbers   1 digit$\overline{)}$4 digit, 1 digit$\overline{)}$5 digit, 1 digit$\overline{)}$6 digit

5 Division with internal zeros in the answer

6 Extension of division to give decimal fractions rather than remainders

7 Strategies for estimation, mental computation and calculator use.

8 Introduction of two-digit divisors

Note: The use of ÷ can be phased in with the use of calculators for division while the alternative concept of division as repeated subtraction should be introduced as an aspect of problem solving with division when the sharing concept and the basic facts are secure.

The division concept

Division relies fundamentally on a good understanding of multiplication, both in the form of visualising the initial concept and in the use of multiplication facts to provide answers for the steps involved in the formal process. For this reason, it is essential for multiplication to be secure before division is formally introduced. Nonetheless, young children acquire an informal knowledge of division through everyday experiences involving sharing and this lays the foundation for the formal study of division from around the fourth year of school. Materials can be used to model these situations although, since division involves forming equal-sized groups, the same difficulty that occurred with multiplication will arise: groups can be formed in two quite different ways, raising the issue as to how the two conceptions are related. When the number of groups is known, the size of each group may be found by a process of sharing, known formally as partition.

There are 18 bananas in a bunch.
Three people will share them.
How many for each person?

3 sixes; 6 bananas for each person

When the size of each group is known, the number of groups may be found by a process that is essentially repeated subtraction. This repeated subtraction approach is also known as quotition.

There are 18 sunflowers.
Three flowers will be placed in each vase.
How many vases are needed?

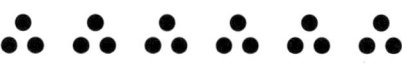

6 threes; 6 vases are needed

Both forms relate to multiplication and it is this relationship that provides the fundamental notion of division. Placing the objects into arrays also leads into a meaningful form of recording for the operation:

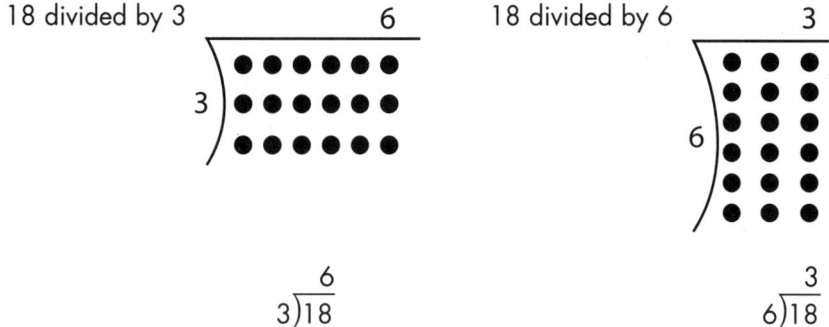

A complete conception of division requires both the sharing and repeated subtraction conceptions; however, the sharing approach provides the better basis for understanding. It is the one young children are more familiar with and enables a consistent development from the initial concept through the links to multiplication facts to the algorithm when the sharing process is related to the arrays used in multiplication.

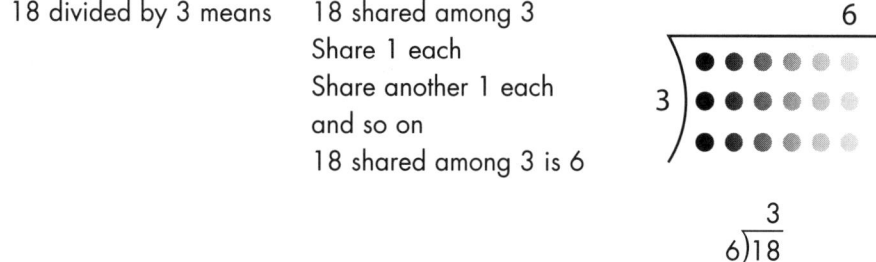

Children will need several experiences with everyday problems that require sharing to obtain an answer. One motivating way of providing activities is to make a number of boards with simple stories and cards that provide different numbers to match them. A child reads the story and shares Unifix or other markers onto a division mat to form an array which answers the question:

One child places a card on the story and reads the story with its numbers. The next child takes a number of Unifix to match the number in the story (24) and shares them onto the division mat. The next child uses the result shown on the array to read the whole story with its answer. The important point to be gained is that the number in each row provides each share.

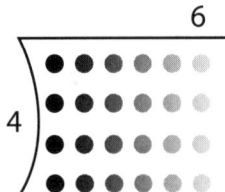

24 . . . shared among 4 . . . is 6 each

To make the activity more game-like and independent, pairs of children can play on different boards. Each pair places a card on their story at the same time. The first pair to obtain a correct answer reads their story with its answer and gains 1 point. The first pair to get 5 points wins that round.

The use of sharing also allows a meaningful interpretation of the processing used in completing the algorithm. Traditionally, an example such as $7\overline{)398}$ has begun with '7 into 3 won't go; how many sevens in 39?'. Yet children with a sound basis of numeration realise that the 3 represents 3 hundreds and that 300 can be divided by 7. In contrast, a sharing approach asks that 3 hundreds be shared among 7, which is not possible, then asks that 39 tens be shared among 7 and so on.

Similar difficulties arise with examples like $7\overline{)3.98}$. The only realistic answer to the question 'how many sevens in 3?' is none. In other words, the 'how many' language that is derived from the repeated subtraction process suggests that such division is not possible. Yet a realisation that division is the inverse of multiplication shows that an answer must exist. A sharing conception of division enables a meaningful answer to be given: '3 ones cannot be shared among 7, so 39 tenths are shared among 7'.

The emphasis in developing the division concept, then, needs to focus on a sequence of experiences involving sharing. The use of arrays in a manner paralleling that used to develop multiplication minimises the difficulties associated with two different forms of grouping and leads to the development of a meaningful language that can be associated with division. This sharing language can, in time, be replaced by a more formal language of 'divided by', but it is not appropriate to use 'goes into' or 'how many . . . in'. While these forms can be regarded as growing out of the repeated subtraction conception, this is not the fundamental basis for division and it can be readily conceived of in terms of sharing through the array model. These forms also have little relationship to the problem situations in which division occurs and are likely to lead to a rote interpretation of division, unrelated to the fundamental notion of sharing.

SEQUENCE FOR THE DEVELOPMENT OF THE CONCEPT OF DIVISION AS SHARING

1 Introduce the division concept through everyday problems that involve sharing.
2 Guide children to use counters in arrays to match the problem situations. At first a language in terms of sharing should be used, but in time this needs to give way to a more formal language. These activities also develop an awareness of the link between division and multiplication.
15 shared among 3:

$$3\overline{)\begin{matrix}5\text{ each}\\ \bullet\bullet\bullet\bullet\bullet\\ \bullet\bullet\bullet\bullet\bullet\\ \bullet\bullet\bullet\bullet\bullet\end{matrix}}$$

3 Gradually introduce the division symbol $\overline{)}$ by relating the recording to the way in which the array has been constructed.

15 shared among 3 is 5 $3\overline{)15}^{\,5}$ 15 divided by 3 is 5

4 Use of $)$ from the outset, rather than ÷, has proven to be beneficial in developing the algorithm in the same way that vertical recording assisted in building up addition, subtraction and multiplication. Moreover, it clearly distinguishes the number to be shared from the number sharing:

number sharing$\overline{)\text{number to be shared}}^{\,\text{number each gets}}$

5 While the fundamental understanding of division is in terms of sharing, it is the link to multiplication that allows division with larger numbers. This grows out of the use of arrays but later there needs to be an explicit stress on the thinking these reveal:

$6\overline{)54}$ means 54 divided by 6; 6 nines are 54 so $6\overline{)54}^{\,9}$

6 Use of 'goes into' or 'how many . . . in' for division is not helpful. The informal language of 'shared among' should be replaced by 'divided by' as the formal expression for division in due course, but this needs to be understood as another way of expressing the notion of sharing.

Introducing remainders

Division with larger numbers will involve remainders and renaming from the outset. For this reason, it is important to stress that sharing means that each person is given the same amount. In particular, it is helpful to avoid phrases such as 'share equally' since one share must be equal to any other, and it also leaves open the question that it may be possible to share and each person get a different amount. Instead, when the sharing is done, there may be some objects that remain after sharing and it is for this reason that the word 'remainder' is used to describe them. This idea of remainders needs to be developed with the division concept, using the same activities used earlier.

One child places a card on the story and reads the story with its numbers. The next child takes a number of objects to match the number in the story (17) and shares them onto the division mat. The next child uses the result shown on the array, together with the remaining objects, to read the whole story with its answer.

$$3{\overline{\smash{\big)}\,\begin{matrix}\bullet\bullet\bullet\bullet\bullet\\\bullet\bullet\bullet\bullet\bullet\\\bullet\bullet\bullet\bullet\bullet\end{matrix}}}\;\bullet\bullet$$

17 . . . shared among 3 . . . is 5 each; 2 remain

To make the activity more game-like and independent, pairs of children can play on different boards. Each pair places a card on their story at the same time. The first pair to obtain a correct answer reads their story and gains 1 point. The first pair to get 5 points wins that round.

This use of arrays with some remaining can then lead to a meaningful recording to show remainders:

15 divided by 3 is recorded $3\overline{)15}^{\,5}$, and read 15 divided by 3 is 5

17 divided by 3 is recorded $3\overline{)17}^{\,5r2}$, and read 17 divided by 3 is 5 remainder 2

Extending the division concept

When the concept of division as sharing has been established, the alternative form of division as repeated subtraction can be introduced and related to both the use of the array model and the fundamental sharing idea. Problems of this form require several objects to be taken at a time and put into some special form of arrangement. These can readily be set out in rows so that the answer to the problem is given by determining the number of rows. In this way, not only is the corresponding multiplication idea drawn on to find or check the answer but the resulting arrangement immediately suggests an interpretation of the problem in terms of sharing.

Melissa is making fabric flowers
Each flower needs 6 petals
There are 18 petals altogether
How many flowers can she make?

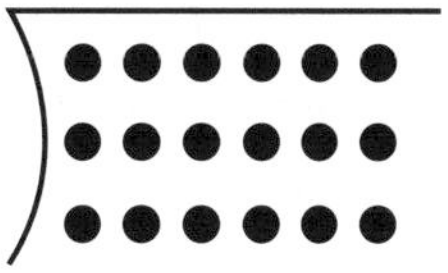

There are 3 rows of 6.
The answer is given by 3 sixes are 18 or 18 divided by 6 is 3

At a later stage, the use of the ÷ symbol can be introduced. This is not necessary for some time because of the unambiguous meaning portrayed by the $\overline{)\quad}$ symbol. However, just as the = symbol needs to be introduced to enable computations involving several operations to be written and to allow the meaningful use of a calculator, so in time this alternative way of showing division needs to be taught. Indeed, it is helpful to introduce the new symbol ÷ through the use of a calculator as it draws attention immediately to the fundamental difficulty with its use:

42 divided by 7	is written $7\overline{)42}$ and read from right to left
or	is written 42 ÷ 7 and read from left to right

This difference has been the source of much difficulty for children and at the very least they need to be made explicitly aware of the different ways of reading and interpreting these two ways of recording the same division. Providing several examples for children to match and then use a calculator to check whether their answers are correct is one good way of allowing them to build in this understanding.

Match the division example to the corresponding expression:

28 ÷ 7	is the same as	$28\overline{)7}$ or $7\overline{)28}$
$7\overline{)63}$	is the same as	7 ÷ 63 or 63 ÷ 7
35 divided by 5	is the same as	35 ÷ 5 or 5 ÷ 35
56 divided by 8	is the same as	$56\overline{)8}$ or $8\overline{)56}$
36 divided by 4	is the same as	$36\overline{)4}$ or 4 ÷ 36
8 divided by 2	is the same as	8 ÷ 2 or $8\overline{)2}$

Likely difficulties

Difficulties with the concept usually occur because of a focus on the symbols and answers rather than the meaning associated with sharing. Consequently, a child may be unable to give a story to match a picture or symbols, may rush to give an answer rather than read the division, and may not be able to show the situation using materials.

- When asked to read an example such as $4\overline{)20}$ a child may not be able to read the expression appropriately, saying '4 divided by 20' instead, which is likely to lead to difficulties with division for larger numbers.
- A child may say '4 into 20' or 'how many fours in 20' indicating that he or she has no meaning for division other than as a command to provide an answer.
- When asked to use materials to show division, the child may simply provide objects to match the answer rather than the parts involved in the story or written expression. For example, a child asked to show $4\overline{)20}$ with materials put out

    ●●●●   ●   ●●●●●   ●

    and explained his result   'four   times   five   equals   20'
- The division may be recorded in a similar fashion to the other operations: $\div 5$ instead of $5\overline{)20}$.
- When asked to provide a division story to match a written expression or picture, a child may simply provide a story which matches the numbers but does not involve division, often confusing what is required with addition or multiplication.

Investigate and discuss

1. Record each of these situations with its answer. Which are sharing and which are repeated subtraction?
    a. Murray has 30 roses. How many bunches of 6 roses can he make?
    b. Melanie picked 32 ripe mangoes. She picked the same number from each of 4 trees. How many mangoes did she get from each tree?
    c. Miranda grows herbs in pots to sell at the market. She has 75 pots to pack into boxes. If she puts 9 in each box, how many boxes will she have to take to the market?
    d. Max needs 28 tomato plants for his garden. The plants are sold in punnets of 6. How many punnets must he buy?
2. Once children are familiar with the idea of sharing and the $\overline{)}$ symbol, the formal word division and the associated language, divided by, can be introduced.
    How could you use materials to give meaning to the statement '27 divided by 6 is 4 remainder 3'?
3. Provide a division story to match each of these expressions: $4\overline{)36}$, $6\overline{)25}$, $48 \div 8$, $37 \div 7$

Linking division to the nearest multiplication fact

Division introduces a new way of thinking that has not been present in any of the other operations. Simply recalling a basic fact will not provide a complete answer. There will usually be a remainder and the task will <u>not</u> be to provide a 'division fact' but to use the nearest multiplication fact. For example, $4\overline{)976}$ uses 4 twos are eight to get as close as possible to 9 and 4 fours for the remaining steps—$4\overline{)976}$ is 244. Situations such as $4\overline{)848}$ or $4\overline{)832}$ where multiplication facts can be used directly are rare.

Sharing objects in arrays to show division with and without remainders has established the basis for this thinking. It now needs to be developed to the point that when a division statement is met, a multiplication fact is automatically brought to mind. A secure knowledge of the multiplication facts is obviously a necessary prerequisite and these facts should be reviewed before an automatic link to division can be built up. Division examples such as $4\overline{)24}$ or $6\overline{)42}$ can then be shown with arrays and the corresponding multiplication fact linked to the result:

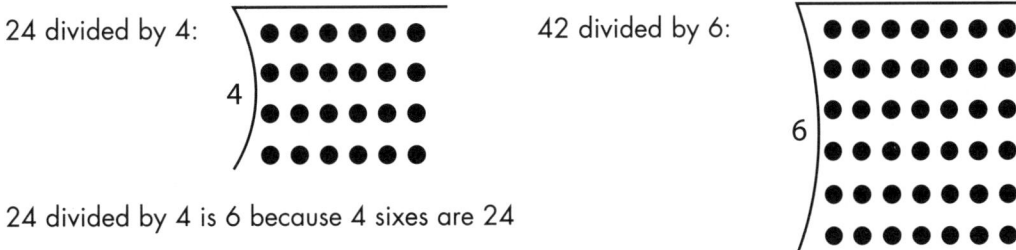

24 divided by 4 is 6 because 4 sixes are 24

42 divided by 6 is 7 because 6 sevens are 42

Arrays can also be provided for children to write the corresponding multiplication and division:

shows $\begin{array}{r}7\\ \times 3\end{array}$ and $3\overline{)21}$

This can be extended to writing all four multiplication and division statements associated with an array:

shows $\begin{array}{r}4\\ \times 5\end{array}$ and $5\overline{)20}$ or $\begin{array}{r}5\\ \times 4\end{array}$ and $4\overline{)20}$

Use can also be made of a number fact wheel to link multiplication and division in a similar way to that used earlier for addition and subtraction:

A number fact wheel to develop the relationship between multiplication facts and division is simply constructed by inserting a wheel with 10 equal parts (to accommodate the digits 0–9 on the multiplication side) between a double-sided cover of the form indicated in this illustration.

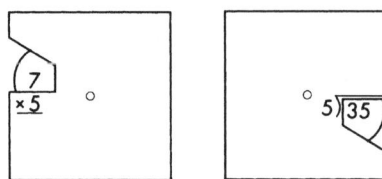

For this number fact wheel based on multiplying and dividing with 5, first have one child nominate a number between 0 and 9, for example 7, then write this in the space on the wheel on the multiplication side. Since the answer to 5 sevens is 35, this answer can be written on the space on the other side of the wheel. The division side now reads 35 divided by 5 and the answer 7 is found on the multiplication side of the wheel. The multiplication fact has been used to generate the answer to the division fact. By repeating this for the remaining digits between 0 and 9 chosen in a random order, the activity brings out strongly the link between multiplication facts and division.

Another means of making this link explicit is to use an understanding of multiplication in terms of factors and product to interpret division as calling for the provision of a missing factor when the product and one factor are known:

Write a multiplication fact on the board.	6	6	...
Ask the children what the product is.	× 4	× ...	× 4
Then cover one of the factors and ask what it must be.	24	24	24
Repeat for the other factor.			
After some introductory examples,	8	?	
the activity need only focus on the missing factors:	× ?	× 7	
	72	56	

Games can be used to encourage children to answer division examples by thinking of the corresponding multiplication fact directly:

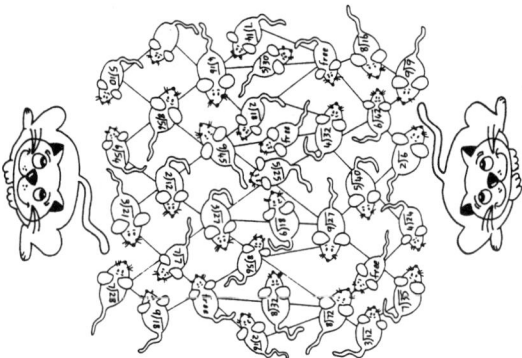

In order to move a counter to the first position, the number rolled on the dice has to match the answer. At first a player will complete all of the divisions and then check to see whether one of the answers corresponding to a possible move corresponds to the roll of the dice. As playing continues, players realise that it is more efficient simply to try multiplying the number on the dice by the number that is dividing to see if the product matches the number being divided.

Later, this can be changed to a game which focuses on the remainder, where there are examples on the board of the form $5\overline{)37}$, $6\overline{)53}$, $7\overline{)55}$. In order to move a counter, the number rolled on the dice has to match the remainder given when the division is completed, including rolling 0 for examples like $8\overline{)64}$. Now players have to focus on the nearest multiplication fact, using multiples of the number dividing to find the remainder.

Division with one and zero

These links between division and multiplication provide almost all the thinking that will be needed for problem situations and the algorithms. However, particular consideration needs to be given to situations that correspond to multiplication by one and zero. An array readily demonstrates that a number divided by itself is 1 and a number divided by 1 is that number:

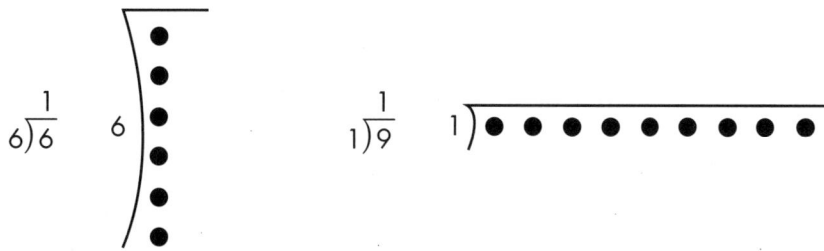

However, situations corresponding to multiplication with zero are not so straightforward and have often been confusing to children and their teachers. Since there are generalisations to further mathematics that depend on an understanding of division involving zero, it is important for these to be developed to form a complete conception of division. If zero objects are shared out, each of those sharing must receive no objects—in other words, zero divided by any number is zero. This will be particularly important in division such as $3\overline{)4202}$ or $7\overline{)49053}$ where zero will need to be recorded in the tens or hundreds place to show that none are shared.

When a number is divided by zero, many children confuse this with multiplication involving zero and expect the answer also to be zero. Indeed, there is widespread confusion in the community about this result and the most frequently given answer is infinity. Yet, infinity is not a number and it is necessary to go back to the initial concept to determine unambiguously what the actual answer is. Division by zero asks for a number of objects to be shared to no one. A little thought shows that this means that sharing does not in fact occur; sharing among zero and thus dividing by zero is not possible. Further, for division by zero to be possible, it must also be possible to find a number that can multiply by zero to give a particular number. For example,

For $0\overline{)8}$ to have an answer, that answer must multiply by 0 to give 8.
This is not possible as multiplying with 0 is always 0.
The same situation will occur for any number at all so division by zero is not possible.
Mathematically, division of any number is said to be *undefined*.

Note that zero divided by zero is also said to be undefined. Otherwise, any number would provide an answer (any number multiplied by zero is zero), implying that zero divided by zero would have an infinite number of correct answers, which clearly could not be correct.

It is interesting at this point to compare the use of the repeated subtraction conception of division for zero with this reasoning. Division of zero by any non-zero number, for example 8, would mean that 8 would have to be subtracted from zero to give an answer. This is not possible so repeated subtraction suggests it is division of zero that is undefined. Yet consideration of multiplication immediately says that since 8 zeros are zero, zero divided by eight must be zero. The repeated subtraction interpretation is incorrect while the sharing conception gives the correct result.

When division by zero is considered, for example $0\overline{)8}$, repeated subtraction suggests that zero be taken away from eight as many times as possible. Clearly this can happen indefinitely, leading to the erroneous view that division by zero gives infinity. Not only does this show that division with zero is not answered by the repeated subtraction conception of division, it also underlines the initial reasoning that sharing is the essence of division. It is certainly the easiest approach to use in developing the concept, the link with multiplication and the development of the basic facts. But it is also the only approach that allows for the consistent development of division from beginning to end.

Once again, this example has shown that the essential understanding for division is based on sharing and the relationship of division to multiplication. Sharing gives meaning to the process of division while an ability to think readily of the associated multiplication facts allows the process to proceed. Being able to link division to multiplication means that appropriate factors can be found at each step of the algorithm, be it a pencil and paper, mental or estimation process.

Likely difficulties

Difficulties usually occur because a child cannot connect division with corresponding multiplication facts, especially when remainders are involved.

- Simply guessing or trying a number of multiplication facts, often by reciting them in order, may be the only means the child has to try to obtain answers.
- The underlying multiplication facts themselves may not be known so that incorrect answers are given.
- Division involving 7, 8 or 9 may not be attempted because a child recognises that these multiplication facts are difficult for them.
- When asked to answer a division involving a remainder, such as $7\overline{)46}$, a child may say they 'don't know' because there is no matching multiplication fact or use the ones digit as a cue and answer 8 by thinking of 7 eights are fifty-six.

Investigate and discuss

1. It is often said that since multiplication and division are inverse operations, they should be taught together.
   a. At what point in the development of multiplication and division would you introduce the division concept?
   b. When would you commence division to multiplication facts?
   c. What is the role of 'families of facts' such as

   $$\begin{array}{cc} 9 & 4 \\ \times 4 & \times 9 \\ \hline 36 & 36 \end{array} \qquad 4\overline{)36} \qquad 9\overline{)36}$$

2. Make sets of division number fact wheels as shown on p. 312 and use them with children at about Year 4. Which facts would you use to initiate thinking that links needed division to known multiplication facts?
3. Make the division games shown on p. 313. Play them with a partner and consider what makes them easy or difficult. What thinking did you use to determine whether you could move when the dice was rolled?
4. Change the division on the board to allow for remainders. How readily were you able to determine the moves indicated by the dice when playing this game?

The division algorithm

The development of the division algorithm is a straightforward extension of the sharing process used for the concept and basic facts. Materials can be used to introduce the steps in the algorithm and develop a meaningful language to govern the recording for the written form in the same way as addition and subtraction were built up. Indeed, this direct use of materials to establish what is needed at every stage of the algorithm should make the learning of division so much easier than multiplication. Although there may be many steps, sharing place by place and renaming where necessary provides a cycle of thinking that governs the whole process, so that the division algorithm should almost be the easiest of all to learn.

It would seem unlikely that many children could experience difficulty with division until common practices, particularly the ill-conceived language that accompanies any recording, are examined. Much school practice involves meaningless 'crutches' to support a rote procedure, a variety of recorded forms chosen in an *ad hoc* manner, and a range of misleading names for what is only one algorithm. In particular, there really is no such thing as 'short division'; the length of the process is always governed by the size of the numbers involved, only the amount of recording varies. This form of division is expected to be done without any recording of the steps involved in the process, instead requiring that they be kept track of mentally. The higher level of thinking that this demands means that any form of division without recording can only be 'short' for those skilled in division. It is highly inappropriate to emphasise such an approach at the introduction of division and only reasonable to expect such a facility at the end of some long developmental time, if at all.

$$\begin{array}{r} 721 \\ 6\overline{)4329} \\ \underline{42} \\ 12 \\ \underline{12} \\ 9 \\ \underline{6} \\ 3 \end{array}$$ shows what has occurred at each step, whereas $6\overline{)43^{1}29}^{\,7\ 21}$ requires learners to keep the meaning in their mind, with no guidance from the small digits written inside the number being divided

For these reasons, no mention is made of division without recording here. Instead, the emphasis is on replacing inappropriate practices with the form of language, use of materials and meaningful recording based on an understanding of place value and renaming that have ensured success with the algorithms for addition, subtraction and multiplication.

The major stumbling block for children is that there are no basic facts that can be used directly. At each step, children need to find the nearest multiplication fact to the number being divided and also use the amount that remains to form the next number they need to divide. This means that there will be a need to rename thousands as hundreds, hundreds as tens, or tens as ones from the outset. Division without renaming occurs infrequently and even then demands no different thinking, simply the sharing on which division relies, so will be considered among the renaming situations. Later, difficulties caused by internal zeros in the answer will also need to be considered and division continued so that the remainder is also divided and decimal fractions are introduced into the answers. It is only a small step to build up the division of decimal fractions from this, but division by decimal fractions as such need not be dwelt on. As division with common fractions really has no practical applications whatsoever it is no longer considered appropriate for primary school mathematics programs. Most uses within mathematics itself can be carried out from an understanding of the division concept, while any further processes that might be helpful can be developed in the context of the situations for which they are needed.

SEQUENCE FOR DEVELOPING THE DIVISION ALGORITHM

1 Once the concept and link to multiplication facts have been learned, extend this thinking to share tens and ones.
2 Division includes renaming and remainders from the outset to introduce the form of the algorithm, the need to begin with the largest place, and the particular use of materials, language and recording.
3 Division with no renaming and no remainders is included among examples with renaming as the thinking involved is identical.
4 This understanding is broadened to include division where there are internal zeros in the answer.
5 The algorithm is continued to include division of the remainders to give a decimal fraction as the answer. This leads to the division of decimal fractions by whole numbers.
6 This process and understanding is extended to include division by two-digit numbers, estimation, mental computation and calculators.

Division with tens and ones

As division is extended from the basic facts to building up the initial algorithm, the only new thinking involved is to share tens then share ones, rather than treat the number solely as ones to be shared. Bundling sticks or Base 10 materials are useful in aiding this transition from

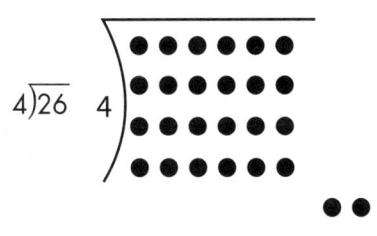

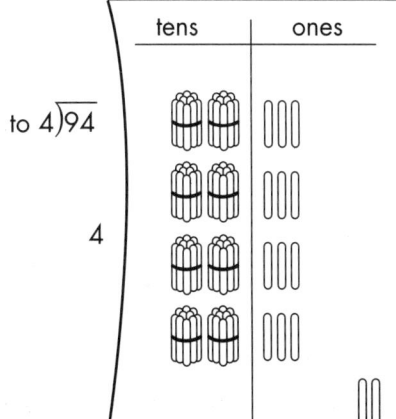

26 divided by 4 is 6 remainder 2   94 divided by 4 is 23 remainder 2

While the use of materials may encourage some children to continue to share one at a time, sharing the materials in arrays on a place value chart will suggest that the tens be shared first. Any child who does share one at a time can then be led to rename their answer into tens and ones. A discussion centred on the efficiency of sharing one at a time when the answer is actually tens and ones can be used to develop an understanding of sharing the larger place first.

Dealing with the tens before the ones is contrary to all of the other algorithms. This conflict with the earlier thinking may not only cause uncertainty with using the materials, but there may also be confusion about where to record the result of sharing the tens. Should it be written above the tens place or at the right in the ones place where the recording for addition, subtraction and multiplication began? It is important that experiences with materials establish this new thinking that tens are shared, any renaming is carried out, then ones are shared before the recording of division is introduced. The recording can then match the use of materials directly to give meaning to the steps involved:

Language	Materials	Recording
94 divided by 4 **Can you share** 9 tens among 4? **Yes** 2 tens each 8 tens are shared out 1 ten remains **Can you share** 1 ten among 4? **No: rename**—14 ones to share	4 { tens | ones (8 bundles of tens, empty ones)	$\begin{array}{r} 2 \\ 4\overline{)9\,4} \\ \underline{8} \\ 1\,4 \end{array}$
Can you share 14 ones among 4? **Yes** 3 ones each 12 ones are shared out 2 ones **remain**	4 { tens | ones (8 bundles of tens, 14 ones, 2 remain)	$\begin{array}{r} 2\,3\ \text{r}2 \\ 4\overline{)9\,4} \\ \underline{8} \\ 1\,4 \\ \underline{1\,2} \\ 2 \end{array}$

94 divided by 4 is 23r2

Using examples with renaming to introduce the division algorithm minimises any tendency to provide answers to division without the associated recording and so avoids any use of so-called 'short' division. It is much more appropriate to show all recording from the outset as it is designed to match the use of materials and to provide a means of governing the many steps involved in the division algorithm. At some later stage, it may be possible to dispense with the recording, but only an individual child will know whether the process is sufficiently secure to allow this.

Note that there has been no reference to 'bringing down the 4' in this process. In the first instance this is because quite literally the 4 is not 'brought down'. It is there to indicate the number that is being divided and remains written at the top of the recording! In actual fact, what has happened is that the remaining 1 ten has been renamed as 10 ones giving 14 ones altogether. The 4 has then been written alongside the 1 to indicate that there are now 14 ones to be shared. Rather than a meaningless—and incorrect—statement about bringing down, the algorithm focuses on what needs to be shared at each step. In the same way, now that there are reasons behind the steps in the process, there is no need to adopt a custom of drawing an arrow from the digit in the number being divided to where it is rewritten on the line where the next amount to be shared is considered. It is language that gives meaning to the processing of division, not a number of ritualised statements about 'bringing down' accompanied by the drawing of arrows.

Division with three-digit numbers

After the idea of division as sharing has been extended to two-digit numbers so that the reasons for beginning sharing with the tens have been accepted, the process can be continued to three-digit numbers where the hundreds are shared first, then the tens and last of all the ones. Base 10 materials should be used with the two-digit numbers before being used to extend this process to three-digit numbers, sharing the objects place by place, beginning with the largest place:

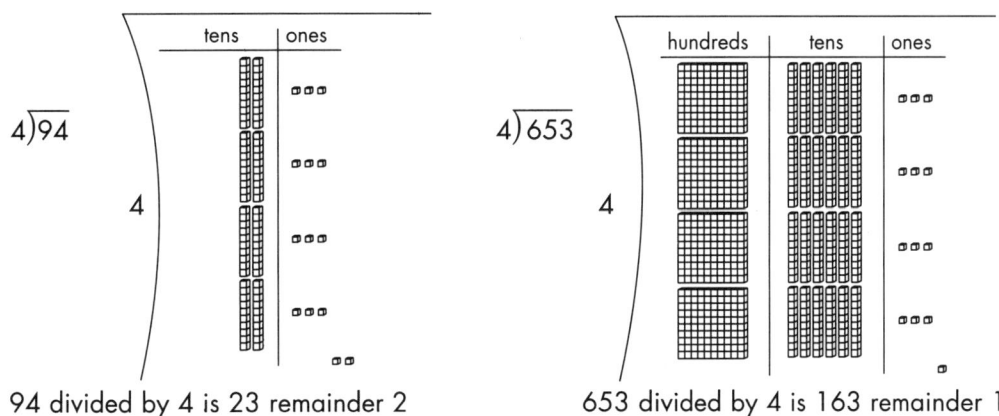

94 divided by 4 is 23 remainder 2   653 divided by 4 is 163 remainder 1

The process of sharing with materials leads directly to a meaningful recording process. It is largely a matter of verbalising what is occurring and, by reference to the underlying numeration ideas, simply evolving a recording step to match each action with the materials.

Not only does place value indicate how the recording should occur, hundreds in the hundreds place, tens in the tens place and so on, but it clearly spells out how one step succeeds another.

Language	Materials	Recording
653 divided by 4 **Can you share** 6 hundreds among 4? **Yes** 1 hundred each 4 hundreds are shared out 4 2 hundreds remains **Can you share** 2 hundreds among 4? **No: rename** — 25 tens to share	hundreds | tens | ones	$\,1$ $4\overline{)6\,5\,3}$ $\underline{4}$ $2\,5$
Can you share 25 tens among 4? **Yes** 6 tens each 24 tens are shared out 4 1 ten remains **Can you share** 1 ten among 4? **No: rename** — 13 ones to share	hundreds | tens | ones	$\,1\,6$ $4\overline{)6\,5\,3}$ $\underline{4}$ $2\,5$ $\underline{2\,4}$ $\,\,1\,3$
Can you share 13 ones among 4? **Yes** 3 ones each 12 ones are shared out 4 1 one **remains**	hundreds | tens | ones	$\,1\,6\,3\,\text{r}1$ $4\overline{)6\,5\,3}$ $\underline{4}$ $2\,5$ $\underline{2\,4}$ $\,\,1\,3$ $\,\,\underline{1\,2}$ $\,\,\,\,\,\,1$

653 divided by 4 is 163r1

At first, examples should involve sharing of each place, but when this is familiar it needs to extend to division where it is not possible to share the hundreds and these need to be renamed to tens before the division can proceed. Materials should be shared to show how at first, but later a number expander can be used to show how many tens the number being divided has and link directly to the written process:

Language	Materials	Recording
653 divided by 7 **Can you share** 6 hundreds among 7? **No: rename**—65 tens to share **Can you share** 65 tens among 7? **Yes** 9 tens each 63 tens are shared out 2 tens remain **Can you share** 2 ten among 7? **No: rename**—23 ones to share	hundreds | tens | ones (number expander showing 7)	$$\begin{array}{r} 9 \\ 7\overline{)653} \\ \underline{63} \\ 23 \end{array}$$
Can you share 23 ones among 7? **Yes** 3 ones each 21 ones are shared out 2 ones **remain**	hundreds | tens | ones (number expander showing 7)	$$\begin{array}{r} 93\,r2 \\ 7\overline{)653} \\ \underline{63} \\ 23 \\ \underline{21} \\ 2 \end{array}$$

653 divided by 7 is 93r2

The realisation that hundreds cannot be shared and need to be renamed as tens is a crucial step in the building up of division understanding. There is a tendency to say that, for example, 1 hundred could be shared among five, but since after sharing they would all receive 2 tens it is not correct—tens were shared if those sharing receive tens. It is also important not to record zero when it is not needed, for instance above the largest place when it is the next place that is shared, and to leave it for when it is significant, such as within the answer to show that there are no tens or ones shared.

Many children, when coming to terms with the new left to right recording for division, think that the first digit that results must be recorded in the left-most place. A stress on sharing place by place, coupled with understanding which place is being shared fostered by a number expander, ensures that this tendency will be minimised if not avoided altogether. In contrast, a rote procedure in terms of 'goes into' or even 'divide, multiply, subtract, bring down' provides no guidance, leading to the suggestion that a zero be written in the first place to stop children writing another digit there. But an answer will then begin with a zero, so what whole number does it refer to? The only numbers that begin with a zero are those decimal fractions between 0 and 1 such as 0.4 or 0.75. At best the writing of a zero above the initial digit of the number being divided will be accepted at face value as another technique the teacher wants adopted, but it may confuse. Even the child looking for meaning may misread the number, especially two-digit numbers where 05 is often interpreted as fifty. For this reason, zeros should not be written into the division algorithm at any point other than where they signify that there are none of a particular place. Similarly, for examples where there are no remainders, the division can simply conclude without recording a zero at the bottom of the division or as the remainder:

Zero is significant—it shows no ones were shared:

```
                          130r2
                        5)652
Share the hundreds         5
Rename and share the tens 15
                          15
Rename and share the one   2
```

Zeros are not recorded when there is no remainder:

```
                          125
                        5)625
Share the hundreds         5
Rename and share the tens 12
                          10
Rename and share the ones 25
                          25
```

Zeros are not recorded when no hundreds can be shared, or when all tens are shared out:

```
                    71r3
                 5)358
Share the hundreds  35
Rename and share the tens  8
                    5
Rename and share the ones  3
```

Division with larger numbers

While materials can be used to extend the sharing of materials place by place for an example such as 8)6348, it should be necessary only as a first step to show that the thinking and recording for thousands continues the same pattern. First share the thousands, then share the hundreds, then share the tens, then share the ones. When it is not possible to share the thousands, as in 8)6348, a number expander is useful in showing that the first step in this case must be to share the 63 hundreds. This also emphasises that since it is hundreds that are being shared, the result must be recorded in the hundreds place.

Step	Language	Recording
Share the thousands	6348 divided by 8 **Can you share** 6 thousands among 8? **No: rename**—63 hundreds to share	8)6 3 4 8
Share the hundreds	**Can you share** 63 hundreds among 8? **Yes** 7 hundreds 56 hundreds are shared out 7 hundreds remain **Can you share** 7 hundreds among 8? **No: rename**—74 tens to share	7 8)6 3 4 8 5 6 7 4
Share the tens	**Can you share** 74 tens among 8? **Yes** 9 tens each 72 tens are shared out 2 tens remain **Can you share** 2 tens among 8? **No: rename**—28 ones to share	7 9 8)6 3 4 8 5 6 7 4 7 2 2 8
Share the ones	**Can you share** 28 ones among 8? **Yes** 3 ones each 24 ones are shared out 4 ones remain	7 9 3r4 8)6 3 4 8 5 6 7 4 7 2 2 8 2 4 4
	6348 divided by 8 is 793 remainder 4	

This same thinking readily extends to five-digit and six-digit numbers:

Can you share the ten thousands?	7857 6)47142	Can you share the hundred thousands?	87927r1 3)263782
No: Rename and share the thousands	42	No: Rename and share the ten thousands	24
Rename and share the hundreds	51	Rename and share the thousands	23
	48		21
Rename and share the tens	34	Rename and share the hundreds	27
	30		27
Rename and share the ones	42	Share the tens	8
	42		6
		Rename and share the ones	22
			21
			1

However, there will be occasions when there are not enough of a particular place to share, so zero will need to be recorded in that place to show this. Children frequently have difficulties with examples of this kind, so zeros occurring within an answer need to be developed carefully after the cases where there are enough of each place to share have been built up.

Division with internal zeros

Since renaming occurs with division in a similar manner to subtraction, it is reasonable to anticipate that difficulties might arise with zeros. Situations such as those discussed above where there are not enough thousands or hundreds in the first place provide one such example. Rather than record a zero in that place, the total number of hundreds or tens needs to be shared and the digit that arises placed accordingly. However, when there are not enough hundreds, tens or ones to share later in the division, zero needs to be written in that place to show this, otherwise children may continue to omit that place and give an answer such as

```
    6 6r1           661            606r1
  7)4237    or even  7)4237  instead of  7)4237
    42              42                 42
    37              37                 37
    36              36                 36
     1               1                  1
```

Using materials to develop this thinking with three-digit numbers is essential, as the amount shared on the division place value chart clearly shows the answer that results, the place where it was not possible to share and why zero needs to be recorded in this place:

Language	Materials	Recording
925 divided by 3 **Can you share** 9 hundreds among 3? **Yes** 3 hundreds each 9 hundreds are shared out No hundreds remain	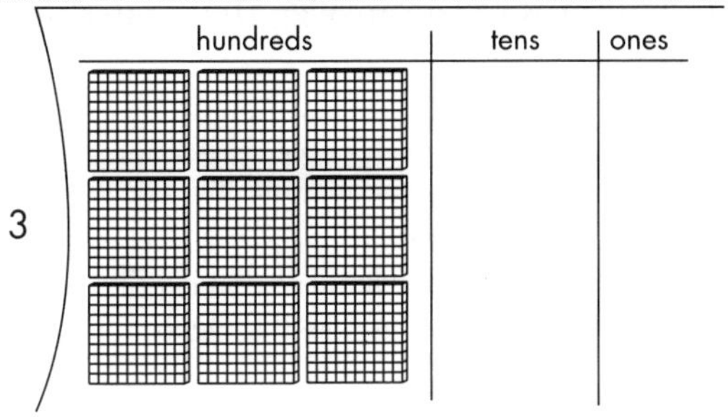 hundreds | tens | ones	3 $3\overline{)9\,2\,5}$ $\underline{9}$ 2
Can you share 2 tens among 3? **No: rename**—25 ones to share	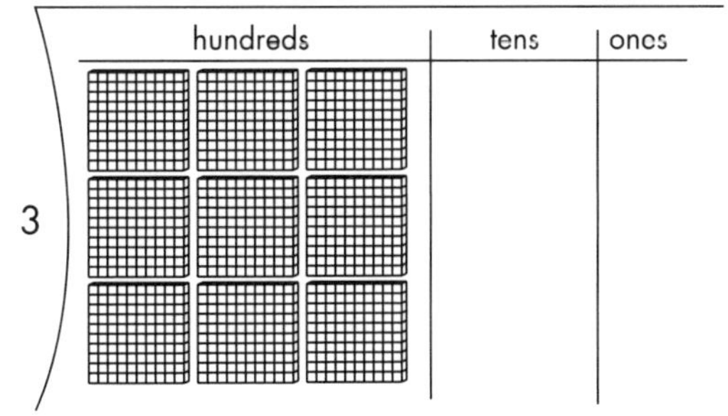 hundreds | tens | ones	30 $3\overline{)9\,2\,5}$ $\underline{9}$ $2\,5$
Can you share 25 ones among 3? **Yes** 8 ones each 24 ones are shared out 1 one **remains**	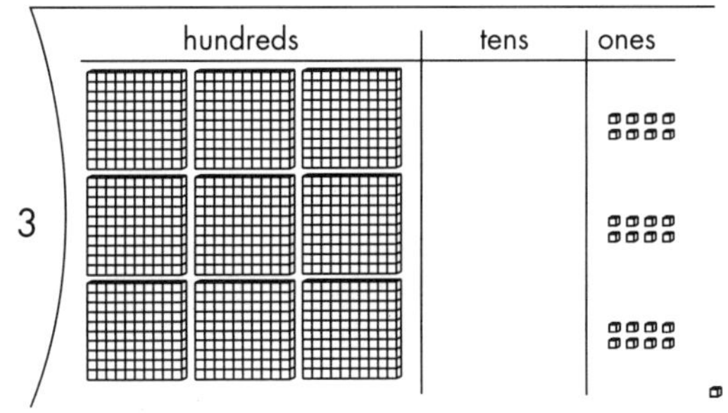 hundreds | tens | ones	$308\text{r}1$ $3\overline{)9\,2\,5}$ $\underline{9}$ $2\,5$ $\underline{2\,4}$ 1

925 divided by 3 is 308r1

Note that it is not an internal zero within the number being divided that causes any difficulty. These are no different from any other division:

905 divided by 7

Share the hundreds:
Rename and share the tens

Rename and share the ones

$$\begin{array}{r} 129\text{r}2 \\ 7\overline{)905} \\ \underline{7} \\ 20 \\ \underline{14} \\ 65 \\ \underline{63} \\ 2 \end{array}$$

While it is not really practical to use materials to show division with four-digit and larger numbers with internal zeros in the answer, the same thinking is used, based on sharing and recording each share.

Step	Language	Recording
Share the thousands	4237 divided by 7 **Can you share** 4 thousands among 7? **No: rename**—42 hundreds to share	7)4237
Share the hundreds	**Can you share** 42 hundreds among 7? **Yes** 6 hundreds 42 hundreds are shared out	6 7)4237 42
Share the tens	**Can you share** 3 tens among 7? **No:** 0 tens each **Rename**—37 ones to share	60 7)4237 42 3
Share the ones	**Can you share** 37 ones among 7? **Yes** 5 ones each 35 ones are shared out 2 ones remain	605r2 7)4237 42 37 35 2

4237 divided by 7 is 605 remainder 2

Division with five-digit and six-digit numbers simply extends this thinking, although zeros within the answer, and the subsequent likely confusion, may occur more often. A particularly difficult situation occurs when the ones cannot be shared and a zero needs to be recorded in the ones place with the ones recorded as a remainder:

	7 060r3		70 650r4
Can you share the ten thousands?	6)42 363	**Can you share** the hundred thousands?	5)353 254
No: rename and share the thousands	42	**No: rename** and share the ten thousands	35
Can you share the hundreds?	3	**Can you share** the thousands?	3 2
No: rename and share the tens	36	**No: rename** and share the hundreds	3 0
Can you share the ones	3	**Rename** and share the tens	25
No 3 ones remain			25
		Can you share the ones	4
		No 4 ones remain	

Likely difficulties

Difficulties with the algorithm usually have their origins with inappropriate recording of their working which does not refer to place value, sometimes in the form of 'short' division, other times because of a meaningless procedure related to 'into' or 'how many in' and the 'bringing down' of digits. There may also be particular misconceptions when zero occurs in an answer, or with the concept of remainder. There may also be errors due to an inability to access the nearest multiplication basic facts but these need to be addressed separately from the algorithm.

- *Inappropriate recording of the algorithm:*

$$3\overline{)639}^{\,213} \qquad 4\overline{)761}^{\,110} \qquad 85\overline{)761}^{\,000}$$

Asking 'how many in' each digit in turn without any recording of what is done leads to responses which make sense in terms of the questions asked, but not in terms of the whole division:

$$6\overline{)4236}^{\,7\ 6} \qquad 4\overline{)2438}^{\,6\ 9r2} \qquad 9\overline{)36\ 457}^{\,4\ 5\ r7}$$

When a place cannot be divided, the division is considered in separate parts.

- *Place value:*

$$4\overline{)287}^{\,71} \qquad 8\overline{)439}^{\,54r7} \qquad 6\overline{)520}^{\,860r4}$$

The result of dividing the first place that can be shared is simply recorded in the largest place. Sometimes a zero might also be placed to fill up the space above the last digit as in the third example.

- *Zeros in the answer:*

$$3\overline{)4223}^{\,14\ 7\ r2} \qquad 6\overline{)2439}^{\,4\ 6\ r3} \qquad 4\overline{)86\ 423}^{\,21\ 653}$$

Instead of a zero recorded when a place cannot be shared, the digit from sharing the next place is simply written in that place. Sometimes the remainder may be written to fill up space above the last digit as in the third example.

- *Renaming:*

$$4\overline{)7^36^21}^{\,1\ 1\ 0r6} \qquad 5\overline{)58^36}^{\,11\ 1r4} \qquad 7\overline{)8^102}^{\,1\ 00r3}$$

The small digits recorded alongside a digit in the answer have no meaning and are interpreted as remainders, rather than showing renaming for the next place.

◆ *Remainders:*

```
      925                758r12
 7)6478             7) 5318
    63                 49
    17                 41
    14                 35
    38                 68
    35                 56
     3                 12
```

The answer may be given without reference to the remainder

The remainder may be larger than the number dividing because an incorrect multiplication fact was used

◆ *Recording:*
Numbers may be confused when the recording does not place digits in the correct places. Use of quad-ruled paper or lined paper turned on its side to allow the correct placing of the long lines of recording is helpful.

Investigate and discuss

1 Use Base 10 materials and a place value chart to complete:

6)968   4)823   6)425   3)285   5)604   4)921   7)678

2 Record how these would be answered using the place value language outlined in this section.
How could you use the materials to provide insight into the occurrence of zeros in the answers?

3 Answer these examples:

7)8045   6)3452   8)4862   5)3524   3)6237   4)2802   9)6488

In which order would you introduce these division examples to children?

Division of the remainder

While many applications of division require remainders, there are other situations where the amount needs to be completely shared out to include a decimal fraction. In particular, this is the form given by a calculator so that there will also be a need to reconcile the two types of answers. The process established so far leads naturally into this extension: after thousands have been shared, then hundreds, then tens and then ones, it is clear that the process could continue to share tenths, then hundredths and so on. This continuation of the sharing process has been termed 'division of the remainder' as the ones remaining to be shared are renamed as tenths in the same way that tens that could not be shared were renamed as ones. In the past, this process has been termed 'write the answer as a decimal' and children have responded literally, changing their answer of 23r4 to 23.4. Sharing gives meaning to this extension of division and so inhibits the development of such an inappropriate response.

There is no need to use any materials in this process as it builds directly on the thinking developed for whole numbers. However, it is best to begin this process with numbers where tenths and hundredths result in the answer, using four-digit or larger numbers so that the place value basis for the sharing and the manner in which this governs each step is most obvious. Quite frequently, if only tenths result, the answer that is given by simply writing a decimal point in place of the 'r' for remainder gives a correct answer. For example, 9)8742 is 971r3 and 971.3 when the division continues to give tenths. When this is changed to 8)8742, the answers are 1092r6 and 1092.75 so that simply replacing 'r' with a decimal point is clearly seen to be inappropriate. Later, examples such as 8)6826 or 5)7652 can be included where only tenths result in the answer.

Step	Language	Recording
Share the thousands	6826 divided by 8 **Can you share** 6 thousands among 8? **No: rename**—68 hundreds to share	8)6 8 2 6
Share the hundreds	**Can you share** 68 hundreds among 8? **Yes** 8 hundreds 64 hundreds are shared out 4 hundreds remain **Can you share** 4 hundreds among 8? **No: rename**—42 tens to share	8 8)6 8 2 6 6 4 4
Share the tens	**Can you share** 42 tens among 8? **Yes** 5 tens each 40 tens are shared out 2 tens remain **Can you share** 2 tens among 8? **No: rename**—26 ones to share	8 5 8)6 8 2 6 6 4 4 2 4 0 26

Share the ones	**Can you share** 26 ones among 8? **Yes** 3 ones each 24 ones are shared out 2 ones remain **Can you share** 2 ones among 8? **No: rename**—20 tenths to share	853 $8)6826$ 64 42 40 26 24 20
Share the tenths	**Can you share** 20 tenths among 8? **Yes** 2 tenths each 16 tenths are shared out 4 tenths remain **Can you share** 4 tenths among 8? **No: rename**—40 hundredths to share	853.2 $8)6826$ 64 42 40 26 24 20 16 40
Share the hundredths	**Can you share** 40 hundredths among 8? **Yes** 5 hundredths each 40 hundredths are shared out	853.25 $8)6826$ 64 42 40 26 24 20 16 40 40

6826 divided by 8 is 853.25

Recording the 2 tenths is simply a matter of writing a full stop to show that the whole numbers have ended, then writing the 2 in the tenths place, which follows the decimal point. There is no need to write '.0' or '.00' in the number that is being divided because the sharing process says that tenths is the next place to be shared and the answer, 2 tenths, tells how it will be recorded. The 5 resulting from sharing the hundredths is simply written in the next decimal place.

Division invariably continues beyond two places for divisors and often produces non-terminating decimal fractions. This generates a need to round answers to *significant places* if the process is not to continue indefinitely and if the answers are to match problem situations realistically. The problem context will suggest how many places are to be considered significant; for instance, for the measurement of length it might be two places corresponding to centimetres or for mass it might be three places corresponding to grams. However, it will be necessary to continue to divide for one further place than the answer actually demands in order to determine what these digits will round to. Considerable understanding of division and the nature of decimal places in general, including the ability to round decimal fractions, will be needed before it can be meaningfully introduced. Nonetheless, the language and recording related to sharing is consistent with earlier division and the process can be developed relatively easily at an appropriate time.

Likely difficulties

- Confusion with the division process may lead to the r from the remainder simply being replaced with a decimal point.
- When the tenths cannot be shared, the hundredths digit rather than a zero might be written in the tenths place.
- When the answer is required to two or three significant places, the division might simply stop at two or three places, rather than continue to the next place to round the result.

Investigate and discuss

1. Answer each of these examples to obtain a decimal fraction using the recording and language outlined in this section. If the division continues beyond the hundredths, write your answers to three significant places.

    $8\overline{)5972}$ $6\overline{)7900}$ $4\overline{)3562}$ $8\overline{)5965}$ $3\overline{)5698}$ $6\overline{)9865}$ $7\overline{)4783}$

2. How would you assist a child to come to terms with the need to keep dividing for one place further than that asked for in order to obtain an appropriately rounded decimal fraction?

Estimation with division

Being able to readily complete the written division algorithm based on a meaningful process is critical to using it in many problem situations. However, a calculator will be used for most division with larger numbers and an awareness of whether an answer is reasonable is crucial in interpreting and using the results of such calculations. Estimation will be the key to assessing the reasonableness of these answers and will also be called on when exact answers are not needed and an approximation will be enough. An estimate is sufficient to see how many cartons would be needed to collect the bottles from the recycling area or how many strips of turf would be needed to cover a new lawn. Measurements involving area and volume are often approximate rather than exact; for example, the amount of paint to buy to provide two coats on all of the walls or the amount of pine bark to cover a garden bed to a particular depth. They often involve estimation of smaller sections before combining them using addition or subtraction. Thus, estimation for division demands a secure understanding of the division concepts and basic facts as well as an ability to integrate any partial results with estimates involving the other operations. As with multiplication, a good visual sense is needed to see what will be involved in practical situations.

Not only do estimation and approximation skills enhance an ability to deal with everyday quantitative situations, they are also crucial within the algorithm itself once divisors with more than two digits are met. With a single-digit divisor, multiplication and division facts are all that is needed to determine each step, but with divisors like 57 an estimate needs to be made first, then checked by exact multiplication and adjusted if necessary. Developing skills and understanding with division is not only feasible once the process has been meaningfully established, it is essential to further development, as well as in everyday applications.

Since division has always been carried out from left to right, an initial use of a front-end strategy is clearly going to be helpful, although this needs to be combined with the understanding of division in terms of sharing. Knowing that sharing the hundreds gives hundreds, for instance, is almost all that is needed to allow good estimates to be made.

6724 divided by 9

$$\begin{array}{r}7\ldots\\9\overline{)6724}\\\underline{63}\\42\end{array}$$

Can you share 6 thousands among 9?
No share 67 hundreds
about 7 hundreds each
42 tens to share among 9
Almost 5 tens each
750 would be a closer estimate

Later, rounding assists in the process in the same way as it did for division with two-digit divisors:

24 273 divided by 37 about

$$\begin{array}{r}6\ldots\\40\overline{)24\,273}\end{array}$$

Can you share 24 thousands among 37?
No share 242 hundreds
37 is close to 40;
about 6 hundreds or 600 each

Similarly, rounding can be very useful when division with decimal fractions is to be considered:

173.46 divided by 5.3 about

$$\begin{array}{r}3\ldots\\5\overline{)170}\end{array}$$

5.3 is about 5, 173.46 is about 170
Can you share 1 hundred among 5?
No share 17 tens;
about 3 tens or 30 each

The use of a front-end strategy commonly leads to an underestimate, however, in most cases continuing to share the next place to make an adjustment is almost as difficult as carrying out the exact calculation itself. To get a closer estimate it is often more practicable to consider the next multiple of the divisor after that given by sharing the largest place possible.

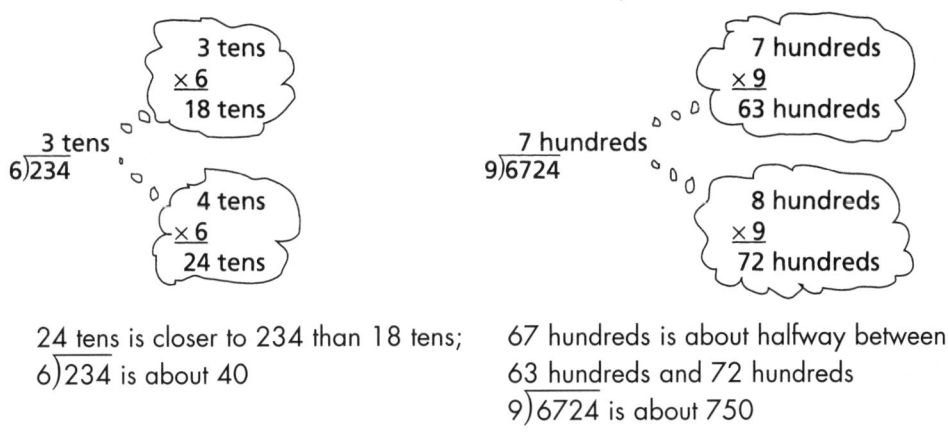

24 tens is closer to 234 than 18 tens;
6)234 is about 40

67 hundreds is about halfway between 63 hundreds and 72 hundreds
9)6724 is about 750

Estimation strategies for division build directly on the thinking developed as an integral part of the written algorithm in a similar manner to the way estimation of multiplication grew out of the understanding required for multiplying large numbers. In contrast, for addition, new thinking involving a front-end strategy proved helpful, as much to build up an acceptance of the idea of finding results close to what was needed instead of exact answers, as to allow a specific new method. Subtraction built in the rounding-based thinking that was also the key to multiplication. Thus, while computational estimation is very important in contemporary society, it is not a topic that should be developed independently of exact computation. Nor should it be treated in the detail that in the past accompanied the development of pencil and paper processes. It is as much dependent on the earlier learning of numeration, basic facts and the thinking that underpins the various algorithms as it is on new ways of thinking. Estimation needs to be seen as a process that builds smoothly from the initial front-end considerations given to addition through the rounding-based thinking that allows estimates to be made for subtraction and multiplication to the use of both front-end and rounding for division. Then its power will be appreciated and it can become an integral part of the thinking surrounding the four operations.

Dividing numbers mentally

Although approximate answers are sometimes useful for division, there is frequently a need for accurate answers calculated mentally—for instance, in sharing a restaurant bill among several friends, or in deciding on a monthly schedule to meet the demands of a lay-by transaction. This requires easily managed processes building on the ways of thinking developed with the written process or even the use of multiplication to determine a result rather than division. There may also be ways of breaking down the problem into easier steps or of combining smaller steps through the use of particular factors and then adding or subtracting. When sharing a restaurant bill of $368.00 among five people, the nearest whole multiple can be obtained first, then the remaining amount can be broken up into successively easier divisions and the results added:

$$5\overline{)368}$$ 5 seventies are 350: $70
$$5\overline{)18}$$ 5 threes are 15: $3
$3 shared among 5 is 60c
Each person's share is $73.60

This thinking can also be extended to dividing by the factors of a number in turn, rather than simply attempting division by the full number. For example, in dividing 6585 by 45, first divide by 5, which readily gives 1317; then divide 1317 by 9, 146r3; the answer to 6585 divided by 45 is 146r15 since the remainder after dividing by 9 is 3 fives and so must then be multiplied by the first factor 5. Use can also be made of particular factors that multiply to give 100 (4×25, 5×20) or 1000 (4×250 or 8×125) similar to the ways they have already been used in multiplication.

There are many ways of dividing mentally and all demand a good understanding of the division process, its relationship to multiplication and an ability to use any of the four operations to obtain shortcuts that are more easily remembered on the way to the final result. Discussion of the merits and ease of use of a variety of self-constructed strategies is essential in building an inclination to choose to use mental arithmetic in everyday situations. This will aid the development of confidence as well as establishing a range of methods adapted to suit particular number situations, individual strengths and preferences. It will also help children to view mental computation as a way of thinking that applies to all operations as part of an expanding number sense rather than a collection of separate strategies to be memorised.

A willingness to attempt to calculate mentally and to do so efficiently is most important and should lead in the longer term to facility with mental computation comparable to the automatic recall and use of number facts. By this time, children should be in a position to choose whether to use mental, pencil and paper or calculator processes to fit a given situation and have confidence in their ability to derive and interpret answers with them.

Investigate and discuss

1 Finding an approximate average provides a common need for estimation with division. How would you determine which of these estimates is reasonable?
   a The total attendance for 7 games was 162 817. The average attendance was about 2000, 20 000, or 200 000?

b You have 48 months to repay a loan of $8500. It will cost you about $20, $200 or $2000 per month without even considering interest.

c The city has 7389 pupils enrolled in 9 primary schools. There are about 80, 800 or 8000 students in each school?

What difficulties might be expected in trying to calculate these answers mentally?

2 List any other situations where estimation and dividing mentally might be required and any techniques of your own that you use in coping with them.

Using calculators in division

The use of a calculator for division can accompany the development of the written algorithm soon after understanding is available. However, if introduced too soon a calculator may inhibit some children from coming to terms with the cycle of steps involved and the manner in which the sharing concept and basic facts are integrated with numeration ideas to develop a coherent process. There is also the issue of balancing a need for understanding how remainders arise in division yet are not readily shown with a calculator. The means of moving from one form to another can be built up using a calculator and this is addressed in the following section on division with fractions. A calculator is almost always used when there is a divisor with two or more digits as the practicality of all of the steps required for division of this form is both time consuming and difficult to manage. It will be sufficient to see how it links to the earlier division with a one-digit divisor using both pencil and paper and calculator algorithms.

Using a calculator often enables answers to be obtained more quickly and means that the user can focus on coming to terms with problems and even finding ways of solving problems that are different from those that might have been used with a pencil and paper or mental method. For example, consider this problem.

658 children and teachers went to the ballet. They travelled in 13 buses, each carrying about the same number of passengers. How many were on each bus?

Division is obviously needed, but the answer given when the numbers are keyed into a calculator, 658 ÷ 13 = 50.615384, needs further analysis. Thinking about the problem and this answer suggests that a solution would be that each bus would contain at least 50 passengers. Perhaps some would have to be left behind, but it is more likely that the additional passengers would be assigned to other buses. If the problem were solved by a pencil and paper method, the remainder would show how many additional students there would be. But use of a calculator provides another method: multiplying 13 by 50 and subtracting this product shows that there were eight additional passengers. So, five buses could carry 50 passengers and eight could carry 51 passengers or it would also be possible for nine buses to take 50 students and four buses to take 52, and so on. The calculator can be used to provide possible answers but thinking about the problem and its solution is provided by the user of the calculator not having to focus on the actual computations that might be required.

Estimation is helpful in introducing calculators where children are asked first to estimate an answer then compare it with that given by a calculator or to use estimation to assess the reasonableness of calculator answers, justifying the results to other members of the class. Discussion should also focus on ways in which use of a calculator might give rise to incorrect answers, for instance when dividing by many numbers a continuing decimal fraction results and the answer might need to be interpreted as a whole number. When 45 ÷ 135 × 3 is keyed into a calculator, the answer 0.999999 is shown, but the actual answer should be 1. Discussion of how these sorts of errors might arise and how they can be corrected will not only develop a good and competent facility with calculators but also assist in consolidating understanding of the concepts, allow children to distinguish between the operations, and promote their ability to apply operations to problem situations.

Investigate and discuss

1 Calculator games can be used to consolidate an understanding of all four operations and to allow children to distinguish among them. Play this game with 2 or 3 other partners, using only one calculator.
   a 100 is keyed into the calculator to start. Each player then selects a number from 1–9: his or her playing number.
   b The calculator is passed around and each player in turn keys one of the operation keys (+, −, ×, ÷), their playing number and =.
   c The winner is the player who makes the display as close as possible to 1000.
2 A calculator is also useful in allowing children to see the relationship between division and multiplication.

a Use a calculator to find the missing numbers:

```
    .. ..           46          693 r53         .......... r29
  ×  ...         ×  ....      782) ......     58) 69745
    ____           ____
    413            414
                   322
    .. .. ..       .. .. ..
    ____           ____
    3953
```

b Comment on the operations required to complete each answer and the notions that might be introduced and reinforced with these types of questions.

Division with fractions

Division of decimal fractions by whole numbers is a natural extension of the language and recording used for division of the remainder. In fact, the existence of decimal places in the number being divided simply makes the earlier process even more explicit.

34.58 divided by 4	8.645
Can you share the tens?	4)34.58
No: Rename and share the 34 ones	32
Rename and share the tenths	25
	24
Rename and share hundredths	18
	16
Rename and share thousandths	20
	20

However, division of decimal fractions by decimal fractions is a more complicated process because of the need to adjust the divisor to a whole number before division can proceed, and then adjust the number being divided in a corresponding manner:

2.4)591.6   is the same as   24)5916

4.38)117.384   is the same as   438)11738.4

Sharing allows a reasonable justification of this since there must be whole numbers of shares (the number of rows in the array). It follows that the divisor needs to be a whole number to allow sharing to literally occur. Sharing 20 objects among 5 gives the same result as sharing 200 objects among 50, 2000 objects among 500 and so on. This can be generalised to: when the number sharing is increased by a power of 10, to receive the same share the amount being shared must also be increased by that power of ten.

Once the adjustment has been made, the resulting division usually involves two-digit or three-digit divisors and, since division by these is invariably done with a calculator, there seems little reason to focus on a pencil and paper process for division with decimal fractions. Justifying the adjustments is an aspect of developing a full appreciation of division and leads to a means of estimating or checking answers. But division with decimal fractions rarely occurs outside of specialist technical areas where calculators would be an integral part of the work. With a calculator, the numbers involved can be keyed in without being changed and the answer given directly, so it is sufficient to expect that this form of division will always be done with a calculator.

On the other hand, the thinking involved in the process of adjusting both numbers to determine the corresponding whole number division also underlies division by powers of ten. As a number slide shows, dividing by 10 means that the digits move one place to the right; the number of hundreds becomes the number of tens, the number of tens becomes the number of ones and so on. Note that it is not the decimal point that moves. The decimal point marks where the whole numbers end and the fractions begin so it cannot move. Rather, the digits decrease in value by a power of ten and occupy a place of lesser value.

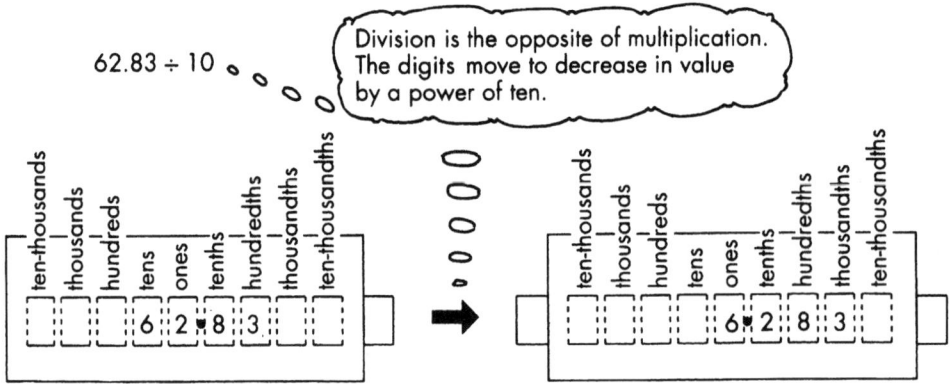

The use of a calculator for division with decimal fractions raises another issue. What is the relationship between division with remainders and division where a decimal fraction results? There needs to be a focus on the meaning of remainder and its connection to the two different forms an answer can take. The number dividing, the number divided and the two parts of the answer, the number shared out and the remainder, are related in the following way:

$$\text{number dividing}\overline{)\text{number divided}}^{\text{number shared out + remainder}}$$

so that   number divided   = (number dividing × number shared out) + remainder
and   remainder   = number divided − (number dividing × number shared out)

When division is continued and the remainder is divided:

remainder = number dividing × decimal part of the answer

These two understandings allow the determination of answers involving remainders using a calculator. For example:

Vickie and 24 of her workmates won $278 768 in LOTTO. They decided to share the money so that each took only whole dollars and to give the remaining dollars to the person from whom they bought the ticket. How much did each win in LOTTO? How much did they give to the ticket seller?

Twenty-five share the winnings   278 768 ÷ 25 = 11 150.72
Each share is $11 150
The remainder is given by   number dividing × decimal part of the answer
25 × 0.72 = 18
$18 is given to the ticket seller
or by   number divided − (number dividing × number shared out)
278 768 − (25 × 11 150) = 18

Division with common fractions is also rather complicated although the actual process is deceptively simple. Justifying this process and ensuring that it is understood fully so as not to interfere with the skills involved with other common fraction operations is a difficult task. The most commonly used rule is to 'invert and multiply', which often leads to difficulties for children who do not understand the ordering implicit in this concise statement:

$\frac{2}{3} \div \frac{5}{6}$ is often interpreted as $\frac{3}{2} \times \frac{5}{6}$ or even as $\frac{3}{2} \times \frac{6}{5}$

It is as likely that children will do things in the order they hear them (invert the first fraction then multiply) or invert both fractions and multiply. A more meaningful instruction is to note that dividing by a fraction is the same as multiplying by its inverse, but even this will cause difficulties for young children and the means of establishing it are not straightforward. The vertical recording for addition and subtraction of common fractions helps in coming to terms with the different-looking as well as differently processed division process when this is introduced as with the multiplication of common fractions. Fortunately, these fundamental difficulties have lately been acknowledged and, since there are virtually no real situations at all that ask for the division of common fractions, it is a topic that is now disappearing from the school syllabus.

Investigate and discuss

1. What new difficulties might arise when division of decimal fractions of the form $5\overline{)3.267}$ and $4\overline{)0.352}$ is introduced? How would you explain placing zeros in these answers?

2. **a** Answer the division $0.4\overline{)31.52}$ using the language and recording developed in this section. List the steps and understandings that you required.
   b How would you justify adjusting the divisor from 0.4 to 4?
   c How might the sharing notion be helpful?
   d How would you develop the idea that the number being divided must also be changed?

3. Calculators can divide with decimal fractions without any prior adjustment to the numbers.
   a How could you use this capability in developing the pattern for adjusting both numbers in division with decimal fractions?
   b What need would there be for such an understanding when children always have access to calculators?

4. **a** Rename these common fractions as decimal fractions and per cents using the division process developed in this section.

   $\frac{4}{5}$ $\frac{3}{8}$ $\frac{5}{6}$ $\frac{7}{12}$ $\frac{2}{7}$

   b How would you explain to a child how to use a calculator to do this renaming?
   c What difficulties might there be in interpreting the answer given by the calculator?

5. Can you think of any situations where division with common fractions might be required? Complete the following examples and list any understandings that are needed:

   $\frac{4}{9} \div \frac{6}{15}$ $\frac{8}{3} \div \frac{12}{6}$

   What place should the development of this computation have in contemporary mathematics programs? Consider whether it might already be part of the process of building up skills in numeration, addition and subtraction for common fractions.

The relationship between division and the fraction concept

One important use of division is to allow the renaming of common fractions as decimal fractions and per cents. The procedure used simply asks that the numerator be divided by the denominator but it is difficult to establish and justify why this is appropriate and why the resulting decimal fraction is equivalent to the common fraction. Unfortunately, many children are just given this rule and then find difficulty in applying it through lack of any understanding as to why it works. In particular, they are likely to confuse their rule and divide the denominator by the numerator as they generalise that the larger number is always divided by the smaller number.

On the other hand, some common fractions can be renamed as decimal fractions by a simple multiplication process corresponding to that used to rename equivalent common fractions. For example, fourths can be changed to hundredths by multiplying both the numerator and the denominator by 25, calling on a regional diagram to justifying what is being done:

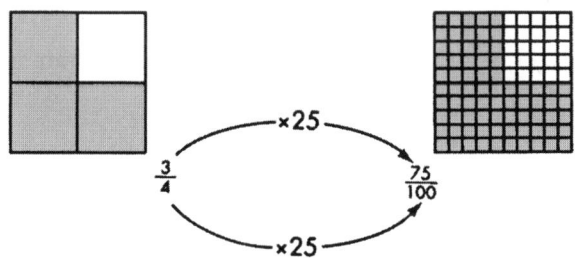

$\frac{3}{4}$ is the same as 0.75 or 75%

While this same multiplication process could be used for renaming many common fractions, it clearly will not work for all cases as there is often no factor in common between the denominator and a multiple of ten. Nonetheless, it can be used to show how some common fractions are renamed as tenths or hundredths, allowing this background of understanding to be called on to assist in understanding the basis for a division process:

3 divided by 4:

$$\begin{array}{r} 0.75 \\ 4\overline{)3.0} \\ \underline{2\ 8} \\ 20 \\ \underline{20} \end{array}$$

$\frac{3}{4}$ is 0.75, so is the same as $4\overline{)3}$

When sufficient examples have been discussed to suggest the connection between division and common fractions, a justification can be made by appealing to their underlying concepts. The key to this is showing that sharing is a fundamental idea common to both the fraction and division concepts.

For common fractions with a numerator of one, a diagram can show this readily. For example, 1 fifth is shown as:

While this diagram shows 1 fifth, it can also be interpreted to show 1 item shared among 5. Each share is 1 fifth since 5 fifths are 1.

1 shared among 5 is $5\overline{)1}$ The diagram shows that each share is $\frac{1}{5}$
$5\overline{)1}$ must be $\frac{1}{5}$

However, when more than 1 item is shared the thinking is not so clear cut: $\frac{3}{5}$ is shown by:

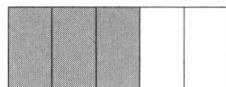

but this does not show $5\overline{)3}$. What is needed is to draw on an expanded notion of the fraction concept. While 3 fifths means that 1 one is partitioned into five equal parts and 3 of the fifths are considered, it can also be viewed as 1 fifth out of each of 3 ones:

This shows how 3 ones can be shared among 5; each share is 1 fifth of 1 one, 3 fifths altogether: $\frac{3}{5}$ is the same as 3 shared among 5 or $5\overline{)3}$.

$\frac{3}{5}$ is the same as $5\overline{)3}$ or

$\frac{3}{5} = 0.6 = 60\%$

$$\begin{array}{r} 0.6 \\ 5\overline{)3.0} \\ \underline{3\ 0} \end{array}$$

A mix of examples that can be represented by diagrams showing parts of several wholes as well as by region diagrams that draw on the multiplicative relationship can be used to justify the general division process for renaming common fractions as decimal fractions. Similarly, the use of real-world situations such as the sharing of pizzas, cakes, fruit and so on is also helpful in generating experiences where wholes need to be shared part by part and where a language of division is linked to this process. For instance:

If 3 pizzas are to be shared among 5 persons, it is commonplace to talk about dividing the pizzas among the five. The pizzas are cut into 5 equal pieces so that a share is formed by taking one piece from each. In this way each person receives 3 fifths of a pizza.

Diagrams can continue to be used for many examples, but their use should give way to reasoning about division as soon as the link between division and common fractions has been grasped. For instance:

$\frac{3}{8}$ is the same as

```
      0.375
   8) 3.0
      2 4
        60
        56
         40
         40
```

$\frac{3}{8} = 0.375 = 37.5\%$

One point that will emerge though is that for some common fractions the division would continue indefinitely. However, these involve a repeating sequence of digits and can be shown to match exactly the common fraction by reasoning with place value. For example, what common fraction is equivalent to 0.33333 . . . ?

multiply by 10	$10 \times 0.33333\ldots$	$= 3.33333\ldots$
from each side, subtract $1 \times 0.33333\ldots$	$9 \times 0.33333\ldots$	$= 3$
then divide by 9	$0.33333\ldots$	$= 3 \div 9$
so	$0.33333\ldots$	$= \frac{1}{3}$

All terminating or repeating decimal fractions are equivalent to common fractions and this has led to them being called rational numbers. On the other hand, there are decimal fractions that are non-terminating, non-repeating, such as:

$$0.101001000100001\ldots$$
or $\quad 0.17265199010034\ldots$

These cannot be renamed as common fractions and are called irrational numbers. Other examples of irrational numbers include the square root of 2, $\sqrt{2}$, and the number $\pi = 3.14159265358979\ldots$

Investigate and discuss

1 a The process for renaming common fractions as decimal fractions has always been deceptively easy to follow but difficult to understand. Briefly summarise the steps used in the discussion above that justify the commonly used rule 'divide the numerator by the denominator'.
  b How convincing is the role of real-world problems in this justification?
  c How would you show that any renaming of common fractions as decimal fractions can be done simply using a calculator?
  d How would you incorporate the renaming between common fractions and per cents into this thinking?
2 The notion of rational and irrational numbers was first discussed by Greek mathematicians, reputedly Pythagoras. Find out the background to this discovery. How might you use this knowledge in teaching children in the upper primary school?

Division with two-digit divisors

While the development of the division algorithm to this point has only included single-digit divisors, the previous section on division with decimal fractions has already shown how division by two-digit and larger numbers can arise. Division by larger numbers is also likely to occur in other everyday situations and in processes associated with other mathematical techniques such as finding averages. Even though calculators can be used in many of these situations, in order to use them appropriately and to develop a feeling for what is likely to occur, some experience with the formal algorithm is needed so that children develop insight. There will also be situations in further mathematics where the use of calculators may not be appropriate or helpful. In algebra, for example, where division may be needed to simplify expressions or to find common factors, it would be rather limiting and even confusing for calculators to be called upon to complete this aspect when they offer no help to the algebraic processing.

The algorithm for dividing by larger numbers is identical to that for one-digit numbers, simply requiring that each place be shared in turn. However, simple multiplication and division basic facts are no longer sufficient to determine each share. Instead, estimation is needed to suggest what the share might be and it is necessary to use the multiplication algorithm to find out how much has been shared. Rounding the divisor to the nearest ten or hundred is often a necessary first step in making an estimate of the share and this estimate may need to be adjusted after multiplication by the divisor has checked that as much as possible has been shared.

This requires a very sure understanding of the rounding, estimation and multiplication processes that turn a first attempt into the correct answer. Consequently, the straightforward extension of the division algorithm tends to be quite complicated in practice. The use of a working space in which the estimation and checking can be set out systematically assists children to see and control this complex thinking. However, the main way to allow children to develop a meaningful and secure process is to build up carefully the underlying processes one at a time, ensuring that these link to the understandings developed with the earlier processes for division by single-digit numbers.

To this end, two-digit divisors close to a multiple of ten where the rounding involved is obvious should be investigated first so that the estimation, checking and adjusting is given most attention. If divisors such as 11, 12 and 13 are used initially, it is possible to begin this development sharing Base 10 materials in the same way as the algorithm for one-digit divisors was established. At this point, the division may result in a remainder or be continued to give a decimal fraction:

846 divided by 13			**Working**	
Can you share 8 hundreds among 13?	65 r 1		13	13
No: rename and share the tens	13)846		× 7	× 6
Try 8. Too large, try 6	78		91	78
Can you share 66 ones among 13?	66			
6 is too large, try 5	65		13	
	1		× 5	
			65	

846 divided by 13 is 65 remainder 1
or 846 divided by 13 is 656.076; 656.08 rounded to hundredths

When the estimation and adjusting steps are secure, larger divisors involving more complex rounding and multiplication can be introduced, leading to the following form for the algorithm for division with larger divisors:

Step	Language	Recording	Working	
Share the thousands	8734 divided by 34 **Can you share** 8 thousands among 34? **No: rename**—87 hundreds to share	34)8734		
Share the hundreds	**Can you share** 87 hundreds among 34? **Yes** 2 hundreds 68 hundreds are shared out 19 hundreds remain **Can you share** 19 hundreds among 34? **No: rename**—193 tens to share	2 34)8734 68 193	34 × 3 102	34 × 2 68
Share the tens	**Can you share** 193 tens among 34? **Yes** 5 tens each 170 tens are shared out 23 tens remain **Can you share** 23 tens among 34? **No: rename**—234 ones to share	25 34)8734 68 193 170 234	34 × 5 170	34 × 6 204
Share the ones	**Can you share** 234 ones among 34? **Yes** 6 ones each 204 ones are shared out 30 ones remain **Can you share** 30 ones among 34? **No: rename**—300 tenths to share	256 34)8734 68 193 170 234 204 300	34 × 7 238	
Share the tenths	**Can you share** 300 tenths among 34? **Yes** 8 tenths each 272 tenths are shared out 28 tenths remain **Can you share** 28 tenths among 34? **No: rename**—280 hundredths to share	256.8 34) 8734 68 193 170 234 204 300 272 28	34 × 9 306	34 × 8 272

... continue in the same way to share hundredths, thousandths and so on as required

8734 divided by 34 is 256.9 rounded to tenths

Since most calculations involving division can be readily done with calculators, the extent to which facility with the written algorithm should be developed needs to be re-examined. While materials show the basis of division in sharing, a full understanding of any process requires reflection on what is involved not just an ability to carry it out automatically. The recording of the algorithm is as much about setting out clearly the steps involved so that the process can be considered and analysed as it is about providing a technique for division itself. If calculators are to be used to their full advantage and their results to be applied and interpreted, a good understanding of what division involves is essential. In many ways, this needs to be the major purpose in extending the division algorithm to larger divisors. By the end of primary school, it is reasonable to expect that children will have sufficient understanding of division to choose for themselves whether to use a pencil and paper algorithm, a calculator or, in time, a mental computation.

Investigate and discuss

1. Investigate how feasible it is to use Base 10 materials to introduce division with small two-digit divisors. Use these examples:

   $12\overline{)649}$ $13\overline{)506}$ $11\overline{)723}$ $12\overline{)510}$

   Are there any differences in the use of language, recording and thinking from that used for one-digit divisors?

2. Discuss how rounding the divisor can assist in making an initial estimate in division with larger numbers.
   How might this help in showing where the first digit of the answer is to be placed?

3. **a** Answer each of these examples using the recording and language outlined in this section:

   $28\overline{)965}$ $32\overline{)7841}$ $62\overline{)64\,529}$ $59\overline{)189\,346}$

   b Comment on how the language, thinking and use of a working space as used above can assist in making division a straightforward and meaningful process.

   c How does this compare to the way of doing division that you were introduced to at school?

4. In the overview to this chapter, it was stated that division simply involved the application of the following cycle of steps:

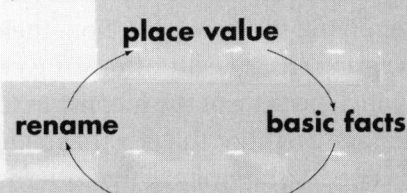

   a Discuss how this applies to division with whole numbers and decimal fractions.
   b What aspects of number understanding are required of each step in the cycle?
   c What is the role of the question 'Can you share . . .'?

> **5** Solve the following problem, using written computation, a calculator and a mental method. What role did understanding of the division concepts and the array model linking multiplication and division play in your solutions? How would you assist a child to come to terms with what is required? Use the general model of problem solving given in Chapter 2 to organise your approach.
>
>   546 children were taken on a 3 week bush adventure. They travelled on 19 buses, each carrying about the same number of children. How many went on each bus?
>   Note: There are several possible answers!

Section V: Errors and misconceptions with computation

WORK with children experiencing difficulties in learning mathematics has led to an awareness that these difficulties have most often arisen because of discrepancies between the materials used to develop mathematical ideas and the language that is then used to direct and control the symbols later used to convey meaning. A major factor in this is the timing of the introduction of the experiences on which mathematical learning needs to be based. In the past, many aspects of mathematics were introduced too soon or too quickly while there were some aspects, particularly numeration ideas, that were introduced too late or too slowly. There were also ideas, such as the renaming shown by the number expander, which were not introduced at all. Meaningless 'rules' such as 'move the decimal point' have also contributed to children's difficulties in learning mathematics. Most of these difficulties occur as a result of what has been done at school so the child can hardly be held responsible and termed 'learning disabled'. It needs to be acknowledged that many children have learned difficulties rather than learning difficulties.

The approach to teaching number understanding and the computational algorithms advocated in this book has grown out of an analysis of these learned difficulties. It is designed to develop interconnected skills and understandings in a systematic manner so as to minimise uncertainties in using and applying mathematics. Nonetheless, it is still possible that some children will form misconceptions as a result of missing crucial steps in a developmental sequence. This can occur as much because of the inconsistencies inherent in and between the algorithms as in the way they are taught or the fact that children have been absent at crucial times. Consequently, there is a need to be aware of the patterns of errors that children are likely to make, the underlying reasons for them and the means of overcoming or avoiding them.

One form of behaviour that should not be overlooked in any analysis of children's errors is that of no response at all to a given question or problem. One reason may be that the child simply did not have enough time, perhaps because he or she works slowly but perhaps also because of factors quite unconnected with the learning situation at all, such as stress, emotional upsets or illness. Other reasons are that children recognise that they cannot do that particular type of activity as they have not met it before or have never come to terms with the thinking involved. In these cases, doing nothing is not so much a sign of inability as a signal that they are aware of their limitations. It could also be that they once knew how to do this particular type of problem but no longer remember the method.

Nonetheless, if children have built up meaningful computational concepts and processes, given a computation to carry out or a problem to solve, they will be able to obtain correct answers using a standard process. That is, they will be able to:

- select an appropriate process
- perform steps in a correct sequence
- use correct basic facts
- apply underlying place value, renaming or other structural concepts
- record any working appropriately

It is also possible that a child will be able to obtain correct answers using non-standard processes. This is appropriate when the child understands what is being done and when it is possible to generalise the particular method to fit future needs. However, many of the *ad hoc* procedures that are observed are learnt by rote and cannot be readily applied or used in different circumstances.

For example, when a child multiplies without reference to place value, this may appear to cause no difficulties when multiplying by a single digit:

```
   5
  4 7      8 sevens are 56      one digit is recorded in the bottom line, one at the top
× 8        8 fours are 32       one digit is simply placed next to the first digit in the bottom line
 3 7 6     and 5 is 37          the other digit is joined with the one at the top and placed next
                                to the digits in the bottom line
```

When 2 two-digit numbers are multiplied, this often results in the result from multiplying the tens digit with the ones digit in the ones place, since that is what is said, 6 sevens are 42, and not 43 tens:

```
   4
   5
   4 7     8 sevens are 56
 × 6 8     8 fours are 32 and 5 is 37
 3 7 6     6 sevens are 42                  one digit is recorded in the bottom line, one at the top
 2 8 2     6 fours are 24 and 4 is 28       one digit is simply placed next to the first digit in
 6 5 8                                      the bottom line, the other digit is joined with the
                                            one at the top and placed next to the digits in the
                                            bottom line
```

Of more concern is the child who is able to obtain correct answers using incorrect thinking. It is possible that in rewarding a child for obtaining a correct answer we are also reinforcing the use of improper methods or reasoning. For instance, a child may be getting all of the answers correct for addition with zero by treating it as nothing, but then find that subtraction is sometimes answered incorrectly:

$$\begin{array}{r} 60 \\ +38 \\ \hline 98 \end{array} \qquad \begin{array}{r} 68 \\ +30 \\ \hline 98 \end{array} \qquad \begin{array}{r} 68 \\ -30 \\ \hline 38 \end{array} \qquad \begin{array}{r} 60 \\ -38 \\ \hline 38 \end{array}$$

A child who simply treats zero as nothing will get the first three examples correct, but will be perplexed as to why the last example is incorrect

When children make errors, regardless of the background to their problems in learning mathematics, they tend to occur systematically, indicating that an inappropriate way of thinking has been established. These patterns of errors have been observed across all manner of diverse groups and dominate the misconceptions that children have about mathematics. Nonetheless, children do sometimes make random errors that are not consistent from one occasion to another. These are very difficult to come to terms with for they may be due to motivation factors, various distractions in the classroom situation or, indeed, actual carelessness. But it must be remembered that a careless error can only be a mistake that the child recognises immediately when asked to reconsider and is unlikely to be repeated at another time or circumstance. Indeed, children's errors are rarely careless; they are deliberate, formed as the result of a misguided attempt to find correct answers. It follows that teachers should be aware of these patterns, both to bear in mind as the initial ideas are established and to alert the children to the possible confusions so that they are less likely to occur.

Origins of errors in mathematics

Many of the errors that children make are sourced in an inadequate set of meanings for number and the operations. Indeed, a lack of numeration understanding rather than difficulties with the operations *per se* is the major cause of errors in computation. Not only is it necessary to understand the full structure of the number system, children also need to learn to interpret numbers flexibly through renaming.

Lack of place value understanding:
Adding the digits as if they are separate numbers

$$\begin{array}{r} 42 \\ +71 \\ \hline 14 \end{array}$$

Adding each place separately with no awareness that the result must be too large through not renaming

$$\begin{array}{r} 36 \\ 47 \\ 62 \\ 58 \\ 39 \\ +25 \\ \hline 2337 \end{array}$$

Lack of renaming:

In subtraction, this often leads simply to finding the difference between the digits in a place	6 4 3 −3 5 9 3 1 6	In division this may lead to dividing the digits or pairs of digits as if they were separate numbers	201 $3\overline{)625}$ or 56 $7\overline{)3542}$
In addition, the additional ten may not be recorded full sized and be overlooked	7 8 + 2 6 9 4	In multiplication, this may lead to multiplying and recording each place separately	3 7 × 5 1535 or 3 7 × 4 5 1235
Common fractions may be added or subtracted without reference to the meaning of the fractions involved		$4\frac{3}{4} + 5\frac{2}{3} = 7\frac{5}{7}$ $8\frac{7}{8} - 6\frac{3}{5} = 2\frac{4}{3} = 3\frac{1}{3}$	

A failure to understand a concept in terms of its underlying meanings often leads to an inability to choose the correct operation to match a problem situation. For instance, many children have to seek confirmation as to whether addition or subtraction is required regardless of the sense of the situation being considered or even despite the occurrence of the operation symbol within a calculation they are attempting to complete. Understanding the part/part/whole model enables appropriate choices to be made as it allows an analysis of the meaning implicit in a problem. Further, the symbol for each operation can only be introduced meaningfully when the action it represents has been built up. When children do not have well-understood, full meanings for a concept, they are forced to rely on word cues in problems that will often lead them astray. For instance, both addition and multiplication problems use the expression 'how many altogether', leading to children deciding to choose the operation based on the size of the numbers in the problem or even how many numbers occur, reasoning that only addition involves more than two numbers. Similarly, since addition and subtraction situations both involve the word 'more', children will often choose to add when subtraction should be used.

Without a secure knowledge of the basic facts, particularly one which builds on powerful thinking strategies rather than rote learning, carrying out the algorithms is usually impossible. There will either be mistakes with the individual steps or the effort of determining these intermediate results will cause a child to go astray in moving from one step to the next.

Incorrect basic fact	58 + 80 130	Uses basic fact from another operation	63 × 2 125	Uses steps from another operation	^{1}64 + 7 141	Incorrect operation	^{2}64 + 7 448

Even when the correct operations and processes have been selected, errors can arise in the sequencing of the steps involved:

Steps are carried out in an incorrect order	$\begin{array}{r} \overset{4}{3\,6} \\ \times\ 7 \\ \hline 4\,9\,2 \end{array}$	The process is incomplete	$\begin{array}{r} 6\,8\,7 \\ -\ 5\,3 \\ \hline 3\,4 \end{array}$	or $\begin{array}{r} 12 \\ 7\overline{)842} \end{array}$

Finally, even when the correct operation or algorithm is chosen and when all steps are performed in the appropriate order, the way in which the working is recorded can cause difficulties when information is translated from a problem situation. A child may

- incorrectly form and read number symbols
- misalign digits within an algorithm

However, the difficulties that arise are not all sourced in the child's lack of understanding of the underlying concepts for the operations or the numbers themselves. Many errors occur through inconsistencies in mathematics itself, in its teaching, in the level of abstraction that is demanded or even in the conflicting forms of language and materials that are required. These too need to be examined for the way in which they might lead to learned difficulties so that children may be made aware of them as they arise instead of falling prey to the errors they engender.

Sources of learned difficulties in mathematics and its teaching

While mathematics is often held to be the epitome of logic and reasoning, there are many inconsistencies in the way in which it is written and read that cause children serious misunderstandings. For instance, while written words are always read from left to right, this does not always occur for written numbers. Teen numbers are read differently to all other two-digit number, for instance 64 is 'sixty-four' but 16 is 'sixteen'. Other numbers involving teen numbers such as 417, 6015, 2.13 or 7.012 are also read differently. This can lead to difficulties with renaming in addition or multiplication:

$$\begin{array}{r} \overset{6}{3\,7} \\ +\ 4\,9 \\ \hline 1\,3\,1 \end{array} \qquad \begin{array}{r} \overset{5}{6\,3} \\ \times\ \ 5 \\ \hline 3\,5\,1 \end{array} \qquad \begin{array}{r} \overset{7}{5.08} \\ +\ 6.09 \\ \hline 1\,1.71 \end{array} \qquad \begin{array}{r} \overset{8}{4.6} \\ \times\ \ 3 \\ \hline 3\,6.1 \end{array}$$

If zero is simply seen as a 'place holder' rather than as a number which represents 'none of something', its significance will not be grasped and it is likely to be overlooked or used inappropriately:

```
    3 3
    3 5
×   6 7
    2 4 5
    2 1 0      A zero is 'put down' to correspond to the tens place, but the next step,
    4 5 5      6 fives, gives rise to another zero, and this is seen as already being placed.
```

```
        36
    3)927
        9
        02     02 is read as 20, and 20 divided by 3 is 6
```

Further inconsistencies occur in the way new ideas are introduced. For addition and subtraction, numbers are dealt with by proceeding down each place from top to bottom while in multiplication this order is reversed. There are good reasons for having different ways of operating when the steps involved are so different: for addition and subtraction the algorithm simply demands that each place be operated on in turn whereas for multiplication there is a crossing of places, ones multiply with tens and so on. This difference is not always so apparent to young children and the method that has given so much success for addition is often continued to multiplication:

```
           ⁽¹⁾                         ⁽⁴⁾
           ⁽⁶⁾7                        ⁽⁶⁾7
         + 3 6                       × 3 6
Since     1 0 3      why shouldn't    3 0 2
```

While addition, subtraction and multiplication are always carried out from right to left, division occurs in two different forms that are read differently and division is completed from left to right:

Why do both $7)\overline{5294}$ and $5294 \div 7$ mean 5294 divided by 7?

Further difficulties arise because of the abstractions intrinsic to mathematics. Symbols necessarily provide concise representations of ideas that first need to be internalised. If this has not been allowed for in the building up of concepts and skills, then children will have little understanding when the symbols are used. For example, the use of exponents builds on the idea of factors occurring several times in a product. If small numbers have not been reserved for exponents but have been used in the recording of algorithms and the word 'times' has been used in multiplication, many children interpret this notation in their own way:

4^3 is 4 times 3 which is 12 rather than 4^3 is $4 \times 4 \times 4$ or 64

Processes that subsequently depend on the meaning associated with the use of mathematical symbols can only be weakly established when the symbols are not strongly linked to the ideas and actions they represent. For example, determining which number is greater or lesser is simply a matter of using place value ideas, yet expressing this symbolically has caused children much distress when expected too early and through the confusion introduced by asking children to choose between two symbols, > and < . There is really only one symbol that indicates both greater than and less than—it can be read or written from right to left. In renaming common fractions as decimal fractions a division process is called on, but if a child does not link the meaning of a fraction with equal parts to the concept of division as sharing, he or she will be uncertain as to which of the two numbers used to describe the fraction is divided by which:

To write $\frac{5}{8}$ as a decimal fraction, do we use $8 \div 5$ or $5 \div 8$?

Finding an average also involves division, but it needs a clear understanding that it is the total of all the numbers concerned that is divided by how many numbers there are. For instance, if a runner has an average of 123.7 seconds after 8 trials, then completing 3 more trials with times of 117.9 seconds, 125.3 seconds and 121.8 seconds does not mean that a new average is found by taking the sum of 123.7, 117.9, 125.3 and 121.8 and dividing by 4. The total time taken in the trials and the total number of trials have to be used:

8 trials with an average time of 123.7 seconds would take 989.6 seconds
3 more trials would take 117.9 + 125.3 + 121.8 which is 365 seconds
All 11 trials would take 989.6 + 365 or 1354.6 seconds
Her new average is 1345.6 ÷ 11: 123.15 seconds

Any difficulties with the abstractions inherent in mathematics are further exacerbated by the level of maturity of the children when the ideas are first introduced. Not only are abstract ideas in mathematics introduced at an earlier age than comparable ideas in other subject areas, the maturity level of the children in a class is not uniform and what seems reasonable for some children may be inappropriate for others. If teachers elect not to use materials and models to introduce ideas after the first few years of school, many children are unable to connect new concepts and processes to the earlier number understanding, concepts and processes on which they are built. Children's fear and unwillingness to participate in division and fraction processes are often due to their inability to see connections to earlier ideas and consequent failure to make sense of what to them seem to be very complex and arbitrary series of steps.

One way to reduce difficulties associated with mathematical abstractions is the use of language that matches the way the ideas are first introduced with materials and then used to direct the recording in formal processes. It is for this reason that the development in this chapter has made such a careful use of language to give meaning to the operations rather than simply accept traditional patterns that accompany the recorded steps without providing any meaningful direction. Notions of renaming and placing digits in the appropriate places provide reasons for the way the operations are carried out, whereas talking about 'putting down', 'carrying' or 'bringing down' provide no direction whatsoever. On the other hand, the formal and concise language of mathematics can create new difficulties if its use is expected too soon, while there are certain words in use in mathematics the meaning of which conflicts from situation to situation as well as with their use outside of mathematics. Perhaps the word 'more' is the most contentious. Often it is associated with notions of 'bigger' yet it frequently occurs in subtraction situations such as 'If I have 5 tennis balls and you have 8 tennis balls, how many more tennis balls do you have than me?'. If addition is also described in terms of more and bigger, many children will be induced to add in these situations rather than subtract. Thus, while language is of great help in alleviating or avoiding difficulties in mathematics, it is also paradoxically a source of confusion and great care needs to be taken with its use.

Similarly, although materials are essential in building up meaning for mathematical concepts and processes, there are situations in which their use can create difficulties for children. For instance, in addition both numbers are shown with materials:

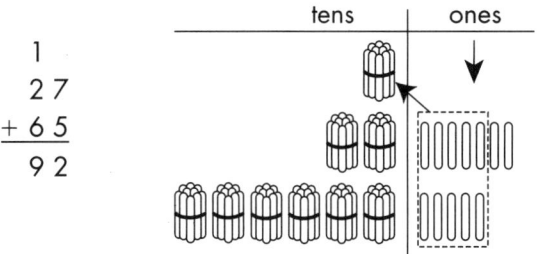

This suggests to many children that they should make both numbers for subtraction. Yet the very concept of subtraction says that only the top number is made, the bottom number is taken away:

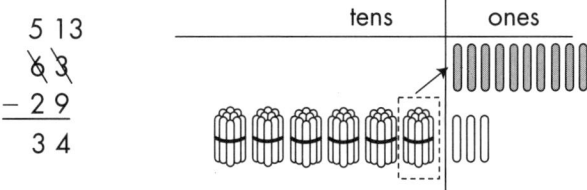

If both numbers were made, then the bottom number would have to be removed before it could be taken away!

Later, it would seem natural to use materials for all larger numbers but for those examples with internal zeros it would then be made more difficult again:

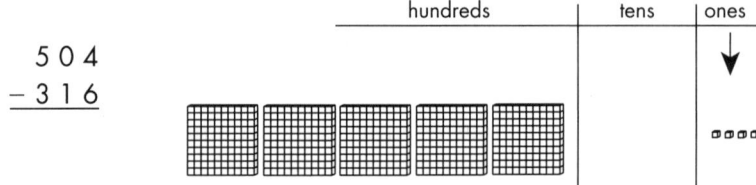

```
 5 0 4
- 3 1 6
```

It looks as if there are no tens, yet consideration of the number alone says that there are 50 tens and that subtraction is done in the same manner as before. The use of materials at this point encourages at best a very different form of thinking and recording:

```
        9
    4 ⩗0̸ 14
    5̸ 0̸ 4
  - 3 1 6
```
and may lead to difficulties with renaming
```
        3
    ⩗ 10 14
    5 0 ⩗4
  - 3 1 6
    0 9 8
```

These difficulties, which can arise out of the nature of mathematics and its teaching, can be minimised if the subject is viewed as a way of thinking that is to be gradually subsumed rather than a series of steps or isolated pieces of information to be committed to memory. It needs to be seen as a means of expressing relationships and solving problems that has grown out of real experiences with materials and models through a language that not only makes sense of these experiences but also is at a level appropriate to those learning it for the first time. It is not simply a system of abstract symbols to be learnt and associated with a set of formal rules. Instruction in mathematics needs to be carefully sequenced to allow the construction of the fundamental understandings on which subsequent processes depend. Abstractions that might inhibit the learning of mathematics need to be delayed until children are ready for them and are able to cope with any associated ambiguities. Above all, this analysis of the mathematical difficulties that cannot be avoided has shown the need to use language and materials carefully and consistently to allow the visualisation of concepts and processes and to lead into the final mental or recorded forms.

Intervention in mathematics

While it is to be hoped that difficulties arising either through an incomplete grasp of the fundamental concepts for numbers and the operations or through the ambiguities created by the nature of mathematics and its teaching will be minimal, there will always be a need to provide reteaching for some children. Ideally, their need for assistance will be noted from the outset rather than left until error patterns are deeply ingrained. In particular, it is through observing children working rather than looking at their completed work that provides information about what they can or cannot do, whether they have internalised appropriate processes or formed their own misconceptions. If erroneous ways of thinking are to be countered as soon as they arise, it is more important to watch children working, to infer their thinking and to determine how answers were obtained than to gather in and correct workbooks. Written work may show the extent of children's difficulties, but it rarely shows how they came into being, or even, necessarily, the real source of the errors. Nonetheless, whether resolving difficulties as they arise or repairing them after they have become firmly established, the following process of intervention has proven to be beneficial in assisting children with difficulties.

THE PROCESS OF INTERVENTION

1 Identify and describe the mathematical error.
2 Determine the source of the difficulty: What thinking is being used? Why?
3 Lead the child to see the inadequacy of this thinking in order to appreciate the need for change.
4 Assist the child to (re-)construct an appropriate way of thinking.
5 Provide sufficient motivating practice to allow the new process to become secure and to provide for the generalisation of the strategy to more complex situations.

Children experiencing difficulties with a particular topic are often simply given more examples of the type where the error first occurred because of a natural belief that more practice is needed. Unfortunately, for many children, such an approach merely serves to reinforce a belief that they are not very good at mathematics or even to cement home the error more strongly as the inappropriate thinking is practised. Pinpointing the actual error and describing the processes being followed ensures that the error is attacked at its source. In turn, this allows the origins of the error to be seen; it shows whether there is difficulty with numeration, for example, or whether there has been confusion with another process.

However, the crucial step in this process is leading children to see their errors. Simply telling them what to do is unlikely to be sufficient. After all, they have probably at least been told a correct method at some earlier stage or may even have been exposed to its development through the use of materials. Yet they have built up for themselves a procedure that seems to make as much sense. It may even provide correct answers some of the time or have applied to a whole class of examples in the past. This previous success is likely to encourage them to keep using an inappropriate procedure until they can see its limitations. These children will first need to see the erroneous thinking involved before they will give up an inappropriate procedure and adopt a more suitable method. Several strategies can help to expose errors in the thinking or processing:

- cognitive conflict—when children are confronted with what they see occurring as opposed to what they think should occur. The use of materials is crucial in bringing about this conflict and in resolving the differences that arise:

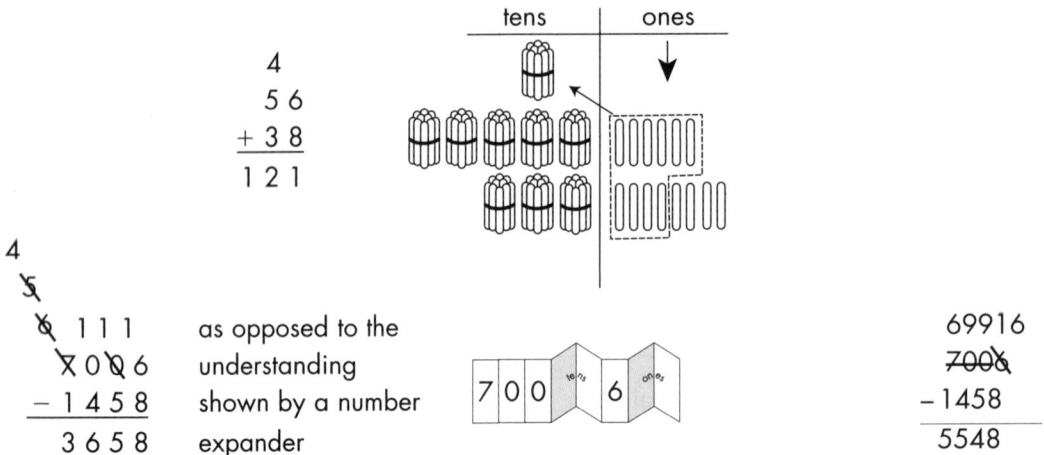

4		
5 6		
+ 3 8		
1 2 1		

```
  4
  ₅
  ₆ 1 1 1       as opposed to the              6 9 9 1 6
    ₇ 0 ₈ 6     understanding                  7̶0̶0̶6̶
  − 1 4 5 8    shown by a number   |7|0|0| |6|  − 1 4 5 8
    3 6 5 8     expander                        5 5 4 8
```

- counter examples—when the use of the same method or way of thinking leads to an 'obvious' error.

Multiplying only the	3 7		3 1	
ones and tens places	× 8 4	suggests	× 2 4	Since 2 by 31 is 62, this is clearly too small
	2 4 2 8		6 4	

- estimation—involves rounding and a good understanding of the underlying concepts and processes. Care needs to be taken that the child with the misconception can, nonetheless, find and accept the estimate that calls into question the earlier way of thinking.

Recording the result	5 7		6 tens	
of multiplying by the	× 6 9	about	× 7 tens	
tens in the ones place	5 1 3		42 hundreds or 4200	
	3 4 2			
	8 5 5		The answer is far too small	

Use of a calculator or the inverse operation to check whether an answer is appropriate, for example adding to check the result of subtraction or multiplying to check division, are other ways that might show children that their methods are problematic. However, these tend to be more abstract than the approaches outlined above and may end up simply being another way in which a teacher tells a child that he or she is wrong. Many children also reject the idea of completing another example to check the first on the grounds that they are being forced to do twice as much work as they first thought they had to do.

An approach to the teaching of mathematics that makes intervention an integral part of classroom behaviour demands fundamental changes in the roles of both teacher and learner. A teacher then becomes more of a facilitator than a director of children's learning, listening to and accepting what children say and assisting them to work out what they are doing in an atmosphere of psychological safety. The child is at the centre of learning, responding to situations that arise rather than passively taking in information. Errors can then be seen not just as something to be avoided, but as an intrinsic part of the learning situation to be confronted, discussed and reconciled. In this way, a teacher can assist learners to develop autonomy in constructing their own mathematics and in determining how and when to apply it.

Investigate and discuss

Analyse each of the errors discussed in this section in terms of the intervention process. That is:
a Identify and describe the mathematical error.
b Determine the source of the difficulty and the thinking that is being used. Suggest reasons why this way of thinking has developed.
c Lead the child to see the inadequacy of this thinking or strategy, using cognitive conflict, counter example or estimation. If you can think of another approach, give some justification for its advantages.
d Outline how you would build up an appropriate way of thinking.

Teaching computation: concluding remarks

THE approach to teaching the four operations that has been developed in this chapter has stressed a number of important points. These have focused on the role of materials in building up processes, the importance of a language that grows out of this use of materials to direct the formal algorithms, and the need for a meaningful recording of the various processes. Consistency between the algorithms and the various forms of number that they apply to has also been paramount in this development. The realisation that children construct their own mathematics has been a guiding principle in all of this with the final discussion on errors highlighting the importance of coming to terms with the different possible and plausible alternative constructions. Yet the knowledge that is required for all of these diverse processes and understandings is essentially only numeration, which governs the behaviour of numbers, the concepts on which the operations are based and the basic facts that allow computation with all numbers—large and small—simply by building up the results place by place. All children should be able to become competent with all of the mathematics discussed in this chapter for whole numbers and fraction ideas. They should then have the confidence as well as the capacity and understandings to apply their mathematics to problems in both real-world and theoretical situations.

Initial uses of these processes and understandings are with the straightforward types of problems that have traditionally been termed 'word problems'. These require the translation of a given situation into the operations that allow it to be solved and have always been a major focus of mathematics in primary school. Children need to be able to identify the operations described in the story situations, to carry out the associated calculations using the appropriate information and to do this in a correct order. A knowledge of the meanings of the operations and their alternative verbal expressions is needed to decide what must be done, while the application of existing knowledge or the translation of this knowledge to a new situation is all that is required to complete the solution.

Successful problem solvers are usually aware of the structure of a problem and use this knowledge to organise their solution strategies or match the problem to similar ones completed previously. Analysis of the structural aspects of word problems allows them to be classified by the number of operations and the number of steps required for their solution. One-step problems involve only one operation and thus require a simple translation of the wording in which the problem is expressed to enable the appropriate operation to be selected. While an understanding of the concept for each operation helps this to occur, aspects of the wording in the problem can be misleading. For instance, a subtraction problem may involve the word 'more' and thus induce some children to think of addition. This confusion is magnified if the children do not meet the full range of addition, subtraction, multiplication or division concepts when they are developed or if they are not provided with a variety of expressions for the operation.

More complex problems require several steps and may involve more that one operation. Two-step problems involving two applications of the same operation are generally quite straightforward. If the operation is addition or multiplication, the order will not matter but if subtraction or division is involved the numbers will need to be used in the appropriate order to give a meaningful answer. When two different operations are used, however, realising the sequence in which they are to be used will always be crucial. Problems with more than two steps magnify these difficulties considerably.

Thus, one further aspect of computation that needs to be considered is the use of several operations concurrently and the order of processing that this demands. The first new notion to be built up is a means of recording several operations. Throughout the development of computation, the stress has been on a recording that matched the form used with large numbers and decimal fractions. In this way, the initial ideas could flow easily into the written algorithms. However, when more than one operation is involved, it is no longer possible to base the recording on these familiar forms and it is customary to use horizontal expressions utilising the equals symbol. In this way, all of the numbers and operations involved can be shown in the one expression and the various rules that govern mixed operations, as well as any use of underlying numeration or computational knowledge, can be applied to obtain answers more efficiently. The first step in this process is to introduce the alternative horizontal form for recording the operations:

$$6 + 7 = 13 \qquad 13 - 7 = 6 \qquad 4 \times 5 = 20 \qquad 20 \div 4 = 5$$

This horizontal form can then be used to express situations that involve more than 1 step:

$$8 + 9 + 12 = \qquad 5 \times 3 \times 4 = \qquad 18 - 8 - 7 = \qquad 32 \div 8 \div 4 =$$

When the same operation is used, the thinking is fairly straightforward. For addition and multiplication, the order in which the calculations are carried out does not matter:

$8 + 9 + 12 =$
- $8 + 9 = 17$; $17 + 12 = 29$
- or
- $9 + 12 = 21$; $21 + 8 = 29$

$5 \times 3 \times 4 =$
- $5 \times 3 = 15$; $15 \times 4 = 60$
- or
- $3 \times 4 = 12$; $5 \times 12 = 60$

However, for subtraction and division, completing the operations in a different order gives different answers.

$18 - 8 - 7 =$
- $18 - 8 = 10$; $10 - 7 = 3$
- or
- $8 - 7 = 1$; $18 - 1 = 17$

$32 \div 8 \div 4 =$
- $32 \div 8 = 4$; $4 \div 4 = 1$
- or
- $8 \div 4 = 2$; $32 \div 2 = 16$

To resolve this confusion, the first convention for completing several calculations in the one expression is to *work from left to right*.

A further difficulty arises when there is more than one operation involved. It is always possible to express the calculations required by a real problem in more than one way. Should an answer be determined by working from left to right or should some operations take precedence over others? For instance:

For his work making deliveries at the local newsagent, John was given $4 a week and $6 for each day he worked. What was the maximum amount he could earn in 1 week?

This could be expressed as $4 + 6 \times 7 =$ or $6 \times 7 + 4 =$
Working from left to right: $4 + 6 \times 7 = 10 \times 7 = 70$
but $6 \times 7 + 4 = 42 + 4 = 46$

Both answers cannot be correct and an analysis of the problem shows that the answer is given by multiplying 6 and 7 and adding 4. In order to accommodate this confusion, a second convention has been adopted: *always complete any multiplication and division before addition and subtraction.*

Situations of this kind pose a very real problem with many calculators where simply keying in the numbers and operations in the order in which they occur in a written expression or in a written problem can result in incorrect answers. Only those calculators with a special logic function taking account of the conventions used for combining several operations will give correct answers without the user specifically entering the information in the order in which it must be processed.

Nonetheless, there are situations where it might be necessary to add or subtract before multiplying or dividing. In order to make this clear, such calculations are placed in parentheses or brackets, and the third convention is that *calculations within brackets are carried out first*. For instance, $(3 + 4) \times 5$ means 7×5 while $3 + 4 \times 5$ means $3 + 20$. Whenever confusion is possible, brackets should be used to indicate the order in which the calculations are to be done.

When working with mixed computations, then, there are three rules that allow answers to be determined unambiguously:

1 First complete any calculations within brackets.
2 Then do any multiplication or division before addition and subtraction.
3 Then work from left to right.

Some examples will show the simplicity of these conventions and how they are applied in practice:

$68 - (24 \times 7) + 49 \times (20 - 8)$ $= 68 - (168) + 49 \times (12)$ Do the parts in brackets
$= 68 - 168 + 588$ Multiply before adding
$= 488$ Work from left to right

$(70 - 20) \div 5 \times (6 + 4)$ $= 50 \div 5 \times 10$ Do the parts in brackets
$= 10 \times 10$ Work from left to right
$= 100$

$70 - (20 \div 5) \times 6 + 4$ $= 70 - 4 \times 6 + 4$ Do the parts in brackets
$= 70 - 24 + 4$ Multiply before adding
$= 50$ Work from left to right

Many people 'learnt' to accommodate these conventions by being told to follow the steps suggested by the acronym BODMAS (Brackets, Of, Divide, Multiply, Add, Subtract) or BOMDAS, which appeared to alter the order for some steps. Since no meaning was given for either ordering, it was often misapplied or simply forgotten altogether. Consequently, many children made errors in problems or calculations involving two or more steps. Yet the thinking outlined here not only demonstrates the need for some convention to remove any ambiguity from arithmetic calculations, it also shows the sensible origins of the conventions themselves.

If children have a secure knowledge of these conventions on calculating with several operations together with a full awareness of the concepts for the operations, then problem solving involving direct applications of their computation knowledge should be straightforward. However, the information that is or is not provided in a problem adds some further complexity. If the wording is suggestive of the operation involved, the translation from the problem to the children's computational knowledge is relatively easy. When the wording is not so familiar, the translation is more difficult. The amount of information given in the problem may also cause difficulties: if there is too much information, the child needs to sort out the useful from the unnecessary before commencing a solution; if there is not enough information, the child must seek out what is necessary before a solution strategy can be formulated.

Solving word problems successfully also requires a good plan of attack and access to a repertoire of strategies, or general processes, which will help solve a variety of problems. These have been discussed in detail in Chapter 2, but all problem solving requires general guidelines to structure the analysis of the question, the formulation of a plan to solve it and an evaluation of any answer that eventuates. If there is a strong base of understanding for numeration and the operations, the foundations are established for good problem solving skills. Seeing that they can readily apply this knowledge also provides children with the self-confidence that allows them to take risks with the more complex problems that are at the core of contemporary mathematics programs.

Moving to algebra

Algebra has its historical roots in the study of general methods for solving problems. Systematic experience with patterns can build up to an understanding of the idea of function and experience with numbers and their properties lays a foundation for later work with symbols and algebraic expressions.

Principles and Standards for School Mathematics,
2000, p. 37

Algebra developed through the efforts of Arabian mathematicians intent on finding methods for solving more complex problems than could be easily dealt with using the arithmetical methods of the day. Its power in showing once and for all that a particular method of solution, and hence the solution of certain classes of problems, was needed only once when words or symbols related to general numbers and operations, rather than specific instances, was also paramount in its rise to the forefront of mathematical thinking. As algebraic notions were further developed and spread to the emerging European mathematics of the seventeenth century, it became the backbone of scientific inquiry into the laws of nature and, in due course, of the techniques of the calculus brought together by Newton in England and Leibniz in Germany.

A defining moment was the introduction of alphabetic symbols, first by the French mathematician Viete and formalised later by Descartes in his *Discourse on Method*. From the point of view of young students coming to terms with algebra, it is unfortunate that lower case letters from the Latin alphabet were chosen as the symbols to represent these ideas. If a new set of symbols, perhaps based on the Greek or some other alphabet had been chosen, many of the confusions between everyday language conventions and the results of arithmetic that were being generalised would have been much diminished.

However, since this was not the case, thought instead needs to be given as to how number ideas and processes can be used to create meaningful ways of thinking about algebra. Many children experience difficulties in coming to terms with algebra because of the conflicts between earlier ways of thinking in numeration and computation and the ways demanded for algebra. This may not only be a problem of adapting to new processes. Often it indicates that understanding of numeration and operations was insufficient, relying on rote memorisation of rules rather than a full meaning of the concepts involved. In particular, it is possible to complete all of the arithmetic of primary school, including quite complex problem solving, while thinking additively. Multiplication may then be thought of in terms of repeated groups so that division, seen as the reverse of multiplication, is related to subtraction. Such an additive view of mathematics is one of the major causes of children's difficulties in accepting and applying the processes of algebra, especially in factorising and in ratio. In reality, the mathematics of high school is based on multiplicative rather than additive ideas and an understanding of multiplication as an array is essential to developing the function concept central to higher mathematical processes.

Algebra as generalised arithmetic is readily introduced using letters to signify numbers and number patterns as in the concluding section of the numeration chapter. When this use of letters has been accepted by children, it can be extended to situations where patterns of numbers are given and from which a general rule needs to be determined. While the process is similar to the replacement of specific numbers by letters indicating a general number, once the four operations are involved the patterns described become more complex and the advantages of using letters become more apparent and powerful. In this way, children can be led to appreciate the need and value of the use of algebraic symbols, using letters to help to explain patterns and determine the outcome of particular sequences of operations as in the following activity.

FIND THE RULE

In these examples, the initial number and the result are given for a range of patterns.
Fill in the missing numbers and then state the general rule:

1 Initial	result
10	5
9	4
16	11
20	
27	
39	
n	
n	$\rightarrow n - 5$

2 Initial	result
6	20
15	29
26	40
30	
58	
100	
n	
n	$\rightarrow$

3 Initial	result
2	10
5	25
12	60
15	
20	
34	
n	
n	$\rightarrow$

4 Initial	result
3	10
5	16
8	25
12	
20	
33	
n	
n	$\rightarrow 3 \times n + 1$

5 Initial	result
3	24
5	44
9	84
12	
20	
100	
n	
n	$\rightarrow$

6 Initial	result
2	21
4	39
9	84
12	
22	
50	
n	
n	$\rightarrow$

7 Initial	result
4	25
10	10
20	5
25	4
50	
100	
n	
n	$100 \div n$

8 Initial	result
5	115
8	112
12	108
20	100
48	
83	
n	
n	$\rightarrow$

9 Initial	result
50	950
260	750
400	600
330	
666	
999	
n	
n	$\rightarrow$

In particular, the children are able to see that the symbols become a means of determining answers, rather than just a way of expressing them. Thus, when an expression of a general rule has been determined, the implications of that rule can also be investigated, leading to explorations of number patterns themselves rather than simply looking at their outcomes or starting points. Other activities are given by:

THINK OF A NUMBER PUZZLES

In these examples, the puzzle can be generalised using the letter in place of the unknown number that the person is asked to choose for him- or herself.

1. Choose a number. Double it. Add 9. Add your original number. Divide by 3. Add 5. Subtract your original number.
   Try several numbers. What happens?
   Can you explain your result?
   If the number you first chose is n, then the sequence of steps is:
   $n \to n + n \to n + n + 9 \to n + n + n + 9 \to n \div 3 \to n + 5 \to 8$

2. Choose a number. Triple it. Add the number 1 more than your original number. Add 7. Divide by 4. Add 5. Subtract 2.
   Try several numbers. What happens?
   Explain your result—if the number you first chose is n, then the sequence of steps is:
   $n \to$

3. Three consecutive numbers total 60. What are they?
   Call the first number n. Then the other numbers are $n + 1$ and $n + 2$.
   $n + n + 1 + n + 2 = 60$ so, $n + n + n = 57$ and $n = 19$
   The numbers are 19, 20, 21.

4. Three consecutive numbers are such that the sum of the first and last is 118. What are they?

Activities like these help to develop an awareness of the power and advantages of using letters to stand for numbers and to express patterns and ideas. Later, as algebra is established as a central part of mathematics in its own right, the manipulation of objects can be used to establish general laws in a similar manner to their use in building up computational ability and understanding. The meaning of the ways in which symbols are operated on will be expanded and notions of number will be extended to include negative and complex numbers as the solutions to different types of algebraic equations are investigated. Such investigations are one of the major topics in the high school curriculum whereas in primary school it is sufficient that children come to think of the letters as standing for unknown numbers and representing general patterns or results. The operations on these unknown numbers are simply those of the computation that has been developed so far, evaluated when the 'missing' numbers are known. If this type of thinking is built up prior to the formal introduction of algebra, the use of letters in this new topic will seem no less reasonable than their use in labelling points in geometry. When their use in solving problems once and for all and expressing results succinctly yet meaningfully is also established from the outset, the few processes that need to be learned in algebra will be welcomed as an extension of problem solving and computational ways of thinking.

CHAPTER FIVE

Shape and space

I f children take up the invitation to investigate . . . then they are more likely to do some worthwhile mathematics for themselves, whereas the invitation simply to classify shapes or to look for squares around the school building is likely to confine children to the more trivial and unrewarding aspects of shapes and space.

Delaney, 1979, p. xvii

People have been intrigued by and have learned geometrical ideas for many hundreds of years. In more recent times, curriculum documents informing mathematics teaching have given more emphasis to important geometrical ideas that can be learned by children in the primary school years. Children can be intrigued by geometrical ideas and can develop a wonderment of shape and spatial relationships, patterns and examples in their environment and culture. The source of this intrigue resides with interesting and dynamic teaching.

The overall aim of shape and space teaching in the primary school, along with the development of a positive attitude, interest and enjoyment, is to equip children to use spatial ideas and knowledge to complete practical tasks and solve a wide range of everyday practical problems. Many problems require shape and space knowledge and understandings in their solutions. Allied to this overall aim are the ideas of helping children to describe and understand their environment, to find their way within that environment, and to construct a variety of objects. Geometry is also a necessary foundation for much of the mathematics that older children will be learning later in high school. This is so for those children studying analytical geometry, calculus and trigonometry in Years 11 and 12, but all children will need to be able to apply some simple shape and space knowledge if they are going to succeed at high school mathematics. Many of the basic concepts and rudimentary processes are established in the early and primary years. Teachers need to be aware of the negative impact that may occur when children complete a boring or tedious worksheet in shape. Children expect the work they are completing to be interesting and exciting.

Spatial ideas have considerable historical and cultural significance as well as modern applications in, for example, design problems and document layout for computer presentations. The inclusion of geometry under the less formal name of shape and space continues to the present day with all Australian state curriculum documents, syllabus outlines and recommendations in reports highlighting the need for children to study geometrical ideas. This chapter will discuss the significance of shape and space in primary school, methods suitable to teach shape and space ideas, important ideas to be learned by children, essential background knowledge for teachers and selected research findings that impinge on teaching and learning shape and space ideas.

Section I: What is geometry?

THE word geometry is a combination of two Greek words, *geo*, meaning earth, and *metron*, meaning a measure. Over the subsequent years it has developed to be much more than 'earth measuring' and is now considered to be the 'mathematics of space' so that mathematicians 'search for mathematical interpretations of space' (Bishop, 1983). More recently, curriculum documents have labelled geometry as shape and space—a name that is reflected in the title of this chapter—a useful name for teachers to understand as it highlights two areas of knowledge that form the basis of geometrical study. Bishop (1997) also noted a link between shape and space and two of his constructs of the cultural roots of mathematical knowledge, namely 'locating' and 'designing'. The name shape and space also suggests links to other areas of knowledge such as graphics, design, art and geography.

Historical development

Many of the principles and theories that constitute knowledge of geometry can be found in the writings from Ancient Greece. Euclid, Pythagoras and other Greek geometers considered the learning of geometry to be important and spent many hours with their students in its study. It was they who developed the accumulated knowledge of shape, space and measurement which had been acquired over centuries in Egypt and Babylon. These ideas had been developed from practical situations like measuring fields and observing the motions of stars and planets. The Greeks turned it into an organised, logical and coherent system now known as geometry. Pythagoras and his colleagues used geometry to explain music, art and science. Ideas such as the Golden Rectangle used in painting and architecture, as well as the description of the movement of planets, stemmed from geometry. Geometry has been central to the understanding of the nature of mathematics itself through the consideration of such notions as abstraction, generalisation, reasoning and proof.

The language and vocabulary of shape and space

The language of shape and space plays a vital role in developing spatial sense. Language has a dual function for children learning shape and space ideas. It allows them to describe concepts and communicate ideas and findings so that others might hear their thinking and reasoning (and vice versa) and helps ideas to form in their minds as they grapple with words and images. Language in the form of discussion and explanation allows children to connect symbols, actions, words and pictures to develop a richness of meaning, relationships and connections for a particular shape. Square becomes much more than an isolated fact and the name for a diagram of a shape seen in a textbook. It is imbued through the use of language with so much more meaning for the learner. In this sense, children literally talk themselves into understanding.

Materials and manipulatives are used with children to represent concepts but materials themselves do not carry meaning. They are used to provide experiences for children that can be discussed and reflected upon so children can build a meaningful system of mathematics. It is through discussion that learners establish links and concepts, relate what they have experienced, filter the input based on their pre-existing schema, and assign features to symbols and names. In this way embedded ideas emerge as mathematical ways of thinking. Teaching is designed to highlight connections. It is not a series of statements made by the teacher. Children need to conduct their own dialogue rather than only hear and repeat someone else's words and phrases. Mathematical activity occurs when there is activity with dialogue. Pseudo-conceptual understanding is developed through name learning and a reliance on rules and rote memorisation. It is also through discussion that teachers gain insight into students' views and see how these views are matched to those of the teacher.

It is at the level of the specific vocabulary of shape and space where many learners develop misconceptions, particularly if they have not used the language of shape and space to develop connections. Most of the words and names associated with geometry are derived from Greek and Latin roots.

Unfortunately there are some anomalies in mathematical language that can lead to misconceptions, for example the teen numbers and the naming of polygons. When dealing with the naming of teen numbers it has been suggested that larger two-digit numbers be introduced first so that children become familiar with the pattern for naming numbers prior to encountering a breakdown in the pattern. Likewise, when naming polygons (literally many angled shapes), it is simpler to name shapes such as the decagon (deca meaning ten i.e. ten-sided polygon), nonagon, octagon, heptagon, hexagon, pentagon and so on before naming quadrilaterals and triangles, which do not follow the naming convention—otherwise we would speak of 'quadragons' and 'trigons'.

The established vocabulary and conventions for shape and space give order to the area and so are vitally important in building knowledge. However, if they become the major focus, difficulties may arise. There may be confusion with the way words are used in everyday life and their more precise use in mathematics. For example, for some children the word 'regular' describes the size of drink or 'fries' that they wish to buy. For others, it means 'found everywhere' in the sense of regularly or usually found. In mathematics, it is the opposite of irregular, that is, it has the same length. Hanbury (2000) noted difficulties experienced by teacher education students with 'opposite', 'adjacent', 'diagonal' and 'right triangle'. The latter was translated by some students as the 'correct triangle'. Anecdotal stories abound of children who, having found a 'right angle', proceed to look for a 'left angle'.

The formal language of shape and space can be overwhelming in its scope and precision for young students. Teachers in all years of the primary school need to be wary of introducing too many new words at the one time if they are difficult or likely to lead to confusion, such as the word perpendicular. In many cases, simple but accurate everyday phrases will suffice. It is desirable to listen to and use children's language in the first instance, gradually refining it and introducing new key words as they become necessary. For instance, the words *similar* and *congruent* have precise mathematical meanings that are confusing for many children. However, the confusion usually lies in remembering which of the two words to use in a particular situation rather than the idea or meaning of the concepts themselves. Neither word is necessary for primary children since there are simple, meaningful everyday phrases that convey the required meaning. Children can note that the same shape means shapes are similar; when shapes are congruent they have the same size and shape. Understanding and using the concept appropriately is important rather than discriminating the difficult terminology.

Forms of geometry

There are two aspects to the study of shape and space in primary school: visual geometry, sometimes referred to as spatial awareness or spatial sense, and formal geometry (Davey, 1998).

Visual geometry

Visual geometry is informal and is the use of space, shape and form at an intuitive, personal and unstructured level, such as in interpreting a map, sketching a picture or rearranging objects. Sometimes descriptive or explanatory language may not be necessary, such as when playing with blocks, solving a jigsaw puzzle or using tangram pieces or pattern blocks. Language may be used in this case in a sub-vocal fashion as the 'arranger' or 'sketcher' talks to him- or herself during the task. However, when language is required to describe and reason, it tends to be natural and informal. Visual skills are required in many everyday tasks, for example in designing gardens, play areas and theatre scenery.

Formal geometry

In contrast, formal geometry is more objective, demanding greater accuracy in language and representation and more rigour in logic. Formal geometry is important, for example, when measuring, such as in making a dress or furniture or when justifying a statement in a logical way. It is concerned with an extensive and exact vocabulary and agreed modes of reasoning. It is highly structured, being made up of a series of provable theorems and facts. It is much less intuitive, with little scope for opinion. Statements are right or wrong and, when right, can be proved to be so. It is an important prerequisite for engineering, science, technology and other mathematically based disciplines.

Section II: Why shape and space are important in the primary school

GEOMETRY is not concerned so overtly with facts and processes to be learned and practised in the same way that many people see arithmetic. However, delving deeper, it can be seen that geometry is useful when, for instance, shape ideas and principles are applied to the environment in the form of plans and maps. Bridge designers and the architects of buildings use geometrical information in their plans and constructions. Over the years, ideas and concepts of shape and space have been vital to navigators and have underpinned the recreational, game-playing and aesthetic activities of many cultures.

Many tasks call for the use of visual skills, as most people need them in day to day living. Spatial skills are also vital to architects, designers of landscaped gardens, golf courses and sets for stage and screen, civil engineers concerned with such things as traffic flow, painters, weavers, furniture and cabinet makers, dress designers and layout artists.

> We use spatial ideas for a wide variety of practical tasks. We describe our surroundings, find our way around and mark out and construct living spaces. Spatial ideas are basic to the solution of many design problems: producing inexpensive but sturdy packaging, laying out a page for maximum appeal, arranging pattern pieces to minimise waste or developing the orbital engine. (*A National Statement on Mathematics for Australian Schools*, 1991, p. 78)

Geometry, in the form of shape and space, has connections to modern definitions of numeracy. Spatial understanding and competence are significant components of numeracy and are essential to functioning effectively in modern times. Recent curriculum documents have charged teachers with the task of developing children who not only have number sense but also spatial and data sense.

> Numeracy is the effective use of mathematics to meet the general demands of life at home, in paid work, and for participation in community and civic life . . . They [the National Numeracy

Benchmarks] will incorporate the development of students' understanding and competence with number and quantity (ie measurement), shape and location and the handling and interpretation of quantitative data (National Benchmarking Taskforce, 1997, cited in *Australian Association of Mathematics Teachers, Numeracy — Everyone's Business*, 1997, p. 13).

Geometry when combined with number is also important in the teaching of measurement and provides a vehicle for many problem-solving situations. Teaching geometry is not simply a matter of teaching children to classify and name a few shapes. It is more important that they learn to discriminate between shapes, to visualise what a shape would look like if it changed in some way and to apply geometric ideas in problem-solving situations.

Geometry provides a useful vehicle for developing logical thought and mathematical ways of working, such as classifying, hypothesising, justifying and generalising. It is also a natural place for children to interpret and describe physical environments. Geometrical ideas are useful in representing and solving problems, developing arguments and analysing characteristics and properties of shapes and situations. It is useful as via exploration it can help children and later adults go beyond the learning of isolated ideas to develop relationships and by using these connections find many ways to solve problems. It is for these reasons that geometry has formed part of a 'sound education' throughout history.

Learning shape and space in the primary school

Primary school geometry should not just be the study and accumulation of a body of facts such as the names of shapes, their properties and sets of theorems concerning them. Rather, the emphasis should be on developing spatial intuition ranging from the appreciation of designs in wallpaper, tiling patterns, Celtic and Islamic decoration, traditional quilt designs and company logos, to the interpretation of maps and diagrams and relatively simple observations such as the diagonals of a square divide it into four identical, right-angled triangles. Building this awareness requires the active creation by each student of knowledge, meanings and understandings about two- and three-dimensional shapes, transformations and location. However, meaning, understanding and appreciation cannot be given to children by a teacher-dominated transmission approach to teaching. Students need a wealth of practical and creative experiences in solving problems by observing, analysing, describing, exploring and drawing a variety of shapes, arrangements, patterns, maps and other geometric structures. They need to engage in mathematical inquiry and investigation in much the same way as in a science experiment.

This content for shape and space can be organised into fundamental strands. For instance, *A National Statement on Mathematics for Australian Schools* (1991) suggested:

1 Shape and structure—dealing with the properties of two- and three-dimensional objects and the relationships between shape, structure and function.

2 Transformation and symmetry—dealing with the mathematical equivalent of changes of position, orientation, size and shape and with symmetries in shapes and arrangements.
3 Location and arrangement—dealing with the representation of position and arrangement including the use of coordinates.

These were further amplified in the *Numeracy Benchmarks Years 3, 5 & 7* (Curriculum Corporation, 2000), which offers guidance for the shape and space knowledge to be acquired by children at various year levels:

Year Three
Students are able to recognise and name a range of common two-dimensional (2D) and three-dimensional (3D) shapes and objects, and describe some of their features using everyday language. They are beginning to be able to use simple grids, maps and plans to locate items or landmarks.

Year Five
Students are able to recognise and describe an increasing range of 2D and 3D shapes and objects and use more formal geometric language to describe their features. They are becoming more aware of spatial patterns and tessellations as well as symmetry in 2D shapes. The students pay more attention to directions about placement of objects and routes when reading straightforward maps and plans.

Year Seven
Students use geometrical language to describe features of 2D shapes and 3D shapes and objects when they compare and classify these. They recognise different 2D representations of 3D shapes and objects, lines of symmetry in common 2D shapes, and basic angles (such as 90°, 180° and 360°). Students use shapes to make tiling or repeating patterns, and identify shapes that match exactly in different arrangements. They use simple coordinate systems, scales and basic compass directions to interpret maps and describe locations.

Other state and national documents also include substantial reference to shape concepts and have reflected this emphasis, while some have added further sections. For example, the Western Australia *Outcomes and Standards Framework* (Education Department of Western Australia, 1998) has four strands—reason geometrically, represent shape, represent transformation and represent location, while in the UK and USA these are:

England and Wales	Understanding patterns and properties of shapes		Understanding properties of position and movement	Using and applying shape and space and measures
USA	Characteristics and properties of 2D and 3D shapes	Apply transformation and use symmetries	Specify location	Visualisation and spatial reasoning to solve problems

Investigate and discuss

1 Note how the benchmark of each year level develops in complexity and sophistication as it builds on the previously established knowledge.
2 Note how there is an emphasis on the language of shape and space throughout each of the three statements.
3 How do the benchmarks outlined above match the Van Hiele levels described later in this chapter?

Ways of working in geometry

A classic article by Hoffer (1981), 'Geometry is More than Proof', and subsequent writing by, among others, Davey and Pegg (1992) have highlighted five ways of operating that need to be emphasised in teaching and subsequently acquired by children. These are directly related to shape and space, are quite general in nature and are designed to help children solve problems in mathematics, in daily life and post-school employment:

- visualising spatial arrangements
- communicating orally and in writing
- drawing and making models
- thinking logically
- applying geometrical concepts and knowledge

Visualising spatial arrangements

Geometry is very visual, and visual comprehension and understanding underpins much of what people do in their daily life. Information, often presented in a visual format, has to be processed. People, for example, routinely engage in tasks such as assembling models or furniture bought in a kit form, using patterns, diagrams and drawings, decorating homes with wallpaper and carpet tiles, and attempting to follow a wiring diagram for a new computer, television or audio system. More often nowadays newspapers, magazines and television present information in the form of diagrams and graphs that need careful and thoughtful interpretation.

Two different aspects of visual skills have been identified by Bishop (1983):

- **Visual processing**—this is the ability that enables students to manipulate and transform spatial images in the mind. It also includes the aptitude for visualising non-spatial relationships and information such as constructing a flowchart or visualising operations on fractions.
- **Interpreting figural information**—this refers more to understanding the conventions used in drawing diagrams, graphs, maps, signs and charts and is concerned with interpretation and reading.

Within the classroom context, the process of visualising may be developed via the following activities with pictures and materials noted by Davey (1998) and Hanbury (1999):

- recognising something from a picture
- picking out the properties of a geometric figure
- seeing similarities and differences in plane and solid figures
- interpreting diagrams and sketches especially of simple three-dimensional objects, but also of two-dimensional situations like floor or street plans
- reading a map
- recognising objects that have been turned around, inverted or viewed from a different direction
- imagining what something looks like after some transformation, such as a rotation or distortion
- seeing if a line is straight
- estimating the size of an angle
- recognising embedded figures
- fitting jigsaw pieces together
- visualising a situation from a written or oral description
- completing a geometric pattern
- visualising the cross-section of a solid object
- visualising a solid from its net

Communicating orally and in writing

As children experience more activities with shape it becomes more important for them to use precise and unambiguous language. Shape and space has a wide and rich vocabulary and a set of conventions with which children will need to engage as they develop. It is important for children to describe and explain ideas and relationships between shapes as part of learning in shape and space. It is acceptable for children to describe shapes and situations as accurately as they can in their natural language and for their particular level of development. This may include some everyday words; however, children will eventually need to develop a more accurate and precise way of communicating ideas in shape and space via teacher scaffolding and example, but there should not be such a heavy focus on correct terminology that it restricts the development of important geometrical ideas and the motivation and positive attitude of children. Further, the difference between everyday words and their precise use in mathematics causes difficulties for many children. Nevertheless, by the upper primary years children are expected to be more exact with their terminology so that ambiguity is avoided. For example, 'faces' would be expected to be used rather than 'sides' when children are describing a dodecahedron. Activities and situations for developing more accurate and fluent communication include:

- discussing similarities and differences between figures
- learning to use geometric language in oral and written forms
- naming and describing a figure or spatial situation accurately
- interpreting an oral description accurately
- describing the properties of an object that the children are holding
- describing a concept
- interpreting written instructions of geometric situations accurately
- describing spatial relationships
- describing the location of something in relation to other objects

Drawing and making models

Along with talking and writing, sketching, drawing and model making are other forms of communicating that are an important part of learning in mathematics. There can be some flexibility allowable in the degree of accuracy of a drawing or diagram. For example, in some contexts and for some problems a 'rough sketch' might be sufficient to communicate an idea or solution. At other times, however, a more precise and accurately drawn figure might be needed.

The idea of representing via sketches, models, diagrams, drawing or computer packages as part of communicating is regarded as more important for primary school children than traditional constructions with a pair of compasses and ruler. Less formal methods fit more closely one of the main aims of mathematics teaching, that of solving problems. Drawing is much more than representing with a pencil, it can be expanded to include constructing, modelling and using computer drawing packages. Drawing, modelling and representing can be developed via activities such as:

- drawing simple plane shapes
- drawing scale diagrams
- cutting out cardboard shapes
- sketching isometric diagrams of figures based on the cube
- drawing very simple perspective sketches

- sketching simple diagrams of cones, pyramids, cylinders
- making 3D models with clay, plasticine, or cardboard
- making plane shapes on a geoboard
- constructing shapes using pattern blocks

Thinking logically

For many people in the past the reason to study geometry was for its logic and rigour. Logic was central to the ability to construct a proof and to follow the work of mathematicians such as Euclid. Logical reasoning remains an important aspect of work in shape and space at all levels but it has taken a wider role to include classifying, identifying similarities and differences, and reasoning. Thinking and reasoning abilities at all levels can be developed through the following activities:

- finding similarities and differences between shapes
- understanding that shape is independent of orientation, position and size
- understanding that certain properties can be used to identify shapes while others cannot
- classifying and sorting—realising that shapes belong to various families and that some families are included in others
- being able to discern and continue a pattern
- conserving
- formulating and testing hypotheses
- making inferences

Applying geometrical concepts and knowledge

Everyday life involves problems that have to be solved and situations that have to be resolved. One of the reasons for studying mathematics in general and shape and space in particular is to help children develop skills, processes, knowledge and ways of working to solve problems. There are excellent connections between shape and space knowledge and the design and function of objects that come from worthwhile activities and discussion. Several applications of geometry are:

- making curtains, patchwork quilts
- creating designs for fabrics, decorations, wallpaper, tiles
- model making, sculpture, architecture, building
- painting and drawing
- making and reading a map
- making cupboards, toys, shelves, drawers
- marking out tennis courts, running tracks
- setting out a car park
- planning a delivery run around a group of streets
- designing the shape of industrial objects to be manufactured for sale

Investigate and discuss

1 How does the curriculum document you would use to plan the teaching of shape and space incorporate Hoffer's five ways of working and the suggestions by Davey and Hanbury for teaching?

How children learn shape and space concepts

It is important for teachers to be aware of some of the models that explain how individuals acquire shape and space concepts and processes. Research from people such as Piaget (1972), Labinowicz (1980) and van Hiele (1986) indicate that children appear to pass through a number of levels or modes of cognitive functioning as they move from infancy to childhood and adolescence. While it is fair to say that not all educational theorists believe in the notion of stages, the concept is a useful one for teachers as they can use it to understand and chart pupil growth and to plan their programs according to student needs. The van Hiele theory, which developed from the work of Piaget, provides significant insights into children's growth and can help classroom teachers plan and assess learning specifically in shape and space (Fuys, Geddes & Tischler, 1998). It should be noted, however, that the use of the term level in this context is different from and unrelated to the levels referred to in many curriculum documents to indicate progression through particular content. Further, the levels are not age related but are based on learning that children have experienced and gained.

The van Hiele theory

An outline for the development of children's learning of spatial concepts was proposed in the late 1950s by two Dutch teachers, Dina and Pierre van Hiele. They delineated five levels of understanding—visualisation, analysis, informal deduction, formal deduction and rigour. Their model describes the characteristics of the thinking process at each level, not the shape and space or geometric content. The levels are sequential and the van Hieles asserted that no student can successfully miss any level and that ideas implicit in one level are explicit in the next.

In subsequent writing, there is variation in the numbering system and names associated with the levels. For example, the original work of the van Hieles started at level 0 and proceeded to level 4 while more recent writers tend to begin with level 1 and progress to level 5. Pegg (1997), cited in Owens & Mousley (2000), has suggested that the second level be separated into two parts—part A, where figures are identified in terms of a single property, such as the length of sides, and part B, where several but independent properties are known. Horne (1998) noted a non-van-Hiele level of precognition suggested by Grouws that comes before the starting level. Here children only attend to a subset of characteristics in their perception of shape. She offered an example of children on one occasion not differentiating between circles and squares, as they are both closed shapes. They may, however, suggest that they are different on another occasion when they consider straight and curved lines. For the purposes of this discussion, it is proposed to keep to the original description but to adopt the modern numbering system starting at level 1.

Van Hiele level 1: Recognition/visualisation

At this basic level a shape can be recognised and named without reference to properties, that is, children operate according to appearance. Children can describe a shape as a whole but will not refer to attributes that make it unique. For example, an oblong may be described as 'like a door' or 'like a window'; a cuboid is a 'box'. Children refer to objects with which they are familiar. At this level children can learn shape vocabulary and may be able to identify specific shapes, for example a square. They may be able to reproduce the figure but would not be able with any certainty to note the right angles or that the opposite sides are parallel. Problems are solved by visual means or by trial and error. Language is imprecise. Words like 'corner' and 'slanting' may be used to describe shapes. A number of children in older primary classes may be operating at this level (Clements & Battista, 1992).

Van Hiele level 2: Analysis

At this level children start to notice and acknowledge attributes and properties of shapes. The awareness of properties is now extrinsic. Initially, a square may be described as a shape with 4 sides or 4 equal sides and an oblong as a shape with 2 long and 2 short sides. Later, additional properties, for example 4 right angles, will be added. Children can, for example, distinguish between triangles and talk about the properties of these shapes. However, children at level 2 cannot understand that some properties are a necessary consequence of others or that there are relationships between them. For example, they may be able to note that an isosceles triangle has two angles of the same size and two equal sides but will not connect the two ideas. Teacher-imposed definitions cannot be understood, and cannot be applied. Simple, accurate descriptions of shapes listing all known properties are appropriate at this level.

Van Hiele level 3: Ordering/informal deductions

At this level, the relationships between properties and shapes begin to emerge. Properties are seen to be dependent on each other with certain properties following logically from others. For example, when lines are parallel, corresponding angles must be equal. This understanding comes from reflecting upon concrete experiences with materials such as geostrips and by drawing and measuring. Level 3 could also be called the *relational* level since children are now seeing relationships between the various shape and space concepts. Children can now give the properties of a shape that will uniquely define it. So an equilateral triangle can be defined as having three equal sides, three equal angles each of 60 degrees, the angles will add to 180 and so on. Not all children at primary school will function at level 3.

Van Hiele level 4: Formal deduction

Children now have an ability to reason abstractly and use logic to develop a proof. They will be able to construct their own proof rather than remember proofs. They will know that an equilateral triangle can be determined by saying that all its sides are equal. They will know that this is sufficient as they 'see' the relationships and roles of other terms. Few students reach this level while at high school.

Van Hiele level 5: Rigour/metamathematical

Students are able to reason analytically about systems dealing with the relationships between formal constructs. This is beyond most students at high school.

As with any model it is important to be aware of strict definitions of levels. The model is useful, however, in that it helps in organising tasks and teaching, and supports the development of understanding of the properties of shapes.

Implications for teaching

The level at which children operate cannot be related strictly to age as they progress at different rates. Development is more related to amount and quality of experience. The original work by the van Hieles offered a teaching approach for each level of:

- *information*—experiences here acquaint children with the content and language of shape and space
- *guided orientation*—children explore objects and investigate relationships
- *explication*—children talk and elaborate their findings, learn and use mathematical language as they discuss with the teacher and other children
- *free orientation*—children solve problems involving the ideas recently encountered
- *integration*—children reflect on how this learning connects with and builds on other knowledge they have

From this approach there are connections to what many people would call effective teaching practices. The teacher has to structure tasks to contain relevant ideas and knowledge as well as guide children in their discussion of the outcomes of the task. The children need to experience a range of tasks that embed the same learning and use their own language in discussion and explanation of their work. Children are required to reflect on what they have done and found. This knowledge is then connected in explicit ways to what they already know. Children also have to use their knowledge in problem and investigative situations. They are not remembering the work and ideas of other people but are building their own ideas in a guided and structured way. They are neither just remembering rules, words and processes nor being left to 'discover things for themselves' in fun activities.

Textbook use when working with children to develop shape and space ideas presents difficulties in relation to the van Hiele theory for at least two reasons. First, textbook pages are limited and tend to offer one or at most two activities relating to an aspect of shape and space learning rather than multiple experiences. Second, textbooks tend to move forward and back between van Hiele levels and are generally based on the author's view of the median age characteristics of children. They often lead to the rote acquisition of names and procedures and thus have no van Hiele level at all, since the van Hiele levels refer to thinking, not memorising.

It is interesting to note that language plays a very important role in this theory. The van Hieles claim that each level has its own linguistic symbols so that children at different levels have difficulty in communicating clearly with each other in their discussion of shape. Hence, it is important that teachers familiarise themselves with the language actually used by the children in their classes to talk about shape ideas, and that they become aware of the way children think and reason at each level. Again two-way communication between teacher and children should be stressed but the language of communication needs to be that of children, not solely that of the teacher.

By the upper primary years most children should be operating consistently at van Hiele level 2, that is, they see objects with properties rather than just visually. Van Hiele level 3 is the point where logical argument is noted; however, it is important that teachers of younger children are aware of this and encourage it, as even young children can begin to reason and offer justification for their choices and comments. The ability to reason and argue a case will develop with expectation and experience and most curriculum documents now require spatial reasoning for children of all ages.

Investigate and discuss

> **1** Find a recently published textbook series and compare its treatment of shape and space learning in relation to the van Hiele levels and suggestions for teaching.

Likely difficulties in shape and space learning

Research in recent years (Atweh, Owens & Sullivan, 1996; Owens & Mousley, 2000) has highlighted a number of instances of misconceptions held by children relating to shape and space. An awareness of possible misconceptions will allow teachers to be prepared and vigilant for their emergence. Thoughtful and careful preparation of rich mathematical shape and space tasks can expose misconceptions and sensitively bring them into conflict making the situation ready for discussion and exploration. Discussion and explanation will expose and make explicit children's thinking and reasoning and allow teachers an insight into children's minds and understanding. It will also allow children to hear other reasons and arguments as their peers talk about their work. Teachers can then match key shape and space understanding with the likely misconceptions and difficulties experienced by children. Such knowledge may also inform teachers of flaws in their practice. For example, are the images portrayed in textbooks and on the board always the same? Are triangles always presented with a base parallel to the bottom of the page or board? Likely misconceptions can be categorised as follows.

Limited examples

Children develop a limited concept of polygons and often only recognise a portrayal of a regular shape in one orientation. For example, when asked to draw a hexagon, children will invariably show a hexagon with all sides of equal length and in an orientation with a side parallel to the bottom of the page.

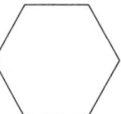

Rarely will they acknowledge the existence of other members of the hexagon family or hexagons placed in different orientations.

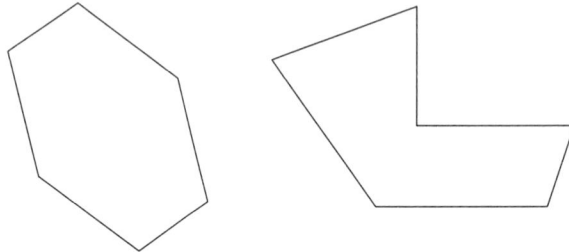

Limited orientations

Allied to this is a similar situation where children are able to recognise and name a square when it is presented in a 'normal' orientation but are unable to identify it when it is rotated and presented in a different position. For some children this shape now becomes incorrectly a 'diamond' rather than a square.

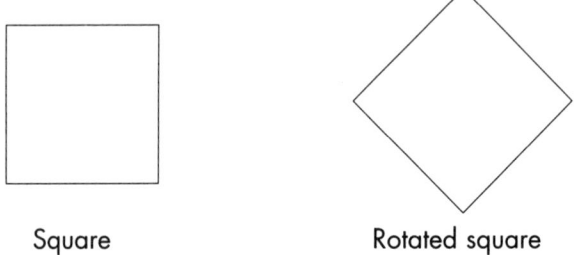

Square   Rotated square

Most young children can recognise circles and squares but they are less comfortable with triangles and rectangles. They have difficulties with 'long thin rectangles' and triangles where the 'point' is not at the top. Further work in identifying and discussing the critical and non-critical attributes of shapes along with exposure to examples and non-examples of particular two-dimensional shapes is needed. These beliefs are difficult to change and children are likely to continue to believe the same thing despite the efforts of teachers and textbook writers to offer a contrary perspective.

Mathematical attributes

Many children when they are asked to describe three-dimensional objects often do not attend to the critical and mathematical features of the object being viewed and comment on other non-mathematical features such as colour. Other difficulties include manipulating from two-dimensional to three-dimensional and in interpreting two-dimensional pictures representing three-dimensional objects. These difficulties may be due to a lack of experience with such activities and also suggest that children need to have their attention focused on the mathematical features of the task or problem. A variety of experiences using imagery and visualisation will lead them to appreciate alternative perspectives. For example, as children begin to investigate situations using turning, flipping, superimposing, folding and drawing their learning and understanding will develop and they will begin to think dynamically.

Symmetry

Many children are comfortable with symmetrical objects displaying a vertical axis or line of symmetry but have difficulties when a shape has other line symmetries. Older primary children may continue to experience difficulty with lines of symmetry that are at an angle rather than in vertical or horizontal. They may complete a shape reflected in a vertical axis of symmetry but may not be able to do this with a horizontal mirror line or a mirror line at an angle. Other shapes may at first sight seem to have line symmetry, for example the diagonal in an oblong.

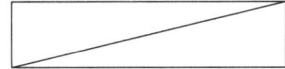

This is particularly likely when consideration of an oblong follows children finding lines of symmetry in a square. Discussion of ways to test line symmetry, then checking via at least one of them, such as folding, help children to overcome this difficulty. In turn, this discussion could lead usefully into the concept of rotational symmetry.

Tessellation

Children are likely to experience difficulties when work with tessellations uses only pencil and paper. Exercises requiring children to complete a tessellation where one starter tile is given often leads to children almost randomly drawing and fitting the shape. There appears to be little attention paid to pattern and regularity of the tile placement. Soon the larger tessellation breaks down and children become frustrated. A second difficulty with drawing round a tile template highlights the need for a degree of accuracy. As children draw round the template the subsequent tile is slightly enlarged and sometimes will not fit into the allocated space, suggesting that the tile will not tessellate when in reality it does. A change in task planning will help to overcome these difficulties. This is not so much a learner misconception but a consequence of the task. Where children are presented with a number of pre-cut or commercially produced tiles, they are able to construct physically a tessellation and easily make adjustments to patterns as difficulties arise. A digital camera nowadays can record a tessellation quickly and easily.

Position

Most teachers are aware of the difficulties children have with ordered pairs or coordinates when attempting to name position. The usual confusion occurs when children confuse or are unaware of the convention of naming the horizontal component before the vertical component. Providing activities involving games such as 'Battleships' or transferring a drawing from one grid to another via coordinate references helps to overcome these difficulties.

Angle

Mention has already been made of the issue of 'right' angles and 'left' angles, which seems to offer for some children a reasoned development. Orientation and length of lines in a shown angle also present difficulties for children who have had limited exposure to them. For example, some children will consider angle B to be larger than angle A because the lines are longer or because there is a larger gap at the end.

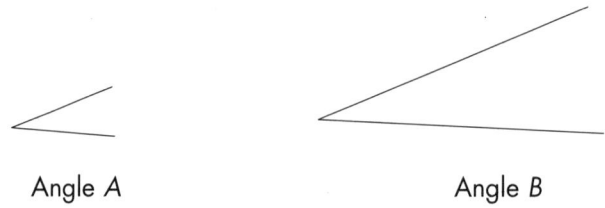

Angle A   Angle B

The concept of angle is discussed later in this chapter and also in the chapter on measurement. Children typically have difficulties with using and reading the scale on standard semi-circular or 'half-moon' protractors. Here they confuse which scale to use and the value of the gradations on the scale. The use of these protractors without sufficient development of the angle concept should be avoided.

Section III: How to teach shape and space

Shape and space teaching and constructivism

Recent reports and curriculum documents (*A National Statement on Mathematics for Australian Schools*, 1991; *Principles and Standards for School Mathematics*, 2000) have advocated styles of mathematics teaching that are quite different from traditional methods of exposition, textbook work, memorisation and practice. Many of these changes to teaching style can best be understood from a constructivist viewpoint (Chapter 1, pp. 10–11), which suggests that:

- ◆ knowledge is actively created by the individual learner rather than passively received
- ◆ learners create new knowledge by reflecting and thinking about their actions both physical and mental. Through the act of reflecting, learners integrate new knowledge into their existing structures

- learning is a social process involving social discourse, explanation and evaluation
- mathematics learning is viewed as sense making rather than purely learning set procedures
- the role of the teacher is neither to transmit the 'correct' procedure, fact or way of working nor to allow a free-for-all discovery. A constructivist teacher plans appropriate tasks and time for talk, guides and focuses children's attention and, thus, unobtrusively directs and guides their learning

Typically constructivist teachers would:

- use and encourage children's thinking, questions and ideas
- use open-ended questions and tasks as well as encourage children to expand their questions and responses
- encourage children to predict the consequences of action
- encourage children to challenge gently and politely the ideas and statements of their peers
- use collaborative teaching and learning strategies
- allow adequate time for reflection and analysis

Hoffer and van Hiele

Hoffer identified five basic ways of working that are needed to understand geometry needed by people to solve problems in mathematics, other learning areas and everyday work, leisure and domestic life. Combining Hoffer's ways of working and the van Hiele levels provides a guide for planning in shape and space teaching.

	Level 1 Recognition Visualisation	Level 2 Analysis	Level 3 Ordering Informal Deduction	Level 4 Formal Deduction	Level 5 Rigour
Visualising					
Communicating					
Drawing and modelling					
Thinking and reasoning					
Applying					

Hoffer–van Hiele planning grid (after Hanbury, 2000)

Source: Reprinted from *Structure and Insight* by Dina & Pierre Van Hiele. Copyright 1986, with permission from Elsevier.

The shaded areas in the table are those that typically apply to the primary school.

Investigate and discuss

> **1** How does constructivism fit with Hoffer's ways of working and the van Hiele approach to teaching?

The inquiry model

This model incorporates important features of constructivist teaching and emphasises children exploring, explaining and making decisions. It encourages children to engage in mathematical thinking by exploring an activity or situation and then, with the explicit aid of the teacher, reflect and think about what they have done and found out. Reflection and discussion with a teacher is vitally important in developing understandings and knowledge embedded in the activity.

Often with activities in shape and space children have fun and enjoy themselves, which is good for motivation and developing a positive attitude to mathematics. However, it is even more important that they also learn something—'having fun' is not sufficient. Children need to build new connections to existing knowledge and develop new ways of thinking as a result of engaging with tasks and activities. This approach is used in the teaching examples throughout this chapter.

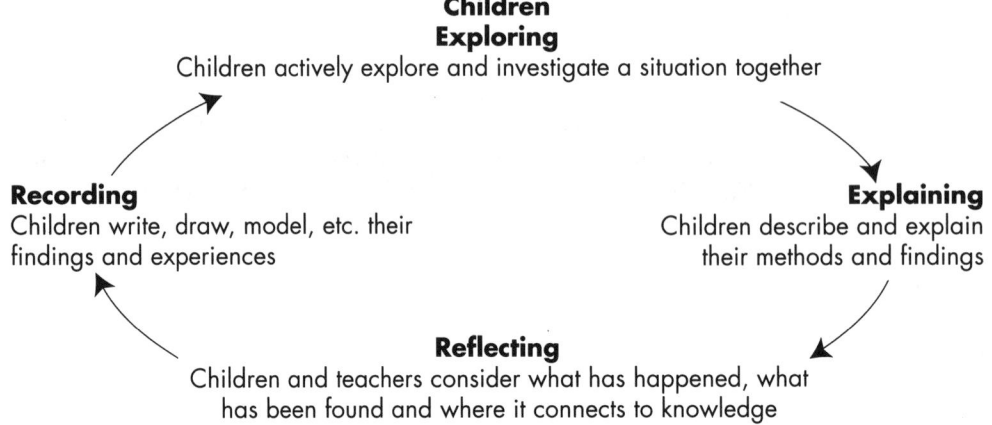

An inquiry approach to teaching

Principles for teaching shape and space

Shape and space in primary school mathematics should be a rich source of information, mathematical activity and enjoyment for children. Recommendations from reports (*A National Statement on Mathematics for Australian Schools*, 1991; *Curriculum Framework for Kindergarten to Year 12 Education in Western Australia*, 1998; *Principles and Standards for School Mathematics*, 2000) advocate:

- children exploring and investigating
- children experiencing a variety of activities
- children seeing a range of sizes and orientations of shapes
- children using a wide range of materials
- children describing activities and relationships

Children exploring and investigating

A National Statement on Mathematics for Australian Schools (1991, pp. 80, 88) noted that with regard to shape and space

> ... the formal teaching and testing of the vocabulary of space should not be the focus of work in this strand ... [rather] the approach should continue to be informal and practical and emphasise the investigation of features of objects.

In other words, the emphasis on remembering and repeating found in the approach of many textbooks needs to be shifted to one of experiencing, exploring and thinking by children. An activity folding a square of paper and investigating the possible shapes that can be formed as corners are folded or left unfolded illustrates this approach.

FOLDING THE SQUARE

Exploring
Children are shown how to fold a square of coloured paper into half through the centre of the vertical side. They then fold in each of the corners to the centre line:

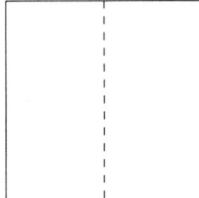 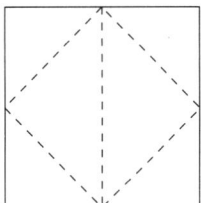

A smaller square is formed and can be recorded by drawing its silhouette onto a separate piece of paper. Children are then challenged to investigate, in pairs, how many different silhouette shapes they can make by folding on the original lines.

Explaining
Children *explore* the situation and think about the consequences of their actions and the significance of their findings. As they work together folding the square they are classifying shapes as they decide if the new folded shape is the same or different from the ones already recorded. During this phase of the task they have to *explain* to their partner, and the teacher, how the shapes are different and therefore acceptable. The pairs can then share with the class shapes they have found so far. The act of sharing allows a short change in working style and others to see what has happened and possibly motivate them to further inquiry.

Reflecting

This phase is as important as exploration, helping children grasp ideas and learning embedded in the task. A teacher still has a role in making explicit this potential learning and helping children connect to knowledge already acquired, so that it is more than an unfocused 'show and tell' session. The teacher already knows the points that can be developed and guides children to make them. For example, children will have met the situation whereby they made a large square (no corners folded) and a smaller square (all corners folded). Some pairs may even have noted the smaller square as a 'diamond' as they have the square rotated through 90 degrees. Discussion can establish that both shapes are squares and that size and orientation have no effect on their 'squareness'.

By looking at the range of shapes made, a teacher can draw the children's attention to the varieties of pentagon and hexagon that may be found. Children will then meet the idea that all hexagons have six straight sides, whereas the regular hexagon, where all six sides are of equal length, is a special case. The teacher can also discuss the critical and non-critical features of shapes as well as the need to decide on rules for classifying. For example, some children may classify the larger and smaller squares as different due to their size, while other children may have them as the same by reasoning that they are both squares and that size and orientation does not matter in their grouping. The attention to sameness and difference as part of classifying is a vital part of working mathematically and a teacher needs to help children attend to these and other important attributes.

Recording

Children can now decide how they will record their explorations and findings. For instance, they could use silhouette drawings again, or make smaller examples of the folded square with a base that can be glued to a recording sheet. This second idea has a dynamic quality of being able to be unfolded to show how it was constructed.

Children experiencing a variety of activities

Using a variety of activities will help children build rich concepts of shapes. It will also help reduce the boredom factor experienced by many children from repetition of the same activity in the same style, for example completing worksheet or workbook pages. Children may also develop a better sense of shape through activities such as fitting two other shapes together to form a new shape. In this way they will begin to see shapes as made up of other shapes much in the same way children with good number sense see numbers in flexible and varied ways. The vital contribution of teacher-led discussion and reflection on the results of the activity, as described earlier, should be noted if the task is to develop from making 'nice designs' to one from which children might make generalisations or connections to other knowledge.

Curriculum and syllabus documents highlight experiences that children should meet and knowledge they should demonstrate. For example, the *Curriculum Framework for Kindergarten to Year 12 Education in Western Australia* (Curriculum Council, 1998) provides a list of verbs that suggests a variety of experiences and tasks that children can undertake:

Visualise, draw, model, predict, show the effect, transform, describe, analyse mathematically, represent, make choices, construct, plan, build, identify, use coordinates and networks, reason, solve problems, justify solutions, generalise, apply knowledge, link, connect, explain, describe, reflect, rotate, translate, enlarge, reduce, explore, use a variety of materials, arrange and rearrange, ask, focus, investigate, stack, fit, balance, handle objects, observe properties, make statements, change shapes, give and follow directions, make spatial patterns, interpret, sketch, sort and classify.

In turn, these allow teachers to plan for the same aspect of learning to be encountered in a series of different tasks. For example, children can experience folding activities, such as 'Fold the square', geoboard tasks, cutting tasks, overlapping tasks and fitting tasks as part of a series of activities to help them develop detailed and connected concepts. The following activities illustrate a variety of approaches and use of materials and also demonstrate how a range of orientations and examples of shapes may be achieved.

OVERLAPS

Children are given two squares of coloured paper. It is important that one sheet has a darker colour than the other and is visible through the lighter piece when this is laid over the dark paper. The situation to explore involves children overlapping the squares of coloured paper and noting the overlapped section.

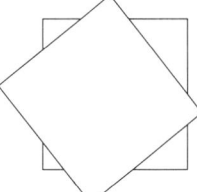

In this case the overlap produces an octagon. Children can investigate other overlaps. They may also use different starter shapes, such as a square and a triangle, two triangles, a hexagon and a triangle.

FOLD AND CUT

Cutting will offer another experience for children and another chance to explain and describe and predict. A sheet of A4 paper is folded in half and a scissor cut made across the folded corner.

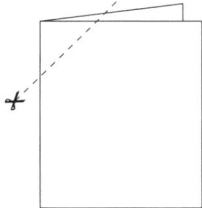

Children have to predict what shape will be formed by the cut both for the smaller piece and later for the larger remaining piece. Visualisation, prediction, reflection and discussion are involved as children think about the shapes and their properties. The activity can be extended when children cut across the fold at different angles, or fold the paper again and repeat the cutting experiments.

The activity is useful not only to add variety to the children's experiences but also to focus their attention when making observations. They will be undertaking fundamental ways of working mathematically by observing, predicting and testing. For many children the act of predicting adds anticipation and an element of surprise to an activity.

Investigate and discuss

> **1** Examine the curriculum document you would use to plan the teaching of shape and space to identify experiences children should be receiving. Select a small area of the document and highlight and list the verbs mentioned.

Children seeing a range of sizes and orientations

For many children their experience and therefore concept of shapes is very limited. A restricted knowledge of, for example, a hexagon is developed because often children only see and interact with one embodiment of the shape. Too often textbooks, teachers and worksheets present only a regular hexagon in the same orientation. A similar restricted view of other shapes is often given—pentagons are regular, triangles are usually presented with their base parallel to the bottom of the page, and squares are rarely seen in a variety of orientations. Computer drawing packages often allow the drawing of both regular and irregular shapes but then restrict children to a stereotyped orientation of the shape unless the rotate function is employed.

FOUR TRIANGLES

An activity for younger children involves fitting shapes together. The traditional *tangram* tasks may be too complex for younger children, but the style of activity is still applicable in a simpler form.

Present children with four card triangles produced from a square. Show them how the original square may be reformed by placing the triangles.

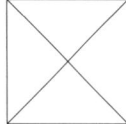

Four triangles cut from a square

The task for children working in pairs is to fit the triangles together to form new and different shapes. These are recorded in silhouette form for checking new combinations and later discussion. In this reflective discussion children will meet a variety of hexagons, different-sized quadrilaterals and different ways to construct triangles.

Children using a wide range of materials

When materials are used they often tend to be limited in scope and variety. Commercially produced, plastic examples of regular shapes are useful but only give one example for children to abstract the features of a particular shape. For example, pattern blocks, while very useful in a number of ways in helping children learn mathematics, can establish a narrow understanding of shapes if they are the only material used to support children's learning. Thinking that all triangles are equilateral and green, whereas all hexagons are regular and yellow may be an unfortunate result of using a narrow range of materials. Nonetheless, although there are only six shapes—a hexagon, a square, a trapezium, a triangle and two rhombuses—the blocks are so versatile that activities can be devised for every year level. Some examples of their use include:

- continuing or creating linear patterns
- solving problems—for example, making a particular shape with exactly seven blocks
- making tessellations—a tessellation is a patterned arrangement of geometric shapes rather like floor tiles. There must be no spaces between the shapes and the pattern must show complete regularity and predictability
- introducing area, perimeter, symmetry and fraction concepts
- creating mosaics, pictures and other artistic designs

Boxes of brightly coloured pattern blocks ought to be available in every primary classroom.

Simple, inexpensive materials can be used to add variety to the range met by children. For example, card tangram-type pieces, coloured paper for folding, scrap paper for cutting, geoboards, geostrips, drinking straws and supermarket packages are all useful to help children develop a broad and rich concept of a particular shape. A range of commercially produced equipment can supplement general, everyday materials but much excellent work can be achieved without huge expense.

THREE SQUARES

Offer the children three of the orange pattern block squares in the format shown below and tell them that this is half the shape.

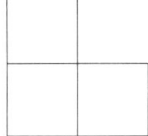

Children have to use this starter to construct 'the whole shape' and justify why and how it is the whole.

Investigate and discuss

1. Complete the activity with pattern blocks and record your solutions.
2. Identify any learning embedded in the activity and the possible difficulties children might encounter.
3. How would you extend this activity for more able children?

Children describing activities and relationships

It is important that children talk about, describe, write about and explain what they have done and found out during their shape and space activities. By having to use language, children begin to clarify and connect what they have learned from the activity with what they already know. Reflecting through the use of written and oral language is an important feature of developing understanding of shape and space concepts. It will move the often unconnected, party tricks of some shape tasks from the realm of occasional amusement to mathematics.

Part of the discussion and explanation by children could be to classify, that is, to say how something is the same, different or has changed in some way (for example the variety of hexagons and pentagons produced with 'Fold the square', p. 387). A teacher's questions help children to observe and focus their attention so that they can analyse shapes and shape situations.

The vocabulary of shape is important for children to know but it is not the main purpose for studying geometry. It should grow naturally from the experiences undertaken by children and from discussion and scaffolding by the teacher. Children appear to learn vocabulary best by using it. In this way they acquire a more precise use of language rather than just obtain more words. Teachers who provide more open tasks and questions tend to use language for reasoning and explanation rather than for memory checking.

A development for teaching shape and space

A general development for primary years should start with offering children rich experiences which begin to develop informal ideas of shapes, allowing children to become familiar with real examples of shapes, their representations, names and connections to the environment. A variety of cultural connections should also be embedded in the curriculum. Gradually, a capacity to name shapes can be extended to an understanding of the defining characteristics of different shapes, their properties and the relationships between them.

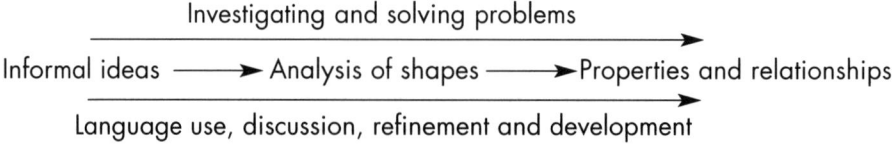

A development for teaching shape and space in the primary years

From recognising of shapes, children can begin to analyse shapes to identify how they are the same and how they are different, noting specific properties and attributes. The use of shape language will also become more precise and sophisticated, so that by the time most children reach the upper primary years they will be working at or near van Hiele level 3. For many of these older children, teaching will focus on building knowledge of properties and concepts of shapes. They will begin to acknowledge the relationships of properties within a shape but also between different shapes. For example, they will develop ideas that a square has four right angles, two pairs of parallel sides, equilateral sides, four lines of symmetry and will tessellate with itself and with regular octagons. They will begin to realise that a square is part of the family of rectangles and is similar to but also different from a rhombus.

Throughout this gradual development there will be a focus on and integration of language and inquiry as children and their teachers engage in discussions, explanations, solving problems and investigations. Work with younger children should focus on all aspects of growth, cognitive, social, physical and affective. Some tasks should be of more substance and depth, and engage children for longer to balance more easily finished tasks. Some work should be in the form of play and be less teacher structured although children should still be challenged to think about and describe what they are doing and to explore new possibilities and directions.

As children become older and have had more experiences their tasks can become more abstract. They can be expected to be more systematic in their exploration and to move from outlining general characteristics to more precise descriptions. There is, however, no clear and definite progression as many children will not follow the same route through these ideas as they learn. Questions remain of where to start, what to do first and what comes next.

Where to start

A class of children in a typical school is usually of a similar age. While they may be chronologically similar, they may not, however, have experienced the same things, understood the same ideas and interpreted the same experiences in the same way. They will be different in relation to their knowledge and understanding of shape and space ideas. Therefore, their needs will be different and teachers have the difficult task of identifying and planning for these needs in a class of 25 or more children.

'Tuning in'

One way to find out what children know is to ask them. This often produces enlightening and sometimes surprising results for the teacher when it is used at the start of some work and also at the completion of a teaching sequence. Asking children to tell, draw and write out all the things they know about a square or a diagram on a page is so much more informative of children's knowledge than, as Clarke (1988) pointed out, a test question. He illustrated this point with the following questions:

Name these shapes.

The answer given by a child to question one was quite correctly *Savoury*, while the second answer, as all snack food lovers will know, is *Barbeque*.

Another way to find out not only what children know but also how they are connecting ideas and relationships is to use concept maps (Mansfield & Happs, 1991). In this technique children start with an idea or statement or fact. From here they begin to write in other things that can be related to it.

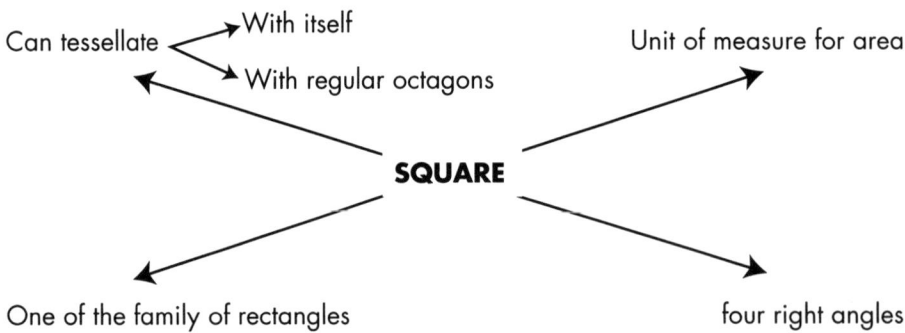

Beginning of concept map for SQUARE

It is also a useful exercise for teachers to attempt when planning learning in a particular mathematical area. Teachers can compare their concept maps with those of the children to identify missing sections. The missing sections can then form the focus for teaching and learning.

The results of either of these 'tuning in' techniques can be used to put children into no more than three rough groups—those children with limited knowledge and connections of the content area, those with extended knowledge, and the rest who fall somewhere between the first two groups. A teacher can then focus questions and tasks more appropriately to the needs of children and achieve a closer match and degree of challenge.

Extending

Open-ended questions and tasks (Sullivan & Lilburn, 1997) provide a good vehicle for children exploring and discussing, as well as learning shape and space ideas. They work in a mathematical way by doing their mathematics rather than repeating the mathematics of someone else. Two methods (*What if...* and *Find more*) for extending tasks and accommodating different knowledge of shape and space as identified via a 'tuning in' exercise are illustrated in the following activity.

SLICE THE SQUARE

Use one straight cut to slice a square into two parts.

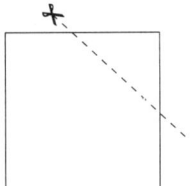

Exploring
In pairs children explore cutting prepared squares to see how many different shapes can be formed. The example offered shows a triangle and a pentagon. Some children may prefer to continue the exploration by drawing lines onto pre-drawn squares rather than cutting each square.

Explaining
Children explain to each other, the teacher and, in a sharing session, their peers how shapes are similar and different. Reasons and justifications will be needed as to why shapes have been recorded as 'different' before they are allowable.

Reflecting
Discussion with the teacher can begin to consider the different sorts of triangles possible, different forms of pentagons, whether it is possible to cut a hexagon, and how a square can be formed from combinations of other shapes such as two isosceles, right-angled triangles.

Recording
Children devise a way to record, describe and explain their findings and any patterns and relationships they have noticed. They may use cut squares of coloured paper, or an A4 sheet of squares onto which they draw the cutting line.

An activity can be extended as follows:

- **Find more** With this approach teachers can target not only the amount of work but also the complexity of the thinking undertaken by children. The technique allows a teacher to vary the task according to children's needs and abilities. Children initially find one example, then gradually find another, several examples, or, where possible or appropriate, find all cases. In the example above, children could initially be asked to find an example of a triangle. Some children could be asked to *find more* triangles, while some could be asked to *find all* triangles.

- **What if ...?** This approach allows the activity to be extended for those children who have found most of the examples applicable to them. The teacher could pose the question to the children starting with the stem *What if.* For example:
  What if you could use two straight cuts?
  What if the cut was not straight but could have one bend in it?
  What if you started with a triangle rather than a square?

Section IV: Shape and space in the primary school

THE major mathematical content and examples of how it might be embedded in learning activities are presented in three strands: shape and structure, transformation and symmetry, and location and arrangement.

Shape and structure

The *Numeracy Benchmarks Years 3, 5 & 7* (Curriculum Corporation, 2000) for shape and structure state that as part of the minimum set of achievements in spatial sense, *Year 3* students are expected to:

> recognise and name familiar 2D and 3D shapes and objects (ie triangle, square, rectangle, circle, cube and pyramid); identify where those 2D and 3D shapes and objects occur or are used in everyday life; and use everyday language such as *flat, round, side, corner* and *curved* to describe common shapes and objects and their properties.

As part of the minimum set of achievements in spatial sense, *Year 5* students are expected to have achieved the Year 3 benchmark standard and, in addition, are expected to:

> recognise and name 2D shapes (ie pentagon and hexagon) given descriptions or drawings of them; recognise and name common 3D shapes and objects (ie rectangular prism, cylinder, cone and sphere) given descriptions or realistic drawings of them; describe and compare 2D and 3D shapes and objects according to their important features (eg say why a shape would be a cone rather than a cylinder); and use conventional terms such as angle, face, edge and base to name parts of these 2D and 3D shapes.

As part of the minimum set of achievements in spatial sense, *Year 7* students are expected to have achieved the Year 5 benchmark standard and, in addition, are expected to:

recognise, describe and name common 2D shapes (ie right-angled and equilateral triangles, quadrilateral, parallelogram and octagon) (eg identify a triangle as equilateral because it has three equal sides), and 3D shapes and objects (ie rectangular, triangular and hexagonal prisms, tetrahedron and square-based pyramid) (eg identify a chocolate bar as a triangular prism because it has two triangular faces and three rectangular faces), and representations of these; use geometrical language (ie 2D (two-dimensional), 3D (three-dimensional), diagonal, right angle, parallel, perimeter, circumference and degrees) to describe, classify and compare shapes and objects (eg describe the crowd as lining the perimeter of the playing field); and recognise basic angles (ie 90°, 360°) and describe them as corners of shapes or rotations.

Two-dimensional (2D) shapes

The study of 2D shapes forms the bulk of the shape and space content for many primary school children. Plane shapes, as they are sometimes known, have two dimensions—length and width. They lack depth, the third dimension, and as such cannot be picked up. It is this lack of depth that puts them into conflict with normal, everyday language and usage in the classroom. Here it is usual for teachers to ask children to draw round the yellow hexagon from the pattern block collection or the thick triangle from the attribute set. In strict mathematical terms the children are tracing round not a hexagon but a hexagonal prism to draw their hexagon. They draw their hexagon from the 'face' of the shape. Furthermore, teachers and children often refer to coins as circles and commercially produced plastic geometric shapes found in most classrooms as squares and triangles. While we can note that these are cylinders, square and triangular prisms, such rigidity would be very confusing for children.

The usual 2D shapes encountered within the primary classroom can be classified by the attribute of their side into three groups:

- those shapes bounded only by straight lines, for example triangles
- those enclosed by curves, such as circles and ellipses
- those bounded by a combination of straight lines and curves, for example semicircles

The first group are known as polygons and include most of the shapes familiar to primary children. The name polygon is derived from the Greek words *poly* for many and *gonia* for angle and means *many angles*. All three groups of shapes—polygons, curved lines and mixed straight and curved lines—have the important defining features of being *simple* and *closed*. They are said to be simple because they do not cross over themselves and closed because there is no break in their boundary.

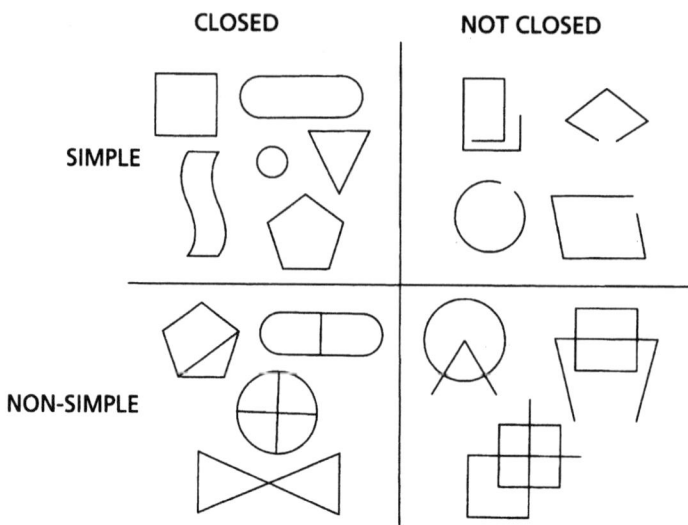

Closed simple shapes examples

Two other attributes are important for children to meet in their study of shape and space. These features allow the shapes to be classified into *convex* and *concave* (having an interior angle greater than 180 degrees) groups.

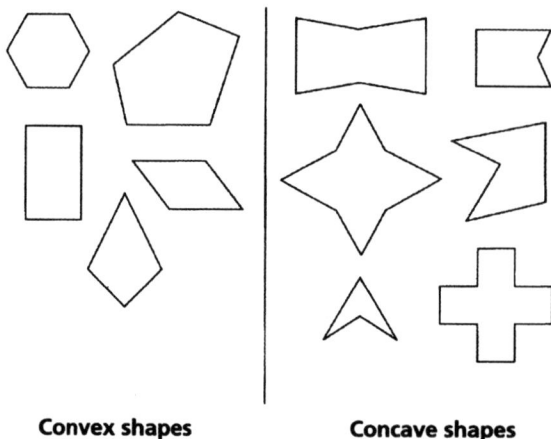

Convex shapes   **Concave shapes**
Convex and concave shape examples

Sorting and classifying activities, apart from being a fundamental way of working mathematically, are most useful to help children notice the features necessary for the three groupings mentioned above. The sorting activities can be:

- open instruction—sort these shapes into a group with one special feature and a group without that feature

◆ teacher directed—sort the shapes that have straight sides into a pile and make another pile of the rest

With open instruction, children have to think and make decisions for themselves. In this way they act in a mathematical way. A teacher-directed style requires children to hear and follow instructions. It also allows the teacher to focus children's attention on a particular attribute that might be missed in a more open style of activity.

CARD SORT

Blank playing cards are a useful piece of equipment that can be used to overcome the problem of the *coin as a cylinder* issue. Two-dimensional shapes from all three groups (straight sided, curved and mixed) can be drawn onto the cards. The shapes will vary in size, regularity and orientation. The cards can then be easily manipulated by children into piles as they apply their sorting criteria.

POLYGON OR NOT?

This general style of activity is based on *Wollygoggles and other creatures* (O'Brien, 1980) and is applicable to other content such as triangles, parallelograms or concave and convex shapes.

Present the children with a sheet that has examples and non-examples of polygons and a mixed group of shapes that they have to sort according to the feature or features they established for polygons.

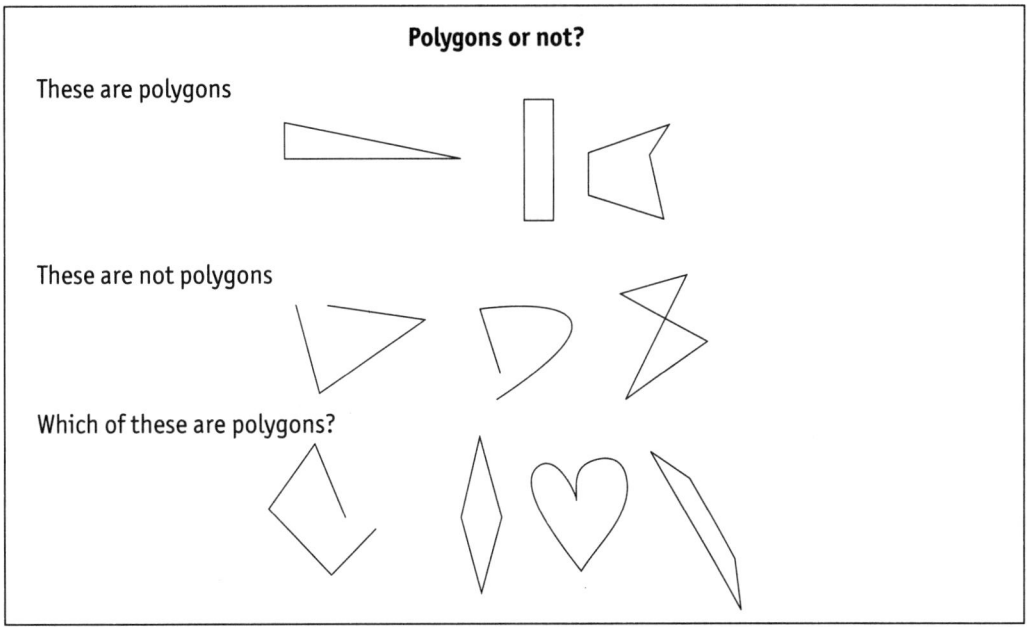

With both activities it is vital that children reflect on what has happened and explain their reasons for sorting so that the features of polygons (simple, closed and straight sided) are drawn out and made explicit.

Naming two-dimensional shapes

Polygons

Children quickly become familiar with recognising and naming common shapes such as circle, square and triangle. Earlier discussion noted the need to develop children's concepts of these words by offering them in a range of orientations, sizes, contexts and materials, rather than just concentrating on shape recognition and naming of stereotypical examples. The classification system for polygons is easily accessible for most children and mostly follows a consistent pattern, based on the number of angles in a shape, though many see it as based on the number of sides. The numbers are generally derived from the Greek word for that number, and this is usually combined with *gon* from the Greek word for angle. So a polygon with eight sides is an 'eight-gon' or octagon. Using the English word 'eight' initially allows children to connect to the number of angles/sides before they have to use and remember the Greek prefix *octa*. Confusion occurs for many children with the use of Greek words if strong connections and images have not been established. The simple pattern does not, however, begin with the same logic. A shape with three angles is not a 'trigon' but a triangle, while a general name for a polygon with four angles is not a 'tetragon' (using Greek words) but a quadrilateral. This word combines the Latin words of *quad* for four and *lateral* for side. From this point the naming system is more logical and regular:

- 3 angles (and 3 sides)   triangle
- 4 angles   quadrilateral
- 5 angles   pentagon
- 6 angles   hexagon
- 7 angles   septagon or heptagon
- 8 angles   octagon
- 9 angles   nonagon
- 10 angles   decagon
- 11 angles   undecagon
- 12 angles   dodecagon

It is useful to begin by naming those polygons that are consistent with the main pattern of number and angle (such as 8-gon) before special cases such as quadrilaterals are mentioned. This allows the main pattern to be established first and the irregular names to be learnt in contrast to them, in the reverse of the typical teaching approach.

Regular polygons

When all angles are equal and all sides are equal in length, the polygon is said to be *regular*. An *equilateral triangle* is a regular triangle because it has three equal 60° angles and three equal length sides. A *square* with four 90° angles and four equal sides is also said to be regular. On the other hand, pentagon and hexagon are general terms and refer to a wide variety of five- and six-sided polygons respectively which are only regular when their angles and sides are equal. Children need to develop concepts of these polygons that note the differences and similarities contained in the name. Activities such as 'Slice the square' and 'Fold the square' illustrated earlier will offer children experiences of this idea not gained from textbook diagrams or commercially produced shapes.

Classification of shapes

The complete system of classification is more complex than the simple system described above. It involves cases where some shapes are included in the group and under the name of another shape. For example, a square is included as a special case of the group rectangles. So a square is a rectangle. This is a difficult idea for children (and adults) to grasp. The use of the word *oblong* for the shape many children call a rectangle may help with the inclusive idea. It will only gradually become clear to children through many experiences and much discussion and as they move from van Hiele level 3.

Triangles

As children work with triangles in different situations and tasks, they gradually notice that while all examples can be grouped in the family triangles, there is a difference between triangles when one considers the angles. Triangles can be classified as *scalene, isosceles* and *equilateral*. A scalene triangle has no angles (and consequently no sides) that are the same size. An isosceles triangle has at least two angles, and two sides, that are the same size. The word isosceles is derived from the Greek word *isos* meaning equal, and the word *skelos* meaning leg. An equilateral triangle has three angles and three sides of the same size. An equilateral triangle is therefore a special case of an isosceles triangle as it also has 'two equal angles'. Triangles with right angles can be both scalene and isosceles.

	Scalene	Isosceles	Equilateral
Acute angles	✔	✔	✔
Right angled	✔	✔	
Obtuse angled	✔	✔	

The seven types of triangles

WHY NOT NINE?

This activity is developed from Pinel (2002) who asked why there are only seven types of triangle and not nine.

Exploring
In groups of three, children draw triangles on pre-cut, playing-card-sized pieces of paper. They are challenged to see how many different triangles they can generate.

Explaining
After a short time, they are asked to sort their collection into piles that show different triangles. Each group then explains to the teacher and class the criterion they used for sorting, how they are triangles and how they are different. The teacher can then set a further task to sort the triangles into the 'different' pile but concentrate on the angles and ignore size, colour of pen, who drew it, and other attributes children may use for classifying. Children record the attribute/criterion they have used each time. Establish the number of different triangles in each group's pile (note that there are no more than seven different triangles) and ask them to consider the attribute of angle size.

Provide the following table and ask them to place their cards with triangle drawings in the appropriate spaces and be able to justify their selection.

Equal angles / The third angle	Zero	Two	Three
Acute angles			
Right angle			
Obtuse angle			

Reflecting
Children note which spaces are not filled and hence which triangles are impossible to make (the ones for the cells of obtuse/three and right/three).

Recording
Children complete a record of what they did, what they noticed and what they now know. The recording could include the sorts of triangles drawn in the appropriate box in the table.

WHAT GOES IN THE MIDDLE?

Exploring
Present the children with an envelope containing a number of pre-drawn and pre-cut triangles. They should be of different sizes and contain examples of isosceles, equilateral, scalene and right-angled triangles. In small groups, children play the 'Guess my secret rule' game. One child decides on a secret attribute for the triangles and sorts them according to this rule into two piles. Pile 1 has the secret attribute while pile 2 does not have the attribute. The others in the group have to establish the secret sorting attribute. This is recorded and another member of the group repeats the secret sorting with a different attribute which is also recorded to avoid repeating and to use later in discussion.

Explaining
Children explain to the rest of the class the attributes they used for sorting. The teacher records these on the board. If they are not included add *right angled, three equal sides, only two equal sides*.

Reflecting
Provide the following Venn diagram for sorting.

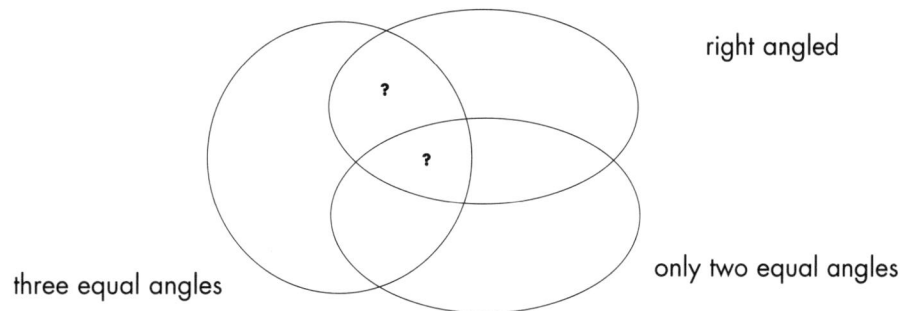

Name the circles individually as *right angled, three equal angles, only two equal angles*. Ask the children to place the pre-cut triangles in a correct place on the diagram and justify their placement. They need to consider the blank sections (marked ? on the diagram) and what might be put in these sections. They may even try to draw an appropriate triangle for one of these sections if one is not available in the envelope.

Recording
Children could draw or make and glue examples of each sort of triangle onto the diagram. The triangles can then be described in terms of their properties.

PAINT THE TRIANGLE

Exploring
This activity uses properties of triangles and develops properties of other polygons, as well as an ability to visualise.
- Ask children to close their eyes and imagine a large equilateral triangle.
- Then take a large paintbrush and paint the triangle completely with white paint.
- Next imagine a pile of three smaller black triangles at the side of the large white triangle.
- Take one of the smaller black triangles and place it into one of the corners of the white triangle. Repeat with the other two triangles, placing them in turn into the remaining corners of the white triangle.
- Now focus attention on the large triangle and the white area that remains uncovered by black triangles.
- Write down (but do not draw) the name of the shape that remains white.

Explaining
The teacher asks for and records on the board the names of the remaining white shapes. A range should be noted to include equilateral triangle, hexagon, square, oblong and nonagon, according to whether the black triangles meet, overlap or are separated. If one or more are not submitted they can be used for a further challenge after the next section. Children are asked to explain and justify how they managed to produce, for example, a triangle. The rest of the class have to listen to the reasons and note the drawing to see if they agree that it is possible.

Reflecting
The teacher can offer further challenges to see if the children can find a way to make, for example, a white oblong. Discussion can highlight the fact that some children interpret the word triangle in different ways and how some children have a limited view of a triangle. For example, many will only imagine equilateral triangles.

Recording
Variations on the answer can be recorded by drawing the types of triangles used and shapes produced along with comments and explanations.

Quadrilaterals

The key to the system for naming these shapes is determining the minimum number of attributes that the general set must possess. For example, the *parallelograms* are a set of 2D shapes with two pairs of parallel sides. So squares, oblongs and rhombuses are all parallelograms. The set that includes the others is the one with the least number of attributes. Squares and oblongs have all the family properties of parallelograms and additional, special properties. For this reason, students must have experience with a wide variety of each shape. Otherwise false properties based on limited experience, such as orientation or 'slantiness' in the case of parallelograms, may be included in the children's criteria for a particular shape.

Children are often confused by the fact that some properties are dependent on each other, while some are not. For example, constructing a shape with four right angles results in it having to have two pairs of equal sides. However, it does not follow in reverse that by creating a shape with two pairs of equal sides a shape with four right angles will always be produced.

A *quadrilateral* is a closed shape with four straight sides whose meaning originates from the Latin *quad* for four and *latus* for 'side'. This shape need not have sides of equal length.

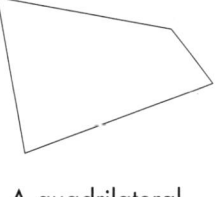

A quadrilateral

A *trapezium* is a quadrilateral with at least one pair of parallel sides. Within this group are also special cases such as a *right-angled trapezium,* an *obtuse-angled trapezium*, and an *isosceles trapezium*.

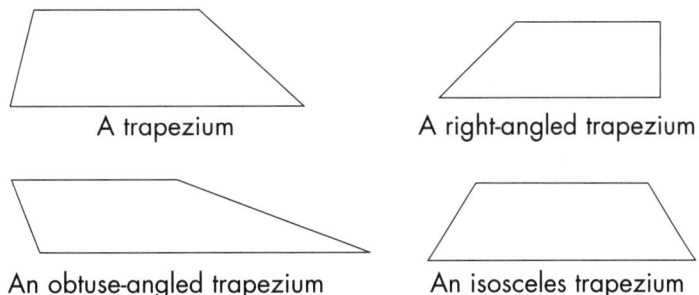

Books written in the USA often seen in libraries within Australia refer to the following terms in a different way from Australia and Europe. This can lead to further confusion if exercises and activities are taken without thought from these resources.

Australia and Europe	USA
Quadrilateral	Trapezium
Trapezium	Trapezoid

A *parallelogram* is a quadrilateral with two pairs of parallel lines. A special case arises when the sides are of equal length. This produces a *rhombus*. Another special case of a parallelogram is a *rectangle* where four angles are right angles. Two other cases of this inclusion idea should be mentioned now so teachers are better able to prepare their pupils to meet it at a later stage. A *rectangle* is a four-sided plane shape with four right angles. The four right angles are the only requirement. No condition is imposed on the length of the sides. Hence, a square with four right angles must be a rectangle. In a similar way a square is a *rhombus* because a rhombus is a plane shape with four equal sides. It should be noted that in this case no condition is imposed on the size of the angles. A square is both a special sort of rhombus and a special sort of rectangle.

Children who are used to calling a shape with four right angles and two pairs of unequal sides a rectangle will find it difficult to accept the notion that a square is a rectangle when they meet this idea at some later time. It is advisable to call a figure with four right angles and one pair of sides longer than the other an *oblong* rather than the more general term rectangle. At some later stage children can be led to see that both squares and oblongs can be grouped together and called rectangles.

There is a strong feeling among many children to use the general, everyday word *diamond* to describe shapes in mathematics. There is no mathematical term diamond. The confusion is further compounded when children see the icon for diamond and the image of a diamond on playing cards. Each of these is, in fact, a rhombus. Added to this are the blue and brown pieces in a pattern block set which look like a diamond and are often called diamonds but in fact have a face of a rhombus. A 'baseball diamond' adds further confusion as it is a special rhombus called a square.

   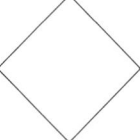

A playing card diamond (rhombus)   A baseball diamond (square)

There is, however, some debate in the area with some mathematicians wishing to use the term diamond to designate a special rhombus. Thus, a shape with four equal sides but with angles not 90 degrees could be called a *diamond* so that the general term rhombus can be applied later on to both squares and diamonds, both of which have four equal sides. All of these shapes have two pairs of parallel sides, which is the only criterion for parallelograms. Again parallelogram is a general term. A parallelogram with two long sides and two short sides should be called a rhomboid and it is advisable that children meet the general term parallelogram as a name for the entire class made up of rectangles and rhombuses.

There are two other four-sided figures that teachers should know. These are the *kite* and the *arrowhead*. Both of these have two pairs of adjacent sides that are equal. The kite has all angles less than 180 degrees while the arrowhead has one angle greater than 180 degrees.

   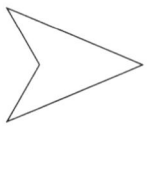

A kite   An arrowhead

The difficulty and complexity of this area lies between the idea of properties of a shape and its defining features. A rectangle, for instance, is defined as a quadrilateral with all its angles right angles. It has many other properties that can be deduced from its definition. Full understanding of the classification of shape comes with the understanding of the relationship between shapes and the relationship between properties within a shape. In order to ensure a full understanding and recognition of a particular shape, children need to see and discuss non-examples as well as examples and the idea of inclusion.

The table below offers a start with attributing properties to quadrilaterals. The list can be extended to include aspects of angle size, parallelism, symmetry and further properties of diagonals.

	Oblong	Square	Rhombus	Kite	Arrowhead	Parallelogram
Opposite sides equal	✓	✓	✓			✓
Adjacent sides equal		✓	✓	✓	✓	
All sides equal		✓	✓			
All right angles	✓	✓				
Diagonals at right angles		✓	✓	✓		

Some properties of four-sided shapes

The shape and space content discussed above is not intended to be taught to children presented in this way. It is offered as background content for teachers so that they are aware of connections between these ideas and can develop the ideas and relationships with children in a confident and competent manner. A *geoboard*, a board with equally spaced rows of nails on which rubber bands can be placed to show shapes, is a valuable way to help children experience the variety of quadrilaterals. Many commercial geoboards are of limited suitability for most primary aged children as they have too many nails, thus offering too many possibilities for answers, and the nails are usually too close together. Home-made geoboards are much better as they can be limited to 3 × 3 or 4 × 4 or at most 5 × 5 'nails'. Have children use a 4 × 4 board and the equivalent 4 × 4 dot paper to produce and record as many different four-sided shapes as possible.

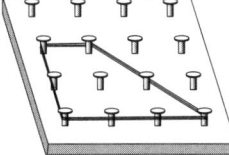
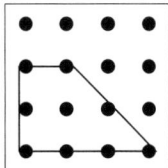

The different quadrilaterals can then be used for sorting by the children in the style of *secret sorting*. Here the attribute selected is not revealed by the sorter until the other children have deduced it from perusing the *have the secret* and *do not have the secret* piles of quadrilateral drawings on dot paper. The teacher can keep track of the criteria used by children for sorting and these can form the basis for discussion of the properties and special features of quadrilaterals.

Circles, ellipses and ovals

A second group of 2D shapes are defined by the attribute of curved rather than straight sides. This group contains the circle, the ellipse and the oval. With young children it is useful to have them distinguish 2D shapes with curved sides from those with straight sides through discussion of similarity and difference. Children can begin to see the connection between 2D and 3D shapes by drawing round 3D objects with a circular face or cross-section, and finding examples of circles in real objects. This is a technique that can also be used to connect polygons to 3D objects and vice versa. Children can quite easily connect the circle to everyday objects such as a wheel, a dollar coin and lids from jars. It is also important that they experience and see large versions of circles, for example garbage bin lids and bicycle wheels, so they have wider examples than the small textbook drawn versions.

Children also need to be familiar with the properties of a circle so that when later work with abstractions like pi (π) is introduced they will have a visual reference to aid their thinking. The act of drawing a circle with a radius, for example a fixed length of stiff card with anchor point and place for pencil, or a piece of string kept taut, or a commercial 'safety compass', is important in establishing this reference. These pieces of simple equipment are much better to use initially as they more easily direct the children's attention to the fixed length of the radius and to the fact that the edge of the circle is a constant distance from the centre of the circle. The string and card techniques for drawing a circle are also useful as they can be used to construct circles ranging from quite small in size to huge. The usual school pair of compasses, with a metal point, a place for the pencil and two often loose and floppy arms, are more suitable for older children after the idea of radius has been established.

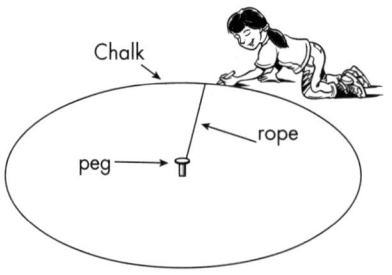
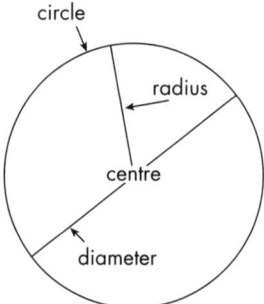

There are other properties of circles that older primary children can start to experience. It is important that they have established these properties via activity and reflection before they are faced with using them in abstract ways via various formulae with pi. The relationship between the radius and diameter is important for children to know—the diameter is double the radius and the radius is therefore half the diameter. Asking children to fold pre-cut circles with a fold line passing through the centre point of the circle illustrates bilateral and multi-line symmetry.

There is also the relationship between the diameter and the circumference and subsequently that of the radius and the circumference that can be established. The connection between the two—the circumference is approximately three times longer than the diameter—can be grasped in a very practical way long before the abstractions are introduced. The circumference and diameter of various circular objects can be measured using a tape measure and then recorded on a chart.

Object	Diameter	Circumference	Ratio C/D

The calculation of the circumference measurement divided by the measurement of the diameter can be performed quickly and accurately using real data and a calculator. The answer is recorded and forms the focus for reflection and discussion with the teacher so that the relationship is made explicit and established. The Greek letter π used to symbolise the constant ratio was proposed by the Swiss mathematician Euler in the eighteenth century because it is the first letter of the Greek word for perimeter (perimeter in a circle is also known as circumference). While the values of $\frac{22}{7}$ and 3.14 have been common approximations for π, they need not be introduced until children have worked with the more general idea of the circumference being a little more than three times the diameter and six times the radius. Indeed, many calculators have a π key that provides a greater degree of accuracy than the values $\frac{22}{7}$ and 3 used in pre-calculator days.

Before the end of primary school, children can begin to explore and compare other plane, curved shapes such as the ellipse and the oval.

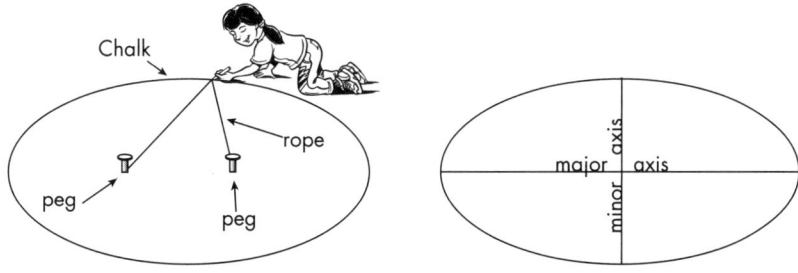

To construct an ellipse, two fixed points (again nails or pegs driven into the ground or drawing pins pushed into thick card) are chosen and one end of a piece of cord attached to each. A piece of chalk or pencil held taut traces out an ellipse. The properties of this curve need to be compared and contrasted with the circle so children develop a greater understanding of the circle's properties.

The word oval comes from the Latin word *ovum*—an egg. An oval is egg-shaped, that is, it is 'pointier' at one end than the other. It has one axis of symmetry unlike the ellipse, which has two, and the circle, which has many.

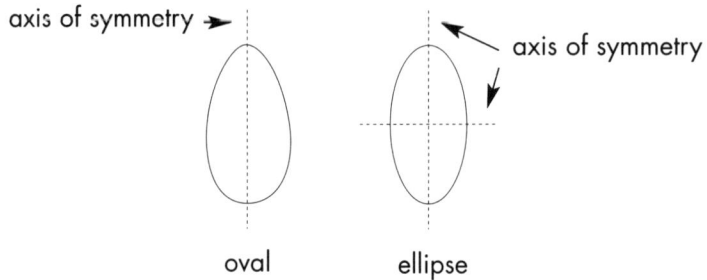

oval   ellipse

In general usage, however, oval is often incorrectly a synonym for ellipse, as in the term 'football oval', which may lead to initial confusion in relation to its properties.

Further properties of 2D shapes

Two-dimensional shapes can be analysed further by considering their angles, diagonals, parallelism and their symmetries. While these are presented separately, it is important to note the connection of angle to measurement and location; of symmetry to symmetry and transformation; and the integrative nature of all of them when describing shapes and solving problems.

Angles

Angle is a form of measurement and its development follows a general teaching and learning approach of the identification of an attribute to be measured by comparison, the use of non-standard units, and finally the use of standard units as discussed in Chapter 6. For many children the concept of angle is difficult to grasp. It has a number of distracters and difficulties, such as the orientation of the drawn angle and the length of the lines enclosing the angle. Teachers need to present children with materials, for example angle arms, angle wheels and Rotagrams (where two circles of clear semi-stiff plastic are combined so they are able to turn to show via drawn lines the size of the angle), and images to highlight the feature of angle and its dynamic nature. This is not possible with static drawings on textbook pages.

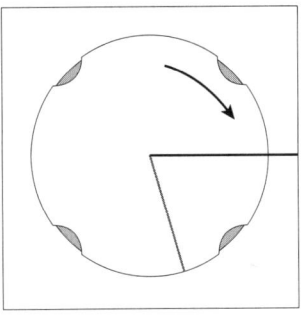

Rotagram

Children can develop the basic idea that an angle measures the amount of turn from one direction to another. From this understanding, teachers can help children understand how angles are measured. Angle arms and Rotagrams can be used to show the connection between turn and the conventional representation of angle on paper, where two lines radiate from a common point.

The three aspects of angle, angle as a corner, angle as the shape made when two lines radiate from a common point, and angle as a measure of turn, need to be integrated into a common idea of turning. This will allow the general concept of angle to be established before reference is made to any specific angle such as 'right angle'. The use of right angle with its connection to corners in rectangles and other objects will come later. It will be included as just one of a series of special angles rather than something that is separate from the general idea of turn.

A teaching development can begin much in the same way as any other measurement concept. Initially, angles can be judged by means of direct comparison where one angle is superimposed on the other. This situation will require children to understand what exactly is being compared and avoid the distracters of length of the lines or sides marking the angle. Later comparisons may be made by means of an instrument such as a Rotagram which shows the angles turned. Other non-standard instruments can be used to help children compare and order angles on real things, polygons, polyhedrons and pre-drawn angles. An angle measure constructed from a circle of paper is quite versatile and easily transportable. The circle has a full turn of 360 degrees and can be folded in half and half again to give non-standard measures for comparison of a 'half turn' and a 'quarter turn'.

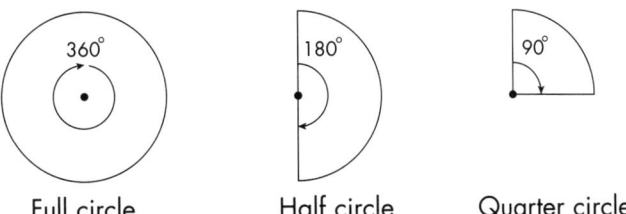

Full circle   Half circle   Quarter circle

A further halving of the circle paper by completing another fold provides the more acute comparative of an 'eighth turn'.

As children begin to compare angles in terms of the standard unit of degrees, a relatively simple division of 360 can provide 180 degrees (half turn), 90 degrees (fourth or quarter turn) and 45 degrees (eighth turn). When working with aspects of direction and location the use of 'turn' can be related easily to the physical act of turning by children and a connection made to bearing. Throughout all this work children should be estimating the size of angles in terms of 'turn' and building personal referents or benchmarks for a quarter and eighth turn.

The final stage of the measuring phase is the introduction of the degree as the standard unit of measuring angle. It is advisable that children use a full-circle protractor in the first instance. A full-circle protractor with a movable line is even better since this instrument emphasises the link between angle size and rotation. The full-circle protractor also connects to the Rotagram of the earlier stage and avoids the confusions and difficulties most children experience with the half-circle protractor. This instrument should be left until later or avoided altogether in primary school.

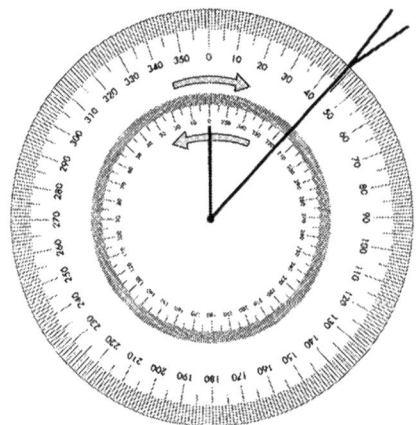

Circular protractor

In addition to these activities, analogue clocks can also be used to explore angle size. Each five-minute interval corresponds to 30 degrees (a twelfth turn). This gives a ready reference for estimation purposes. It would also be useful to have a wall chart displaying various angles, especially angles from 90 degrees up to 360 degrees with their connections to earlier benchmarks of quarter and eighth turns. The clock also enables children to see that angles greater than 360 degrees do exist, for example, the minute hand turns through 720 degrees every two hours and so on.

Angles in polygons

Measuring angles accurately with a protractor is very difficult for many primary school children. There will need to be some allowance for error in measuring and scale reading. This tolerance for error presents difficulties when working with children to establish the sum of angles in polygons. For example, a tolerance of a few degrees on each angle measurement could lead to the incorrect conclusion that the angle sum of a triangle is 189 degrees rather than the actual 180 degrees.

An action proof, however, retains the investigative and active principles of teaching while providing an accurate and visual representation. The teacher asks the children to draw a triangle on the provided paper—make sure that they produce a variety of triangles with a range of sizes and angles. The children then mark each corner with the number 1 or 2 or 3.

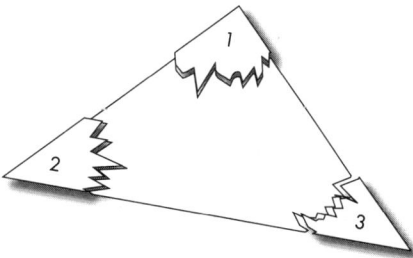

Each corner is then torn from the triangle. The torn and numbered corners are fitted together with the numbered points touching. This should form a straight line which children can measure with a folded circle measure. Many should, however, be able to relate the angle to a half turn (180 degrees).

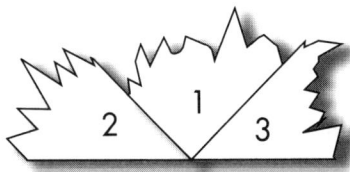

A similar activity can be undertaken to establish that the angle sum of any quadrilateral is 360 degrees (a full turn).

Investigate and discuss

> **1** Draw a range of quadrilaterals, including the arrowhead. Use the corner-tearing method to convince yourself that all quadrilaterals have an angle sum of 360 degrees.

The same approach may be used with other polygons but not in such an effective and convincing way due to the angle total being greater than a full turn. Instead, polygons other than triangles and quadrilaterals are more effectively seen as the sum of triangles.

CHOP INTO TRIANGLES

Exploring
Ask children to draw round a regular pentagon, a regular hexagon or other regular polygons that are to hand. Mathematical Activity Tiles (MATs) are particularly useful here as they include octagons, decagons and dodecagons. Demonstrate how to 'chop' the shape into triangles by selecting a corner (vertex) and drawing lines to each of the other corners without allowing the lines to cross each other.

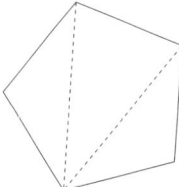

A pentagon 'chopped' into 3 triangles

Have children explore 'chopping' other regular polygons into triangles in a similar way.

Explaining
Children can report to the class how many triangles they have for each polygon. Some children may have drawn extra polygons and may offer insights into them. Some children may be able to describe and explain the relationship between the number of sides of the polygon and the number of triangles constructed.

Reflecting
A table of results is useful to help children see the pattern that emerges from chopping into triangles.

Shape	Number of sides	Number of triangles	Sum of angles
Quadrilateral	4		
Pentagon	5		
Hexagon	6		
Octagon	8		
Decagon	10		
Dodecagon	12		

Through discussion and thinking draw out the following relationships:
- The number of triangles for a polygon is always two less than the total of sides for the same polygon. For example, a pentagon has three triangles: 5 − 2 = 3.
- The angle sum of a polygon can be calculated by adding the angle sum of each triangle drawn in the polygon. For example, a pentagon has three triangles so the angle sum of a pentagon is 180 + 180 + 180 = 540.

◆ The size of each angle of a regular polygon can be calculated from the angle sum. If the shape is regular then all the angles are the same size. Thus, for a pentagon the angle sum of 540 is divided by the number of angles, 5, to give 108 degrees for each angle.

Children could investigate 'chopping' non-regular polygons including both convex and concave examples into triangles in the same way.

The knowledge gained here can be added to the growing concept of each polygon as well as being related to work with regular and semi-regular tessellations.

Investigate and discuss

> **1** Construct a table similar to the properties of four-sided shapes table p. 407 and title it *Properties of polygons*. Include columns for triangle, pentagon, hexagon, octagon and other polygons and rows to note the relationships and information about angles.

Diagonals

A diagonal of a polygon is a straight line drawn between two non-adjacent vertices. That is, it connects two corners that are not next to each other. While triangles have no diagonals, all quadrilaterals have two. A special case with polygons is a concave figure. In this case one of the diagonals will be outside the figure. For example, in quadrilaterals the arrowhead is concave and will have one diagonal outside.

Diagonals in a quadrilateral

Quadrilateral (arrowhead) with exterior diagonal

Diagonals are properties of 2D shapes and provide a further way to classify polygons.

Investigate and discuss

1 Construct a range of quadrilaterals using geostrips and paper fasteners. Use elastic or rubber bands for the diagonals. Make statements about each set of diagonals and add the properties to the table of properties of polygons p. 415. Consider whether each of the diagonals is cut in half by the diagonals crossing, if all pieces are of equal size, and if the crossings are at right angles. For example, with a kite the diagonals cross at right angles.

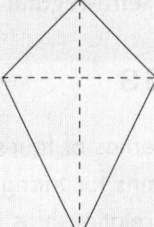

Diagonals in a kite cross at right angles

Check also diagonals in a variety of pentagons, hexagons and octagons.

Parallelism

The development of children's concepts related to parallel lines also follows the van Hiele theory. Initially, children recognise parallel lines but are unable to explain their relationship or outline their properties. At this level, as with all shapes, it is important to provide a range of examples of parallel lines in various orientations, and with lines of different lengths between and within examples. Viewing lines in this way will help children recognise them in quadrilaterals when the sides are presented in orientations other than in the 'usual' way. For example, the opposite sides of a kite are parallel but many children who have only experienced oblongs find this difficult to understand.

Variety in the presentation of parallel lines

Sufficient 'hands on' and 'heads on' (where children think about the activity) experiences with a variety of materials, shapes, sizes, orientations, reflection and discussion will help children develop a wider and more stable concept of parallel lines. The stereotypical, real life examples of railway tracks and roads may confuse some children as they see tracks and roads bend and curve as they go round corners. While the tracks remain the same distance apart, they do not match the need for parallel lines to be straight. Better examples to use are louvres on doors, Venetian blinds and, for examples of parallel planes in three dimensions, the floors and walls of normal rooms, and packages.

Children will then build the concept of parallel lines as straight lines that are the same distance apart and hence never meet or cross. They point in the same direction and have the same slope. Parallelism can be checked in regular polygons when children meet pre-formed static shapes and dynamic models constructed from geostrips or thick cardboard strips with paper fasteners. With the dynamic models children can observe the effect of movement on pairs of sides and the way, for example, a square changes into a rhombus. Discussion and reflection on shapes can lead to children seeing the relationship between parallel lines and angles.

Investigate and discuss

1 Consider parallelism in a range of quadrilaterals and other polygons. Add a further row called parallelism to the properties of polygons table p. 415 and complete statements about the various polygons.

Three-dimensional (3D) shapes

Children meet and interact with 3D objects from their earliest days as they move around, pick things up and play. It is important for schools to build on these experiences and help children connect their world with the more abstract mathematical representations. The real world is three dimensional. Young children come to school from home, pre-school and kindergarten—imaginative environments where they played with building blocks, cans, balls, containers and boxes, all real objects. Everything children see and touch has three dimensions: width, height and depth. Shape and space learning begins with children exploring these solid shapes, working out how to move them, if they will sit flat on the floor, how they can be stacked, whether one will fit inside another and so on. The basic shapes to children are not flat, artificial, 2D figures such as triangles, squares, oblongs and circles, but rather the prisms, cylinders and other 3D shapes that occupy their world. Teachers need to build from these 3D experiences and connect them to the 2D world of books and diagrams. Rather than a strict treatment of 3D solids and then a consideration of 2D figures, it would seem appropriate to interweave the ideas in rich situations of exploration and discussion.

Conventionally, 3D shapes are called 'solids' with 2D shapes being referred to as 'figures'. In their classroom presentation 3D 'solids' may in fact be empty with a hollow middle, for example when constructed from Polydron, or skeletal with just their edges used when they are made from tooth picks and small balls of Plasticine (or frozen peas) or Geoshapes.

Three-dimensional solids can be classified into three general categories:

- those with curved surfaces only, for example spheres
- those with all their surfaces or faces flat, for example prisms and other polyhedrons
- those with a combination of flat and curved surfaces, for example cylinders and cones

Naming three-dimensional solids

A 3D solid with all its faces or surfaces flat is called a *polyhedron*. The word is derived from the Greek words *poly* meaning 'many' and *hedron* meaning 'base'. Polyhedrons, therefore, are solids with many bases or faces, although some older books refer to them as polyhedrona. As with 2D figures, it is important that children eventually acquire the vocabulary and names of 3D solids but this should not be the focus via simple memorisation. Children will gradually learn and use the mathematical names and terms as they experience and discuss shapes and their properties. A more precise and sophisticated working definition of shapes and ideas will emerge as children experience rich and thoughtfully prepared activities. It is suggested that children develop spatial language in much the same way as they learn to talk about various animals and objects—by hearing it used appropriately by others (their teachers and peers) and being encouraged to use progressively more exact language in describing their experiences. As children become familiar with more advanced concepts such as vertex, edge, face and angle, their descriptions will contain additional information. By the end of primary school, a description of one particular shape should be sufficiently specific as to exclude all other shapes.

Many of the names of 3D solids are similar to those used for 2D figures and discussion can help children see the connections and similarities. For example, an eight-sided polygon is called an octagon while a polyhedron with eight faces is an octahedron.

Platonic solids

An interesting subclass of polyhedrons is the Platonic solids. They were identified and developed from the writings of the Greek philosopher and mathematician Plato. He concluded that they must be the fundamental building blocks (the atoms) of nature. He assigned each of them to the essential elements of the universe. The tetrahedron was allied to fire, the cube to earth, the octahedron to air and the icosahedron to water. The remaining dodecahedron was assigned to the cosmos as it was so different from the rest having (12) pentagonal faces.

All faces of a particular Platonic solid must be identical in shape and size, all edges must have the same length, all angles must be the same and all vertices must be identical. This means that all vertices must be surrounded by exactly the same plane figures in the same order. These properties give each figure extraordinary symmetry. Any face can equally be the top or the bottom. Sometimes they are described as regular or perfect. There are only five such figures and they include one pyramid and one prism. Note how the pattern of language breaks down with the commonly used name cube. Another name for the cube is a hexahedron, which connects into the pattern derived from the Greek prefixes.

Number of identical faces	Shape of face	Name
4	Equilateral triangle	Tetrahedron
6	Square	Hexahedron (cube)
8	Equilateral triangle	Octahedron
12	Regular pentagon	Dodecahedron
20	Equilateral triangle	Icosahedron

 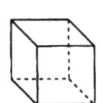 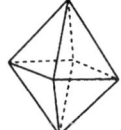  

Tetrahedron   Hexahedron   Octahedron   Docecaheron   Icosahedron

Pyramids and prisms

Children often have difficulties distinguishing pyramids from prisms and tend only to meet limited examples of each. Typically, the pyramid is presented as being similar to the 'Egyptian pyramid' with a square base, while the prism is given as a triangular prism similar to ones seen in science and the refraction of light and the spectrum. There are many others that fall into each category and children need to experience them and note their similarities and differences.

Prisms have two identical, parallel faces joined to one another by rectangles. Boxes, bricks and cereal cartons are commonplace examples. Children may also be familiar with Toblerone chocolate boxes. These provide motivation and a useful connection to real things. The end faces may be of various shapes not just triangles. The name of the end face determines the name of the prism. When a prism is cut in cross-section parallel to its base or end it would form two smaller versions of the same prism because the end shapes would remain the same. Children can experience the faces of different prisms in a tactile way by having an example of a prism hidden in a canvas bag or 'feely box' and then having to describe it to a partner who tries to select it from a group of 3D shapes.

Pyramids have one face with at least three straight edges; the faces joining these edges are all triangles which meet at one common point. The common pyramid from Egypt is typically shown with a square base. The tetrahedron is a pyramid with a triangular base. Pyramids are often named after the shape of their base, thus there are pentagonal-based pyramids.

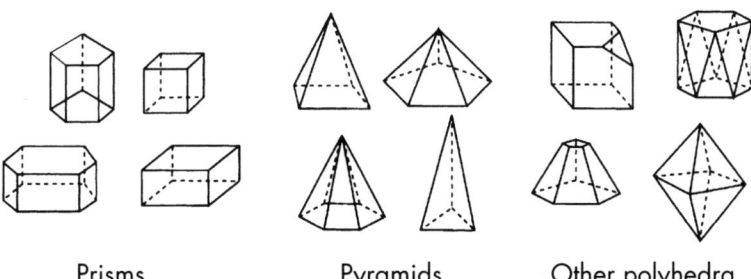

Prisms   Pyramids   Other polyhedra

Children can often see the similarities and differences between the two and between each other when they are shown in net form.

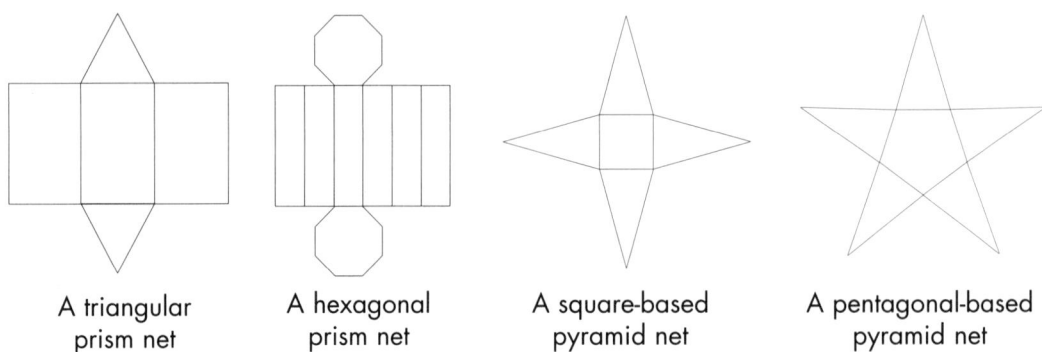

A triangular prism net   A hexagonal prism net   A square-based pyramid net   A pentagonal-based pyramid net

Spheres, cylinders and cones

One property of polyhedrons usually found by children as they play and interact with collections of 3D solids is that they do not roll. They can only slide along a tabletop or down a small incline. Those with both flat and curved surfaces, such as *cylinders,* which have two identical, flat surfaces joined to one another by a single curved surface, can slide and roll. The two flat surfaces are bounded by curves, not necessarily circles. Connections are easily made to tins and containers. *Cones* have one flat surface whose boundary must be curved and a single curved surface extending to a point. Again, connections can be made to witches' hats, traffic or ice cream cones and other similar objects. Children can be asked to investigate what happens when sectors of increasing sizes are cut from a starting circle and then folded to form a cone. *Spheres* with a complete curved surface can be compared to table tennis balls. Beware though of some of the newer design soccer balls as they use a different system for their construction and, while they are 'sphere like', they are not true spheres. These 3D shapes will roll or slide depending on how they are placed on a table or floor.

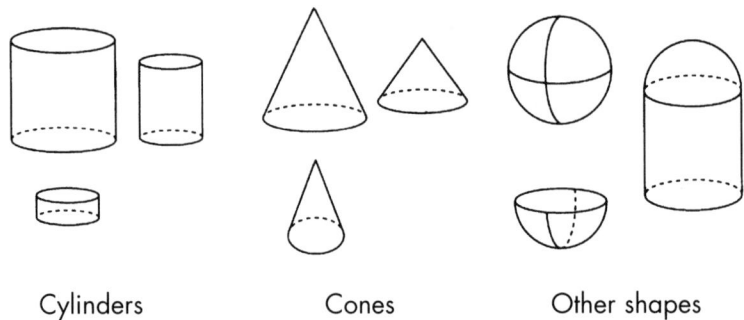

Cylinders   Cones   Other shapes

Other solids

It is possible to have 3D solids other than the Platonic ones and the examples of prisms and pyramids described so far. Often they are quite complex in structure and present problems for children when they attempt to make them from card and a 'net' diagram. With the advent of newer construction materials in shape and space such as *Polydron, Clixi* and *Geoshapes* that have interlocking edges, some of the larger and more spectacular polyhedrons are accessible to older primary-aged children. A modern soccer ball is constructed of a combination of panels that are regular hexagons and regular pentagons. Careful observation and discussion will establish that the panels follow a pattern at each vertex (corner) of hexagon, hexagon and pentagon. This can be given a coding of 6, 6, 5 to represent the 2D shapes after the Schlafi system (Cundy & Rollett, 1981). So long as children check that this pattern exists at each vertex they will be able to construct a soccer ball polyhedron which happens to be known as a truncated icosahedron and has 32 faces made of 12 regular pentagons and 20 regular hexagons.

A truncated icosahedron made from Geoshapes

The Schlafi system of coding works as follows. Each regular 2D shape is given a code number that corresponds to the number of sides it contains. For example, a square has four equal sides. It has a code of 4. A regular hexagon has a code of 6 and a dodecagon a code of 12. It is possible to describe polyhedrons by their faces and how these faces meet at a vertex. When this system is applied to a cube, for example, a code of 4, 4, 4 is generated which acknowledges the three squares that come together at a corner. The code can be checked at any vertex on the cube. A code of 3, 3, 3 will produce a tetrahedron.

Other exciting polyhedra may be constructed following the codes given below.

Code for each vertex	Official name
3, 3, 3	Tetrahedron
4, 4, 4	Hexahedron (cube)
5, 5, 5	Dodecahedron
3, 6, 6	Truncated tetrahedron
3, 4, 4	Cube octahedron
3, 8, 8	Truncated cube
4, 6, 6	Truncated octahedron
4, 6, 8	Great rhombicubeoctahedron
3, 5, 3, 5	Icosidodecahedron
3, 10, 10	Truncated dodecahedron
5, 6, 6	Truncated icosahedron
3, 4, 5, 4	Small rhombicosidodecahedron
4, 6, 10	Great rhombicosidodecahedron
3, 4, 4, 4	Small rhombicubeoctahedron
3, 3, 3, 3, 4	Snub cube
3, 3, 3, 3, 5	Snub dodecahedron

Two dimensions to three dimensions and back again

It is important for older primary children to be able to visualise what a 2D representation of an object will look like if it is to be constructed. Equally important is the ability to represent a 3D object in a 2D way. Adults move freely between 3D objects and 2D representations of them in a stylised mathematical way. For example, the following is a 2D stylised representation of a cube.

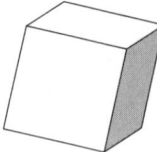

Children need to experience these conventions and representations and discuss them. One way to help with these connections between 2D and 3D is via the use of nets.

Nets of 3D solids

From early years onwards children often work with nets—the 2D form of a 3D shape—to construct 3D solids. Much of the work, however, does not involve mathematics as it is often merely a cutting and gluing exercise. It is important to add aspects of mathematics to this exercise by involving young children in recognition and discussion of faces, edges and corners. Older children can begin to consider other net configurations that might be used to construct a cube or other simple polyhedron, for example a tetrahedron as in the following activity.

MORE THAN THE CROSS?

This activity is suitable for older children and involves them in investigating other combinations of six squares that can be folded to form a cube. A similar activity with triangles can be used for investigating nets for a tetrahedron or octahedron.

Exploring
Show the children the typical cross-shaped net for a cube made from Polydron or Geoshapes (or similar) squares linked together. Carefully fold the Polydron net into a cube. Challenge children, in pairs, to find and record more Polydron combinations that will fold to make a cube.

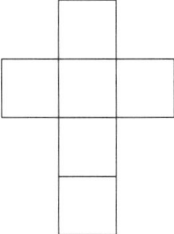

Cross-shaped net for a cube

Explaining
In pairs, children discuss the similarities and differences of the new Polydron combination they have found. This is recorded on square-centimetre graph paper to represent the Polydron. Children explain to each other, the class and teacher how each combination is different and how it folds to make a cube. Rotations and mirror images of examples already recorded will need to be identified and removed.

Reflecting
Some children will attack the problem in a haphazard and random way; others may take a more systematic approach by considering the movement of only one square at a time around a stable base.

Recording
Combinations of Polydron may be recorded on graph paper or smaller versions of the net may be made from paper that can be glued to a chart by a base square and then folded to show how it makes a cube.

It is also important for children to explore transformations from 3D to 2D as well as the other way round. This can be done via asking children to visualise and then draw an object as in the following example.

OPEN A TOBLERONE

This activity uses a Toblerone box, an example of a triangular prism. Other commercial packages, such as cornflakes packets, can be used for similar activities.

Exploring
Present the children with a Toblerone box and ask them to visualise it if it was cut along some of the edges and opened out flat on the table.

Explaining
Children describe how each piece of their opened-out box will re-form to make the original box. They might suggest how other nets might also be formed to make the box.

Reflecting
Children can begin to connect the flat 2D figures of their net with the faces and edges of the 3D box.

Recording
In a similar way to 'More than a cross?' children can glue their solutions and describe the 2D object's connections to the 3D version.

Printing nets

Younger children's first experiences with shape will be three dimensional, so they will need special activities that focus their attention directly on the individual faces so as to make the distinction between a 3D shape and its 2D surfaces. This will also allow them to explore how 3D solids can be represented in a 2D way. For example, everyday, familiar packages for products such as cornflakes, toothpaste and teabags can be carefully cut along selected edges and folded out flat to show the net. Children can explore cutting along different edges and discuss the resulting slightly different nets. These can be folded back to form the original 3D shape or package.

Another approach is to have children print the net from the original package and then compare the two, although a little more teacher guidance may be needed. Children use a wooden 3D shape, for example a triangular prism, as the vehicle for printing. One by one the faces are painted a different colour (or coloured with chalk). After each face is painted it is used to make one print on a large piece of paper. The next face is painted and printed next to and joining by edge to the previous printed face. Eventually all faces are printed and the 'flat' print can be compared to the 3D shape. Colours and face shapes should match to assist children make connections between the 2D representation of the 3D solid.

Sketching 3D shapes

Older children can begin to represent 3D objects in 2D sketches and drawings. Commercially produced materials, such as the DIME Build Up pack and its associated booklets, form an excellent basis for children to develop the skills of drawing 3D shapes on isometric dot paper.

The materials, based around arrangements of cubes, engage children in both visual processing and visual interpretation.

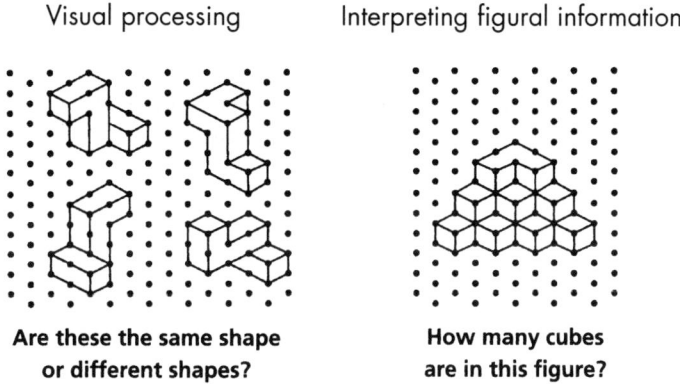

Visual processing

Are these the same shape or different shapes?

Interpreting figural information

How many cubes are in this figure?

Interpreting figural information skills can be improved through exercises and problems involving reading, understanding and the interpretation of graphs, charts and diagrams and by teaching children something about simple 3D drawing including isometric and perspective sketches.

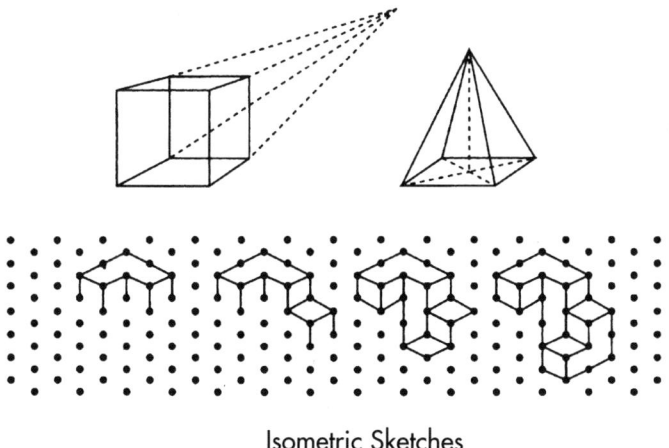

Isometric Sketches

Towards the end of primary school the students should be able to visualise, sketch and describe 3D shapes adequately. Drawing cubes and other shapes based on rectangular solids can be done on isometric dot paper in a landscape format. Drawing such isometric views is a useful skill that is easily acquired and students find it enjoyable.

An activity for older children, 'Bird's-eye view', offers interesting challenges, develops ways of representing 3D shapes in a 2D format, and requires the interpretation of 2D drawings to construct 3D shapes.

BIRD'S-EYE VIEW

This activity asks children to solve the problem of representing three-dimensions in a 2D way that is easily understood by others.

Exploring
Give children a small 4 × 4 square grid and some small wooden or plastic cubes. Have them make the following shape from seven cubes. There is only one layer of cubes.

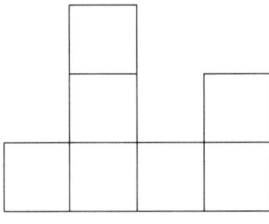

Front view

Challenge children, in pairs, to devise ways to show the 'building' from the top as if they were flying over it in a helicopter.

Explaining
Children show and describe their methods for showing 3D in a 2D way to the rest of the class.

Reflecting
The teacher keeps a record on the board of the different solution methods. Typically these involve:
- Some form of colour coding or shading with a key, for example red is one layer of cubes high, green is two layers, etc.
- The use of smaller squares within a larger one to represent an extra layer.

This shows two cubes high

- A numbering system whereby a number in a square represents the number of cube layers.

| 1 | 3 | 1 | 2 |

Children can discuss the advantages and disadvantages of each method. They then use their preferred method to show a bird's-eye view of a building designed by them on the grid paper using 10 cubes. Further challenges can be set for children requiring them to draw their building in 'silhouette' form from each side (left, right, front and back). Other children then have to construct the cube building and draw its bird's-eye view using the agreed representation system.

Recording
Children draw their 10 cube building on grid paper and describe their method for showing 3D in a 2D format.

Transformation and symmetry

This aspect of shape and space emphasises learning associated with transformation (movement) of shapes—translation, rotation and reflection—and associated concepts of symmetry and tessellation. While symmetry and tessellation are grouped here for convenience of discussion, they, like many other areas of shape and space, are not separate and isolated from each other. Both, for example, are connected to properties of polygons and form an important feature of discussion about 2D and 3D shapes.

The *Numeracy Benchmarks Years 3, 5 & 7* (Curriculum Corporation, 2000) note the following achievements for the designated year levels in this strand organiser:

Year Three
- recognise and continue simple patterns based on repetitions of common shapes;
- continue a linear pattern based on repetitions of common 2D shapes or common 3D objects (eg a pattern beginning with . . .; or a pattern comprising a can, ball and cube);
- recognise a single flip or slide of a simple 2D shape (eg use flips or slides to see whether shapes are identical);
- recognise 2D shapes that have an obvious line of symmetry (eg fold a drawing of a butterfly to show the line of symmetry).

Year Five
- recognise and continue simple patterns based on repetitions of common shapes;
- recognise and continue repetitions of a shape embedded in simple patterns or tessellations;
- construct a simple tessellation or linear pattern using common 2D shapes (eg continue a 'brick wall' pattern);
- recognise or draw the result of a single slide, flip or turn of an object or shape.

Year Seven
- identify symmetrical 2D shapes and recognise line symmetry in 2D shapes (eg pick out triangles that have line symmetry from those that do not);
- describe single movements of 2D shapes (ie flip (reflection), slide (translation) and turn (rotation), and use combinations of these to create patterns;
- identify symmetrical 2D shapes;

- pick out the 2D shapes that have at least one line of symmetry (eg pick out the triangles that are symmetrical from those that are not);
- use practical activities to find symmetrical 2D shapes (eg use mirrors to check whether triangles are symmetrical; fold a rectangle of paper to find all lines of symmetry);
- identify and describe transformations and use combinations of transformations of shapes—flip (reflection), slide (translation) and turn (rotation)—to create patterns;
- identify identical 2D shapes including when they are in different orientations (eg identify repetitions of identical non-symmetrical shapes in a fabric pattern);
- describe single transformations of simple non-symmetrical shapes;
- use a sequence of transformations to create a simple repeating pattern or tessellation (eg from this starting point create a simple border pattern using a slide and a quarter turn).

Transformation

This aspect of shape and space focuses on the process by which an original shape is transformed into a new one, rather than on the individual shape *per se*. The three main processes by which shapes are changes are *translation* (slide), *rotation* (turn) and *reflection* (flip). Associated with these are other processes of *enlargement* and *dilation*.

Transformation is used extensively in pattern making and design in many and varied cultures throughout the world. Common examples are where shapes are repeated or turned in some way to form linear patterns especially in fabric printing and weaving. Ancient Greece and Rome made extensive use of transformational geometry in their designs for borders and decoration.

This border pattern of triangles is a common one.

Translation

The simplest transformation is that of translation, or slide. Here the original shape is moved in a straight line to a new position where it forms another image. Typically, as many patterns are linear in format, the movement would be from left to right. However, this does not always have to be the case as the original can be translated to the left, to the 'north' or 'south' of the original, or at any angle so long as it is completed in a straight line and does not change direction.

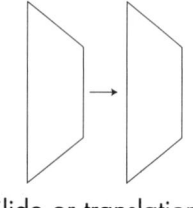

Slide or translation

Everything in the new image is the same as the original. It is only different in its position. The process of translation can be repeated a number of times in the same direction to produce a linear pattern.

Reflection

Another process for transforming an original shape is by reflection, or 'flipping', about a 'mirror line'. Connections need to be made with the children to ideas possibly presented separately in an activity involving symmetry. As with translation, all properties of the new image, its size of lines and angles for example, are the same as the original. The only difference to be noted is that of orientation—the left-hand side is now the right-hand side and vice versa.

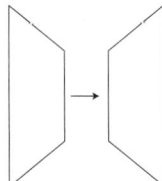

Flip or reflection

The reflection may be about any of the sides of the original. Usually, however, it is on the right-hand side to form a linear pattern. The pattern when extended shows examples of vertical or bilateral symmetry.

Rotation

The third simple transformation in which all the properties of the object are retained in the image is the rotation. In this case the object is rotated through an angle about a point to give rise to the new image. The point is called the centre of the turn. The angle of the rotation can be anything from zero degrees to 360 degrees. However, interesting patterns are generated by multiple turns through angles that are factors of 360 such as 30, 45, 60 and so on. Such patterns also illustrate rotational symmetry. Standard drawing packages on most computers include flip, slide and turn options. Children could use such packages to explore the changes made to a shape or a letter from the alphabet when these tools are used.

Combinations of transformational processes

Interesting linear patterns may be formed by combining two or more of the methods for transformation. For example, a linear pattern can be constructed by using a combination of reflection and translation.

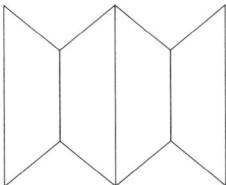

A linear pattern using reflection and translation of the original image

Pattern block pieces are useful, in this instance, for creating patterns using the different forms of transformation. Children can quickly manipulate the pieces to construct their pattern. Provide children with a quantity of each piece to make the pattern rather than using one piece as a template. The trapezium is particularly useful here as the change resulting from the transformation is very obvious (not so with the regular hexagon). Once examples of the various forms of transformation are shown and recorded then children can begin to explore, describe and record the effect of combinations of transformations.

Investigate and discuss

1. Obtain copies of books by Claudia Zaslavky *The Multicultural Mathematics Classroom: Bringing in the World*, (1996) and the series of large-format multicultural mathematics books by Calvin Irons, James Burnett and Stanley Wong Hoo Fong *Mathematics from Many Cultures*, (1994) or similar books that illustrate designs for cloth, pottery and other artifacts from different historical and cultural groups around the world.
2. Identify and copy examples that illustrate patterns made by different forms of transformation.
3. Outline how these might be used as discussion starters and examples for pattern making and be used with children to explore the idea of transformation in linear patterns to construct border patterns.

Other transformations

Enlargement is another form of transformation. In this case the original shape is enlarged by a factor of a number greater than one, for example it is multiplied by 2 or doubled. *Dilation* transformation is similar to enlargement but here the factor used is between zero and one, for example 0.5 (a halving). Thus, the original shape is reduced in size. In both cases the shapes are *similar* but not *congruent* to the original. Similar changes can be made when the second grid is not uniform or distorted in some way, for example the grid lines are not straight but follow a wave pattern. Children can produce designs similar to the effect of fairground distortion mirrors.

There are connections here to the measurement and location strands. One method that can be used with children for enlarging is to use grids. Children draw a simple square grid onto the original shape. A second grid with the same number of squares but with each dimension twice the size is prepared for the transformed image. Children use the grid reference points (coordinates) to fix the features of the original image to the larger second grid to produce the enlargement. Connections and relationships, for example while the side length doubles the area quadruples, can be made via discussion to the size of sides of the original and the enlargement as well as the area of each.

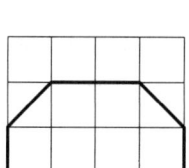
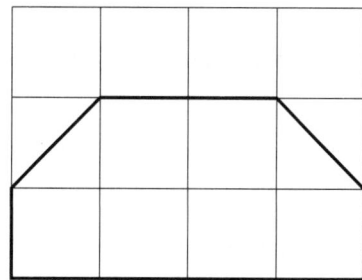

A doubled enlargement

Patterns

Pattern making using shape and space has been a feature of numerous cultures and civilisations for thousands of years. Patterns help people understand the world as well as help them predict and anticipate. They form a vital part of the aesthetics of many cultures. The definition of mathematics noted in *A National Statement on Mathematics for Australian Schools* (1991) is 'The science of pattern'. The skills of recognising, using and generating patterns are important tools in solving problems and provide an effective vehicle for exploring and understanding aspects of shape and space concepts. Patterns developing from work in symmetry and tessellation will be highlighted later in this section.

A pattern can be described as the repeating of a unit or element in a regular way and can range from the very simple—*red, blue, red, blue, red, blue*—to complex and involved Islamic designs. At all levels of learning children should be encouraged to:

- see a pattern
- say the pattern
- record the pattern in diagrams, or symbols and words

Once children can see the regularity of the repeated feature of the pattern, they can begin to copy a given pattern using different mediums, for example copying a clap, clap, finger click pattern and showing it using multi-link cubes. Next, children can extend or continue the pattern, and, finally, they can be expected to create patterns of their own. These patterns can also develop in complexity from a linear, one-dimensional pattern to those requiring two dimensions and involving aspects of reflection, rotation and symmetry.

Multi-link materials are a good medium for pattern making as they fit together on each face and have a variety of colours. At a very basic level, colour can be used to draw children's attention to the 'unit' of the pattern, which is then repeated. Children can be expected to copy and extend the same pattern in a number of different ways, for example with various objects and manipulatives found around the classroom, in words or sounds, with musical instruments, in diagrams and drawing, and with themselves and their peers. Pattern is something that is used continually in the classroom and should not be restricted to isolated sessions. Children should be recognising and having their attention drawn to patterns in many different situations throughout their years in primary school.

Multi-link patterns

Younger children can be shown a pattern in multi-link with an AAB unit such as:

> red, red, blue, red, red, blue, red, red, blue, red, red, blue

They discuss the pattern and, with teacher guidance, identify the unit of the pattern. Pairs of children can then be set the task of recreating the same AAB pattern structure with pattern blocks while others use an empty number line. Towards the end of the lesson children share their interpretations of the AAB structure and explain how their pattern in a different medium still shows the original structure.

Islamic patterns

This activity is adapted from Delaney and Dichmont (1979) and is suitable for older children. It is more complex in structure and embeds elements of diagonals in polygons, symmetry and transformation within the basic unit of the pattern.

Provide the children with centimetre or two centimetre squared dot paper. Have them outline a square unit of 3 dots by 3 dots (it will contain 4 smaller squares). They then lightly draw in the lines of symmetry and join the midpoints of each side to form an inner rotated square.

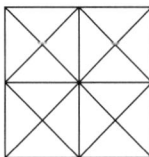

Some of the lines are then picked out and emphasised. This will form the tile design or template that will be translated horizontally and vertically on the large dot paper to form the overall Islamic-type pattern.

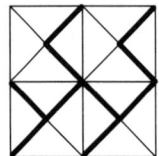 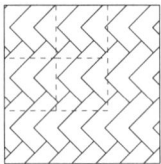

The lines on the final version can be drawn in colour. Discussion of the finished design will identify the unit of the pattern, the features of its design and the effect of it being combined with other translations. Comparisons can be made with other finished designs from the same starting square but with different lines emphasised. Children can search the Internet for examples of Islamic patterns from the Alhambra in Spain or other famous Islamic sites. Other children might like to consider the effect they can achieve by starting from a different base such as a regular hexagon or adding more lines to a larger square that is composed of 5 × 5 grid of smaller squares.

Tessellation

Tessellation can be an exciting and motivating area of shape and space for children. There seems to be an affinity for children with the fitting of shapes and the colour and pattern of the finished tessellation pattern. The word tessellation comes from the Latin *tessara*, which was a square tablet. Smaller square stone versions used for making mosaic floor coverings were known as *tessella*, hence the connection to tessellation and surface covering. There are many excellent examples in books of Roman mosaics which may be a useful discussion starter for work in tessellation. However, many of the Roman mosaics became pictures rather than the stricter mathematical meaning of tessellation—to cover the plane with a pattern in such a way as to leave no region uncovered. Through discussion children should understand that mathematical tessellation uses pattern rather than a random placing of shapes or the making of pictures which belong in the art domain.

Teaching of tessellation follows the general principles of shape and space teaching outlined earlier of inquiry, discussion, reflection, using a variety of materials and relating new knowledge and understanding to that already acquired. Often, however, work in tessellation becomes a low-level, 'busy work', art exercise where starter patterns on black line master sheets are completed and coloured. That sort of activity misses the essential mathematical ingredients of discussion and reflection, and relating to the underlying mathematics content of, for example, pattern and angle. There also needs to be a development of work in tessellation from the fitting of regular polygons to find which tessellate and which do not and why this is so. The work of M.C. Escher (see p. 436) provides more sophisticated use of translation, pattern and angle that are well suited to older children.

Tessellation work can be a good vehicle for the application of spatial ideas and solving realistic problems. For example, children in the middle primary years can be set the task of providing brick paving designs for a driveway. Often brick manufacturers advertise their products and show a driveway design outline. This can be used to start discussion on further investigation into possible patterns (tessellations) with bricks (oblongs). Pairs of children are given the task of being brick paving designers and have to produce a letter to a client showing the different designs possible with their product. Red Cuisenaire or other coloured rods can be used as manipulatives to allow children to experiment with possible combinations and designs. This serves as a way of recording children's work and provides a basis for discussion of the patterns completed. Similar tasks can be set using regular polygons for floor and wall tiling designs. It is possible here to arrive at the 3 regular and 8 semi-regular tessellations.

Regular tessellations
Six equilateral triangles at each point (3, 3, 3, 3, 3, 3)
Four squares at each point (4, 4, 4, 4)
Three regular hexagons at each point (6, 6, 6)

These polygons tessellate because when they are placed together their interior angles add to 360 degrees, a complete turn. For example, with squares four right angles (4 × 90° = 360°) meet at a point.

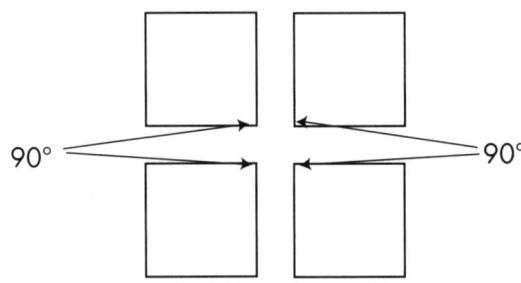

Other regular polygons such as pentagons have interior angles (108 degrees) that do not total 360 degrees, for example 108° + 108° + 108° = 324°, while four pentagons will give 432 degrees, which is too large. This is a useful connection to the interior angles in regular polygons and forms another property for polygons.

Semi-regular tessellations

If children are given piles of manipulatives (MATs are excellent because they contain all necessary polygons) of regular shapes they can be challenged to find which two or more polygons will fit together to tessellate. Having many tiles with the same sized edges is important as it allows children to experiment and quickly see if a combination works or not. Simply drawing around a template often leads to frustration and incorrect conclusions on many occasions.

The eight semi-regular tessellations
Four triangles and a hexagon at each point (3, 3, 3, 3, 6)
Three triangles and two squares at each point (3, 3, 3, 4, 4)
Triangle, square, hexagon and square at each point (3, 4, 6, 4)
Square and two hexagons at each point (4, 6, 6)
Triangle, hexagon, triangle, hexagon at each point (3, 6, 3, 6)
Two triangles, square, triangle and square at each point (3, 3, 4, 3, 4)
Triangle and two dodecagons at each point (3, 12, 12)
Square, hexagon and dodecagon at each point (4, 6, 12)

The use of the Schlafi coding system in a similar way to its use with 3D solids helps to highlight the connections of shapes and angles at a vertex. The order of the polygons as well as their angle size is important as a change in order in some cases may produce a different tessellation. Children may also find tessellations that follow a pattern for part of the tessellation but then change at another point. These are known as semi-regular tessellations. A good example of this is the 3, 4, 3, 12 and 3, 12, 12 tessellation.

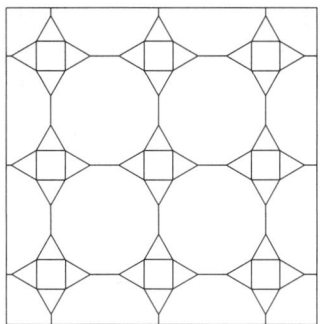

Semi-regular tessellation

Children may also investigate tessellation patterns that may be made using regular polygon tiles and 'gaps'. These 'gaps' are left when two or more polygons come together but do not quite make a complete 'turn', for example four pentagons leave a 'gap' of a rhombus. Other 'gaps' are possible with pentagons. Octagons are another useful shape to investigate, as are shapes that do tessellate such as squares. These tessellations are best recorded with the use of a digital camera.

Tessellation of octagons and squares

Tessellation with triangles and quadrilaterals

Another property of triangles and quadrilaterals is that they will tessellate. Any example of a triangle, not just equilateral triangles, will tessellate with itself. The same is true, though not so obviously so, of quadrilaterals.

Ask older children to produce different quadrilaterals (four-sided figures) on their 5 × 5 nail geoboard and record them on 5 × 5 dot paper. Try to obtain all forms of quadrilateral from the class so that you are able to pose the question for investigation—*Will all quadrilaterals tessellate?* A similar question and task could be set for triangles.

Children then select a quadrilateral (ensure an arrowhead is included) to check and begin to produce a number of tiles of it using their dot paper example as a template. This may be tedious, so the process can be quickened by placing more children into a group or by offering ready made tiles where these correspond, for example the pattern block rhombus. The tiles are then used to tessellate, leading to explanations and discussions focusing on the point where the tiles join in the way demonstrated by squares earlier that show that it is possible for all quadrilaterals to tessellate. Chain link fencing is a good example of the tessellation of quadrilaterals.

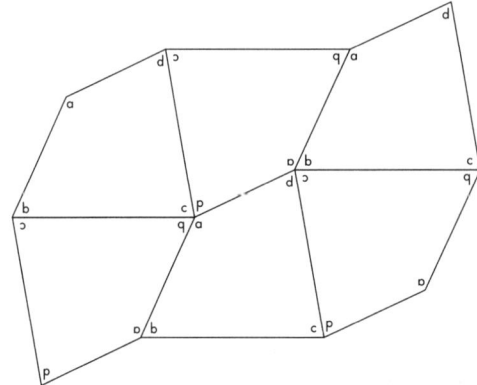

Quadrilateral tessellation

Tessellating and circles

On their own circles do not tessellate. However, if children consider the cross-section end of a pile of drainage pipes or stacked rolls of carpet they will see a tessellation of circles and gaps. There are two forms of gaps between circles.

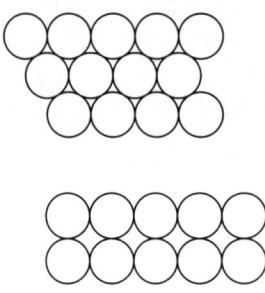

Circle tessellation

Escher tessellations

Many of the art works from the Dutch artist M. C. Escher demonstrate excellent examples of tessellation, albeit at a more complex level. The basic technique can be simplified to make it accessible to older primary children. Some of his work shows tessellations of animals and birds that are formed from a simple base tile.

A square is a good starter base tile. Use square grid paper and have children mark out a 10 × 10 larger square. This will act as a guide for the translations that will take place. Have children draw a piece to be cut from one of the four sides. This is then translated to the opposite side and fixed into place with sticky tape.

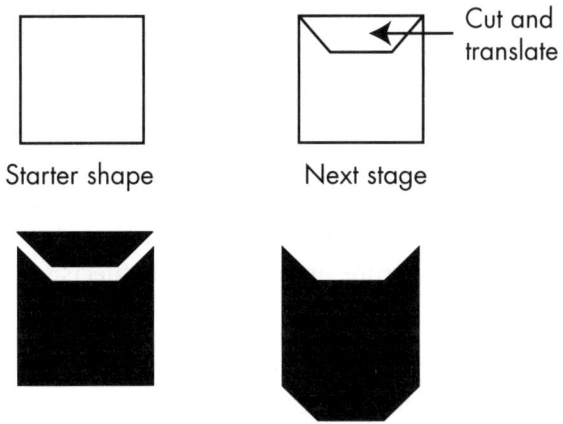

Starter shape   Next stage   Final Tile

The final tile made from squared paper is then used as a template as a new tile is drawn on firm card. This is used to tessellate the piece of paper. Each tile can then be anthropomorphised into a cat or a fish or something else that seems appropriate for the shape.

Tessellation of mice (Kit Sparrow)

Modern computer programs such as 'Tesselmania' or websites (www.worldofescher.com/store/mania) allow children easy access to designing Escher-type tessellations.

Symmetry

Human beings seem to have some innate appreciation for bilateral symmetry. For example, young children's drawings and 3D constructions with blocks and similar materials often reflect the balance that is characteristic of bilateral symmetry. There is also a tendency to avoid or dislike 'lopsidedness' in architecture, art and natural objects.

Types of symmetry

There are two types of symmetry, *bilateral* or *line*, seen in reflections in mirrors, and *rotational*, given by a sequence of partial turns about a central point. Rotational symmetry, while reasonably common in nature and architecture, seems to be less recognised and understood by children.

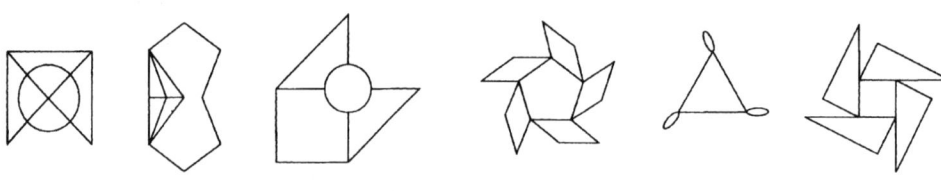

Bilateral symmetry   Rotational symmetry

Bilateral symmetry is commonly shown in classrooms through younger children constructing 'butterfly' pictures from folded paper containing blobs of paint along the crease which are squashed out to provide a print on each face of the folded paper when it is opened. Other typical activities include finding the line or axis of symmetry in a drawing of an object or completing the whole shape where a half is given. All of these activities are helpful in developing children's concepts of symmetry. To be more effective, however, teachers need to have children explaining how they know something is symmetrical and why they placed their line where they did.

Completing the shape activities typically require children to use a vertical line of symmetry. Tasks should develop further aspects of bilateral symmetry so that children are aware of reflection lines that are horizontal to the foot of the page, at an angle to it, and the fact that shapes can have two or more lines of symmetry. This is particularly important for analysis of the symmetrical properties of polygons.

Pattern block symmetry

Children love to use pattern blocks and many immediately begin to make patterns with the blocks that exhibit vertical line symmetry. Teachers can use this motivation to set similar problems to be solved. For example, the pattern must have horizontal line symmetry, or a line of symmetry at an angle to the foot of the page or desk, or it must have more than one line of symmetry. Pattern blocks are very useful in this context as they allow children quickly to experiment with ideas, predict what might happen, test the ideas and change them as necessary. The speed of construction also empowers teachers to ask *What if?* questions, as they observe and interact with children without the fear of setting a huge task, as might be the case with paper and pencil drawing. Once the patterns are made, a teacher can develop the notion that there is more than one type of line symmetry via discussion and children's explanations.

Rangoli patterns

As part of the celebrations for the festival of Divali, Hindu people often construct Rangoli patterns, usually from coloured powders and spices. The patterns are an excellent example of horizontal and vertical line symmetry. The final designs are meant as a sign of welcome to visitors, and to Lakshmi, the goddess of prosperity.

Square dot paper, with either one centimetre gaps or two centimetre gaps, is most useful here to guide the accuracy of children's lines and coloured areas. Have children make a square on the dot paper which is 7 dots by 7 dots. Avoid making the starter square larger as it may become too complex and not allow children to complete the task successfully.

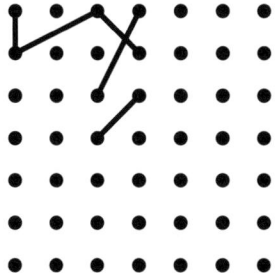

Children then mark lightly the horizontal, diagonal and vertical lines of symmetry. Begin in the top left-hand corner or any corner of choice and mark in some lines to join dots (see diagram above). These lines are then reflected and reproduced across mirror lines into other sections until the pattern is complete. Lines can be drawn in coloured pens and the spaces also coloured to produce the finished design.

The development of a Rangoli pattern

Testing for symmetry

A simple test for bilateral symmetry in a 2D figure is the *fold test*. If the shape can be folded over on a straight line (the mirror line) so that one half of the figure coincides exactly with the other, then the figure has bilateral symmetry. The experience of actually performing this test helps pupils move into the level of understanding where they begin to appreciate the characteristics of bilateral symmetry. Many children experience difficulties with using mirrors to test for symmetry as they misunderstand the role of the mirror and forget to project the image beyond the mirror. The 'Mira', a commercially produced semi-transparent plastic mirror, is better for testing line symmetry as it allows the correct pattern to be reflected but also allows children to see through the Mira to their work for comparison.

To test a 2D figure for rotational symmetry, the *turn test* is used. If a tracing of the figure can be rotated about a central point of the original shape through an angle of 180 degrees or less so that the tracing coincides exactly with the original shape then the object has rotational symmetry. Some shapes will fit inside their outline more than once and it is this feature that can be used to classify their degree of rotational symmetry.

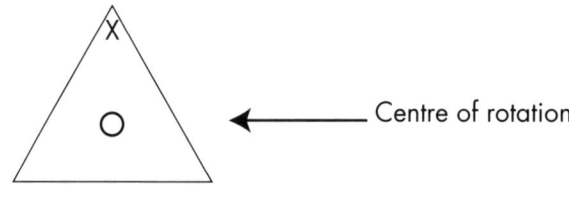

An equilateral triangle

In an equilateral triangle, the X can be rotated to three different positions to fit within the triangle outline. This would categorise it as having 120 degree rotational symmetry or symmetry of order 3 because it can be placed three times in one complete revolution of 360 degrees. A regular hexagon would have order 6, while an oblong will be order 2.

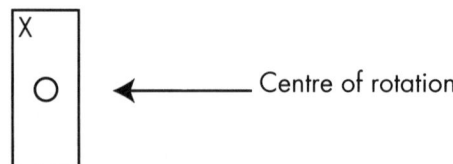

An oblong has order 2 rotational symmetry

A shape that has two or more lines of symmetry always has rotational symmetry, but a shape that has rotational symmetry may not have line symmetry.

Investigate and discuss

1 If a shape with order 2 rotational symmetry turns an angle of 180 degrees how many other orders of symmetry are there?
2 Draw diagrams to illustrate this.
3 Explain how you know you have found them all.

Symmetry in polygons

Another way to classify polygons is by their symmetrical properties both bilateral and rotational. Triangles, for example, can be classified according to line and rotational symmetry. The number of lines of symmetry will vary depending on the type of triangle that is considered.

Triangles / Lines of symmetry	0	1	2	3	Rotational
Scalene	✔				
Isosceles		✔			
Equilateral				✔	✔

Quadrilaterals may be considered in a similar way. Ask children to cut various quadrilateral shapes out of paper and to fold them and rotate them to find their symmetrical properties so that the following chart may be completed.

Lines of symmetry / Shape	Quadrilateral	Oblong	Square	Rhombus	Parallelogram	Arrowhead	Kite
0	✔				✔		
1						✔	✔
2		✔		✔			
3							
4			✔				
Rotational order		2	4	2	2		

Regular polygons exhibit the same number of lines of symmetry as they have sides. For example, the regular hexagon has six sides and six lines of symmetry.

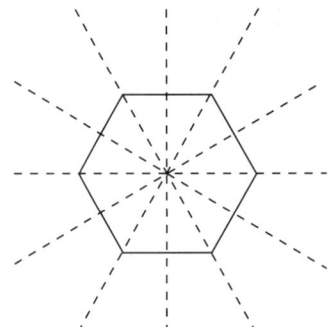

Lines of symmetry in a regular hexagon

Children can often find the lines of symmetry by cutting out a regular polygon shape and using the fold test. This will highlight the 'mirror lines,' which can be drawn in later. It will also allow the turn test for rotational symmetry to be checked.

Geometric shapes, both plane and solid, may have either type of symmetry, both types of symmetry or none at all. For instance, a kite has line symmetry only, while an oblong has both and a scalene triangle has no symmetry at all. The more properties a shape possesses the greater the symmetry.

'Tuning out' assessment of symmetry

Two activities are offered here to demonstrate how *assessment by task* can provide information of children's learning and understanding about symmetry. These tasks are suited to older primary school children. As no specified directions are offered, for example find examples of rotational symmetry, children will react to this situation at different levels of sophistication showing not only an aspect of knowledge but also an inclination of knowing when to use it.

ALPHABET SYMMETRY

Exploring
Provide children with a sheet containing the letters of the alphabet. Select a simple sans serif font (for example Helvetica) from the computer menu and print out a sheet of the upper case letters in a large font size for the children. They are then required to classify these letters according to their symmetry. Pair work is usually good here to encourage children to talk, explain and justify their choices and decisions.

Explaining
Children demonstrate the symmetry they have detected in the letters. They show where they have placed the 'mirror line'. Some may also indicate rotational symmetry—these generally will be children who have a more advanced grasp of the symmetry concept.

Reflecting
Children will note that there are a variety of symmetries shown within the alphabet ranging from no symmetries to multi-lined symmetry and rotational symmetry.

Recording

Letters can be classified and shown in a table:

	Has vertical line	Has horizontal line	Has multi-lines	Does not have line symmetry
Has rotational symmetry			H I O X	N S Z
Does not have rotational symmetry	A M T U V W Y	B C D E K		F G J L P Q R

From here children can begin to investigate symmetry in words; for example:

```
OHIO       O        M
           H        O
           I        M
           O
```

Sentences may also be used, such as the palindromic sentence of Ferdinand De Lesseps, the famous engineer:

AMANAPLANACANALPANAMA

or

MADAM I'M ADAM

THREE PIECES

The following idea, originally proposed by Geoff Giles, the inventor of the Rotagram and DIME solids among many other teaching materials, provides a very useful assessment task either for 'tuning in' or 'tuning out' of a piece of teaching.

Exploring

Provide children with three pieces of card in the proportions shown.

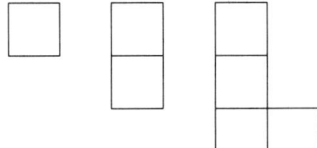

Ask them to fit some or all of the pieces together to form symmetrical shapes and record them on square grid paper.

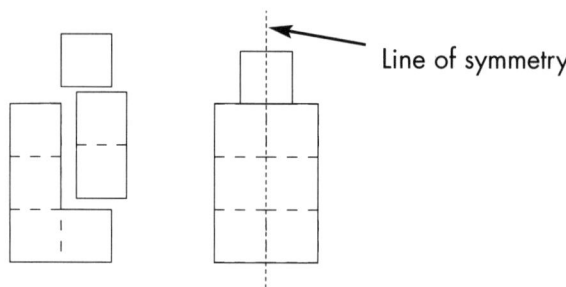

Line of symmetry

Explaining
Children show where the line of symmetry is placed on the shapes or how each shape shows symmetry.

Reflecting
Teachers can note the lines of symmetry children find. Many will only show vertical symmetry, but by questioning and causing them to think and reflect on the situation children may realise that there are other ways to find symmetry within shapes, for example with more mirror lines and rotational symmetry.

Recording
Selected examples can be recorded by children gluing replica coloured paper to a larger A3-sized sheet of paper or poster. Lines or rotations of symmetry can be drawn onto the shapes along with explanations of the symmetry and commentaries on the activities.

Investigate and discuss

1 Collect examples of logos from banks, businesses and cars (for example the Mitsubishi car badge logo) which show symmetry.
2 Identify the symmetries used and classify them in a Venn diagram.

Location and arrangement

The third aspect of shape and space brings together knowledge and understanding relating to position, arrangement and direction. It involves grids and maps. Aspects of angle are associated with direction and need to be connected to these ideas.

The *Numeracy Benchmarks Years 3, 5 & 7* (Curriculum Corporation, 2000) noted the following learning for this strand:

Year Three
- use language which shows they understand position and direction when using a simple grid, map or plan (eg 'The tree is between the house and the fence');
- use position to place objects (eg place a cross between the tree and the house; shade the tree at B3 on the grid);

- recognise and use common positional and directional language in relation to simple grids, maps and plans to find general locations;
- follow directions to move from one place to another on a map, grid or plan (eg use a finger to trace a route from the shop, across the bridge and past the post office to show how to get to the school).

Year Five
- place an object accurately on a simple map or plan and follow or give directions to find a particular place on a map or plan;
- follow or give directions based on order and proximity (eg put a cross to show the basketball ring which is between the taps and the year 2 classroom, but closer to the taps);
- locate an object or pathway accurately on a map or plan;
- follow or give directions about pathways or routes (eg 'Go over the bridge, turn left at the crossroads and keep going until you reach the bus stop');
- locate specific objects using a key or legend (eg find and interpret the symbol in square D5 on the map).

Year Seven
- identify and describe locations and routes using simple coordinate maps (eg street maps) and major compass points (N, S, E, W), and draw simple scale plans of familiar locations (eg draw a sketch plan of the route from school to home showing major landmarks; find the lake that lies east of the highway);
- use a map to locate major features (eg find a street that runs from B3 to A4);
- use a simple map to plan a route (eg say that to get from the school to the swimming pool, you must go south from the school gate until you get to the main road, then turn left, go past the post office, then take the first turn right);
- identify and describe locations and routes using simple coordinate maps;
- describe positions on a map using the major compass points (N, S, E, W) (eg find the lake that lies east of the highway);
- use proximity and a sense of scale when drawing simple plans of familiar locations;
- draw a simple outline diagram on squared paper using a straightforward scale (eg draw an outline of the netball court on centimetre grid paper using a scale of 1 cm : 1 m).

The idea of location begins to develop from an early age as children are asked to place mugs onto tables, stand next to the shopping trolley and are warned not to play behind the car. Over the next few years the concept of location will develop in sophistication and accuracy as children relate location to grids and maps and provide answers to the following questions:

- Where is it?
- How far away is it?
- Which way is it?

As they build their understanding of locations, children develop skills that relate to direction, distance and position. These skills and understandings are built through a range of experiences that include play, physical activity in such things as PE, literature, art, model making, talk and discussion with the teacher.

The language of location

Initial work with young children revolves around the use and understanding of the many words associated with location. Children need to be introduced to and use everyday words for position such as behind, in front, next to, and on top of. Their understanding of the words of position develops through use and experience to include, for example, under, above, near, outside, inside, beside and to the left. Gradually, they are introduced to the idea that movement and distance can be added to the instructions. Children now add words like forward, backward and turn to their vocabulary. Some of the words have the potential to present minor confusions for children as they meet conventions and variations in language use. Children can readily place a pattern block on top of the table or above the painting on the wall. These are in a 3D setting. A 2D context is slightly more complex. Here 'above' the square refers to a drawing of a square on paper and children have to place it to the top of the paper.

Difficulties also occur for children as things change with orientation of the object and the position of the viewer. What is on the left of an object for me as I look at it from the front is on the right for you as you look at it from the rear. Remember conversations that seek to clarify 'my left or your left?' as people attempt to understand location from different viewpoints and differing perceptions. Much of the learning here is incidental and occurs in informal, everyday situations as children carry out requests to 'put the book behind the table' or are asked to 'line up next to the climbing frame'. Simple barrier games are useful to help children give and receive positional instructions, where one child builds a simple shape from joining cubes or pattern blocks, for example, while a second child attempts to recreate it from the instructions and descriptions offered by the first child. Neither child can see the other person's building due to the intervening barrier. Stories are an excellent source of the rich language of position. Most children are soon familiar with the adventures of Rosie in *Rosie's Walk* as she navigates the farmyard and avoids the hungry fox.

Direction

The early uses of directional language arise from the use of positional words and ideas. Children begin to develop navigational words such as left and right, clockwise and anticlockwise. To these are added ideas of distance and measurement as children 'stand by the desk, turn to the left and move forward five steps' or they 'play robots' where class members direct a robot child to 'move forward three steps, turn to the right . . .'. Following this kinaesthetic activity, children in the middle primary years can begin to use programmable toys such as 'Roamer' and progress to computer programs like LOGO.

A 0–99 board can be used to set children problems that require them to devise routes using directional language to move from 25 to 37 (down one square, right two squares) or from 12 to 46. Not only can there be connections to other areas of mathematics such as number, there are also connections within the shape and space strand as children attempt to make the 'Roamer' trace an outline of a hexagon or program in LOGO for the 'turtle' to draw a regular pentagon. There are also strong connections to angle and the idea of turn. Initial instructions for left or right can become quarter turn or half turn which lead to the use of angle and degrees. A quarter turn can then become 'turn 90 degrees' while a half turn is 180 degrees.

Points of the compass, for example north, south, east and west, can be used to develop the directional and navigational language of children. Gradually, the more sophisticated use of all points of the compass, for example NNE, can be incorporated. This idea connects nicely from the full range of points of the compass to the use of bearings. The DIME materials for direction with their use of Rotagrams are an excellent source of teaching and learning support materials to help children develop these ideas.

Maps

Young children have quite sophisticated notions of maps but often emphasise features idiosyncratic to the drawer of the map. For example, the main feature of childrens maps might be their home, which is given a central place and drawn in a larger format. Other maps might show side views of objects along with some top views or combinations of both sometimes on the same building.

Children begin by drawing maps of familiar places which they can see and if necessary check against their map. Gradually, the maps can move in a concentric circle approach from the familiar and immediately viewable in the inner circle, to the familiar but not easily viewable, to the unfamiliar and distant. Children can start with their desk or classroom or bedroom and then move to the next circle out with an example of the school and then to their local area until they finally experience maps of large areas such as their state or Australia. They can also consider maps for special areas and for special situations, for example the Melbourne or Sydney train maps are much more of a diagram and in that sense are unlike road maps.

On their familiar maps, teachers can help children reflect on the order of things and their relative position to other objects. For example, children can be directed to consider if they have things in the correct order on the map—'Is the canteen after the staffroom?'. Are they able to describe the position of key features relative to others, for example the canteen is between the sports store and the staffroom. Gradually, children add to their repertoire of ways of representing, locating and describing features. Their sketch maps of routes to school can develop proportion and scale as well as conventions of top or bird's-eye views.

When progressing through primary school, children start to come to terms with the idea of describing location by the use of grids. Following on, connections can be made to other areas of mathematics, including graphs and data presentation.

Early ideas of position with a grid usually follow an alpha-numeric system (a combination of letter and number), for example B3.

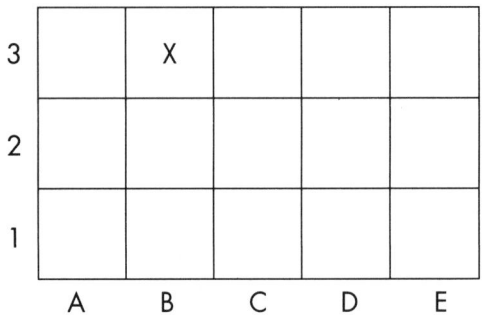

An alpha–numeric grid

It would be useful to begin the letters and numbers from the bottom left-hand corner of the grid. Note here that the spaces between the grid lines are marked with the letters and numbers and the coordinate B3 refers to the whole square. The teacher may also wish to place the letters across the bottom of the grid and the numbers to the side so that children become familiar with the later convention of writing the *x* (across) coordinate before the *y* (up) coordinate. Local area street directories use a similar system to identify the position of a particular street in the suburbs or central business district. A street directory, which includes the school and local homes, can offer a realistic and relevant example of the application for the children and is much more motivating and appealing for the children than stylised examples found in worksheets or textbooks.

The move from using the alpha-numeric system to 'ordered pairs' appears at first sight only a small development. It is, however, a source of considerable difficulties for children.

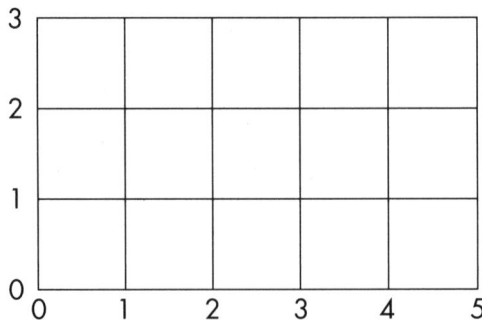

A grid for ordered pairs

The first potential pitfall for children relates to the order to which the pair (2, 3) refers. The mathematical convention is to go across (the x coordinate) first. Then the up (the y coordinate) is considered. This makes the position (2, 3) using the (x, y) convention different from the position (2, 3) using (y, x). The reference relates to the crossing points of the grid lines or the corners of the squares on the grid. Teaching approaches to help children develop these ideas usually involve 'battleship'-type games and pirate treasure maps. Children have to describe and identify position using the conventional method of (x, y) ordered pairs. Later, position, direction and distance at various levels of sophistication are involved, for example 'walk three paces north from the palm tree at (3, 2)'. Towards the end of primary school some children may be introduced to using Cartesian coordinates not just in the positive quadrant but in all four quadrants which will involve them with negative as well as positive integers.

Large-scale maps, such as those suitable for guiding bush walkers and orienteers, need to be able to offer more accurate position fixing. They adopt a six-figure reference system and are able to identify a location 'within the square' so that orienteers, for example, can find a marker by a rock or significant tree. Latitude and longitude are examples of a referencing system being applied to larger-scale examples and a real context. With maps and plans a further factor relating to distance, that of scale, has to be considered. Modern navigation and position finding uses satellites and the Global Positioning System (GPS).

Postscript

GEOMETRY is an integral part of mathematics and one of the basic areas of mathematics that everyone needs in order to function rationally and usefully in everyday life and in many occupations. While the study of geometry on a regular basis has many benefits for primary students, it should not be studied in isolation but integrated with other parts of the curriculum, such as fractions and measurement, and with the arts, language and writing.

As with other topics in mathematics, a major aim in studying geometry is to develop an appreciation of its place in our cultural heritage and its application to the solution of problems. Teaching geometry is not a matter of passing on information. It should assist the child to develop skills and processes of observation, analysis, description and classification, visualisation, representation, exploration and inference. To do this successfully, teachers need to listen to and observe children carefully so they become aware of how children learn. This will enable them to introduce children to appropriate activities suitable to each individual's level of functioning.

Investigate and discuss

State and national documents include substantial reference to shape and space concepts. Websites provide details of current syllabus and curriculum statements.

1 Identify the shape strands in the syllabus or curriculum documents you use. Compare and contrast it to the national statement.

Some important websites are:

Australian Capital Territory:	www.decs.act.gov.au
New South Wales:	www.boardofstudies.nsw.edu.au
Northern Territory:	www.ntde.nt.gov.au
Queensland:	www.education.qld.gov.au
South Australia:	www.dete.sa.gov.au
Tasmania:	www.tased.edu.au
Victoria:	www.bos.vic.edu.au
Western Australia:	www.eddept.wa.edu.au

CHAPTER SIX

Measurement

Measurement . . . has its roots, both historically and in individual development, in significant exeryday activity. Thus it can develop in the earliest years from children's experience . . . Further, it spans and connects mathematics and other sciences and thus can ideally integrate subject matter areas. Finally, it can serve as a foundation for the development of other topics within mathematics.

Clements, 2003, p. xi

Why teach measurement?

Children have a natural curiosity about the world around them and often desire to compare things. They will compare physical attributes such as height and attach language to the comparisons, for example 'I am *taller/shorter* than . . .' or 'My Dad can kick a ball *further* than your Dad'. Over time it becomes apparent that descriptions such as taller and shorter are not appropriate in some circumstances and the question 'how much taller?' arises. This in turn leads to the need to measure and quantify the measurement. *A National Statement on Mathematics for Australian Schools* noted that 'measurement . . . underlines many of the descriptive statements we make' (1991, p. 136). Consider your own language and gestures. How often do you make use of measurement-related words?

Measuring skills are essential for coping with everyday experiences. Measurement provides the link between various aspects of mathematics such as number and space and provides a vehicle for integration across the curriculum. Measurement skills, for example, are used in conducting science experiments, making models or determining the winner of the long jump. Measurement involves the application of number to spatial (and other) qualities of objects and events. Therefore spatial, numerical and measurement concepts generally develop together. Consider some links between number, space and measurement, for example scale diagrams, the discovery of pi as a ratio of the diameter and circumference of a circle and collecting height and arm span data to be plotted on a scattergraph.

Measurement also provides links between mathematics and other school subjects. Measurement skills are a fundamental tool for investigating physical phenomena in all branches of school science and in studying mapping and weather in social studies. The history of the improvements in measuring techniques and of the refinement and rationalisation of units leading to the general acceptance of the metric system provides useful and interesting links with historical investigations in social studies. Measurement skills, especially estimating skills, have an important place in many school physical activities, sports and games. They even play a part in some art and music experiences. Measurement, therefore, provides a medium for integrating mathematics with other topics in the primary school curriculum.

How children learn measurement ideas

An analysis of the process of measuring suggests that children learn to measure by first becoming aware of the physical attributes of objects and hence perceiving what is to be measured. When children have perceived a property of some object they naturally compare it with other objects having the same property. This leads to the need for a standard of measure for the property. Initially the standard referent may be child-chosen from readily available materials. Use of such informal referents leads to the need for standard units for better precision and unambiguous communication.

What is measurement?

The *Principles and Standards for School Mathematics* (2000) define measurement as 'the assignment of a numerical value to an attribute of an object'. Measurement involves comparing an amount of the attribute to be measured with a quantity of that attribute chosen as a unit in order to provide a number. For example, if the amount of the attribute area possessed by a piece of paper is compared with a unit of area, such as 1 square centimetre, and the number obtained is 30, this result is expressed as 'the paper has an area of 30 square centimetres'.

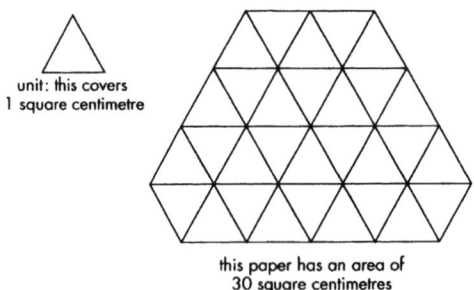

Attribute: area, how much is covered?

In essence measurement is a process that allows people to quantify the world. For example, a trip may be described as being 200 kilometres long or 3 hours' drive. If using a hire car and paying for each kilometre, then the first piece of information would prove useful. However, if a person wished to drive to the airport to catch an aeroplane at a certain time then the second piece of information would be more useful.

In order to measure, students need to be able first to decide on an attribute to be measured and select an appropriate measuring device and unit. If asked to measure an orange where might one start? Perhaps weigh the orange or measure the circumference. Once told that the orange is to be used to make juice then measuring the juice that could be extracted from the orange would become important. Once this decision has been made then it would make sense to extract the juice and pour it into a container. The amount of juice would then be measured in mL.

Before children can cope with this relatively sophisticated process of measuring an attribute efficiently and stating the measure in terms of formal units, there are several preliminary processes that must be acquired. Most children, when confronted with an attribute that they must eventually be able to measure, develop their measuring processes effectively when provided with experiences in the sequence described below. Wilson and Rowland (1993) provide a summary of the research bases for this learning sequence.

LEARNING SEQUENCE FOR MEASUREMENT

1 **Identifying the attribute**—to cope with the measurement processes in a meaningful and effective manner, a child may have to develop new concepts, skills and language. Some attributes to be measured may not yet form part of the child's conscious experience. If so, initial experiences are needed to develop an awareness of the attribute and to introduce the necessary new language. Hence, opportunities must be provided for the child to manipulate objects that possess the attribute being explored, to discuss these experiences with others, and to make comparisons involving the attribute. Considerable time may have to be devoted to these experiences to allow the child to become aware of what can be done to an object without changing the quantity of the attribute that is present. For example, does the length of the ribbon change if it is moved? Does the area of a piece of paper change when the orientation is altered from portrait to landscape? Many experiences of these kinds may be necessary before some children have sufficient maturity and familiarity to be ready to make comparisons between different amounts of an attribute.

2 **Comparing and ordering**—when children are aware of the attribute under investigation, they should be given opportunities to compare and order different examples of the attribute. Activities for comparing two examples should lead into activities for ordering or seriating (that is, placing in correct order or sequence by size) three or more examples, using both direct and indirect methods. These experiences reinforce the development of concepts associated with the attribute and prepare for the next stage when an example is measured by being compared with a chosen amount of the attribute used as a unit. For example, when comparing two heights students might stand next to each other and should be encouraged to make statements such as 'I am taller than my friend' and the alternative statement 'My friend is shorter than me'. Once a third object or person is included then the

descriptions would include words such as tall, taller and tallest, or short, shorter, shortest and statements such as 'Jeremy is taller than Daniel but shorter than Adrian'.

3 **Non-standard units**—the development of formal measurement skills proceeds from initial measurement activities using units from the everyday experiences of the child, for example lengths measured in paces or hand spans, and volumes measured in cupfuls or spoonfuls.

Children should be provided with many opportunities to measure using these kinds of informal non-standard units in order to prepare them for measuring in standard units. Such experiences can be seen to be natural and related to the child's everyday experiences, and although they use familiar experiences in ways that can involve fun and enjoyment, they introduce the child to the use of units to provide numbers that describe a measurement outcome, for instance the room is ten paces across. However, the complications and difficulties associated with formal notation and conversions between units are delayed until children are ready to use their now-familiar measurement processes with standard units. Before standard units are introduced, most of the principles associated with measurement can be developed naturally using non-standard units.

The following *measurement principles* should therefore be established first through measurement experiences using non-standard units:

- Units (even standard units introduced later) are not absolute but have been chosen for convenience. For example, in measuring the length of the classroom, Kim's pace would be a more convenient and appropriate unit to choose than Kim's hand span.
- During a measurement activity, the unit being used must not change. For example, it is not appropriate to begin measuring the room length using Kim's paces and to change during the activity to the use of book lengths as the unit.
- Comparisons are performed conveniently by taking all measurements in the same unit, so that the largest number represents the greatest amount of the attribute being measured.
- The larger the unit used the smaller the number obtained, and vice versa.

A wide variety of experiences in the use of non-standard units also provides opportunities for:

- estimating amounts of attributes before they are measured
- choosing appropriate units to suit different situations
- coping with tolerance for error in judging the degree of accuracy appropriate in a given situation

Students should be led to the discovery that while non-standard units are useful they are not very good for communicating a set of measures. To illustrate this a group of students could make a model seven hand spans long and then instruct an older or younger group to make a similar model. On comparison the students should note a discrepancy in the length of the two objects. Note that the longer the object in terms of hand spans the greater the discrepancy. There are several children's stories such as *The King's Foot* and *Six Feet Long and Three Feet Wide* that develop the idea that non-standard units can lead to miscommunication. Eventually, wide experience with these non-standard units leads to the establishment of the need to standardise units to eliminate ambiguities, to facilitate communication and to enhance precision in measurement in all situations. It is at this stage that the introduction of standard measurement units is opportune.

4 **Standard units**—the standard units used in Australian schools are the metric units (with a few approved exceptions relating to time, temperature and angle) recommended for Australian use by the Metric Conversion Board. These are summarised at the end of this chapter (see also www.nsc.gov.au).

Appropriate learning experiences are needed to allow each child to become familiar with the quantity involved in each unit, the correct language for naming each unit and the correct conventions for writing measures using approved symbols. Children then need much practical experience in using appropriate measuring devices, in making estimations of quantities in real-world situations and in mastering equivalences between units.

5 **Applications**—when children are comfortable and efficient in measuring correctly using appropriate standard units, learning experiences should be directed towards extensive applications of measurement and to the use of measurement formulae as the powerful formal tools for generalisation. For example, simple measurement formulae may be developed and used to generalise methods for calculating areas, volumes and perimeters. Later, these now-familiar formulae provide appropriate starting points for secondary school algebra, for instance for providing experiences in substituting specific values of a variable to find values of the expression, for manipulating an expression to change its subject, and similar introductory algebraic experiences.

The role of language

Children often use imprecise measurement language such as big, small, large, tiny, massive and huge to describe the size of an object but these words beg the question 'how small?' or 'how big?'. Is a huge object larger than a massive object or vice versa? Children begin to use words such as taller and shorter when comparing the heights of two people and eventually learn to compare the heights of several people using comparative statements involving the words tall, taller and tallest or short, shorter and shortest. As a result of participating in measurement lessons children's language should become more precise. Teachers should always take care to model the correct use of language. Discussions in the context of measurement lessons serve several purposes including developing students' understanding of the measurement process and the development of appropriate measurement vocabulary.

Children sometimes become confused by words that have several meanings—one in the classroom setting and another out of school. This issue occurs when discussing the notion of volume, or the amount of space an object occupies. Some children confuse the use of the word volume with the button on the TV remote control that alters the loudness. Adults often confuse the terms volume and capacity or use them interchangeably, whereas the two terms refer to different ideas. Volume refers to the amount of space taken up by an object, whereas capacity refers to the amount it can hold.

Another area where children often experience difficulty is with time concepts. Children are often told to wait a minute when the reality is that it may take half an hour before their wait is over. Expressions such as 'waste time' and 'save time', and 'time dragging' are common and serve to confuse students' understanding of time.

The implication for teachers is the need to encourage students to talk about measurement concepts during mathematics lessons. Teachers also need to initiate conversations about measurement concepts as part of the routine of any measurement lesson. The value of such a routine is well emphasised in the following advice given by the NCTM:

Disc[ussion] builds students' conceptual and procedural knowledge of measurement and gives teachers valuable information for reporting and planning next steps. The same conversations and questions that help build vocabulary help teachers learn about students' understandings and misconceptions. (*Principles and Standards for School Mathematics*, 2000, p. 102)

Constructivism and measurement

Current learning theory suggests that students construct knowledge in a social context. Applying this thinking to the teaching of measurement implies that for students to learn to measure they must become involved in meaningful measurement tasks. Unfortunately, many measurement lessons are basically number lessons given in a measurement context. Students are given calculations to complete involving measurement formulas, which is essentially a substitution exercise or an addition and multiplication exercise. Work with formulas is the endpoint of the teaching and learning of measurement concepts. Prior to reaching this stage students should be engaged in deciding what attribute needs to be measured and what measuring tool and unit are appropriate. Students then need to be given the opportunity to perform the measurement task, read the scale, make decisions about measures that fall between divisions on the scale and choose the measure to assign to the object.

At times teachers can remove all the thinking from a measurement lesson by choosing the measurement instrument and unit most appropriate to the tasks, rather than allowing the students to make these choices. For example, a teacher may provide students with a trundle wheel and ask students to measure the perimeter of the basketball court in metres. Essentially, all the decisions about the appropriate measuring tool and unit have been made and the exercise has been reduced to one of click-counting. If, however, students were given the opportunity to make choices they would be more likely to learn about the measurement process. For instance, students who chose to use a standard 30 cm ruler to measure the perimeter of the basketball court would learn that it takes many ruler lengths (smaller units) to measure the perimeter of a basketball court in comparison to using a metre length. Comparison of measurements made by students should lead to discussion of factors that affect measuring, such as the limitations of the measurement tool, how accurately the students measured and how well they read the scale. When a measure falls between two divisions on a scale students need to make decisions to round the measurement up or down, or perhaps estimate the size of a part unit. Students can only learn these skills by being involved in the measuring process and subsequent reflection.

Materials and equipment for measurement

A lack of equipment is often cited as a reason for measurement lessons becoming just another paper-and-pencil exercise. There are many strategies that may be used to overcome the issue of a lack of equipment. These include setting up workstations, using groups and making use of everyday objects as measuring devices such as 2 L plastic bottles.

It should be noted that in an effort to streamline measurement lessons the learning opportunities may be reduced by nominating and providing equipment. Rather, why not pose a question that stimulates thinking about the unit and measuring device.

In addition to a variety of containers, for example 2 L ice-cream containers, 2 L drink bottles and yoghurt containers, the following commercial resources would assist in the teaching and learning of measurement ideas:

Length 30 cm and metre rulers, dressmaker's tapes, builder's tapes, trundle wheels, height measure.

Weight balances, scales (kitchen, bathroom, spring), set of weights. Balances assist students to compare weights whereas scales allow students to assign a numerical value to the weight of an object.

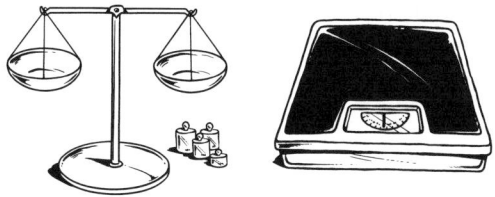

Time sand timers, water timers, tocker timers, stop watches, analogue and digital clocks.

Capacity graduated cylinders, measuring spoons, cups, jugs, centicubes, wooden cubes, 1 cubic metre kit.

Historical aspects of measurement

The development of measurement ideas over history is a fertile ground for discussion. There are many interesting stories to be told that will liven up measurement lessons. A basic search of the worldwide web will provide a great deal of information that may be used to enrich measurement lessons. Try www.nsc.gov.au for a comprehensive background to the metric system and the history of measurement.

Historical examples may be used to explain the origin of words used in mathematics or everyday language. These examples may also be used to prompt mathematical activity and investigation. For example, consider the expression 'to reach a milestone'. Links to the word mile, which was a measure of distance in the imperial system of measurement, may be noted but of more interest is the origin of the term. The origin of the milestone may be traced back to the time of the Romans. As Roman armies marched across various countries they would keep count of their paces and take particular note as they reached 1000 paces. The Romans would refer to this distance as *mille passus*. It is from this expression that the word mile was derived. The term milestone, however, is derived from the practice of the Roman army to leave marker stones, known as milestones, to identify the distance travelled from one point to the next. These marker stones may still be found in some places in England. The expression 'to reach a milestone' literally meant to have covered a certain distance. While the story is very nice and adds historical and cultural aspects to the mathematics being taught it may be used as a prompt for mathematical activity involving pacing. Children could be taken outside and pace length measured over a set number of paces (eg 100) and compared to a single pace. The pace lengths of individual children could be compared and used to examine the problems associated with using non-standard units. Children could be asked to investigate whether short children have short pace lengths and tall children longer pace lengths. (It should be noted that a Roman pace was two steps.)

When dealing with time and the recording of time using a calendar there are many useful historical anecdotes that may be used for motivation in a mathematics lesson. A study of the modern calendar is steeped in history. Students could be asked to explore why some months contain more days than others or why the dates for Easter change from one year to the next. An interesting discussion may be developed around the question 'Why is October the tenth month of the year, when the prefix Oct means eight?'.

Much of the history associated with measurement revolves around the standardising of units of measurement. People realised the need for standard units and those in power recognised the need to formulate some standards. For example, King Henry I (1100–1135) of England made a decree that the yard would be the distance from his nose to his thumb when his arm was outstretched. A class activity involving children measuring this distance would soon reveal issues with this decree. What would happen when the King died? This distance became known as a yardstick (this expression is still used today) and the system on which it was based as the system of imperial measures, because of the links to the royalty. Later in 1215 King John signed a document known as the Magna Carta, which in part included standards for weights and measures. A yardstick, generally made from bronze, was often kept in the town square so that tradespeople could verify their measurements against it.

The metric system we use today was suggested in 1670 but did not really develop until the time of the French Revolution (1790s), when the new government wanted to ensure that the measurement system would be fair to all. The scientists of the day recognised that any measurement system based on physical characteristics of the human body would be subject to change so they decided to use the earth as the 'yardstick'. The standard unit of length was to be the metre, which was to equal the distance between the North Pole and the Equator divided by ten million. The story of how the mathematicians of the day found this distance is fascinating and shows how mathematics may be applied to solving problems that were hindering the progress of society. The system referred to as the Système International d'Unités (SI), or more commonly the metric system, is related to the Greek word to measure—*metron*. Children often have trouble remembering the prefixes attached to the base units in the metric system, that is, milli, centi, kilo. For example, some children think that a millimetre must mean one million metres as they associate milli with million. The orgin of milli dates back to the Latin *mille* meaning one thousand (note the earlier reference to milestones). When used as a prefix, as in the case of millimetre, milli is used to indicate 'one one-thousandth'. This does not explain why one million does not mean 1000. The mille indicates 1000 but the suffix 'on' alters the meaning to big thousand.

Australia owes much of its heritage to the people who travelled here by ship. It is only in recent times that the aeroplane has taken over as a popular means of transport. As an island nation Australia is still reliant on sea transport for carrying goods to and from its shores. Consider the measurement practices used and developed on ships and the words that have crept into our language as a result. Jules Verne, for example, wrote a book called *20 000 Leagues Under the Sea*. Is it possible to travel 20 000 leagues under the ocean? What is a league? How did navigators establish the position of a ship on the ocean? What role did time play in establishing longitude? These are questions that may be integrated across the curriculum, for example into society and environment, in science lessons as well as mathematics. Consider how sailors would measure the depth of the ocean. Basically, they would let out a weighted rope over the side of the ship and wait until they felt it reached the bottom. When they hauled the rope back up they would stretch it across their open arms from fingertip to fingertip. This measurement was called a fathom from a word meaning to spread out, which in turn came to mean the length of two arms spread out. Eventually, the fathom was standardised at six feet long. Over the years the word has evolved in part to mean 'to get to the bottom of'. Children could be asked to collect data to determine the mean fathom length for their year and other school years. Children could be asked to explore the relationship between height and arm span and plot the data on a scattergraph. These data could be used to explore the problems that may have occurred if a short sailor took fathom readings and a tall sailor took fathom readings. There are many other ship-related mathematics that might be explored such as how was the speed of a ship measured and how time was related to the calculation of longitude, which in turn may lead to a discussion of time zones.

The historical and cultural aspects of mathematics add to the richness of the mathematical experience for children. The development of units of measurement and devices for measuring provides teachers with interesting stories to tell but also more importantly may be used to prompt mathematical investigation.

Measurement across the primary years

The following extracts from the *Numeracy Benchmarks Years 3, 5 & 7* (Curriculum Corporation, 2000) provide a guide as to minimum expectations for students in Years 3, 5 and 7. Note the emphasis placed on developing measurement language in the early years, the focus on estimation in the middle primary and the use of standard measurement tools and units in the later primary years. The Year 7 benchmark makes reference to measuring area by counting squares and volume by counting cubes but not the use of formulas to calculate area and volume. Formulas are best left to the domain of the secondary school. Students might, however, investigate the calculation of the area of rectangles by observing patterns when counting squares. Likewise areas of triangles may be related to the area of the associated rectangle.

Year 3 benchmark

Students notice features of objects such as their length, capacity or mass (weight). They compare objects directly and use words like *heavier* or *longer* to describe these objects, and sense the need for accuracy when measuring length. They are beginning to use digital and analogue clocks to tell the time. The students can sequence everyday events and use calendars to locate information.

As part of the minimum set of achievements in Measurement and data sense, year 3 students are expected to:

- use language such as shorter than, holds more than, heavier than and later than to describe or compare objects and events
- decide whether to focus on length, capacity or mass (weight) when comparing two objects or quantities (eg do not go on height alone to decide which of two containers would hold more)
- choose a uniform unit such as straws to estimate, measure and compare lengths (eg say that the verandah is wider than the doorway because it takes 5 more straws to measure its width)
- use direct comparison to work out whether one container holds more than another (eg pour water from one container into the other to decide)
- tell the time in hours and minutes on digital clocks, and hours and half-hours on analogue clocks
- sequence regular activities during a day; say the days of the week and months of the year in order; and find dates on calendars

Year 5 benchmark

Students are developing skill in using metric units to estimate, measure and compare length, capacity and mass (weight). They are able to use clocks to tell the time accurately and use basic timetables and calendars to find information.

As part of the minimum set of achievements in Measurement and data sense, year 5 students are expected to have achieved the year 3 benchmark standard and, in addition, are expected to:

- know that a group of objects can be ordered differently depending on whether the objects are arranged according to length (or height or width), capacity or mass (weight)
- estimate, measure and compare lengths using metres and centimetres
- estimate, measure and compare areas by counting squares on a grid
- estimate, measure and compare capacity using litres (and have some awareness of millilitres)

- estimate, measure and compare mass (weight) using kilograms (and have some awareness of grams)
- interpret measures expressed in decimal form (eg know that a measure of 1.5 L is one and a half litres; know that a jump of 2.95 m is nearly 3 metres)
- tell the time in hours and minutes on analogue and digital clocks
- sequence times in a day, and interpret simple timetables and calendars to find information

Year 7 benchmark

Students use common measuring instruments to measure and compare length, capacity, mass (weight) and temperature, and estimate using a range of standard units. They measure area and volume by counting units. They recognise small and large standard units (eg kilometres, tonnes, millilitres, millimetres), and choose appropriate units for a task. They tell the time accurately using digital and analogue clocks, and read timetables and calendars. As part of the minimum set of achievements in Measurement and data sense, year 7 students are expected to have achieved the year 5 benchmark standard and, in addition, are expected to:

- make reasonable estimates of different measurements—length, capacity, mass (weight), temperature and time—and of area by comparison with a square metre
- use standard measuring instruments and read scales to the nearest graduation (eg tape measure, measuring jug, thermometer)
- use standard units to measure length (ie millimetres, centimetres, metres and kilometres), capacity (ie millilitres and litres), mass (ie grams and kilograms) and time (ie seconds, minutes and hours)
- measure area by counting squares and part squares, and volume by counting cubes

The ideas and concepts embedded in the benchmarks are expanded in the following section. The focus is on the development of measurement concepts and the physical act of measuring.

Measurement topics in primary school mathematics

The sequence used to develop proficiency in measurement involves first knowing what is being measured, that is, the development of awareness and knowledge of the attribute to be measured. Different examples of this attribute can then be compared and ordered, and then can be measured using non-standard units, leading to the use of standard units.

This sequence is quite general and can apply to the measurement of any attribute. Indeed, one of the broad aims in teaching about measurement is to help children develop an overall schema for coping with any measurement situation.

Some attributes are more important than others in preparing children to cope with life needs in measurement and in later mathematical experiences, and these require specific detailed attention in school curricula. Some of these relate to spatial or geometric experiences, for example length, area, volume and angle.

Others relate to common physical measures, for example mass, time, temperature and velocity or speed. Primary school syllabuses generally include all of the above except velocity and similar rate measures. Money and value is commonly included as a measurement category also, together with probability (the measure of chance) and statistics (the measure of data characteristics).

Geometric measures

Length

This one-dimensional concept is related to the geometric concepts of direction and line. Length measures need to be investigated in the many practical situations in which they occur, for example as length, width, height, depth, thickness or nearness of objects. Most initial experiences relate to straight lines, but distances along curves, for example in measuring the length of a curved driveway, are also relevant. Distances around plane shapes, both straight and curved, are of particular practical interest and lead to measures of perimeters.

Area

This two-dimensional concept is related to the geometric concept of region enclosed by a plane shape. Investigations of the amount of surface of plane regions help identify the attribute that can then be compared, measured using non-standard and standard units, and then investigated for regular shapes by the informal development of area formulas. The use of formulas to calculate areas of common shapes is the appropriate final stage of the learning sequence, and not the beginning stage as has often happened in schools in the past. Calculations also provide a medium for developing relationships between the standard units for length, area and volume.

Volume and capacity

The three-dimensional concept of volume refers to the amount of space taken up by the solid shape. Two different practical situations need to be experienced by children when they learn about volume and its measurement. One relates to investigations involving 'how much space a three-dimensional object takes up', leading eventually to measures of volumes of prisms, cylinders and the like in units derived from length units, for example cubic centimetres and cubic metres. At primary school level the term *volume* is usually used to refer to specific measures of these kinds. The term *capacity* is used to refer to the same three-dimensional attribute, but only when it refers to measures of substances that can be poured, that is, fluids. Hence, most school experiences involve investigations of amounts of liquids or solid analogies like sand, sawdust, sugar and the like which can be poured between containers of different shapes and sizes. Occasionally gases may be involved also (for example, in investigating how much air a balloon holds), and experiences lead eventually to the use of the alternative standard units like litre and millilitre, and their relationships to the standard volume units like cubic metre.

Angle

The measurement experiences associated with the geometric notion of angle involve comparisons of the sizes of two-dimensional corners and measures of the amount of turning or rotation from one ray of the angle to the other. These measurement experiences tend to be more closely associated with appropriate geometric and spatial experiences than most other measurement topics. When dealing with angle there can be a tendency to focus on measuring angles with a protractor rather than on developing the concept of angle as a measure of turning. Rotagrams (consisting of a plastic disk that rotates inside another piece of plastic) are a useful device for comparing the size of angles but to measure an angle requires the use of a protractor. Measuring an angle involves aligning the protractor with the vertex and sides of the angle and then reading the correct scale. Students who have developed a feel for angle and who have established benchmarks such as right angles and straight angles are less likely to misread the scale on a protractor.

Common physical measures

Mass

Mass is the measure of the inertia of an object. Inertia is the amount of response to a force or push applied to the object, whether the push is trying to move the object or stop it. If the same push is applied to several stationary objects, the smallest masses experience the largest effects and vice versa.

Clearly, mass may not be proportional to volume. A small piece of lead may have a larger mass than a big piece of foam plastic. This suggests that one of the important issues in teaching about mass is to distinguish it from volume when the term 'amount' is used loosely.

Weight is the force that gravity exerts on an object. While the mass of an object is constant, its weight can vary from place to place. For example, an object weighs less on a high mountain than at sea level because it is further from the earth's centre of gravity, and an object weighs about six times more on earth than on the moon.

Teachers need to be aware of the distinction between mass and weight for several reasons. The terms mass and weight are loosely used in everyday speech to mean the same thing. It is common to refer to the process of 'weighing' an object to find its mass. Masses are usually compared by comparing their weights (for example, when using a spring balance) or the effects of their weights (for example, their moments or turning effects about a pivot point when a beam or pan balance is used). Also, children's early perceptual experiences with mass involve making judgements as a result of 'hefting', that is, experiencing muscle tension or the downwards pressure on hands when objects are held or lifted.

In this case, as in performing 'weighings' with balances, the weight of the object is used to provide evidence about its mass. This is valid because the mass of an object is proportional to its weight at any particular location, and desirable because it provides experiences consistent with weighing experiences in everyday life. Nevertheless, the Australian Metric Conversion Board recommends that mass be used correctly in all situations, and school syllabuses generally reflect this correct use of terminology.

Time

While telling the time may appear to be a relatively simple skill to teach, the measurement of time is in reality much more complex. Time involves much more than reading a scale, such as a clock face or a digital display, it also involves developing an understanding of the passage of time. To a child waiting for a special event such as a party, time appears to drag, but speeds up once the child is fully occupied at the party. Some children who watch the clock develop a feeling that time moves more slowly as the minute hand makes its way from the six to the twelve, in other words the second half-hour. This belief possibly stems from the fact that in most schools lunch time is around twelve o'clock (literally of the clock) and school finishes around three o'clock, hence time seems to drag just prior to these events.

Rather than provide a series of formal lessons on teaching children to 'tell the time' it is recommended each classroom have two or more clocks. One clock should be analogue and the other digital and they should hang next to each other. Throughout the day the teacher should make reference to the time. For example, a teacher might state 'it is nearly lunchtime, it is almost twelve o'clock' while pointing to the clocks. The children may then compare how the time is indicated on the various types of clock.

Another issue associated with telling the time is the way it is stated. A digital clock showing 9.15 is often referred to as showing 'nine fifteen' whereas an analogue clock showing the same time is often referred to as 'quarter past nine'. This reading of the clock assumes a basic knowledge of fractions because all the student sees is the hour hand pointing to nine and the minute hand pointing to three. Once the time advances a further five minutes a different expression, 'twenty minutes past nine', is used to state the time. Imagine the confusion when at 9.45 an adult states the time is quarter to ten rather than three-quarters past nine. Essentially reading a clock face is a scale reading exercise and as such can become quite complex and all the more so because time is not metric, involving a range of bases. To confuse matters even further children see times for races recorded to hundredths of a second when they have been taught there are 60 seconds in a minute and sixty minutes in an hour. Children will meet both forms of clock and both types of reading hence the suggestion to use more than one clock in the room and discussion of the various ways to read the time.

Young children may experience difficulty understanding the passage of time. Referring to decades and centuries can be problematic when a child is only six or seven. To understand better the passage of time students should be given the opportunity to make timing devices.

Using shadow sticks and marking the shadow onto paper at 30-minute intervals of the day will help children link time to the natural rhythms found in nature. Building water clocks and sand-timers from plastic drink bottles will also allow children to observe the passage of time as the sand runs from one container to the next. Making a candle clock and watching it burn along with observing a pendulum swing will show the possibility of integrating mathematics and science, while at the same time allowing children to observe the passage of short periods of time. While designing and making these devices the opportunity to discuss the history behind them will arise. Further background on the topic of clocks may be found by conducting an Internet search based on the key word 'horology', or the science of timekeeping.

Children also need to develop their ability to sequence events in time. This might include activities such as describing the sequence of events in a single day. For example, children may describe waking up and getting out of bed, having breakfast and then brushing their teeth. Simple activities such as cutting up cartoon strips and putting them back in order will help children to sequence events. The children should be encouraged to explain and justify their order or events.

As the children gain experience they will develop techniques for estimating short periods of time. To gauge whether one minute has passed many children employ unusual counting techniques such as 1 elephant, 2 elephants, 3 elephants . . . As part of the daily routine in a classroom reference should be made to the calendar and the date. At the appropriate dates and times teachers could refer to specific events such as the shortest and longest day, phases of the moon and the changing of the seasons. The study of time and the calendar itself has filled the pages of many books. It is recommended that teachers keep a current calendar hanging up in the classroom as well as the calendar for the previous year and the following year. In this way older children can be encouraged to investigate questions as to why their

birthday is not on the same day each year or why the date for Easter changes from year to year. When studying the calendar the opportunity arises to include some cultural aspects of mathematics such as the Jewish calendar (based on the lunar cycle) or the Buddhist calendar.

Time is unique in that it cannot be perceived through sensory experience in the way other measurement concepts can. The effects of passage of time can be perceived through changes in natural phenomena, such as day and night, the changing lengths of shadows, seasonal changes in climate, vegetation and animal behaviour or by the use of mechanical devices, through the audible ticking of a clock, the visible movements of clock hands or the changing of digital clock numerals. However, people are unable to move through time at varying rates or in varying directions of their own choosing. A period of time cannot be isolated and stored for later examination or comparison. Contrasting experiences, for example, when a time change occurs twice as fast, cannot be made available. In summary there are three main ideas to be developed in time

- concept of duration of time
- knowledge of sequence and order in time
- clock face reading

Temperature

This is the measure of how hot or cold things are. Children can perceive large differences in temperature and are commonly exposed to temperature variations for objects like hot food and cold drinks, and locations such as sitting near the fire to get warm, and times (for instance, it is hot today). However, human perception of temperature change can be distorted by an immediately prior experience, and small differences in temperature may be difficult to judge without the use of appropriate thermometers.

An additional complication in learning about temperature occurs because of the choice of numbers used to describe the reference points of boiling and freezing temperatures for water. Arbitrarily locating the zero of the temperature scale at the freezing point of water has resulted in the need for the use of negative numbers (numbers smaller than zero) to describe temperatures that are relatively common in many locations. Arbitrarily assigning the number 100 to the boiling point of water has resulted in the unit temperature size (1 degree Celsius) being such that large negative numbers are required to describe temperatures of very cold situations. These outcomes have implications for the number skills required to cope with determinations of temperature changes, for example what is the change from –10 degrees Celsius to 13 degrees Celsius?

Some ambiguity may occur when discussing temperature. The kelvin is the recommended metric unit for measurement, but most reports of weather will express the temperature in degrees Celsius. A brief consideration of the history of naming temperature intervals will assist in the understanding of temperature units.

Anders Celsius (1701–1744), a Swedish astronomer first proposed dividing the temperature between freezing and boiling point into 100 steps, with zero representing the freezing point and 100 the boiling point. Originally this scale was called the Centigrade scale—literally centi (100) gradus (steps). With the introduction of the metric system and the use of the prefix centi to mean one-hundredth, some ambiguity in the use of the term Centigrade developed. In 1948 a decision was made to change the name from degrees Centigrade to degrees Celsius and use the symbol °C. This was not the end of the matter as in 1954 the definition of the Celsius scale was changed and 0°C was set to 273.15 degrees kelvin. The kelvin scale became the key unit of measurement. Further changes were made in 1967, when the term degree was dropped from the kelvin scale. From this point on when stating a temperature using the kelvin scale it was simply called kelvin (using a lower case k) and the degree symbol was no longer used.

For the most part children at primary school will be more familiar with the term degrees Celsius, rather than kelvin. They may, however, come across temperatures given in kelvin in science. Some children may have heard of a temperature scale known as Fahrenheit. This is a non-metric temperature scale. Some books show formulas for Fahrenheit and Celsius conversions. This is not appropriate content for primary school.

Other measures

Value and money

Money provides the units used to measure the value or cost placed on objects. Formal computations involving money use precisely those number skills required to cope with operations on whole numbers and hundredths, and therefore provide a motivational context in real-world situations that are highly relevant to all children for learning the number operations involved. Nevertheless, money is not just a topic for using number in important life situations. The development of the concepts of value and worth, trading and exchanging, and hence equivalences of value can be developed using the general approaches to the teaching of measurement topics, with their emphasis on manipulation of materials and active child involvement in, for example, class 'shop' activities. Children need to develop the ability to recognise coins and to handle money. It is recommended that real coins be used in preference to plastic coins. Experience with giving change in the class shop will lead to discussion of subtraction as building up.

Probability

Probability involves the measurement of *chance*. Chance deals with the concepts associated with randomness, so probability provides a measure of how likely it is that particular events will occur. Many life situations involve chance and randomness. Adults, for example, commonly make decisions about health and property insurances and many participate in a wide variety of legal gambling activities. Children commonly play games that use dice and spinners and enter competitions where luck determines the winner. Everyone is interested in the likelihood of fine weather for a coming event. Yet there are many misconceptions and poor intuitions about probable situations, and many vain expectations of impending luck. Hence, it is important that school mathematics programs provide experiences that lead to an understanding of random events and the development of processes used to measure their probabilities, using both experimental, or empirical, methods and analytical, or theoretical, methods. Chance processes are dealt with in a separate chapter.

Statistics

Statistics deals with ways of collecting, sorting and summarising information or *data* in order to make sense of large quantities of it. This aspect of statistics is called data handling or descriptive statistics. A second aspect, termed 'inferential statistics', deals with procedures for making statistical inferences, that is, generalisations or predictions about a population based on information obtained from a sample of that population.

Both aspects of statistics involve common experiences in everyday life. Large quantities of data are summarised in the consumer reports, surveys and the like that are discussed almost daily in the media. The voting intentions of the electorate are inferred from the samples of electors investigated by pollsters. Corporations test the effectiveness of their products or their advertising campaigns by collecting and processing data about the reactions of samples of consumers and using the results to draw conclusions about the total population. Hence, for effective informed participation in society today, all adults need to be able to make sense of the huge amounts of data presented daily in reports, surveys, advertising and the like. These statistical concepts are therefore considered to be an important component of numeracy and hence an important basic skill. School mathematics programs should provide extensive experiences in data handling that encourage children to collect, organise, represent and interpret data from situations of interest to themselves in order to answer questions that they and others have asked. These activities should include opportunities to develop skills in measuring and describing some of the basic properties of collections of data including averages or central tendency and spread or variability.

School mathematics programs often treat probability and statistics as topics distinct from the major strands of number, space and measurement, in part because they are seen to involve quite different concepts and processes. Yet probability and statistics do draw on skills and understanding from each of the other areas, and they do involve measurement processes in particular that are central to their study. Chance, for instance, is measured in terms of the probability of an event occurring. Similarly, a set of data has important properties that can be measured once it has been collected and organised. It is therefore appropriate and convenient that probability and statistics are here discussed within the overall framework of measurement. Since 'Chance and data' was proposed as a discrete strand separate from measurement in *A National Statement on Mathematics for Australian Schools* (1991), many school mathematics programs have increased emphasis on this aspect of mathematics. These topics are discussed in more detail in Chapter 7 to reflect this increased importance as discrete strands in primary school mathematics.

Teaching measurement in primary schools

Processes and attitudes

In learning about measurement, children merge important sensory and perceptual ability with the cognitive aspects of geometry and number. Therefore, to develop beyond the beginning stages of perceiving the attribute to be measured and making direct comparisons between examples of the attribute, appropriate number skills need to be available. A full understanding of numbers, including rational counting, must precede the development of more formal measurement activities using non-standard then standard units. As number competence expands from the use of whole numbers to common and decimal fractions, measurement activities in units that require fractional representations become feasible and desirable, and choices of appropriate units become realistic expectations.

Similarly, full understanding of computation for appropriate number operations must precede the introduction of activities to determine areas, volumes and perimeters of shapes, calculations of total costs, temperature changes, times for journeys and the like. Clearly, there are appropriate geometric experiences relative to shapes and their properties that must also precede measurement activities relating to attributes of these shapes, such as areas of triangles and quadrilaterals and volumes of prisms and cylinders.

Although the development of measurement activities depends upon prerequisite knowledge of concepts and skills developed from number and geometry, there is a reciprocation process from measurement back to number and geometry in several ways. Measurement activities provide very appropriate contexts for applying the number skills and geometric concepts to highly motivational activities associated with real life functioning and problem solving situations. Measurement experiences therefore provide ideal settings for demonstrating the usefulness and importance of mathematics in everyday human activities and therefore in each child's personal life. Skill in correctly measuring cooking ingredients has an immediate practical outcome in the quality and taste of the product to be eaten, while reading a clock accurately is important if a child is not to miss a favourite television program. Correctly measuring the cloth used in sewing a garment or the piece of timber used to construct a tree-house requires the use of measuring skills that are important in producing a desired outcome. These circumstances all provide opportunities for a child to feel satisfaction from having done a good job of measuring. Hence, school experiences in measurement should result in at least two affective outcomes:

- children should derive personal satisfaction and self-confidence from the actual process of measuring and enjoy being able to measure for themselves
- children should appreciate the role of measurement in their own lives and in society generally

Because success in measurement activities depends in part on number and spatial competencies, these affective outcomes of personal satisfaction and enjoyment achieved in measurement may extend generally to the mathematics of which measurement forms a part, and the usefulness of mathematics in life is further emphasised. Its importance in society can be given additional emphasis as children learn about the crucial role that measurement plays in great scientific and technological achievements, and as they use their measurement skills in art and physical education and integrate their mathematical skills in science and social studies.

Thus, measurement in a school mathematics program is justified as an important topic in its own right, but also because it provides opportunity for using number and spatial skills in practical situations that offer opportunities for personal satisfaction and for the appreciation of the importance and usefulness of mathematics in general.

Approaches to teaching measurement

Learning outcomes

While each of the specific measurement topics has outcomes pertinent to itself, there are general outcomes that apply to measurement as a part of the school mathematics program. These include:

- **Problem solving**—children should be experienced in using measurement concepts, skills and processes to solve problems from a variety of contexts.
- **Application**—children should apply their measuring processes to real-world situations and common life experiences.
- **Estimation**—children should develop estimation skills and use estimation as the first step in any measuring process.
- **Interrelationships**—children should become aware of interrelationships between measurement and other topics in school mathematics or other school experiences outside mathematics classes, and between the specific attributes investigated in measurement, for example length and volume.
- **Appreciation**—children should appreciate the role of measurement in their lives and in society.

Within the broad study of measurement itself, some specific measurement outcomes apply to all topics. These tend to give structure to the sequence of classroom experiences provided for children and should allow them to:

- perceive the attributes or properties of objects to be measured
- compare objects that have a similar property
- measure properties first in non-standard units
- measure properties later in standard metric units
- use measuring instruments effectively and accurately
- read measuring scales correctly
- understand that all measures of continuous quantities are approximate

- apply their measurement knowledge and skills in drawing and constructing objects of given sizes
- appreciate the role of measurement in their own lives and in society
- experience the actual process of measurement and derive personal satisfaction from it

These measurement outcomes are now applied in developing the teaching and learning sequences for each measurement topic in turn. Length measurement is discussed in detail, to establish the fine detail for the sequence used at each stage of the development. This provides an overall model that may be adapted for any other measurement topic. The other topics are discussed in a condensed summary form.

Length

Perceiving and identifying the attribute length

Early length experiences must develop an awareness of what length is, and of the wide variety of words used in discussing lengths. The following list of some 'length' words illustrates the extensiveness of the vocabulary that must be used and experienced:

long	longer	longest
short	shorter	shortest
wide	wider	widest
narrow	narrower	narrowest
high	higher	highest
low	lower	lowest
thick	thicker	thickest
thin	thinner	thinnest
deep	deeper	deepest
shallow	shallower	shallowest
tall	taller	tallest
broad	broader	broadest
near	nearer	nearest
far	farther	farthest

Also, location and proximity words like the following have relevance in developing distance vocabulary:

above	below	up	down
close	away	distant	
next to	beside	as long as	same length as

These words should be used naturally in class discussions at every opportunity. They gradually extend children's vocabulary to more precise words than the general use of 'big' and 'little'. As the words are used, attention is being directed to the attribute of length. Awareness of the property then grows as comparisons of lengths are made at the next stage.

Appropriate activities to help children develop awareness of length and to use length vocabulary include the following:

- Draw a tall man and a short man, some long lines and some short lines, some wide things and some thin things.
- Cut string, paper strips, ribbon to the lengths of people or objects and stick them on walls. Discuss the lengths using appropriate language, for example the door is high, the table is long but the book is short, the room is wide, Mary is tall, Jimmy is short, play in the shallow end of the pool, that is a thick sandwich.
- Sort objects by length. Name objects that are short, tall, wide, narrow, thick.
- Distances around things should be included in these early experiences. For example, tie string around the parcel, it needs a long piece; run around the oval; walk around the block; cut strips of paper to fit around things and body parts (for example, long strip to fit around my head, a short strip to fit around my wrist).

These activities and discussions soon lead naturally to comparisons of lengths and associated language. Comparisons of the attribute help to reinforce the understanding of it and then prepare for measurement activities using units at the following stage.

Comparisons of lengths

Comparing lengths of objects is a natural second stage to the perception of length. Activities that involve perception of length quite often involve comparisons of lengths as well. For example, when children are discussing things that are long or short, they soon begin to talk about something being longer or shorter, such as 'I made a long train but Maria's is longer'.

The vocabulary of comparison therefore develops almost simultaneously with that of the attribrute. This applies to all attributes, for example 'it is hot today; it is hotter than yesterday'. Comparison activities therefore help to consolidate the growing awareness of the particular attribute as well as preparing for more formal measurement activities using units.

Comparison activities also constitute a measurement process in their own right. People often measure in real life without using units. Direct or indirect comparisons often provide sufficient measurement information to solve the immediate problem. For example, when moving furniture, particularly through a doorway, people often use direct or indirect methods that do not involve the use of units.

It is important that children experience activities in which they compare and order attributes to extend their understanding of the attribute and to introduce them to informal measuring processes. The sequencing of these experiences would generally involve three successive stages. In the first stage, examples of objects for which the attribute is the same or different may be investigated. In the second stage, activities involve comparisons of two examples such as 'which is longer, thicker, shorter?'. In the third stage the activities require three or more objects to be seriated or ordered for the given attribute. For example when comparing the heights of three children the following statements might be made. Amee is taller than David but shorter than Adrian. Adrian is the tallest and David is the shortest.

One early comparison activity that helps to consolidate knowledge of the attribute involves making collections of two sets of objects, one in which the property or attribute is constant and other attributes differ, the other made up of objects that are identical except for

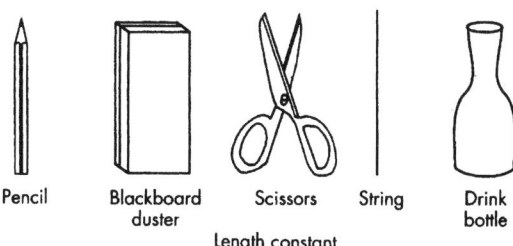

Pencil Blackboard Scissors String Drink
        duster   bottle
Length constant

the attribute under discussion. Children are then asked to examine each set and identify what is the same or different for the objects in each set.

For length comparisons, a set of pencils, pens, books, scissors, pieces of string, ribbon, paper strips and other objects all of the same length could be compared by aligning them so that the constant length can be readily identified.

A second set might include strips of paper of identical colour, texture and width but of differing lengths. Teacher-prompted discussion directs attention to the similarities and differences first to focus attention on length and then to make comparison of this property for different objects.

Early comparison activities should involve two similar objects that differ only in the attribute of length, but with a considerable length difference so that they can be distinguished by eye. The objects should initially be aligned correctly at one end to facilitate the length comparison. Later experiences should demonstrate that accurate length comparisons are possible for any positions of the objects. Comparisons of objects for which the length differences are less obvious then reinforce the need to align the objects correctly at one end.

Examples of appropriate activities for comparisons of lengths include the following.

Direct comparisons
- Is your pencil longer than mine?
- Who is taller, Jim or Bill?

- Which bean plant is higher?
- Which is the longer (wider, thicker) of two sticks, straws, pencils, ribbons, pieces of string, paper strips?
- For a given stick (piece of string, rope) find things that are longer, shorter, as long as it.
- Estimate which tree looks taller, which hole is deeper, which building is longer.
- Look at a stick then put it away and find something that is about as long as the stick. Use the stick to see if you are correct.

Indirect comparisons

Compare lengths through the use of an intermediary, for example:

- Use string, paper strips, arm spans, body height to compare the length of the table with the space between the windows; the width of the table and width of the door; the height of the box and the distance between the shelves.
- Use a stick to compare the depth of the pool and the height of the netball ring.

Comparisons of distances around things

- Use string, paper strips, ribbon to compare distances around head sizes, wrists, biceps, feet, cans, bottles.
- Roll two cans (bottles, bicycle wheels) one turn each to see which rolls further.

Investigate and discuss

> **1** Which is longer, the height of the glass or the distance around the top rim? Are all drinking glasses like this? Are cups, mugs, jam tins similar?

Ordering lengths

When children can compare two lengths and consistently report the result correctly, they should be given opportunities to order three lengths using a variety of situations and objects. The process of ordering three or more lengths is not a simple extension of comparing two lengths, because a transitivity process is involved, that is, if one length is longer than another and this length is longer than a third length then the first length must also be longer than the third length.

Activities involving children's personal dimensions generate interest, for example arrange groups of children from shortest to tallest, mark children's heights on the chalkboard, write names with marks and compare the heights. Appropriate variations that provide relevant experience in seriating lengths consolidate understanding of the comparison process in preparation for the introduction of comparisons with units. They also relate the seriating process to ordinal numbers at various levels of sophistication, for example 'Is the fifth stick the longest?', 'Show me the second longest stick'. A word of caution—some children can be sensitive about taking personal measurements (especially mass). Teachers need to be aware of this issue.

Measuring lengths using non-standard units

There are strong arguments for introducing formal measuring activities through the use of non-standard units. Some of the reasons were summarised in the overview of the measurement process. As well as establishing most of the measurement principles through the use of natural enjoyable experiences in the child's immediate environment, non-standard measures also establish measurement as a commonly used problem-solving process. Not having to rely on the ready availability of a standard measuring instrument is common in real life. For example, the question, 'Will this heavy table fit through the door?' requires a degree of flexibility in problem solving that is not necessarily fostered by having children only measure lengths with a ruler or tape.

Using non-standard units also gets children started on measuring quickly and easily. Not having to memorise new names for the unfamiliar teacher-imposed formal units allows children to focus primarily on the process and hence to begin measuring using units that are common and familiar. Some suggestions relating to the commencement of measuring using non-standard units will now be discussed.

Use body parts

Parts of the body provide interesting units for introductory use. They also provide opportunity later to introduce discussion of the history of measurement, for example the pre-metric inch standardised from various finger widths, the foot from human feet lengths, the ancient cubit as the distance from elbow to fingertip, the yard from various pace lengths. Children should therefore be encouraged to measure a wide range of objects using as units their own body parts as appropriate—width and length of finger, palm width, hand span for short lengths, and foot, pace, stride, arm length, arm spans for longer distances.

pace   fathom

Use other units

Other readily available objects may then be used as units, for example pencils, blocks, paper clips, straws, toothpicks, edges of paper sheets, lines on writing paper, sugar cubes, coins, as well as strings, ribbons, sticks and dowels of various lengths. The arbitrariness of choice of units is demonstrated when different children use different units to measure a length, each with comparable success. However, when different lengths are being compared, use of different unrelated units for each object makes valid comparisons difficult. For example, the board is 15 pencils long; the table is 17 hand spans long. Which is longer? These kinds of discussions can be used to lead to the need for a common unit in such cases, for example 'let's measure them both in Mary's hand spans'. Appropriate activities can be used to reinforce this principle, for example find two fixed objects in the room that are as long as each other. How do you know you are correct? Which is longer, five pencils or five paper clips? These activities also provide opportunities to establish the generalisations that, for a given length, the smaller the unit the bigger the number, and vice versa.

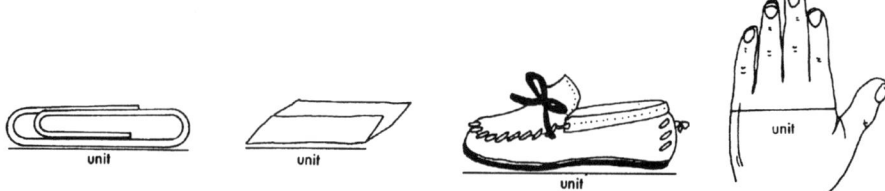

Estimation

From the earliest of these experiences, children should be encouraged to think and estimate first, then measure. Initially, these estimations are no more than guesses, but guessing establishes a prior commitment to the measurement activity and the resulting measurement is a meaningful result that provides a check on the guess. As familiarity with sizes of the units grows, these guesses become somewhat more educated and hence the teacher's request becomes one for the child to estimate first, then measure:

> Estimation activities are an early application of number sense; they focus students' attention on the attributes being measured, the process of measuring, and the value of referents. Thus estimating measurements contributes to students' development of spatial sense, number concepts, and skills. Because precise measurements are not always needed to answer questions, students should realize that it is often appropriate to report an estimate as a measurement. (*Principles and Standards for School Mathematics*, 2000, p. 105)

Tolerance of error

At about this stage, discussion can begin to focus also on the accuracy achieved in measuring, and the tolerance for error that distinguishes most measurement activities from children's experiences with number, where exact answers to exercises are usually expected and valued. Measures of length where exactness is not possible because the length measured lies somewhere between successive integral numbers of the unit chosen provide rich topics for discussion. The focus of these discussions should be on sense making. For example it would make little sense to round a 2.3 m length of wood to 2 m because of following a 'rule' to round down as you may not have enough wood to finish the job.

Reasons for choosing whether to round to the nearer integer or to estimate a fractional part need frequent discussion, so that the influence of the practical circumstances of the measurement process on the degree of accuracy sought might be identified. These more difficult aspects of measurement need to be broached frequently over considerable periods of time to allow children the opportunity to appreciate them and develop some facility in their use.

Perimeter

As well as investigating lengths, heights, thicknesses and distances to things, measures around things should be included naturally in these activities to maintain the development towards establishing perimeters as one kind of length measurement. Children often confuse perimeter and area measurement. Examining the origin of the word perimeter will assist in children's understanding of measuring perimeter. The word perimeter is made up of the words peri meaning 'around' and metre from the Greek 'metron—to measure', so to find the perimeter means to measure around. Consider how children perceive the word circumference and how an understanding of the origin of this word would assist children attempting to find the circumference of a circle. Irregular and curved shapes can be investigated by the use of string and the like, and by rolling appropriately sized circles like coins, bottles and bicycle wheels around the shapes.

Measuring lengths using standard units

Establishing standard units

When children can measure lengths efficiently and effectively using non-standard units, they are ready to move to the use of standard units. The motivation for moving to this next stage will probably occur naturally as a result of discussions of measures obtained using different non-standard units for the same length.

The potential for confusion becomes evident when two people report their results obtained when they used different non-standard units to measure the same length and others attempt to make rational judgements about which is correct. Establishing the need for consistency in the use of units to facilitate communication about measurements leads naturally to the introduction of standard units that permit ready communication in all possible circumstances, for example in everyday life, in industry, business, science and technology, and at local, national and international levels.

Children's school measurement experiences must aim to produce the same kind of understanding and competency in the use of appropriate metric units as is expected from their number experiences. It is not enough merely to memorise the names of units, their symbols and conversion ratios, rather children must seek to achieve a total internalisation of each unit, so that they readily conceptualise its size, they use it confidently without having to refer back to some prior referent, they comfortably estimate using the unit, and they readily convert between units when appropriate. The following stages assist in building this facility.

- **Use a common unit**—class experiences establish the convenience and advantages of all people using a common unit; children identify, construct and use a common unit.
- **Establish the standard unit**—the standard unit is constructed and modelled in as many ways as possible, and its relative size is established; simple measuring devices are calibrated for this unit.
- **Internalise the standard unit**—the unit is used widely to measure personal dimensions and commonly used objects, so that its size relative to familiar objects is established and internalised to the stage where it becomes a reference for estimating.
- **Use the unit with facility**—the unit is used extensively in all possible situations by routinely estimating then measuring, with gradually increasing accuracy.

Standard length units

The usual sequence used in primary schools is to introduce the *centimetre* then the *metre*. These units cope with most classroom length situations. Note, however, it is common for builders to use millimetres when drawing plans and for clothes designers to use millimetres. Later, the kilometre is introduced for measuring very large distances, and the millimetre for very small lengths.

Some school programs introduce the metre first, in part because it is the fundamental unit, and then the centimetre as a convenient unit derived from the metre. The following discussion could be modified in obvious ways to apply to this approach.

The centimetre is commonly introduced as the first formal unit because it is small enough to measure common objects (books, pencils, desks, hand spans) to give a convenient number result, and the 'bits over' are not seen to be of great significance because of their small size. Hence 'about 7 centimetres' is quite an appropriate early measurement response.

The size of the centimetre unit can be established by modelling and constructing it, for example by cutting 1 centimetre pieces from drinking straws of different colours and threading them in groups of 10 on string. Commercially made centicubes provide convenient models also, and 10 centicubes of one colour can be clipped together to make an appropriate measuring stick, then sets of tens of different colours can be used to build longer measurers. Measurers of 10 centimetres can also be constructed by cutting strips from 1 centimetre grid paper and pasting these on cardboard; sets of these can be combined to form longer measurers, eventually up to 1 metre.

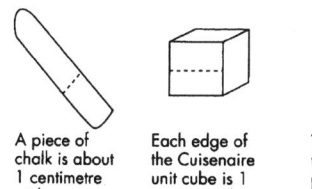

A piece of chalk is about 1 centimetre wide. | Each edge of the Cuisenaire unit cube is 1 centimetre long. | The width of the base of the pointer finger is about 1 centimetre. | The diameter of the head of a thumbtack is about 1 centimetre.

The centimetre unit is gradually internalised through wide use in measuring common objects including personal body measures, for example height, arm, hand, leg, foot, finger lengths and widths, and distances around waist, chest, hips, head, neck, biceps, wrist, ankle and the like. A word of caution, when taking body measurements be sensitive to those children who are obese and who may object to being measured. Children should be encouraged to identify personal body reference lengths for 1 centimetre, 10 centimetres and later 100 centimetres as these may be used as the basis for estimating lengths.

As familiarity with the unit increases, children should be expected to estimate routinely first, then measure. The situations requiring length measures should be extended to include internal and external diameters, depths, thickness of gaps and the like using, if available, a variety of measuring instruments. Commercially available kits include rulers, tapes, callipers, trundle wheels, gauges for measuring depths, thicknesses and internal diameters, micrometer screw gauges and vernier callipers.

micrometer screw gauge | callipers | vernier callipers | depth gauge | trundle wheel | internal diameter gauge | thickness gauge

The correct use of a ruler to draw and measure straight lines requires specific instruction and practice. The correct alignment of the zero of the ruler with one end of the line or object may not be obvious to children in cases where the zero does not occur at the end of the ruler. Other difficulties associated with the correct use of a number line also require attention, such as counting the number of 'spaces' or intervals rather than the marked points on the line.

Metres and millimetres are established and internalised through a similar sequence of experiences in making and modelling the unit, then using it to measure in appropriate circumstances. The kilometre may be established by marking out this distance in the school grounds and having children walk, jog and pace out this distance. Discussions of travelling experiences, for example 'how far to your home?', 'how far to the nearest town?' further consolidate this unit.

Applications

Applications of length measurements apply chiefly to measuring lengths in situations not met every day but still of practical importance. Some of these were suggested in the previous section, for example measuring internal and external diameters, depths and thicknesses or gaps needing gauges. Other applications involve measures of special lengths met particularly in geometric situations, for example bases, altitudes and hypotenuses of triangles, diagonals of polygons, edges of polygons and polyhedra, perpendicular separations of parallel lines, and the like.

Investigations to discover formulae for perimeters of common polygons and circles provide important applications of length measurements. Rolling circular objects or using string to measure circumferences and generalising the comparison of these circumference lengths with the lengths of corresponding diameters and radiuses lead to the discovery of the formula that links these measures.

If perimeters of rectangles and squares are investigated in the course of exploring perimeters of polygons generally, the identification of the generalisations that evolve (namely, measuring around a square involves four sides all the same length; measuring around any rectangle involves two 'lengths' and two 'widths') leads naturally to the identification of the appropriate perimeter formulas. It is not expected that children be taught formulas for calculating perimeters but rather develop an understanding of perimeter as measuring the distance around an object.

The issues relating to the teaching sequence for measurement topics were discussed in detail for length. Other measurement topics are discussed more briefly, but the justifications for the approaches used and the detail of the sequence adopted for length should be applied with appropriate modification to these other measurement topics.

Area

Perceiving and identifying the attribute area

Awareness of area as 'amount of surface' and appropriate language are developed by covering activities, for example wrapping parcels, colouring in, painting, dressing dolls, cutting and pasting one shape to cover another, covering books and greeting cards, covering desktops with paper or tiles of various shapes, using leaves to cover surfaces.

Comparing and ordering areas

- ◆ Directly compare mats made from carpet pieces, cloth, cardboard, paper. First use mat sets of identical shapes but different sizes (squares, circles, similar polygons) so that one compared shape fits inside the boundaries of the other. Initially, overlapping shapes should be used only where estimations are easily judged.
- ◆ Directly compare other areas where superposition is possible, for example book on book, book on table, paper on gift card. Provide ordering experiences by gradually introducing additional items into the sets of mats or objects.
- ◆ Indirectly compare areas by cutting paper to cover one surface then comparing the paper with the second. Include experiences where the paper must be cut and reassembled to test coverage of other surfaces.

- Provide tessellating experiences and use of dissection puzzles, for example tangrams, pattern blocks, jigsaw puzzles.

Measuring areas using non-standard units

- Use a variety of non-standard units (for example hands, sheets of paper, tiles) to measure areas by covering surfaces and counting how many are used.
- Tessellating (from *tessara* (Latin) and *tessares* to tile) with non-standard units (for example, stamps, triangles, quadrilaterals, regular hexagons) establishes the need to cover surfaces without leaving gaps and demonstrates the advantages of using arrays that can be readily counted by using multiplication, for example 6 rows of tiles, 10 tiles in each row, area is 60 tiles.
- A specific tangram piece can be assigned unit area and used to measure areas of shapes covered by the tangram pieces.
- Determine areas of shapes made on geoboards using rubber bands after nominating a square or triangle on the board as unit.

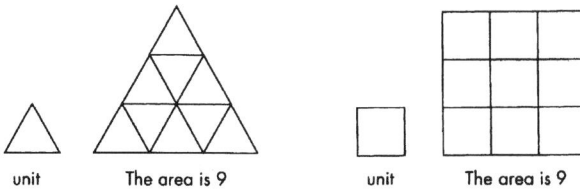

Measuring areas using standard units

- Construct 1 metre squares and 1 centimetre squares from paper, cardboard or cloth. Use sets of these unit tiles to cover the surfaces of regular and irregular plane shapes and count how many are used (estimating part units).
- Use geoboards with 1 centimetre squares and rubber bands to investigate shapes. Perform similar investigations using grid paper and transparent grid overlays. Use these materials and counting techniques to investigate areas of regions inside polygons and simple closed curves. Investigate areas of body parts and personal possessions, for example hands or shoe soles.
- Provide experiences to help conceptualise 1 hectare and 1 square kilometre, for example by researching areas of states and countries, exploring local parks and school grounds.

Applications

- Use investigations with grid paper, grid overlays and geoboards to establish the product formula for determining areas of rectangles and squares. Investigate areas of triangles, parallelograms and trapeziums by using dissecting and reassembling techniques that relate these to areas of equivalent or enclosing rectangles. Investigate areas of circles by dividing circles into a large number of congruent sectors and reassembling the sectors to form an approximate rectangle of length half the circumference and breadth one radius.
- As each area formula is 'discovered' and understood, use it to investigate areas of shapes of interest in the environment, for example which needs more space, a tennis court or a basketball court? By how much? Find the cost of surfacing the school oval.

The role of area formulas

Rather than teach children meaningless formulae or hang up charts containing formulae, relationships between the areas of common geometric shapes may be investigated. After establishing the area of an oblong by counting squares some children may notice that the area of an oblong may be calculated by multiplying the length measurement by the width measurement. Similar thinking may be applied to calculating the area of a square. Other geometric shapes may be related to these two shapes.

Area of a triangle

In order to lead children to the conclusion that the area of a triangle can be found by relating it to half the area of an associated oblong children need to be given the opportunity to explore a variety of triangle types and associated oblongs. For example, it is a fairly simple task to enclose a right angle triangle within an oblong and either count squares to determine the area of the oblong and the area of the associated triangle. Alternatively, the triangle may be cut out and superimposed over the remaining piece to show that the two pieces are the same area. A similar approach may be used for other triangles, although a little more cutting and rotating may be required to show that the areas of the cut out pieces is the same as the area of the remaining triangle piece.

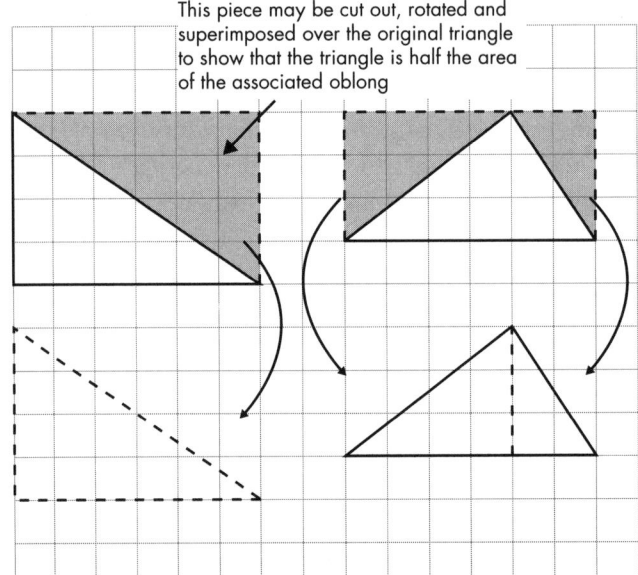

This piece may be cut out, rotated and superimposed over the original triangle to show that the triangle is half the area of the associated oblong

Triangles may also be dissected to produce oblongs with the same base, but half the height.

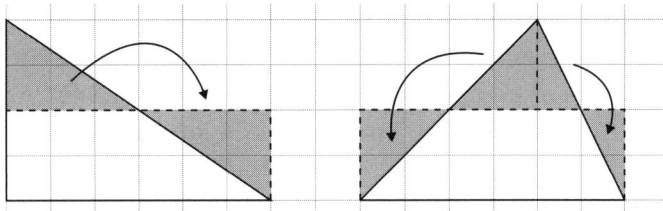

Area of a parallelogram

With a single straight cut and a simple translation a parallelogram may be transformed into an oblong, the area of which may be found. It becomes clear that multiplying the length of the parallelogram by the perpendicular height is similar to multiplying the length and width of an oblong.

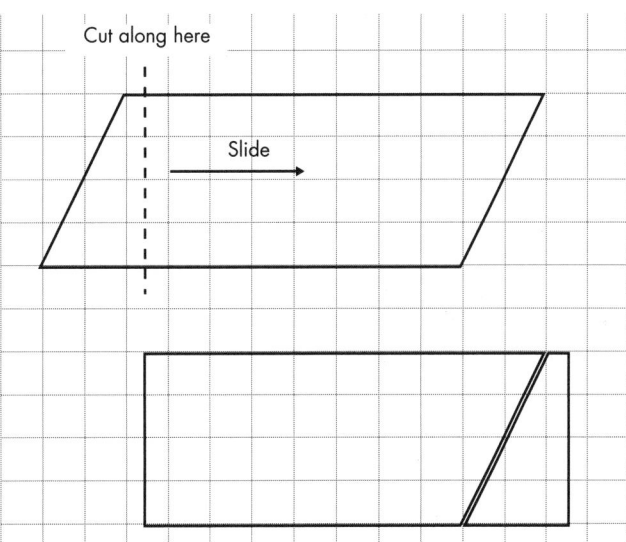

Area of a trapezium

Using a dissection method where one or both sides of a trapezium are cut, the area of a trapezium may be related to the area of an oblong.

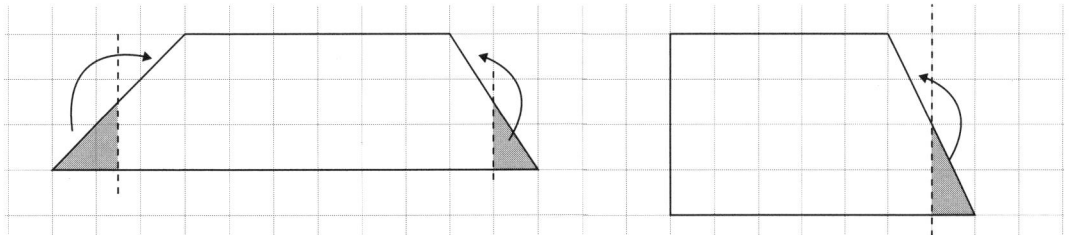

Similar logic may be applied to other trapezia.

Area of a circle

Children often confuse the formulas for finding the area of a circle and the perimeter of a circle because both involve pi and the radius and the number two. The following sequence helps to establish the area of a circle as approximately $3r^2$. First, a circle needs to be enclosed within a square and the square divided into fourths. Second, cut the excess outside the circle from each square. Reform the excess and note it almost forms one complete square. This in turn means that the area of the circle is approximately three of the smaller squares. The length of the side of each of the smaller squares is the same as the length of the radius of the original circle. Given that the area of a square may be found by multiplying the length of each side by itself (squaring the length), the area of the circle must be 3 squares or $3r^2$. Pi is a little over three and therefore the formula may be represented by πr^2.

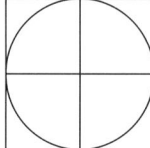

Investigate and discuss

1 Nine tiles are placed in a three by three array. Students are then challenged to remove a tile and maintain the same perimeter. Next students are challenged to remove a second, third, fourth and fifth tile while maintaining the same perimeter. Try the challenge and then explain which misconception is targeted in this activity.

Investigate the activities in the following worksheet on perimeter and area.

INVESTIGATING PERIMETER AND AREA

This exercise is for student teachers and is not suitable for primary students.

a Use square tiles or a geoboard to construct squares of successive sizes, like this:

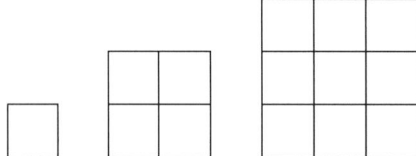

Use the length of the side of the smallest square as your length unit, and its area as your area unit, thus:

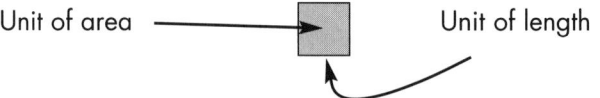

Tabulate your results for as many squares as possible like this:

side length	perimeter	area
1	4	1
2	8	4

and so on.
- What number patterns can you find? Use these patterns to discover the formulae for the perimeter and area of any square.
- Does any square have the same number value for its perimeter and area? How many squares of this kind can you find? Why?
- Draw graphs on the same axes to record your results from your table above like this:

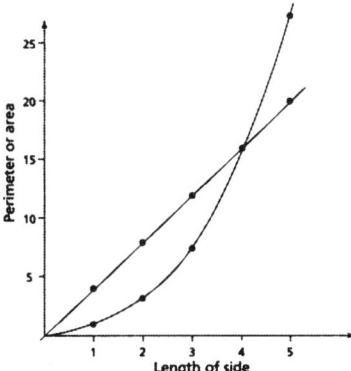

You should draw one graph to show how perimeters change as the length of the side changes, and one graph on the same axes to show these changes in areas. How do the graphs show which squares have the same number value for their perimeters and areas?

b Use 24 square tiles to construct as many different rectangles as possible (or use rectangular dot paper to draw different rectangles with areas of 24 units). Tabulate your results for the perimeters and areas of the rectangles like this:

length	width	perimeter	area
24	1	50	24
12	2	28	24

and so on.
- Do the results show any general relationships between perimeters and areas? For example, is it true that for rectangles the bigger the perimeter, the bigger the area?
- Use a calculator to find the side of a square with the same area as your rectangles. Find the perimeter of this square. How does this compare with your other results? Why?

c Tie the ends of a piece of string 24 centimetres long to form a loop. Use it and pins or plasticine to make rectangles having one side of length 1 centimetre, then 2 centimetres, then 3, 4, 5, . . . cm long until you have made all possible rectangles. Alternatively, use rectangular dot paper to draw all possible rectangles with perimeter 24 units. Tabulate your results for the rectangles like this:

width	length	perimeter	area
1	11	24	11
2	10	24	20

and so on.
- Do your results show any general relationships between perimeters and areas of rectangles?
- What special property of squares is shown by your results?
- What does this tell you about the best shape for a free-standing cow paddock, hen yard or guinea-pig hutch if you want to use the least fencing material?

Volume and capacity

As noted earlier, children often confuse volume and capacity. It is important that 'hands on' activity be accompanied by appropriate discussion.

Perceiving and identifying the attribute
- Play with building blocks and build structures. Stack boxes and blocks, pack things away, fill boxes and cartons with objects, build cubbies and forts, crowd children into large boxes, use the same blocks to make different-shaped buildings. Use clay to make a ball, a long snake, large and small animals.
- Use sand and water play for filling and emptying, pouring between containers, building sandcastles.
- Mix ingredients for cooking, blow up balloons, immerse objects in water and observe the levels. Provide pouring opportunities involving containers that are:
  - similar in shape but varying in capacity
  - identical in capacity but varying in shape
  - of regular and irregular shapes

Comparing and ordering
Use sets of nesting shapes such as boxes, Russian dolls, cylinders and prisms and compare them directly by placing one inside another. Make indirect comparisons and orderings by pouring water, sand or rice between containers to see which holds more. Immerse objects in water to see which displaces more.

Measuring using non-standard units
- Use sets of congruent shapes, for example cubes, matchboxes, marbles or chocolate boxes, to pack larger containers and count how many are used. Use the sets to build shapes of children's choices and count to see which used more. Make different shapes from a given number of blocks.

◆ Use pouring activities to measure how many of a smaller container, for example a cup, spoon, lid, glass or bucket, fill a larger container. Guess first then measure.

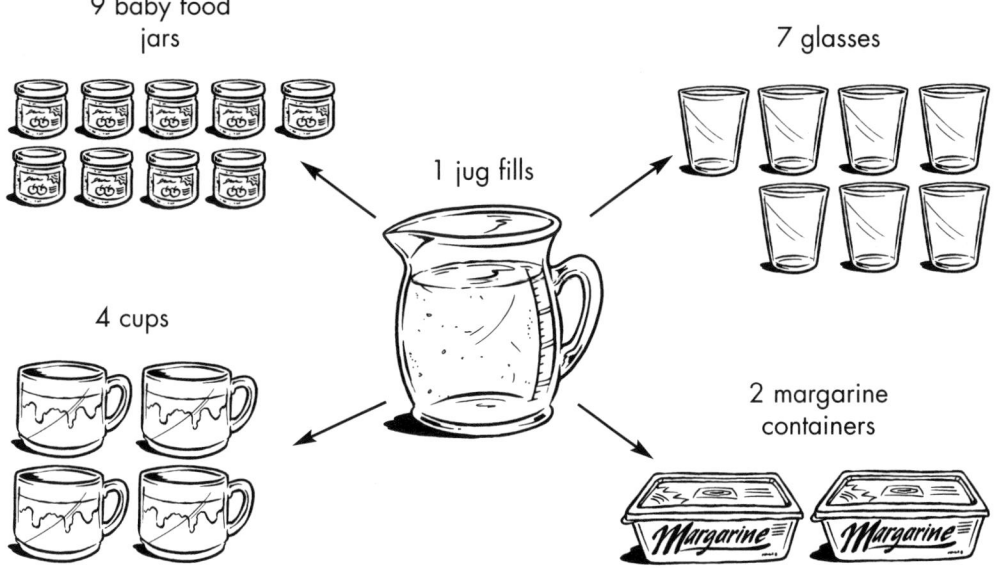

Measuring using standard units

◆ Construct cubes of sides measuring 1 centimetre. Use these, or centicubes, to construct different shapes and count how many. State volume in cubic centimetres. Use 1 centimetre cubes to pack appropriate rectangular prisms (matchboxes, chocolate boxes, cassette cases) to find volumes.
◆ Connect dowelling lengths each of 1 metre to form 1 cubic metre. Experiment with 1 cubic metre, for example how many children occupy 1 cubic metre?
◆ Construct cardboard cubes of sides measuring 10 centimetres. Use pouring activities to check that each contains as much as a 1 litre bottle. Pack each large cube with 1 centimetre cubes to establish relationships between cubic centimetre, litre and millilitre.
◆ Use pouring activities and measuring cylinders to measure capacities of everyday containers, for example cups, jars, jugs, glasses, spoons and shoes, and to test commercial packaging.
◆ Relate kilolitres to litres. Research the use of kilolitres, for example for capacities of reservoirs, and cubic metres, for example for air-conditioners.

Applications

◆ Use sets of unit cubes to construct rectangular prisms, tabulate results for linear dimensions and volumes, and lead to the discovery of the volume formula: volume = length × width × height.

Volume formulas

There is little need for children at primary school to calculate the volume of objects. Cubes may be used to build larger cubes and rectangular prisms. When building these objects children should be encouraged to count the individual cubes required to build the prism and then to look for patterns that would make the task easier. For example, some children will recognise that each layer contains the same amount of cubes as the previous layer. Once the number of cubes in the first layer is established, multiplying by the number of layers will produce the volume of the object. Likewise the number of cubes in a single layer may be calculated by multiplying the number of cubes along one edge (length) by the number of cubes along the other edge (width). Some children may come to the realisation that multiplying the area of the base of a prism by the height will calculate the volume (i.e. length × width × height). This formula along with others for the volume of a pyramid, cylinder, sphere and cone are best left until children reach secondary school. There is certainly no need to display posters containing volume formulas in the primary classroom.

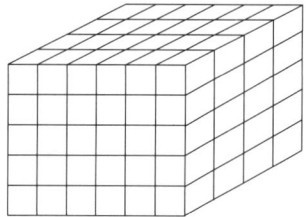

Investigate and discuss

1 In some cases such as when posting a parcel two measures are used to determine the postage cost. The parcel will be weighed and the length, width and height measured to determine the volume of the parcel. Explain why two measures are taken.

Mass

Perceiving and identifying the attribute

- Investigate light and heavy objects through free play by handling, moving and hefting objects. Include large, light objects and small, heavy objects.
- Use seesaw activities and balancing experiences with pan (beam) balances to establish correct interpretations for level and non-level beams.
- Investigate how light and heavy objects stretch springs or elastic.

Comparing and ordering

- Use hefting to compare objects that differ markedly in mass. Extend to seriating three or more objects. Later, use objects of similar masses to establish the limitations of hefting and the need for mechanical balances.
- Use home-made beam balances and spring balances to compare and order everyday objects.

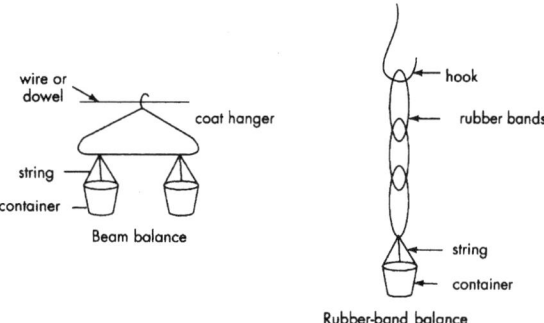

Measuring using non-standard units

Use beam balances and small objects as units (for example, paper clips, marbles, pins, erasers, blocks, stones of similar mass) to measure masses of everyday objects.

Measuring using standard units

- Experience the feel of a 1 kilogram mass by hefting. Compare it with other examples of a 1 kilogram mass with differing volumes, for example 1 kilogram bags of stones, styrofoam, metal, cork, sand and sawdust. Compare a standard 1 kilogram mass with other objects first by hefting then using a balance.
- Use similar activities to experience the feel of a 1 gram mass. Use a balance to measure representative quantities of materials, for example 10 grams, 100 grams, 200 grams, 250 grams, . . ., of sand, stones and so on.
- Use a variety of balances and standard masses to weigh personal and everyday objects. Weigh objects using spring balances. Research the use of kilograms and tonnes in everyday situations.

Investigate and discuss

> **1** In a classroom setting several objects of quite different mass were disguised by being packed with light paper in large ice-cream containers and the lids attached with sticky tape. The children had to guess which object was in each container by hefting it. Consider whether this activity is appropriate for introducing mass. Why?

Applications
Investigate density by measuring masses and volumes of everyday materials. Density is the mass of an object divided by the volume it occupies. Density is expressed in grams/cubic centimetre. The density of an object is often used to distinguish it from other objects.

Time
Because of the difficulties involved in experiencing time as a measurable attribute, it is necessary to develop time-telling using clocks and timers and hence standard units early in the overall development. When telling the time it is recommended that except for o'clock (literally of the clock) and half past children should be taught to read the time digitally, that is, 9:40 rather than 'twenty to'. Children will also need to draw on some of their numeration concepts such as addition and subtraction when telling the time and calculating time intervals.

Perceiving and identifying the attribute
- Use everyday events to introduce the relatively abstract language of time. Assist children to understand what the words mean by encouraging use of the words in correct 'point of time' context, for example children come to school in the morning; there is a holiday tomorrow; Jan arrived early, but Antonia was late.
- Arrange pictures of events in correct sequence. Discuss events that occur in cycles, for example events in the daily and weekly routine, seasonal changes, birthdays, holidays and other important events of the year.
- Record daily weather variations.
- Experience events that take short or long duration of time intervals. Note when things start and finish. Experience a variety of activities that all take the same time. Record plant and pet animal growth.

Comparing and ordering
- Use observation and discussion of everyday events to establish that the duration of an event requires the noting of starting and finishing points in time. Directly compare which activity of two with a common starting point takes longer, for example listening to a music tape or eating lunch, running around the building or writing a paragraph. Indirectly compare activities by timing both using the same techniques, for example pendulum swings, sand timers, water-drip timers and candle timers.

- Use similar techniques to seriate three or more events. Relate events to a common event, for example eating breakfast takes longer than brushing my teeth but less time than walking to school.

Measuring using non-standard units
- Use regular occurrences to time events, for example regular clapping, tapping, hopping, counting; pulse beats; pendulum swings or rocker timer oscillations.
- Make timing devices and calibrate arbitrarily, for example pins at regular distances on a burning candle, water or sand running through a small hole into a calibrated collector.
- Investigate sun dials.

Measuring using standard units
- Introduce minutes, hours, seconds, reading analogue and digital clock faces, twenty-four hour clocks.
- Experience representative time intervals through appropriate activities, for example how many star jumps can you do in one minute? two minutes? thirty seconds?
- Use clocks and stop watches to time events.
- Relate daily events to the times at which they commonly occur.
- Relate hours to days and investigate longer time spans, for example weeks, fortnights, months, years, centuries. Use weekly charts to relate events to longer time spans. Use monthly calendars to record events like birthdays and holidays and to relate days and weeks to fortnights and months.
- The National Standards Commission in Australia (www.nsc.gov.au/) has tried to eliminate confusion in writing dates by adopting an eight-digit format for dates. The system proceeds from the largest to the smallest calendar elements. For example 17 January 2004 would be written 2004-01-17.

Applications
- Use dates of historical events to investigate points of time and periods of elapsed time over the world time-scale. Investigate Greenwich Mean Time, time systems in different countries, daylight saving, time adjustments while travelling, the international date line and time zones.
- Use time computations in problem-solving experiences. Relate time to other physical attributes by investigating rates, for example speed of travel, rates of water usage.

A TEACHING SEQUENCE FOR INTEGRATING TIME CONCEPTS AND TELLING TIME

1. Sequence events: focus on events during the school day. Discuss 'what comes after?', 'what happens next?'.
2. Cycles of events: discuss events that occur each day. Show the days of the week, months of the year and seasons on circular charts. Use the calendar to investigate repeating events and patterns.
3. Associate events with times of the day: sort picture cards of events into morning, afternoon, evening piles. Match picture cards to the appropriate hour on the clock face (when learned).

4 Duration of events: use kitchen timers, egg timers or stop clocks to experience 1 minute, 2 minutes and so on. Estimate and count how many times a particular event can be performed in a chosen period.
5 Telling time on the hour. Identify the hour hand and the minute hand. Identify where the hands point for time on the hour. Practise reading clock times on the hour and associate these times with events throughout the day.
6 Movement of clock hands: use a geared clock to investigate the coordinated movements of the two hands of the clock. Identify how many minutes it takes for the minute hand to move from one numeral to the next, and for one complete revolution around the clock face. Count by fives to determine how many minutes it takes for the minute hand to move from the 12 to any other numeral. Identify how many minutes it takes for the hour hand to move from one numeral to the next.
7 Read clock time after the hour: count by fives to tell how many minutes after the hour when the minute hand points to a numeral. Use the position of the hour hand (if not pointing to a numeral) to tell which hour has just passed.
8 Use digital notation: use digital time to record the number of minutes after the hour. Read time from digital clocks.
9 Refine time-telling skills: associate numerals 3, 6, 9 with 15, 30, 45 minutes after the hour, leading to an ability to read time on the half-hour. Tell time when the minute hand does not point to a numeral by counting by fives to the previous numeral and then by ones to the minute hand. Extend this ability to reading time after the hour: 3:15, 3:30, 3:45, 3:20, 3:50, 3:21, 3:42 and so on.
10 Passage of time: devise appropriate computational methods for determining the time elapsed between two clock readings for example use 'counting on' techniques. Apply these skills in solving real life problems involving time.

Investigate and discuss

1 Would it be possible to have metric or Base 10 units for measuring angle? For measuring time? Are there any 'natural' units for these attributes? Brainstorm a number of interesting ways that children could use to measure angle or time.

Temperature

Perceiving and identifying the attribute

- Experience a variety of temperatures by touching warm and cold objects, making sure you observe appropriate safety precautions. Observe the effects of heating and cooling objects.
- Describe daily temperatures, for example it was cool this morning but it is hot now.
- Investigate pictures of hot and cold objects and extremes of climatic variations.
- Observe body temperature variations, for example Courtney feels hot after running.
- Read stories involving temperatures—for example, *Goldilocks*—about fire brigades, bush fires, icebergs.

Comparing and ordering

- Compare temperatures directly by touch in order to feel the change from one object to another. Use bowls of water at different temperatures to allow comparisons by feel. Compare the feel of objects with body temperature, for example that book feels cooler than my cheek. Compare temperatures by changes in other properties, for example ice melting as it warms.
- Discuss how people test the temperature of milk in a baby's bottle.
- Discuss hotter and colder days and climates. Record and discuss daily entries on the class weather chart.
- Order sets of pictures showing variations in hot and cold situations. Investigate pieces of dull and shiny metal placed in sunlight, to determine which gets warmer in a given time.

Measuring using standard units

- Investigate thermometers and their use. Get a feel of the unit (degree Celsius) by comparing water at room temperature with samples 5 degrees hotter and colder.
- Use a variety of thermometers to measure temperatures of common things, for example ice cubes, boiling water and body temperature.
- Record daily temperatures on the class weather chart.
- Research extremes of climate.

Applications

Investigate temperature measurements in health, science and industry. Use temperatures below freezing point as one approach to investigating negative numbers.

Value and money

Because value is relative rather than absolute and arbitrarily determined, it is necessary to develop money-handling skills early in the overall development to facilitate concrete experiences. The necessary number understanding should be developed first; for instance, facility with one-digit and two-digit numbers is necessary for the efficient manipulation of money amounts involving cents or whole dollars. Conversely, coins and paper money provide convenient concrete embodiments of these numbers. This allows children to model money amounts and engage in class shop activities.

In later school years, knowledge of tenths and hundredths and operations involving decimal fractions may also help to provide number skills needed to cope with money transactions involving dollars and cents. Again, the money experiences provide concrete embodiments of, and real life uses for, numbers involving decimal fractions. Money transactions also provide appropriate opportunities for children to explore and develop alternative computational algorithms, for example the 'making change' algorithm for subtraction and the use of various estimation strategies discussed in Chapter 4. These experiences are relevant in the development of good number sense, and frequent use of money transactions in real-world problems together with other money-based activities and games contributes substantially to the achievement of quality number competence.

Money-handling skills
- Establish a class shop and use for free and directed play.
- Provide coin recognition experiences, using real coins.
- Sort mixtures of coins by size, shape, colour, designs and numerals. Establish coin equivalences and relate to number facts, exchanging activities, shopping, money games.
- Provide similar experiences for recognition and use of notes.
- Extend buying, selling and making change experiences to complement the development of computation and problem-solving abilities.

Perceiving and identifying the attribute
- Discuss things that are most valuable to children and why they are valued.
- Discuss things of little value.
- Sort pictures of expensive and cheap things.
- Play 'The Price is Right', a guessing game where objects are ranked from the cheapest to the most expensive.
- Choose between alternatives by value.

Comparing and ordering
- Compare objects by judging value and investigating prices.
- Seriate objects and sets of pictures of objects by judged value to individuals and by price.

Measuring using non-standard units
Investigate bartering and exchanging in other societies and in children's activities, for example swapping stamps, marbles, comic books.

Measuring using standard units
Develop money-handling skills. Reinforce through practical activities of increasing sophistication in the class shop, and in actual buying and selling situations involving lunch money, the school canteen, making and selling goods for school fêtes, funding school outings and the like.

Applications
- Relate money computation to number operations for whole numbers and decimal fractions, and use in real-world problem-solving situations.
- Investigate interest and discount, lending and borrowing situations and institutions, and managing personal finances.
- Research currencies used in other countries and exchange rates.

Likely difficulties and misconceptions in measurement

Throughout this chapter emphasis has been placed on students being given experience in deciding what measurements need to be taken, what tools should be used and what units should be applied. It was noted that this discussion is vital if students are to develop their understandings of the measurement process and if teachers are to learn about their students' misunderstandings. For example, a common difficulty students experience when using a ruler involves failing to align the zero mark on the ruler with the beginning of the length to be measured.

Students need to develop the understanding of measurement errors and the degree of precision or accuracy represented in measurements that are taken. Some students using crude measuring devices present results to a ridiculous level of accuracy. For example, measuring to the nearest millimetre when using a trundle wheel would indicate a lack of understanding of the precision of the process or the accuracy of the trundle wheel. Teachers may use such instances as opportunity for discussion about levels of accuracy.

Some students experience confusion with the idea that square units are used to measure objects that are not necessarily square, for example measuring triangles in square units. Similar problems occur when measuring volume in cubic units. Introducing formulas too early can contribute to these misunderstandings.

Common misconceptions include the belief that perimeter and area are fixed. Consider the problem of fencing an area of 36 square units. Many different perimeters are possible. A square of length 6 units will use the least fencing. Likewise many students feel that surface area and volume are fixed. Take a piece of A4 paper and produce a cylinder by rolling the paper lengthways and joining the edges with a piece of tape. Take a second sheet of A4 paper and roll it widthways to produce a cylinder. Fill each cylinder and compare how much each holds. These activities help to create cognitive conflict that challenge these common misconceptions. Note, however, a single activity is unlikely to dispel entrenched misconceptions.

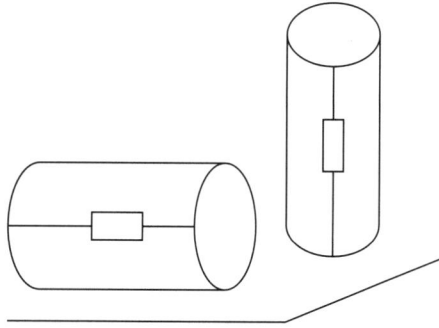

Summary

MEASUREMENT is an important component of any primary mathematics program. It provides rich sources of applications for using number processes and geometry. It provides useful links for integrating mathematics with most other school subjects. It provides many opportunities for demonstrating the relevance of mathematics to everyday activities.

Children learn to measure efficiently by first perceiving the attribute to be measured, then comparing and ordering examples of the attribute. Measuring using informal units develops the measurement principles involved and leads to formal measurement using appropriate metric units. The measurement processes can then be applied widely to life situations and appropriate generalisations can be developed.

Learning about measurement should involve extensive practical experiences involving manipulation of materials at all stages, with important emphasis on developing estimation skills and practical applications. It therefore provides rich opportunities for developing good attitudes to mathematics and a healthy appreciation of its usefulness in life.

Measurement units recommended for Australian use

The *Metric Conversion Act 1970* established a metric system of measurement for Australian use. The units used are those of the Système International d' Unités (SI units), which is a comprehensive and practical system of units of measurement of all physical quantities for technical, scientific and general use. The SI system is founded on seven base units, which include the metre, kilogram and second. All physical quantities may be measured in terms of these units taken singly or in mathematical combinations obtained by multiplication or division. For instance, areas are measured in square metres, which are obtained as a product of two lengths measured in metres. Larger or smaller multiples of these units of more convenient size are obtained by combining the unit with an appropriate prefix selected from a specified series. For example, 'kilo' is combined with 'metre' to form 'kilometre', a convenient unit for large distances.

Units used in school measurement

Physical quantity	SI unit
length	metre (m)
area	square metre (m^2)
volume	cubic metre (m^3)
plane angle	radian (rad)
mass	gram (g)
time	second (s)
temperature	kelvin (K)

Common prefixes and units

Prefix and symbol	Meaning	Example
milli (m)	one-thousandth	a millilitre is one-thousandth of a litre
centi (c)	one-hundredth	a centimetre is one-hundredth of a metre
kilo (k)	a thousand	a kilogram is one thousand grams
mega (M)	a million	a megalitre is one million litres

Length SI base unit: metre (m)

1000 millimetres (mm)	= 1 metre (m)
100 centimetres (cm)	= 1 metre (m)
10 millimetres (mm)	= 1 centimetre (cm)
1000 metres (m)	= 1 kilometre (km)

Area SI base unit: square metre (m^2)

$10 \times 10 \; (= 100)$ square millimetres (mm^2)	= 1 square centimetre (cm^2)
$100 \times 100 \; (= 10\,000)$ square centimetres (cm^2)	= 1 square metre (m^2)
$100 \times 100 \; (= 10\,000)$ square metres (m^2)	= 1 hectare (ha)
100 hectares (ha)	= 1 square kilometre (km^2)
$1000 \times 1000 \; (= 1\,000\,000)$ square metres (m^2)	= 1 square kilometre (km^2)

Volume SI base unit: cubic metre (m^3)

$100 \times 100 \times 100 \; (= 1\,000\,000)$ cubic centimetres (cm^3)	= 1 cubic metre (m^3)
	= 1 kilolitre (kL)

Alternatively for liquids and gases:

1 cubic centimetre (cm^3)	= 1 millilitre (mL)
1000 millilitres (mL)	= 1 litre (L)
1000 litres (L)	= 1 kilolitre (kL)
	= 1 cubic metre (m^3)
1000 kilolitres (kL)	= 1 megalitre (ML)

Mass SI base unit: kilogram (kg)

1000 milligrams (mg)	= 1 gram (g)
1000 grams (g)	= 1 kilogram (kg)
1000 kilograms (kg)	= 1 tonne (t)

Non-SI units

Several so-called non-SI units are recommended for use because of their practical importance or use in specialised fields.

Attribute	Unit	Definition
Time	minute (min)	1 min = 60 s
	hour (h)	1 h = 3600 s
Temperature	degree Celsius (°C)	°C = 1 K
Angle	degree of plane angle (°)	1° = rad = revolution
Atmospheric pressure	millibar (mb) (meteorology only)	
Length	nautical mile (n mile)	1 n mile = 1852 m
Speed	knot (kn)	1 kn = 1 n mile/h

Notes:
- Multiples of the unit of mass are formed from the unit 'gram' instead of the base unit 'kilogram'.
- The prefix 'kilo' is pronounced 'killo' regardless of the unit to which it is attached.
- Hectare is a common unit for land area in Australia.

Rules for using the metric system

Spelling

Spelling and symbols used in Australia are in accordance with international standards. The correct spelling of the common units is metre, litre and gram.

Writing unit names

- All unit names are either written in full or represented by their correct symbols, for example metre per second or m/s.
- Mixtures of full or abbreviated names and symbols including the symbol for 'per' (/) should not be used, for example kg per cubic metre, metre/second and m/sec are not allowed.
- Unit names follow the normal rules of grammar. The plural is used with numbers greater than 1 and numbers less than −1, for example 20 grams, 150 kilometres, −20 degrees Celsius.
- The singular is used for numbers between −1 and 1 inclusive, for example 1 gram, 0.15 kilometre, − 0.32 metre per second.
- The plural is the same as the singular, for example centimetre—cm, centimetres—cm
- All unit names except Celsius are written in lower case letters, for example gram, litre, kilogram.

Writing unit symbols and prefixes

- All unit symbols are lower case letters except the symbols derived from people's names and the symbol for litre, for example metre (m), gram (g), litre (L).
- All prefix symbols are lower case letters except mega (M), for example kilogram (kg), megatonne (Mt).

Spacing of symbols

Prefix and symbols
- The prefix symbol is part of the unit symbol and is attached to it without a space or a dot, for example millimetre (mm), kilolitre (kL).

Compound symbols
- Multiplication of one unit by another is represented by a dot or a space, for example N.m or N m for newton metre, a unit of torque.
- Division of one unit by another is represented by a horizontal line, by an oblique line or by multiplication by a negative power of the unit; for example:
  $$\frac{km}{h} \text{ or km/h or km.h}^{-1} \text{ or km h}^{-1}$$
  $$\frac{kg}{m^3} \text{ or kg/m}^3 \text{ or kg.m}^{-3} \text{ or kg m}^{-3}$$

Numbers and symbols
- The number applies to the whole symbol and not to any particular unit within the symbol. The number is therefore separated from the symbol by a space, for example 5 g, 60 km/h.

Numerical notation

The decimal symbol
- A dot on the line or a comma on the line, for example 27.50 or 27,50.

The thousands marker
- A space should be left between groups of 3 digits to the right and left of the decimal marker, for example 0.321 456; 12 236 412.
- In four-digit numbers the space may be omitted, for example 1234 or 1 234 are both allowable.
- A comma should not be used as a thousands marker. In expressing sums of money, for example where falsification of amounts on cheques and securities needs to be avoided, the numbers should be printed or written without spaces, for example, $10250.75.

Numbers less than one
- A zero should be used before the decimal marker with numbers less than one, for example 0.5; 0.123 45.

CHAPTER SEVEN
Chance and data

Skill in the critical reading of data is a necessity in our highly technological society. In particular, processing information presented in newspapers, magazines, commercial reports, and on television is dependent on a reader's ability to comprehend graphs . . . Graphs provide a means of communicating and classifying data [and] allow for the comparison of data and the display of mathematical relationships that often cannot easily be recognised in numerical form.

Curcio, 2001, pp. i—1

Understanding and facility with chance and data, or probability and statistics as these topics are perhaps better known, are increasingly needed by every informed adult. Many everyday situations involve chance, from complex uses in forecasting traffic demands on roads and freeways, the direction and intensity of cyclones and other storms, to the possibilities and consequences of legal gambling. For example, in the widespread game of Lotto where 6 numbers are to be picked out of 45, most people do not realise that their chances are about '1 in 8 million'. If they buy a ticket with 10 games on it each week, then they can expect to win once every 800 000 weeks or once every 16 000 years! With luck a player might win a prize; however, it is more probable that he or she will lose every game.

Notions of chance are also embedded in everyday experiences and language such as 'my number is certain to come up'; 'it will probably rain on my birthday'; 'it's unlikely that the next bus will come soon'. Even the pronouncements of government and business leaders are now couched in terms of likelihood rather than certainty. To participate fully in life in the twenty-first century, children will need to grow up with a realistic set of expectations and interpretations of these announcements and expressions.

Similarly, abilities related to collecting, recording, describing, displaying and organising data are essential in a society that places great emphasis on communication and technology. A capacity to use this information to make decisions, judgements and interpretations will be increasingly important in a world where statistics, rather than simple computation, is used to determine outcomes. Children need to experience and discuss realistic activities from the early years of primary school to help them to refine and extend their understandings of chance and data in order to make better sense of their everyday experiences.

All state and territory curriculum documents have responded to this need by significantly increasing the emphasis on chance and data—not only to provide children with the mathematical processes and understandings required to analyse these situations, but also to build up an intuitive awareness of the ideas and results that will increasingly influence their lives. These essential processes can be understood as three interrelated notions: *chance*, *data handling* and *data analysis*.

Chance

Chance is concerned with concepts of randomness and the application of probability as a measure of how likely it is that a particular outcome will occur. It is a familiar part of children's lives from their earliest experiences with games and competitions, arising in the play itself and also in the expressions surrounding the games, such as 'my team is certain to win'; 'it will probably be fine for Saturday's game'. At the same time, many of the widespread misconceptions about chance processes and luck also become established while children are quite young, for example thinking that if 3 games in a row have been lost, then the odds are that the next outcome will be a win; or that if every toss of the coin has been to the advantage of the opposing team, then the next toss has to be favourable to the home team. These perceptions are often based on subjective reasoning, including children's own preferences and illogical thoughts, rather than quantitative reasoning which is developed over time. For example, when marbles are drawn from a bag containing 8 red and 4 blue marbles a child may say that a blue marble will be drawn next because 'blue is my favourite colour' or because blue 'hasn't had a turn yet'.

Misconceptions about chance processes are widespread and likely to be encountered by teachers in many classroom situations. Many become established while children are quite young and are difficult to overcome. Mathematical investigations that allow children to develop realistic expectations of chance events and question the fallacies regarding luck and odds of success need to be provided in schools from the earliest stages in order to help students develop more inclusive conceptions. As Jones et al. (1999) point out, children's reasoning about chance grows over time from an initial subjective, non-quantitative reasoning through informal and incomplete consideration of quantitative thinking, to full numerical reasoning. They propose four stages in this growth as a framework to describe children's probabilistic reasoning (p. 150) and discuss how it might be used to construct appropriate tasks, monitor and assess children's reasoning, and enable teaching to adapt to children's needs.

Since chance is essentially a form of measurement, the sequence of steps underlying the measurement process developed in Chapter 6 can be used to develop a full conception of what is needed.

1 *Perceive and identify the attribute*

- Discuss experiences in chance using the everyday language associated with chance events, for example fair and not fair, lucky, no chance, always, sometimes, it might happen, probably, possible, impossible, certain, uncertain, more likely, most likely, least likely.
- Classify events as certain and uncertain, possible and impossible. Distinguish between certain events and those very likely to occur, and impossible events and those very unlikely to occur.

Developing these ideas is not easy for young children. They need to be fostered through discussion around activities such as these:

- Play games using spinners, assorted dice, tossed coins and the like. Identify all possible outcomes in establishing the rules of the game.

- Consider the fairness or otherwise of techniques for random assignment of children's roles, for example draw a task card from a shuffled set of cards, spin the bottle, toss a coin, play musical chairs, draw names from a hat.
- List possible outcomes from simple chance experiments like tossing a coin, rolling dice, spinning a spinner, drawing from a bag of coloured bears.
- Investigate combinations of possible choices from menus, clothing outfits, coin collections, gift lists and so on.

2 *Compare and order the likelihood of outcomes*

- Investigate, by discussion and experimentation, two outcomes that are equally likely, for example head or tail from tossing one coin; drawing a ball from a container with ten red and ten blue balls; spinning a fair two-coloured spinner.

- Similarly investigate which of two outcomes is more or less likely when, for example, drawing from containers with different proportions of red and blue balls, spinning spinners with unequal coloured sectors or choosing two heads or one head and one tail from tossing two coins.
- Investigate situations with three or more outcomes and seriate according to how likely it is that a particular event will occur. Such investigations could include examining the outcomes from tossing two coins, rolling two dice, spinning a collection of different two-coloured spinners, spinning an unequal multi-coloured spinner, drawing from a collection of several colours of balls, or drawing playing cards from a deck.

3 Measure probabilities

Because probabilities are measured as ratios they are expressed as real numbers from 0 (impossible) to 1 (certain) without unit names. These can be common fractions, decimal fractions or per cents as well as whole number ratios. For example, the probability of tossing a head with a fair coin may be expressed as one-half, 0.5, 50 per cent, or as a fifty-fifty chance. Proficiency with the relevant numbers is essential, as is a knowledge of certain concepts and, in due course, terminology such as:

- *experiment*—an activity for which the result is unknown
- *trial*—one performance of the experiment
- *outcome*—the result of a trial. For instance, when a ten-sided dice is rolled, the outcome could be any number from 0 to 9
- *event*—a selected outcome such as scoring 7 on one roll of the ten-sided dice
- *sample space*—the set of all possible outcomes of an experiment. The sample space for one roll of a ten-sided dice is 0, 1, 2, 3, 4, 5, 6, 7, 8, 9. This is often written S = {0, 1, 2, 3, 4, 5, 6, 7, 8, 9}

The likelihood of an event may be determined *theoretically*, by analysing the experiment and listing all possible outcomes, or *experimentally*, by performing the activity a given number of times and calculating the proportions of each outcome. The probability of an event is then the ratio of the number of favourable outcomes to the number of outcomes in the experiment or sample space. In an experimental situation, this will reflect the number of trials conducted:

$$\text{Probability} = \frac{\text{number of favourable outcomes}}{\text{number of outcomes in the experiment}}$$

Where it is possible to analyse the situation theoretically

$$\text{Probability} = \frac{\text{number of favourable outcomes}}{\text{number in sample space}}$$

For example, in rolling a ten-sided dice, there is only one way of scoring 7 from the 10 equally likely outcomes, so the probability is 1 out of 10, or $\frac{1}{10}$ written $\Pr(7) = \frac{1}{10}$.

Simple experiments with coins, dice, spinners and the like can be carried out experimentally or analysed by systematically listing all possible outcomes and assigning simple numerical probabilities based on reasoning about symmetry. Children should be introduced to both aspects by actively participating in investigations so that they develop an understanding of the relationship between the numerical expression of a probability and the events that give rise to these numbers. In this way, they can come to understand that the measure of certainty or uncertainty varies as more data are collected. At the same time, many of the misconceptions and poor intuitions about chance that children have developed can be confronted by asking them to predict outcomes before an experiment is conducted or situation is analysed. Unexpected results may then bring about the examination and rethinking of initial assumptions and intuitions.

For example, a coloured bear can be selected from a collection of 3 red bears and 6 blue bears in an opaque bag so that it can be felt but not seen. Children can be asked to predict how likely it is that a red bear would be selected in 1, 10 or 25 trials. The experiments can then be conducted, the resulting data recorded and organised to provide a measure of the probability and the results related to the children's predictions.

A theoretical analysis could be carried out next based on this experience and understanding. If one bear is selected at random there are 9 equally likely outcomes

{Red, Red, Red, Blue, Blue, Blue, Blue, Blue, Blue}

and of these there are 3 different ways of selecting a red bear and 6 different ways of selecting a blue bear. So the probability of selecting a red bear is 3 out of 9, or $\Pr(\text{red}) = \frac{3}{9}$. Similarly, the probability of selecting a blue bear is $\Pr(\text{blue}) = \frac{6}{9}$. These results can then be compared with the experimental data to enable children to come to terms with their understanding of probability.

Situations that are experimentally simple but that may be difficult or impossible to analyse theoretically as there is no way of listing all possible outcomes should also be explored. Examples of situations that can only be explored experimentally include determining the probability that a tossed thumbtack will land point up as there is only one way that the thumbtack will lie point up but an infinite number of ways that it can lie on its side. However, factors such as the weight of the tack or the length of the shank will influence whether it will readily land point up or not. Similarly, the probability that a dropped plastic drinking cup will land on its side will need to be investigated experimentally because there may be something in its mass or centre of gravity that makes it more or less likely to land on its top or bottom.

It is helpful to have a systematic method for listing and counting the possible outcomes in an experiment. This can range from the use of lists and tallies to more complex tables that capture the nature of the experiment as in the following examples:

- Tossing two fair coins. A *table* can show all of the possibilities by combining all equally likely outcomes from each independent event.

		second coin	
		H	T
first coin	H	H H	H T
	T	T H	T T

This shows that the probability of 2 heads (HH) is 1 out of 4, the probability of 1 head (HT and TH) is 2 out of 4 and the probability of 0 heads (TT) is 1 out of 4:

so Pr(2 heads) = $\frac{1}{4}$, Pr(1 head) = $\frac{2}{4}$ or $\frac{1}{2}$ and Pr(0 heads) = $\frac{1}{4}$

Yet children commonly expect that the three outcomes are equally likely until such a table is drawn up.

- Another way of seeing this outcome is to use a *tree diagram*:

first coin	second coin	outcome	probability
H	H	HH	$\frac{1}{4}$
	T	HT	
T	H	TH	$\frac{2}{4}$
	T	TT	$\frac{1}{4}$

- Rolling two fair six-sided dice. A table can show all of the possibilities by combining all equally likely outcomes from each independent event.

		second dice					
		1	2	3	4	5	6
first dice	1	2	3	4	5	6	7
	2	3	4	5	6	7	8
	3	4	5	6	7	8	9
	4	5	6	7	8	9	10
	5	6	7	8	9	10	11
	6	7	8	9	10	11	12

From the table it can be seen that of the 36 equally likely outcomes, there is only one way to score 2 but six ways to score 7. So

$$Pr(2) = \frac{1}{36}, \text{ while } Pr(7) = \frac{6}{36}$$

The first two investigations that follow these sections on *chance*, *data handling* and *data analysis* fall into this category and produce situations in which children's intuitive ideas may be challenged by the data they produce themselves. In order to come up with winning strategies for the games, they need to decide the likelihood of particular outcomes and not simply hope for the largest numbers possible. The data produced in the playing of the games can itself be used to compare strategies used by different players and the changes to playing patterns after a variety of strategies have been made explicit through class discussion. When the activities are extended to real applications, students need to consider the actions of others faced with these choices and select outcomes that will appeal to their sense of opportunity. In this way, these activities mirror many of the choices that children will face on entering the adult world and provide a base of experiences that can influence their behaviour and tendencies to analyse the situations fully rather than rely on intuition, which is often flawed or unduly influenced by hope.

4 Apply probability to chance events

- Children should investigate applications of probability by researching its use in surveys, Gallup polls, risks in certain situations, weather forecasting and the like.
- They should make and justify subjective estimates of probabilities in familiar situations, for example 'the chance that I can hit 6 out of 8 targets' in a side show alley game.
- They should investigate common games of chance by analysis and experimentation to find the probability of winning, and should research the social implications of public gambling.

Data handling

Data handling is concerned with collecting, organising, summarising and presenting data for ease of communication and interpretation. Large amounts of data are provided in newspapers and in television news reports every day, are transmitted around the world via fibre optic cables and satellites, and are churned out by computers in seemingly endless forms and quantities. If adults are to take control of their own lives they need to be able to take in and make sense of those parts of the data that are impacting on them so as to draw conclusions and make decisions.

To lay the foundation for this independence, all students need to be provided with opportunities to participate in collecting and handling raw statistical data, packaging it for presentation to others, and having practice in interpreting and making inferences from this data and also from second-hand data collected by others. In particular,

[t]he achievement of confidence and competence to make sense of and interpret data which have been collected, organised, summarised and represented by others is a major goal [for this strand]. If they are not to be subjected to the kind of exploitation implied by the expression 'you can prove anything with statistics', students need experience in judging the quality and appropriateness of data collection and presentation for answering the questions at hand. They should interpret tables, charts and graphs which range from simple to complex, and from those which provide very good models of data presentation, through those which are flawed but still communicate in useful ways, to those that are misleading (*A National Statement on Mathematics for Australian Schools*, 1991, pp. 163–4).

The exploration of data involves many aspects of mathematics. Counting and measuring are needed for data collection. Data then needs to be organised through sorting and classification before being represented in the form of pictures, charts, tables or graphs. Calculation and estimation are involved in summarising and comparing data, while appropriate units must be chosen in order to facilitate communication of the data to others. At all times, decisions need to be made about how to count and measure, the degree of accuracy that is needed, and the amount of information that will suffice so that students have continually to make connections between the numbers that arise and the situations these numbers represent.

Collecting and representing data on drinks bought at the tuckshop

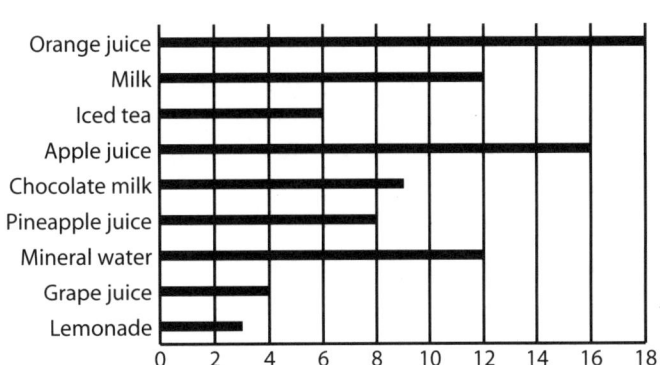

Children also need to be presented with data that has already been put into tables or graphs. This can provide them with opportunities to interact with the data beyond the common activity of reading values in order to answer routine questions. For instance, the children can generate a collection of their own questions that can be answered by the data contained in the tables or graphs. Exchanging questions with other children and discussing each other's answers can lead to significant engagement with the data. Investigative questions can also be posed ranging from simply sorting a table to find the smallest, largest or most common entry to determining the likelihood of particular outcomes based on the data provided. Forming a list of the things that the graph or table does not tell them or for which there is ambiguity or uncertainty is also helpful in building children's awareness of the possibilities and pitfalls in data collected by others. An example of this type of activity is contained in the third investigation that follows the sections on chance, data handling and data analysis. Data obtained from a World Bank database is examined via a simulated activity to bring out its significance to children in a very real way.

The ready availability of calculators and computers in classrooms allows children to investigate situations involving large quantities of real and simulated numerical data and to engage in extended practical investigations using data drawn from everyday contexts. Databases can provide real data to work with; spreadsheets can provide ways of varying the data and its significance; graphing programs can readily allow tables, graphs and plots to be constructed. However, it is important that any computer simulation or use follows active exploration with the experiment and the data it generates so that the greater number of trials and resulting data can show how a particular model or outcome can be strengthened or refined.

Thus, school experiences should expose students to a range of purposes for data collection and handling and help them to understand the processes involved. In turn, these processes can be used to solve problems that are inherently interesting and relevant while providing rich opportunities for significant mathematical investigation. They provide children with many opportunities to ask their own questions and collect and handle data from situations of interest in order to seek answers to questions that they and others have posed. In this way, the collection, use and interpretation of real data can make mathematics an essential part of the world in which children live. Computation can be seen to be purposeful, rather than a series of isolated skills unrelated to any reality. Numbers can tell different stories, revealing aspects of their own lives and community, and bring out the significance of mathematics as a tool for describing, comparing, predicting and decision making.

Representing the data

Early school experiences and play provide opportunities for children to classify and sort objects. These activities also introduce children to the notion that models and pictures can represent real and imagined things. For instance, children use dolls to represent people and toys to portray animals and vehicles. They draw pictures to depict people and things and make models from clay and building blocks. Eventually, during their early number experiences, they are able to use counters, blocks or sticks to represent the numbers of things in different collections. These activities prepare children for early experiences in data handling in which they can collect, sort and organise information in simple ways so as to answer practical questions about themselves, their families and friends and their environment. For example, how many children in class ride bicycles? Are there enough bike racks? How many like different fruits? Does the tuckshop sell these fruits? This discrete data can then be readily represented using objects, counters, pictures and tallying in order to collect frequency information that can be displayed in the form of charts and elementary graphs.

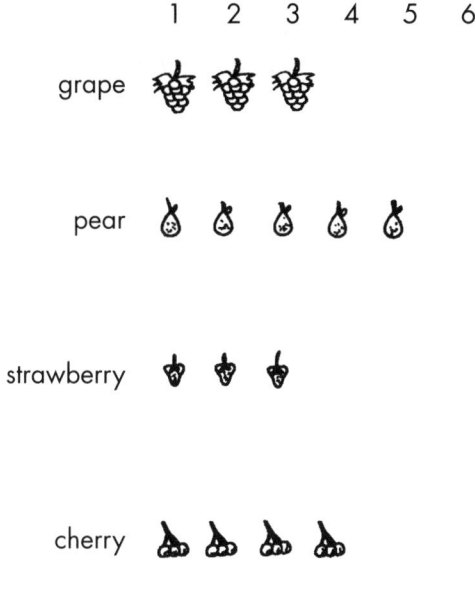

Favourite fruit

Continuous data, for instance children's heights, hand spans or foot sizes, can be treated in a similar manner using paper strips or string to provide a measure.

Once data has been collected and organised, it is important to give some thought as to how it will best be displayed. Over time, many different ways have been invented, each with its own purpose and use. These include horizontal and vertical bar graphs, picture graphs, line graphs, stem and leaf plots, box plots and so on. Each is useful in highlighting different aspects of the data and revealing different points of interest. Consideration of the various graphs and their purposes will determine which graph or table best suits a particular situation. The following section on data analysis describes a range of different graphs that are often used for data presentation and discusses how they can be used to analyse the data. The list is not meant to be exhaustive but rather a guide to the various ways data can be represented in order to be discussed and analysed.

Data analysis

Data analysis, or statistical inference as it is also referred to, deals with drawing conclusions and making predictions based on both data and principles of chance. Coming to terms with data that has been collected and represented requires looking for patterns and trying to understand what these patterns mean. Consequently, discussion and reflection on the meaning of the data is central to data analysis.

In particular, data is *described* by examining the set of values to see where it is centred or clustered and its spread or range. Various descriptors have been constructed for this including mean, median, mode, outliers, quartiles, inter-quartiles and standard deviation. These allow important descriptions given the type of data to be *summarised* so that the shape or feel of the data can be grasped through a few key indicators. Descriptors also make data comparison much easier and more manageable by allowing key values to be contrasted rather than the complete range of values. Sets of data can then be *compared* by contrasting the same variable such as the mean, spread or quartiles. Having an idea of what is typical of each data set is particularly useful in comparing data sets of different sizes. At other times, it might be necessary to *identify relationships* within the one set of data by looking at different variables and trying to discover a link among them.

Thus when we analyse data we are usually doing one of four things:

- describing
- summarising
- comparing
- identifying relationships

Data is often described in terms of average. An average is any numerical value used to identify the middle or centre of the data and there are actually different kinds of average: mean, median and mode. However, many people automatically think of the mean and the terms mean and average are often used interchangeably in everyday conversation. A thorough understanding of all these measures of *central tendency* is important to gain a true understanding of average and its importance in describing data.

- **Mean**—the balance point for a set of data, usually obtained by adding all the values and dividing by the number of values. For instance, adding 3, 4, 6, 8, 9 and 12 gives 42; dividing by 6 gives 7 as the mean. Students often only learn about the mean through this algorithm without understanding anything about how it describes the data. Indeed, it is possible to use the algorithm and obtain the mean without any comprehension of the data at all.

The idea that a single number represents a set of data is a complex one and students first need to build up an understanding of the relationship of the mean to a data set through a variety of experiences with different data. If we think of data set out on a line plot as in the example below, then the mean is the precise point at which it will balance. In other words, the point at which the sum of the differences on one side is the same as the sum of the differences on the other side. In this case, using the algorithm shows that the mean for this data is 35.

```
                    x
                    x
            x       x       x               x
    x       x       x       x       x       x
    x       x       x       x       x       x       x                               x
_____
    31      32      33      34      35      36      37      38      39      40      41
```

The difference of each value from the mean is readily measured, showing that the sum of the differences on each side of the mean totals 17:

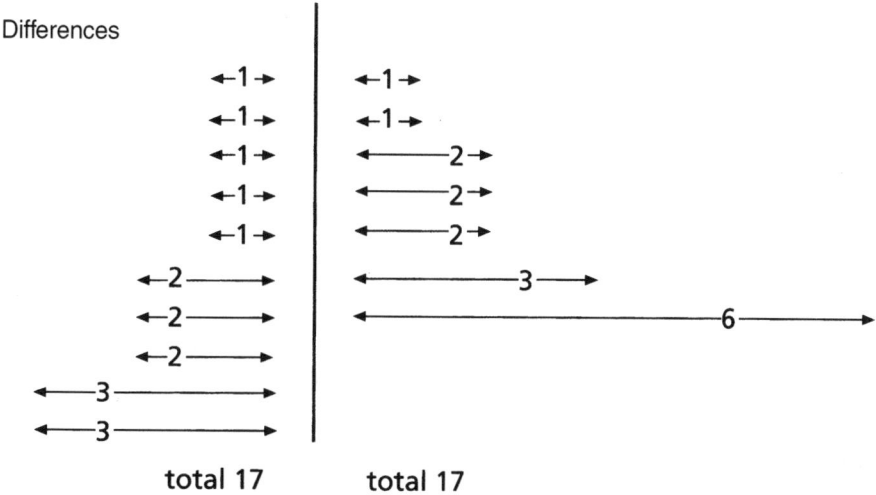

This example not only shows how the mean is a number which describes the middle or balance point of the data, it also indicates how the mean can in fact be estimated without any reference to computation or the formal algorithm at all.

The following activity, adapted from *Used Numbers—Statistics: Middles, Means and In-Betweens Grades 5–6* (Friel, Mokros and Russell, 1992), shows how an intuitive understanding of balance can be used to build up an initial concept of the mean before the formal algorithm is introduced. Data on the amount of soft drink consumed by class members was placed onto a line plot using sticky labels. Once the data had been plotted the class members were asked to examine the data and to think about where they thought the mean would be—in other words at which point the data would 'balance'. Having made their estimates the sticky labels were moved along the line plot to discover where the balance point really was. Using the sticky labels allowed the students to manipulate the data readily and check their estimates. For example, one group estimated the mean for this data to be around 5:

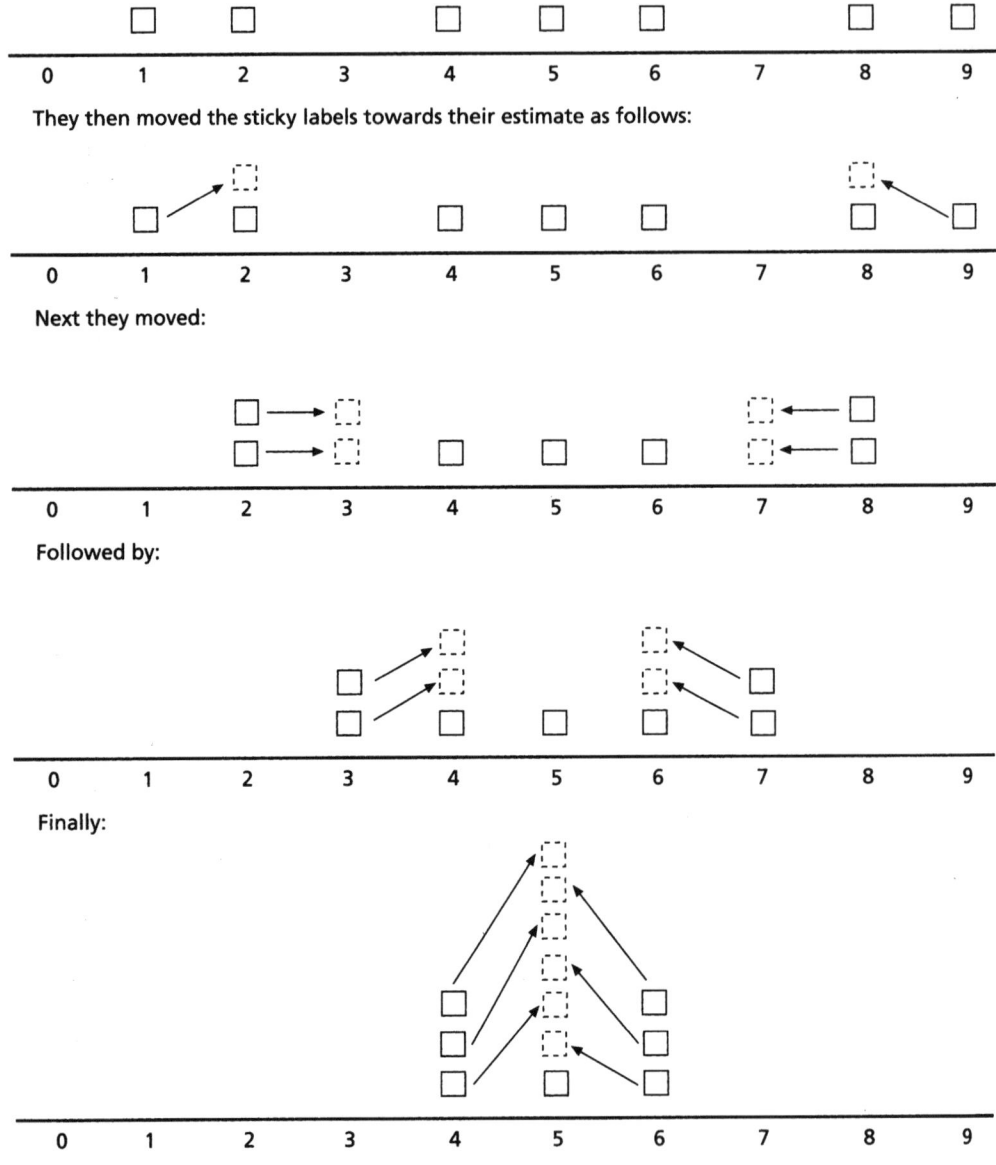

They then moved the sticky labels towards their estimate as follows:

Next they moved:

Followed by:

Finally:

Hence we can see that 5 is the point at which the data balances, in other words, the mean. If the initial estimate had been incorrect, this would have become apparent when the labels were moved as the data would have clustered about a different point, the actual value for the mean.

- **Median**—the midpoint of a set of data. If the data is lined up in order the value of the middle or half-way point is the median. If there is an odd number of values then the median is the middle value. If there is an even number of values then the median is midway between the two middle values, for example 3, 4, 6, 8, 9, 12 has 6 and 8 as the middle values. The midpoint between 6 and 8 is 7 so the median is 7. The fact that the median can be a number that is not actually in the data can seem confusing to children, especially when the number turns out to be a decimal fraction, for example when the median number of children that Australian families have may be 1.9 or less. Sometimes the median and mean will give similar information about a set of data. However, they can also be very different. While the median gives the value that divides the data in half, the mean describes the point about which the data is balanced.

- **Mode**—the value or values that occur most frequently in a set of data. A set of data can have more than one mode. Finding the mode is very easy. However, determining the significance of the mode can be more difficult as the most frequent value is sometimes related to chance and as such may be a point of interest rather than of any real consequence in terms of understanding the data.

The distribution, or *spread*, of data is another important aspect to consider when analysing data. There are a number of ways to describe data in this way with perhaps the most common being:

- **Range**—the interval from the lowest value to the highest value (found by subtracting the lowest value from the highest). The range is heavily influenced by *outliers*, which are extreme values either much lower or much higher than the majority of the data values. In other words they lie outside of the overall shape and pattern of the data. Outliers need to be examined carefully as sometimes they can be errors—someone has counted, measured or recorded incorrectly, or they can be unusual values that have happened for some reason and can generate interest in how and why they occurred.

- **Quartiles**—show the range of data within subsections of the entire data set formed by breaking the results into four equal parts. The lower quartile is the number that has 25% of the values below it. It is determined by finding the halfway point between the least value and the median or midpoint of the data. The upper quartile has 75% of the values below it and is determined by finding the halfway point between the median and the greatest value. As with the median, sometimes a quartile will be midway between two values. The lower quartile is thus the middle number on the lower half of the data while the upper quartile is the middle number on the upper half of the data.

  Upper and lower quartiles are identified by examining the values in rank order (in either ascending or descending order) so that the median of the whole data and then the midpoints which give the quartiles are readily seen. This middle section of the data, from the lower quartile to the upper quartile, is called the *inter-quartile range*, used as a measure of the spread of the data.

For example, putting the data shown on the line plot on p. 516 in ascending order:

32 33 33 34 34 **34** 34 34 35 **35 35** 35 36 36 **36** 37 37 37 38 41
↑   ↑   ↑   ↑   ↑
least   lower   Median   upper   upper
value   quartile   quartile   value

There are twenty items and the middle point lies between 35 and 35 so the median is 35.

The least value is 32. There are ten items in the lower half of the data, the middle number lies between 34 and 34, so the lower quartile is 34.

The greatest value is 41. There are ten items in the upper half of the data, the middle number lies between 36 and 37, so the upper quartile is 36.5.

Data may also be broken down in a similar way into ten equal parts, rather than four, called *deciles*, or 100 equal parts, called *percentiles*. They are frequently used when there is a very large amount of data, for example in the distribution of test scores across a large population.

◆ **Standard deviation**—another measure of spread, showing how far each value differs from the mean. This is used mostly by senior students and usually performed on a calculator with built-in statistical functions.

Often the way the data is actually presented provides a description that is more readily analysed. Thus, in addition to using tables and frequency counts, many types of *graphs* have evolved:

◆ **Picture graph**—data is represented using actual pictures to show how many in each category with labels on one axis and values on the other axis and a title usually at the top. This graph is easy for young children to construct as they simply need to glue the appropriate picture onto a large sheet of paper to make the graph. On the following graph, which shows travelling to school, the students have simply selected the picture of how they travel to school, glued it on to a piece of paper on which the teacher has written the labels, title and number in each column. The end result is a comprehensive graph that enables even young children to analyse, discuss and draw conclusions from the data.

How we travel to school

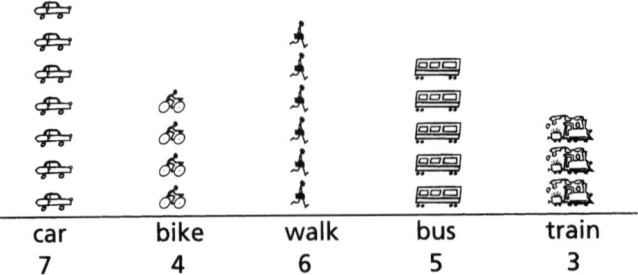

Picture graphs show how many (the frequency of occurrence) in each category

- **Bar or column graph**—this can be horizontal or vertical and has values on one axis and labels on the other axis. Bars or columns are used to show the values. A vertical column graph would have values on the vertical axis and labels on the horizontal axis while a horizontal bar graph would have values on the horizontal axis and labels on the vertical axis. A title, which is usually written at the top of the graph, is always included. This graph is easy to read and clearly shows frequency of values. It is useful for comparison as the two sets of data can be drawn on the one graph, using bars or columns of different colours to distinguish between the different data sets.

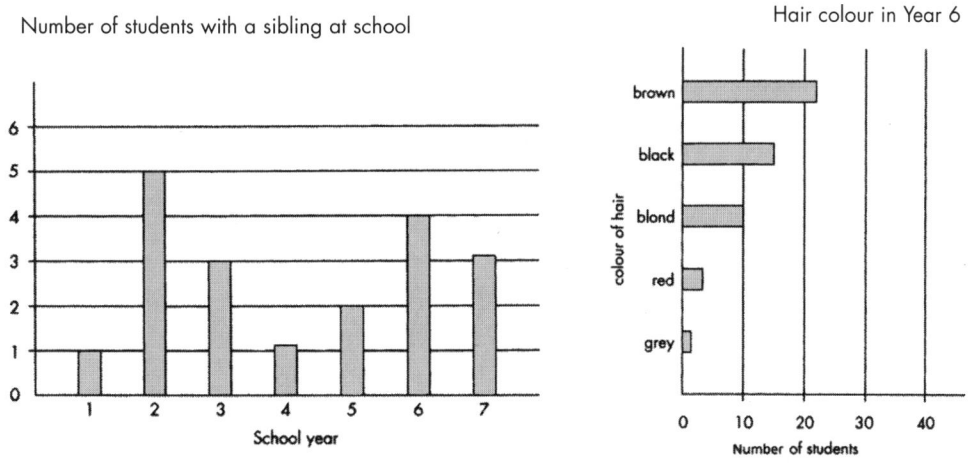

A bar graph (horizontal or vertical) has a scale on one axis only

- **Line graphs**—these are used to show trends and relationships and have labels on the horizontal axis and values on the vertical axis. Values are plotted using dots which are joined to form a line allowing easy identification of rises and falls in the data. Line graphs are often used to show changes in interest rates. If we look at the following line graph we can see that the population of Australia climbed steadily until the early 1940s at which point it sharply increased. This sharp increase has continued and has not dropped off or decreased at any time.

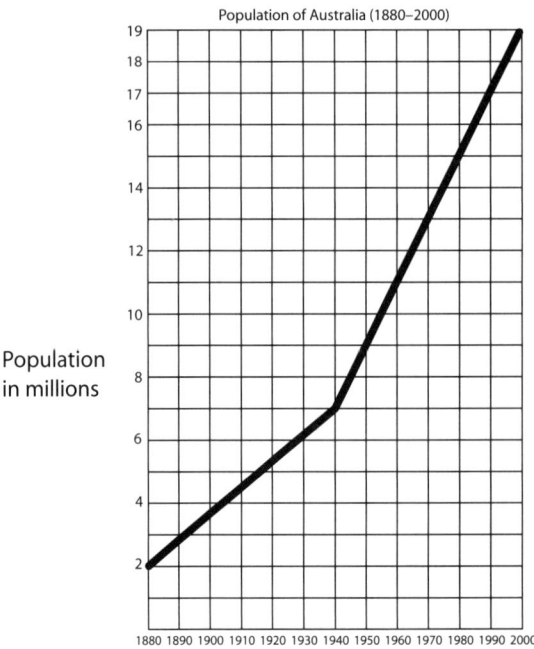

Line graphs show trends and relationships

- **Line plots**—a quick way to organise numerical data, these are often used as a working draft during data analysis. Each piece of data can be recorded directly as it is collected by drawing crosses or dots above a number line to show the frequency. This readily shows the range of the data and how it is distributed over that range. As a line plot is not a formal graph for data presentation, it has no title, labels or vertical axis. Line plots work well for numerical data with a small range and are often used to analyse data at an initial stage as they readily show the mean, median, mode, range, outliers and quartiles.

```
                  x
                  x     x
                  x     x     x     x
            x     x     x     x     x
      x     x     x     x     x     x     x                       x
─────────────────────────────────────────────────────────────────────────
   31    32    33    34    35    36    37    38    39    40    41    42
```

If we look at this line plot it is easy to see where the data begins and ends (range), that it clusters around 34–37, the middle point of 35 (median) and the most frequent values of 34 (mode). If we think of the mean as the balance point it is possible to estimate at which point the data would balance—around 35. If we know the median then it is possible to find the quartiles as the lower quartile is midway between the beginning value and the median and the upper quartile is midway between the median and the last value. In this example, the first value is 32 and the median is 35 so the lower quartile is 34; as the last value is 41 the upper quartile is 36.5; the inter-quartile range would be 34–36.5. It is also possible to see at a glance that 41 is an outlier, a value much higher or lower than the main data set.

- **Stem-and-leaf plots**—another easy way to show the range of data and how it is distributed over that range. In essence, they provide a frequency distribution which retains the data. Usually the plot is organised by tens, so it works best with a range over several decades and with up to about 100 numbers. Each number is divided into tens and ones, the tens are listed vertically and the ones horizontally next to the appropriate tens.

In this way, the tens become the *stem* of the plot and the ones the *leaves*. For example, the values of 104, 122, 126, 127, 133, 138, 142, 142, 144, 146, 147, 148, 149, 150, 157, 158, 158, 165, 176, 179, 183 can be graphed onto a stem-and-leaf plot for analysis. The number of ones alongside each ten show the distribution of the numbers, while each piece of data can be read off very simply; in the first line, 10 | 4 means 104, in the third line, 12 | 2 means 122, 12 | 6 means 126, and so on.

Stem	Leaves
10	4
11	
12	2, 6, 7
13	3, 8
14	2, 2, 4, 6, 7, 8, 9
15	0, 7, 8, 8
16	5
17	6, 9
18	3

This graph is also useful in showing the range, mean, median, mode, outliers and quartiles while at the same time highlighting clusters of data that would be less easy to see in other graphical representations. Looking at this example, we can see at a glance where the data begins and ends (its range), that it clusters around 14 tens, that 104 is an outlier, that 142 and 158 are modes, and that 147 is the midpoint or median. If we think of the mean in terms of balance, then the data would probably also balance somewhere around 147. The quartiles are also readily determined: the lower quartile is 135.5 and the upper quartile is 158.

- **Box plots (box and whisker plots)**—provide a concise summary of data by plotting only five numbers and allow ready comparison of a number of related sets of data. They can be horizontal or vertical.

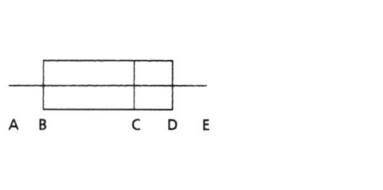 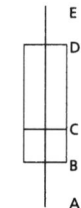

One edge of the box shows the lower quartile (B), the other edge of the box gives the upper quartile (D). Thus 50% of the data is contained within the boundaries of the box. The median value of the data set (C) is indicated by a line drawn within the box (not necessarily the middle). The least number in the data (A) and the greatest number in the data (E) are also plotted onto the graph and a line drawn from A to E, usually referred to as a whisker.

Several important pieces of information can be derived from comparing two or more box plots and they are frequently used to compare outcomes on test data from different groups or related tasks. A comparison of medians gives an approximate indication of the difficulty of a task while the lengths of the boxes provide an indication of the spread of performance for a given task. The following example, taken from Beesey et al. (1998, p. 179), summarises performances for two tasks for the same group of students:

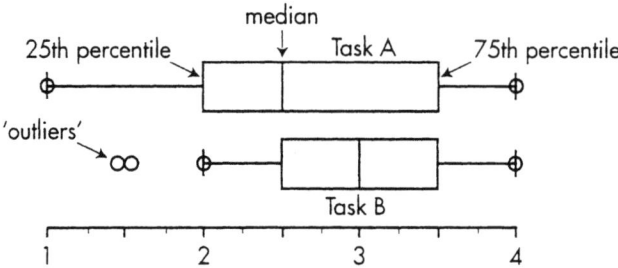

From the diagram, the median score for Task A was 2.5 with 50% of students scoring between 2 and 3.5. Task B is therefore easier for students since its median score is 3. Furthermore, 50% of students scored between 2.5 and 3.5 on Task B. While the top score for each task was 4, and 90% of students scored 2 or above for Task B, there were two students who scored 1.5 for this task.

The values needed to draw a box plot are easily located on a line plot, making it a straightforward matter to construct a box plot from the data that has been arranged on a line plot. Reading from the following line plot, 32 is the least value, 35 is the median, 33.5 is the lower quartile, 36.5 is the upper quartile, 41 is the greatest value.

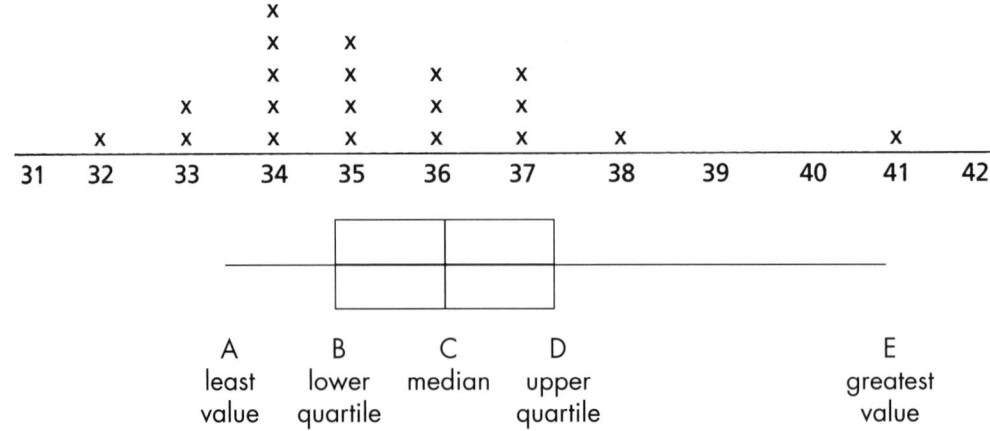

A ruler and pencil can be used to draw vertical lines under the points where the lower quartile, the median and the upper quartile are situated. The box is then completed by drawing horizontal lines above and below the vertical lines. Finally a line is drawn through the box from the least value to the greatest value and the box plot is complete.

- **Circle or pie graphs**—show data as a percentage in which a circle is divided into sections according to the fraction or percentage of 360°. These graphs can be difficult for children to draw accurately. This graph represents data as a whole and divides it into parts of the whole. If we look at the graph above regarding income expenditure we can see that we have a complete set of data—we know exactly what all the money is spent on.

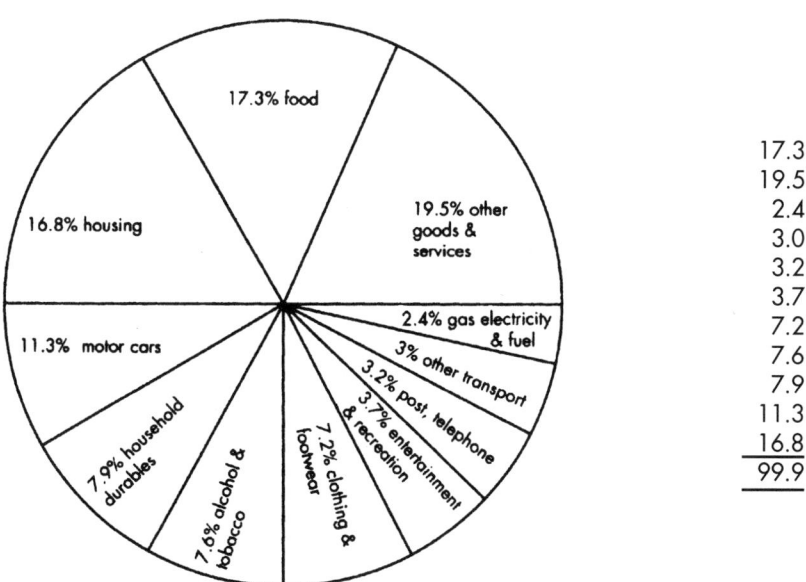

```
17.3
19.5
 2.4
 3.0
 3.2
 3.7
 7.2
 7.6
 7.9
11.3
16.8
────
99.9
```

Circle graphs show proportions of the whole

We could graph this using a column or bar graph but knowing how much is spent is not as useful as knowing the proportion of income spent on each category. A circle graph which shows proportions is much more informative and allows the data to be described and analysed in a more useful manner.

When there are too many values to be graphed and analysed separately, data is usually *grouped* into related clusters. Examples of these are:

- **Frequency tables**—the range of values is clustered into groups or intervals usually of equal space and the count or frequency of each range entered into a table. While these tables can have any number of clusters, between five and seven allows more ready analysis.
- **Histogram**—this form of graph displays data from frequency tables using columns with no gaps. It has a scale on both axes with the horizontal axis showing the data groups from the frequency table. As you can see from the following example the data has been clustered in groups spanning five years.

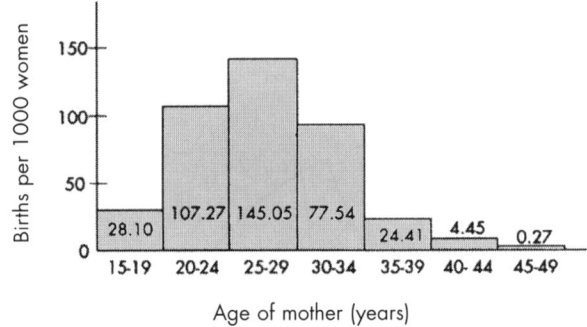

Histograms show frequencies with a scale on each axis

Note that histograms should not be confused with column graphs, which use discrete values or labels rather than clusters.

All of the graphs, tables and plots described so far can also be used to compare and summarise more than one set of data. By drawing two or more graphs *side by side* or *back to back* it is relatively easy to make comparisons between the various data sets and to gain a feel for the different data sets and what they mean.

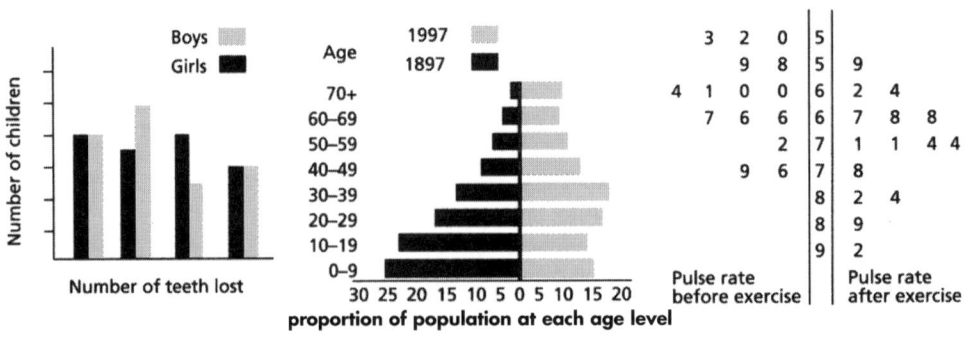

side by side column   pyramid graph   back to back stem

Different forms of graphical displays are required for identifying relationships within the one set of data. This involves looking at all of the values and trying to discover a link between them. Some examples are:

- **Contingency tables**—words or category data rather than numerical data is displayed in tables showing the frequency of different combinations. The tables used on p. 505 to examine the tossing of 2 coins and the result of rolling and adding 2 six-sided dice are examples of contingency tables.

- **Scatter plots**—two sets of numerical data are obtained and plotted as dots of different colours. The scatter of these dots shows where a relationship is likely to exist between the variables.

- **Q-plots**—a set of data is plotted using a scatter plot with the set of data being plotted vertically and then horizontally. A vertical line is drawn at the median of the data plotted horizontally and a horizontal line is drawn at the median of the data plotted vertically thus splitting the data into four regions. Relationships are identified by examining the data in diagonally opposed regions.

A final point needs to be made to link together the two aspects of data handling and data analysis:

> Although children need to learn that data can be displayed in a variety of ways and come to terms with many of these representations, it is equally important that they understand that the choice of display depends on the questions being asked of the data. Graphs should not be regarded as ends in themselves; rather they should serve purposes which are clear to children . . . Little is likely to be achieved by providing a collection of data and having children practise drawing graph types in isolation. Children should begin to locate and interpret data from a variety of sources and incorporate the results of their research into reports of their own. (*A National Statement on Mathematics for Australian Schools*, 1991, p. 168).

Teaching chance and data

In order for chance and data to be built up in a meaningful manner that children can use in their everyday lives and take forward into their adult work and leisure, a number of points need to be borne in mind:

- Chance and data can and should be taught concurrently, not treated as two separate topics taught in isolation from each other. The strand should also be taught in conjunction with other areas of the mathematics curriculum, providing a purpose for the study of particular concepts and processes as well as drawing on experiences in these areas to make chance and data a meaningful part of children's mathematics rather than a set of stage-managed routines.

- Chance and data is better seen as composed of a few large interrelated key ideas rather than partitioned into a host of small, disjointed techniques. Ideas such as natural variation, long-term frequency, the distribution of probabilities and statistical inference via the analysis of data presented in a variety of forms should be central to all activities and learning.
- The learning of chance and data should be contextually based, not presented as abstract pure mathematics. Learning in context can be driven more by students' interests and provide situations in which their uninformed predilections can be investigated and challenged. In this way, children can build their own intuition and sense of chance and data.
- Technology such as computers and calculators (simple, scientific, graphical) should be used to enhance learning at every opportunity. The Internet can be a ready source of topical data.

The three investigations that follow provide examples of how chance and data might be explored in a classroom situation, integrating many aspects of chance, data handling and data analysis within a meaningful context that can allow children to construct their own understanding of the processes involved. The first investigation uses a familiar game from the early study of numeration, building in a need to recognise and respond to the elements of chance involved in rolling a set of numbers on a dice and then utilising graphical representations as a means to improve outcomes and extend the initial activity into a more complex playing situation.

Investigation 1 — Highest Number

Materials: six cards numbered 1–6 per student
one three-digit place value chart for each student
one six-sided dice per group of three students

Introduction

Students first need to be familiarised with the initial game. Each player in turn rolls the dice and puts the card that matches the digit rolled onto his or her place value chart. For example, if a player rolls a 4, then the card with a 4 on it is positioned in the hundreds, tens or ones place. This process is repeated until each player has used three cards to represent a three-digit number on the place value chart. If a rolled number has already been used the player rolls again. Students compare numbers and decide which person in their group has the highest number. Only one dice is provided per group to encourage players to watch each other roll and make a decision on where to place the digit card. In this way, the players can consider a large number of outcomes and reflect on the likely outcomes and best moves.

The activity is then extended by playing five rounds of the game, recording the numbers obtained, and adding the five numbers to obtain a grand total. The winner is now the player with the highest total.

Collection of class data

Scores can now be compared across the whole class to find the overall winner. One way to do this would be simply to record the results of the winner in each group. But students would also be interested to compare their individual results with others in the class. Suggestions from the class would probably first lead to an extended recording of every student's result in some ordered way. But this is not easy to do, and does not readily allow students to compare their performance against the class.

A line plot could be introduced as a means of gaining more information, but it would soon be apparent that the spread of data would make it unwieldy. Other possibilities that could be explored are frequency tables or histograms. This discussion can then lead into a natural way to introduce a stem-and-leaf graph as a means of presenting the data concisely, in which the frequency of results in each band can easily be seen, allowing each individual to compare his or her results with the overall class pattern as well as with other students' scores. Each score needs to be broken into its hundreds (the stem), alongside of which the individual tens and ones (the leaves) can be recorded. The following table shows data collected by one class of 26 students.

Class results

hundreds	tens and ones
30	43
29	54
28	61, 70, 09
27	43, 56, 58, 08
26	35, 11, 91, 43
25	26, 31, 87, 84, 12, 03
24	18, 71, 53
23	74
22	46, 49
21	
20	
19	
18	73

By reflecting on their outcomes, the class can discuss their impressions of the game and the strategies they employed. This should give rise to informed speculation about the possible range of total scores over the five games and estimates of likely outcomes. Since the maximum score possible in any game is 654 and the minimum is 123, the totals will range from about 3000 to 700. In turn, this understanding of the possible results can then be applied to the actual outcomes and the expected distribution of results. It may be that further trials are needed to capture a more likely spread of results or particular possibilities. A focus on the chances involved could be achieved by discussing a game in progress, in which the first roll was a 3 and asking children their move when a 4, 2, 5, 1 or 6 was rolled. Some children might always put the second number they roll in the ones place, in the hope that a 5 or 6 will still come up. Others might relate their choice to a straight numerical value—'If it's smaller than 3, I'll put it in the ones place, larger than 3, I'll put it in the hundreds place'. An investigation of outcomes using a table as on p. 505 can then highlight the reasonableness or otherwise of their strategy.

Once students have experimented with the game and explored the data and chosen methods, a computer can generate hundreds or thousands of simulated results. This follow-up can broaden their understanding and provide opportunities to observe how a greater number of trials can refine their awareness of the underlying probability and the implications for their methods of play. The Curriculum Corporation package, *Chance and Data: Exploring Real Data*, (Finlay & Lowe, 1993) provides a ready-made example, but the children could be challenged to create their own.

Extensions

Further analysis and discussion can be provided by establishing a contextual setting in which the data might be used. For instance, the game could be established as a fundraising activity: 'spend $1 to win $5' at the School Fête where one dollar is wagered on a game to better a particular score for a prize of five dollars. The problem is now one of selecting the score that has to be bettered. The class data displayed on the stem-and-leaf graph can be used to investigate outcomes for several suggested cut-off scores:

Score	Money in	Money out	Result
3000	$26	$5	$21 profit
2600	$26	$65 (13 wins)	$39 loss
2700	$26	$45 (9 wins)	$19 loss
2750	$26	$35 (7 wins)	$9 loss
2800	$26	$25 (5 wins)	$1 profit
2850	$26	$20 (4 wins)	$6 profit

They may realise at this stage that the psychology of game playing needs to be considered. If the winning score is too high, no one will play and the school makes no money. If it is too low, the school will lose money. So there is no absolutely correct answer—only one that makes sense in the context in which it is used.

A further extension could address whether students can have confidence that the results they generated might be repeated on the fête day. This leads to the students playing several times to expand their database and highlights one of the key ideas in statistics and probability—the amount of data that needs to be explored before underlying patterns become apparent. With a computer simulation, profit and loss scenarios could be investigated using the results of many hundreds or thousands of trials.

Investigation 2 – Greedy Pig

This investigation shows how a rich activity can be used to build up chance and data understanding through experimental methods rather than solely through a formal exposition. It also uses a game to build up a sense of chance, contrasting individual probabilities against long-run outcomes. A different form of graphical representation is then used to compare participants' scores within and between rounds of the game after a range of playing strategies have been discussed.

Materials: 1 six-sided dice used by the teacher
1 tally sheet per student

Game	Numbers rolled	Points
1		
2		
3		
4		
5		
	Total	

Introduction

In this game the teacher rolls the dice and the students record the numbers rolled on their tally sheets. At the end of each game the points are added and the winner is the student who obtains the highest total at the end of five games. However, there is a catch—if the teacher rolls a 2 then all points are lost and the players get zero for that game!

All students begin the game standing. The first two numbers the teacher rolls form part of each student's score regardless of their value. After each roll, a student must decide whether to keep playing for more points and risk losing all the points so far if a 2 is rolled or sit down with the total at that time. The game continues until the teacher either rolls a 2 and all remaining students get zero or until there is no longer anyone standing.

Collection of class data

Totals can now be compared across the whole class to find the overall winner, although students would also be interested to compare their individual results with others in the class. Listing every student's result in some ordered way would most likely result in suggestions for a frequency table or a line plot. However, these do not readily allow students to compare performances. A stem-and-leaf graph would be more efficient, as it also represents the data concisely, shows the frequency of results in each band and allows easy comparison of totals. The data collected by one class of 28 students is shown:

Class results

tens	ones
0	0, 0, 0, 0, 0
1	
2	8, 8, 5, 2, 7, 2, 7
3	1, 7, 3, 2
4	5, 7, 0, 9, 2, 4
5	3
6	4, 7
7	1, 0, 9
8	
...	

Students can use this data to explain why they sat down or continued to play and debate when it would be best to sit. These strategies can then be displayed and discussed; for example:

Strategy	Reasons
after a 6	I wait for a six and then sit down as I don't want to lose the points
after four rolls	Because after that I figure a 2 can't be far away
after 10 points	I reckon 10 points a game is pretty good, so I wait for that and then quit
half-class	I wait till half the class is sitting down, then I reckon its too risky after that
two rolls	I reckon if you always sit down after the two free rolls you can never get zero and your total score will build up

The notion that some strategies are better than others can now be raised. For example, all students would probably agree that waiting for one hundred rolls is a poor strategy as opposed to getting 20 points and then sitting down. Various strategies used by different students could be discussed and rated as poor, average or good. Each student could also think about which strategy he or she would judge to be 'best'.

The game can then be played again, with each student now using whichever strategy he or she had decided was most appropriate to obtain a large score. The data obtained from playing these further five games could then be compared with the initial round to see if a change in strategies had influenced the outcomes. Use of a *back-to-back* stem-and-leaf graph is now appropriate, where the second set of data is put alongside the first:

	Second round		First round
	ones	tens	ones
		0	0, 0, 0, 0, 0
		1	
	8	2	8, 8, 5, 2, 7, 2, 7
	4, 7	3	1, 7, 3, 2
4, 6, 0, 9, 0, 6, 3		4	5, 7, 0, 9, 2, 4
	8, 9, 4	5	3
	4, 2, 8, 3, 7	6	4, 7
	4	7	1, 0, 9
	4, 6	8	
	7, 4, 0	9	
	9, 7, 5, 3	10	
		11	

While five students scored zero in the first trial, no one scored zero in the second trial. The students were more aware and made an informed decision as to which strategy they might use. However, given the variability of throwing a 2, a strategy will only help to a certain degree.

Discussion could now centre on the probability of throwing a 2, $Pr(2) = \frac{1}{6}$ as on p. 503, and the notion of long-run frequency, where a particular number is rolled several times in a row or does not occur for many rolls, can also be brought up and examined. For many students, the relationship between these two related outcomes is difficult to see. When a number is rolled many times in a row, a student might think that it is more likely and the dice is 'loaded'. When a number has not appeared for some time, another student might think it is much more probable that it will occur on the next roll of the dice. Patterns within the graphs could be used to investigate these ideas, although once again the issue of the amount of data needed to be able to infer any relationship should be raised.

A computer program could be used to test these hypotheses or to generate a large number of trials in order to determine which strategy is better than others. For example, a program could allow students to test whether it would be better to quit after two, three, four or more rolls by letting the number of rolls before quitting be the variable and simulate each of these many times. Alternatively, a program could simulate quitting after a nominated number of points. (Computer programs that enable this are available through the Curriculum Corporation's Task Centre project.) One teacher who used this approach has reported:

> I found my Grade 6 students could easily follow the logic of the computer testing and hone in on the optimum strategy. ... they can do this empirically whereas to find a theoretical answer by analytic means is well beyond their capabilities at this stage. This illustrates one of the advantages of empirical investigations versus theoretical analysis.

The data gathered from either a larger number of actual games or from a computer simulation is not as easily compared as the above use of a stem-and-leaf plot for two data sets. The issue of which of the measures of central tendency would provide a meaningful comparison could be raised and the investigation could then be used to introduce a box and whisker plot and show its value in comparing many sets of data. After this, a challenge might be to work out the theoretical expected score for quitting after two, three or more rolls.

Investigation 3—Who goes to primary school?

This investigation illustrates how a set of 'second-hand' statistical data can be explored to bring out the significance of particular ideas that are being reported. The data is used as a base for a simulation that can bring the data to life and as the basis for the development and exploration of a variety of graphical representations.

Materials: 1 six-sided dice for each student and the teacher
1 copy of the data table per student

Primary school enrolment (% of age group)

	% of age group enrolled in primary education 1984–86		Primary school enrolment (% of age group) total: 1987	% grade 1 enrolment completing grade 4
	males	females		
Ethiopia	44	28	37	50
Bangladesh	76	49	59	20
China	142	124	132	77
Haiti	86	89	95	32
India	114	81	98	40
Indonesia	121	115	118	80
Sudan	58	41	49	76
Bolivia	97	85	91	60
Philippines	109	106	106	75
Papua New Guinea	79	64	70	62
Mexico	118	116	118	69
Malaysia	102	102	102	99
Brazil			103	22
Algeria	105	81	96	90
Yugoslavia	95	94	95	98
Iran, Islamic Rep.	123	105	114	87
Saudia Arabia	78	65	71	90
Australia	106	105	106	100
United States	101	100	100	100
Japan	102	102	102	100

Source: World Bank, WDI, *Chance and Data: Exploring Real Data*, Finlay & Lowe, 1993, Curriculum Corporation, reproduced with permission.

Introduction

Students examine the data table and discuss any observations, implications or 'stories' that they can draw from it. Class discussion is likely to give rise to issues such as the differences between male and female opportunities in education, types of countries that have more or less educational participation, or the meaning of some of the statistics that arise, such as a 132% participation for China.

Attention can then be drawn to the situation in Bangladesh. Data presented in the table shows that only 49% of girls enter school and that subsequent participation falls so that only 20% of these children graduate four years later. Assuming that an equal number withdraw each year, this means that the initial number needs to be multiplied by the same factor on each of the four years to produce an end result of 20%. The factor obviously must be less than 1 and some experimentation shows that a factor of $\frac{1}{2}$ is too small as $0.5 \times 0.5 \times 0.5 \times 0.5$ is 0.0625 or 6.25%. Similarly $(\frac{3}{4})^4$ is about 32% which is too large but $(\frac{2}{3})^4$ is 0.19 or 19%, close to the correct proportion. In other words, although 1 out of 2 girls is able to begin education, each year only 2 out of 3 of those who are at school will continue for the next year.

This analysis of the data and what it represents provides the basis for a simulation using simple six-sided dice. Children will quickly see that the rate of entry can be determined by rolling a dice to find whether the outcome is odd or even. The drop-out rate can be modelled by breaking the outcomes into three equal groups, such as 1 and 2, 3 and 4, 5 and 6, and accepting numbers in only two of the three groups.

All the children, taking the part of girls in Bangladesh, now stand at their desks, roll their dice and note the result. The teacher rolls his or her dice and announces whether the result is odd or even. All those whose results match the teacher's now sit down as they have not been 'selected' to attend school. The remaining children now roll their dice again and note whether their result is in the group 1 and 2, 3 and 4, or 5 and 6. When the teacher rolls again, a roll of either 1 or 2 means that all of those in group 1 and 2 sit down; a roll of either 3 or 4 means that all of those in group 3 and 4 sit down; a roll of either 5 or 6 means that all of those in group 5 and 6 sit down. In this way, only two-thirds of those in their first year of school have now gone on to their second year.

Repeating this again for those standing gives the number who can go on to the third year. One more roll of the dice in this manner then shows that only a very small number remain, the girls who are able to graduate after four years of school (approximately 10% of the population of school-age girls).

Not only does this make the data in the table come to life for the children, it provides even further impetus to discuss the meaning inherent in the data. It also raises the issue of the amount of data that needs to be collected to depict the situation in Bangladesh accurately. Although the theoretical analysis shows that rolling dice in this way will allow about 10% to 'graduate', in fact there will be discrepancies from this result, reflecting the small number of trials available. A number of further trials not only will show that the accuracy of the simulation will be increased with a greater amount of data, but it will also serve to make the simulation even more personal as different students find that they can or cannot go to or continue at school. Questions can then be raised as to why some children always seem to 'win' their place, or are unable ever to get the chance to go to school. This analysis in terms of the probability model can then be contrasted with reasons that children might advance from a social analysis of the situation.

Many other questions relating to this data table are presented in *Chance and Data: Exploring Real Data* (Finlay & Lowe, 1993, p. 42), but children can easily be challenged to investigate further many of the other situations the data portrays. This could lead to meaningful ways of introducing the use of other graphs. For instance, scatter plots could be used to show the ways in which the proportions of girls and boys participating in education differ and the effects of using different types of graphs to show or hide the message in the data could also be investigated.

Data sense and numeracy

Chance and data have often seemed distant and difficult areas of the mathematics curriculum. The background presented here is intended to shed light on the few, fundamental ideas that are actually involved. More importantly, the investigations have been used to show how these ideas can be introduced in a meaningful and involving way, so that misconceptions can be challenged and new understanding and a sense of what is concerned can be built up. In this way, children can extend their understanding of mathematics beyond its traditional bounds of number, computation, space and measurement to gain the levels of numeracy appropriate for informed participation in the data-filled society in which they will live.

Investigate and discuss

1. Try each of investigations 1, 2 and 3 with a Year 6 class. Use the data gathered by the class in the manner suggested here and write a report on the outcomes of each investigation.
2. Use the data on the line plot on p. 511 to construct a box plot that presents a concise summary of this data.
3. The following data were gathered on four chance and data assessment tasks.

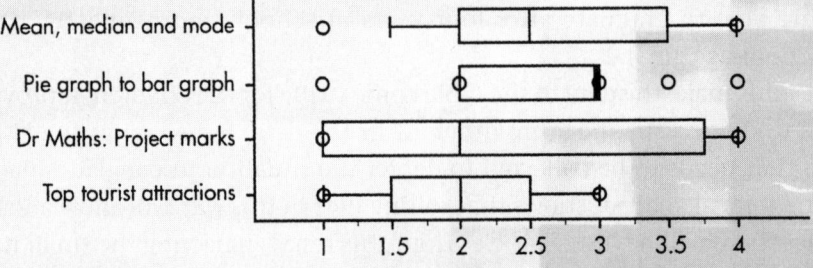

What comments can you make regarding the relative difficulties of the tasks and the performance of the class that completed them?

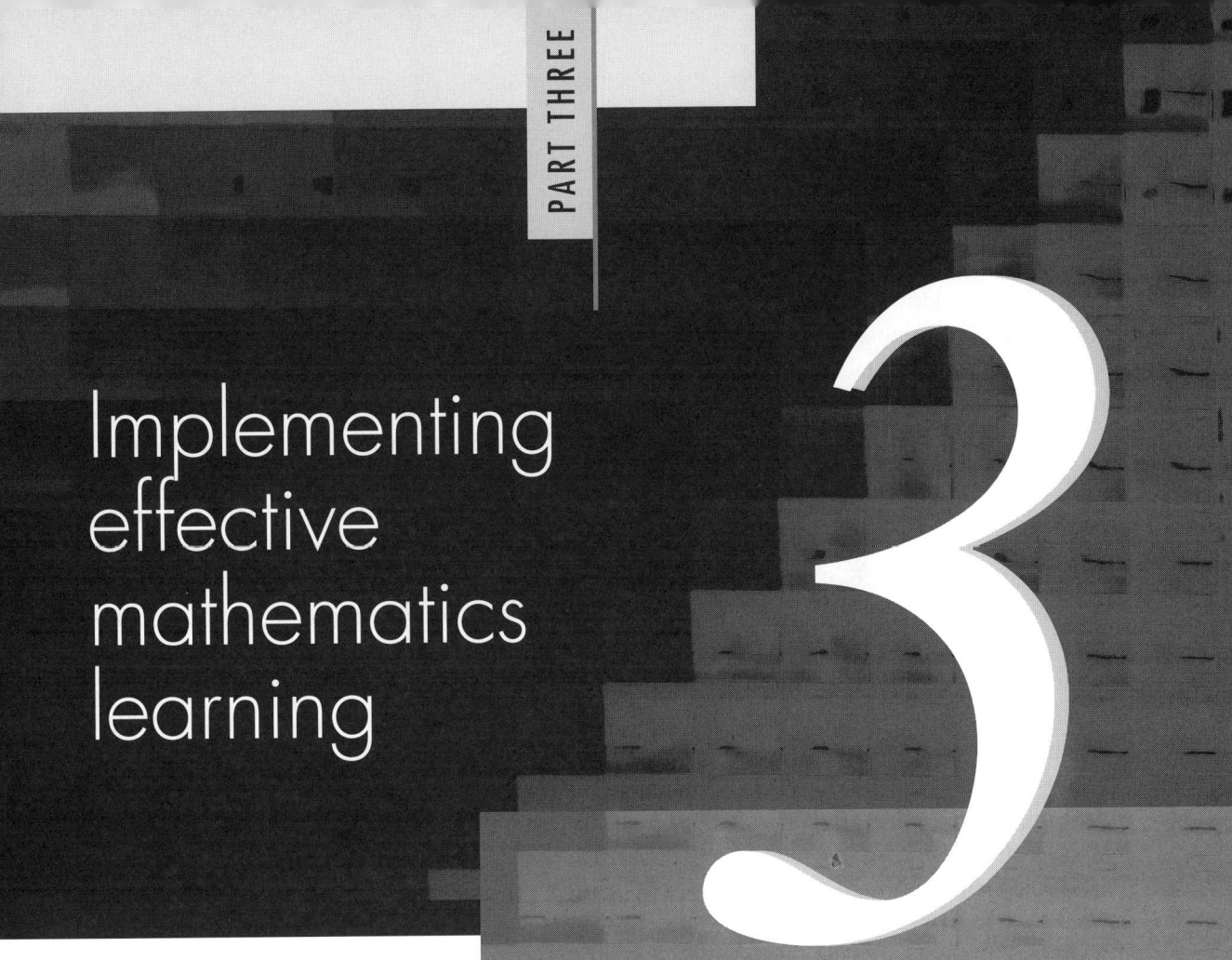

PART THREE
Implementing effective mathematics learning

THE OVERVIEW to Part Two proposed a basis for developing mathematical understanding of concepts and processes in order to build up an ability to use and apply mathematics. This approach analysed each concept and procedure from the perspective of an individual learner so that the chapters and sections that followed developed an order of presentation for each new idea that accorded with children's needs as novices. Not only did this take into account the underlying mathematical understandings, but it also reflected contemporary beliefs concerning the manner in which these mathematical ideas are best developed by children. The use of materials and models out of which a meaningful language could emerge and in turn lead to the recording of numbers and processes with understanding was held to be crucial. However, simply organising each separate topic on this basis is not sufficient. It is at least as important to look to the order and type of development across the entire emerging mathematical framework.

Often the learning of one aspect is crucially dependent on a concept or understanding based in another, even seemingly unrelated, area. For instance, the teaching of perimeter requires an understanding of number to provide a meaningful measure of each side, and of addition to provide the sum of the lengths of all sides. It is also often the case that a concept or skill in one area has already been fostered in earlier learning in another area. If numeration has stressed the idea of counting on from the outset, there is already a large knowledge base when this thinking is called up for mastering many of the addition facts. Similarly, if the initial ideas for numbers have stressed a visualisation of numbers in terms of a ten frame, the basis for the doubles facts and the make to ten strategy has been acquired long before any formal work on addition has begun.

On the other hand, any attempt to develop mathematics as an activity in which thinking strategies are paramount is substantially discounted if much time is devoted to re-teaching material already gained in a different context. For instance, practice in the addition of ten as a basic fact hardly seems necessary or instructive when 'five and ten is fifteen' is obvious from earlier knowledge of the teen numbers. Multiplication 'facts' involving ten are also known from the naming convention for the multiples of ten, '6 tens are 6t', while the twos facts are simply a revisiting of ideas familiar from the addition doubles. Appropriately sequenced, the 'new' material can be brought out to be a re-statement of something already developed, not only reinforcing the connections between the various topics, but also building up a child's confidence in his or her growing mathematical knowledge and understanding.

Consequently, although Part Two has set the scene for sequences of development for each of the fundamental areas of mathematical development in the primary school, the need to consider an organisational framework across these areas is of at least equal concern. A matching of the understandings and thinking needed for and available from other areas must be built in. A further consideration is that if ideas are to be generalised and applied, the development of each concept or process must not be rushed. There needs to be some lapse of time between the initial development of the ideas and their use in relatively straightforward applications. In turn, further time is needed before more complex problem solving is to be routinely attempted. This is not to ignore the need to set initial learning in the context of problem solving, but provides a pathway to a confident ability to attempt to resolve new, unfamiliar situations. Accordingly, a development that goes from

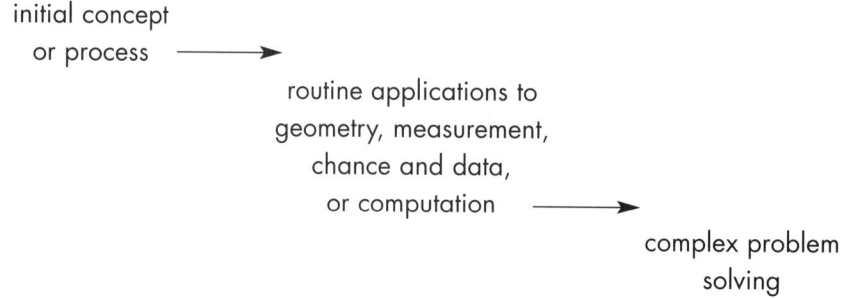

is most helpful. So, in addition to the considerations of sequencing within each area, it is necessary to consider interactions across the areas of number, the operations, geometry, measurement and chance and data along with the requirements of an emerging problem solving focus in organising an individual, class or school mathematics program.

Clearly the development of numeration for the various number ranges needs to be completed before these numbers are called for in processes for the operations. The types of situations required to introduce concepts for the operations need to involve small numbers in order for understanding to emerge clearly. Developing the addition concept, for example, depends on a knowledge of the one-digit numbers. A full understanding of these numbers needs to be internalised before the concept can be meaningfully introduced. The addition facts generally have teen numbers for answers, so learning them does not make sense until the numbers that constitute the answers are secure. In fact, given the sequence leading up to the teen numbers, children should be familiar with all of the two-digit numbers before learning of the full range of addition facts is begun. While this could appear to be delaying the learning of addition facts, the strategies that generate these answers are all based on earlier ideas built up as the underlying numbers were developed. Not only do counting on and using doubles facts involve numeration ideas, but the strategy of making to ten is implicit in the thinking used in grouping numbers in tens and ones. The thinking strategies for addition facts are largely available as a consequence of this interaction between numeration and addition. In the same way, a full understanding of two- and three-digit numbers is imperative for the introduction of the addition algorithm when the renaming associated with the learning activities that lead to place value ensures a ready building of addition with renaming from the outset.

Similarly, it makes little sense to expect facility with subtraction facts if the corresponding addition facts are not known. Subtraction with renaming requires the reverse of the idea that new units are created by grouping 10 of an earlier place to form the next place. Consequently, it is helpful to introduce subtraction with renaming only after addition with renaming is secure. More difficult subtraction demands a good understanding of renaming in terms of the number of tens altogether, particularly for numbers with internal zeros.

The following sequence of development linking numeration, addition and subtraction reflects these considerations to provide a way of planning for the learning of number in the early years.

Level	Numeration	Addition	Subtraction
1 *Establish place value*	one-digit numbers two-digit numbers concept of ten 20–99 teen numbers, 11–19	+ concept + facts	
2 *introduce renaming*	three-digit numbers • internal zeros • number expanders	2 digit +2 digit renaming	− concept − facts 2 digit −2 digit no renaming
3 *consolidate renaming*	four-digit numbers	3 digit +3 digit renaming	2 digit −2 digit renaming 3 digit −3 digit renaming internal zeros

After these initial steps, most of the development of numeration, addition and subtraction is simply a matter of making straightforward extensions to the ideas developed so far.

For the middle years of school, multiplication and division are the focus for computation. Knowledge of larger numbers is essential as even the multiplication concept involves a secure knowledge of numbers to 99. Addition as a concept and as a process must also be well known so that these can be used to assist with the learning of multiplication while at the same time being able to be distinguished from it. Multiplication with larger numbers involves the use of addition facts and the addition of larger numbers. Consequently, learning multiplication does not make sense until both the addition facts and the addition algorithm have been built up. In turn, as a knowledge of multiplication is fundamental at all levels of division, from the concept that shares the use of arrays to the written and mental processes that rely on recall of multiplication facts, the development of division needs to follow the corresponding multiplication concepts and processes.

Level	numeration	multiplication	division
4 second place value pattern	thousands five-digit six-digit fraction concept	× concept—arrays × facts 2 digit × 1 digit renaming	
5 recording fractions	millions decimal fractions • tenths common fractions	2 digit × tens	÷ concept ÷ facts 1 digit)2 digit 1 digit)3 digit renaming and remainders
6 renaming fractions	billions decimal fractions • hundredths • per cents	2 digit × 2 digit	1 digit)4 digit 1 digit)5 digit internal zeros in answer
7 equivalence of fractions	exponential notation	× decimal fractions	divide the remainder decimal fractions

Similar interactions are needed between each of numeration, addition, subtraction, multiplication and division at all levels and the stage is also set for the application of these ideas, along with emerging spatial understandings, to develop strategies for dealing with measurement, including chance and data. It is in this way that the discussion in Part Two was able to focus on the development of concepts and processes at the outset with a diminishing amount of time spent on these as mathematics was built up.

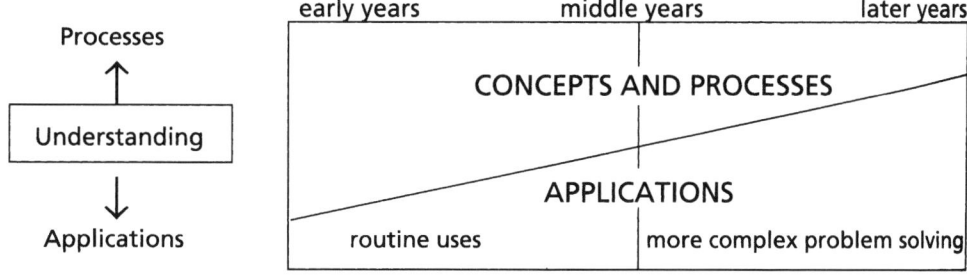

If the connections between the various topics and areas are taken into consideration then not only will the understandings and processes be built up as a cohesive mathematical view, but the time and the underlying content knowledge will be available to develop an ability to think mathematically and solve problems.

A major aspect of the organisation of the teaching/learning situation relates to the sequencing of the ideas within mathematics as a whole. However, all learning takes place in a social context and the way in which the classroom environment is managed is at least as important. Certain activities may be more suitable than others in bringing out the discussion that is needed to allow children to construct ideas for themselves. Particular arrangements of the physical environment or of the children themselves may be more suited to different learning activities. Some work may need to be carried out at home so that parents will need to be informed about the particular approaches being followed if they are to participate in and assist their children's learning.

A measure of the manner in which learning is progressing will also be required. As the emphasis in school mathematics programs has moved to focus on individual outcomes, assessment has become more diagnostic in nature, providing a measure of strengths and weaknesses and the reasons underlying both. Consequently, assessment has become integral to teaching and needs to be planned from the outset as a pivotal part of a mathematics program, particularly in informing children, other teachers and parents how individual learning is progressing.

These issues in the organisation and management required to allow the implementation of an effective mathematics program are the focus of this concluding section.

CHAPTER EIGHT

Planning mathematics

> Effective mathematics teaching requires understanding what students know and need to learn and providing a challenging and supportive learning environment in order for them to learn it well.
>
> *Principles and Standards for School Mathematics, 2000 p. 16*

The way students learn mathematics has been a topic for debate and discussion since classrooms were first devised. Conceptions about how students learn mathematics have changed over time. Mathematics, in many instances, has long been associated with memory and practice. Memorising facts, procedures, definitions and formulas together with pencil and paper work formed the basis of many classroom programs. The way in which the content was organised was seen as paramount to success with the teacher being the means by which the content was transmitted. While an appropriately organised curriculum is important, teaching mathematics has taken a more learner centred approach in which understanding is at the fore.

Much of the literature on how students learn mathematics now centres around the ways students construct their own ideas, processes and understandings rather than simply following someone else's ideas and understandings. Constructivism is an approach to learning which acknowledges the power of students building their own ways of thinking rather than simply accepting a teacher's explanations. In this way classrooms are transformed from places where the teacher has all the knowledge and information which needs to be imparted to the students to ones in which students think, plan, discuss, renew, construct and reconstruct their mathematical understandings.

A National Statement on Mathematics for Australian Schools (1991) highlighted a number of implications regarding the ways in which students learn mathematics:

- learners construct their own meanings from, and for, the ideas, objects and events that they experience
- learning happens when existing conceptions are challenged
- learning requires action and reflection on the part of the learner
- learning involves taking risks

A teacher's ideas and beliefs on how students learn mathematics form the basis of all classroom practice. Although there is no recommended approach or style for the teaching of mathematics, a teacher who believes students construct their own understanding will have a classroom in which mathematical thinking, problems solving, number sense, communicating and investigating is at the fore. Conversely, a teacher who believes rote learning and memorising of procedures are the best ways students learn will have a classroom which reflects this belief.

In 1999 the Australian Association of Mathematics Teachers (AAMT) and Monash University were commissioned to establish a set of standards for excellence in teaching mathematics. The aim of the project was to identify factors that were important and three areas, *professional knowledge*, *professional attributes* and *professional practice*, were identified as being essential for excellence in teaching mathematics. The standards highlight the importance of knowing the students, knowing mathematics and knowing how students learn mathematics. In addition, they acknowledge the importance of the learning environment, of being an enthusiastic teacher, of ongoing professional development, planning and assessment. The complete document together with comments from the teachers involved in the project is available on the AAMT website—<www.aamt.edu.au>.

Planning a mathematics program

A National Statement on Mathematics for Australian Schools (1991) provides a framework for teaching mathematics in Australian schools. This document was a collaborative project of the states, territories and Commonwealth of Australia and as such it 'documents areas of agreement between education systems about directions in school mathematics, the principles which should inform development and the extent and range of school mathematics' (p. 1).

Other Commonwealth documents relating to mathematics include *Numeracy, a Priority for All: Challenges for Australian Schools* and the *Numeracy Benchmarks Years 3, 5 & 7*, both of which were published in 2000. The former outlines the numeracy policy for Australian schools with a particular focus on numeracy in primary schooling. It details the government's commitment to achieving the national goal of numeracy for all Australian students. The *Numeracy Benchmarks Years 3, 5 & 7* followed from a meeting in 1996 in which all the ministers for education agreed that 'every child leaving primary school should be numerate'. The ministers agreed to develop national benchmarks for use in reporting minimum acceptable standards of numeracy achievement.

These documents are particularly useful in providing a general structure for schools or school systems to develop programs of learning experiences and assessment that are appropriate for the unique needs of their student clientele, while allowing for helpful broad consistency across the regions and states of Australia.

The education authorities in each state of Australia have produced their own mathematics curriculum documents or syllabuses. The structure of the curriculum documents varies from state to states to reflect the local nature of each region. Currently most state documents take an outcomes-based approach to teaching and learning and are organised according to various strands, for example number, space, chance and data.

An outcomes-based approach to education defines the end products or outcomes of an education process rather than its inputs. The learning outcomes are used as the starting point for decisions about appropriate student learning experiences. The learning experiences are devised to enable students to demonstrate outcomes from within strands and across strands. Most outcome documents have core outcomes and discretionary outcomes, with the core outcomes being the essential learnings and the discretionary outcomes being up to a teacher's discretion.

In turn, each school usually develops their own school mathematics program based on the state curriculum documents. The school mathematics program identifies the beliefs, goals and core curricula for each local community. It enables a school to develop a program of mathematics specifically situated to their particular area. While programs differ from school to school, a program may contain the school's shared philosophy of how students learn mathematics, the core curriculum for each year level, the language to be used throughout the year levels, for example 'add' not 'plus', the setting out of the algorithms and assessment ideas or requirements. Many school programs take an outcomes-based approach in line with current curriculum documents and educational trends.

The school mathematics program is used by class teachers to plan their individual programs of work. When planning a program, a teacher needs to start with a yearly plan which is then broken down into term plans and finally smaller units of work. Some work would be ongoing throughout the year and other work would build on previous work. Units of work may relate to particular mathematics strands, topics or classroom contexts. Weekly and daily programs are planned from these units of work. During the planning phase, teachers need to make decisions not only about what curricular and how to organise and sequence it, but also about appropriate time allocations, available resources and classroom organisation. Daily and weekly programs enable teachers to fine tune the amount of time allocated to a particular strand, topic or concept.

When planning weekly programs, decisions about continuity and variety must be made. Some teachers plan to include experiences from several strands, such as number, space and measurement, in a particular sequence each week, but modify these plans to suit specific needs when, for example, it is opportune to complete a longer than usual measurement investigation in successive class sessions. Other teachers plan extended programs of several days' duration for each strand in turn, for example space, then number, then measurement, then chance and data, and so on, together with relevant input from other program strands, for example attitudes and appreciation or mathematical enquiry, to cover a three- or four-week unit of class experiences. Several textbook series are available for Australian schools that include work programs which outline specific objectives for each lesson topic. While these resources are helpful for use at the planning stages as well as for later class use, teachers still need to make decisions about sequencing, revision and the individual needs of each child. No textbook series will provide everything needed for a well-balanced rich program of mathematics. In this way textbooks should be seen as a supplement to any class program just as calculators, computers and other resources all combine together to make a well-balanced and varied program.

From this it is possible to see that a teacher's planning involves three aspects—a yearly plan, a term or semester plan and weekly or daily plans.

Yearly plan

- Provides a broad overview of the work to be undertaken in each term or semester across all the relevant curriculum strands of space, number, measurement, chance and data, attitudes and appreciation, mathematical enquiry, choosing and using mathematics and algebra.
- Emphasises the manner in which these different topics are interconnected, showing how ways of thinking in one domain can be drawn on to develop other areas, and including applications that call on such an integrated view of mathematics.
- Highlights the setting out and the language to be used.

Yearly program

Strand	Outcomes
Number	Compare and order whole numbers to 999
	Identify and solve addition problems
	Identify and use addition strategies for basic facts
	Complete two-digit with two-digit addition with renaming
	Compare and order ordinal numbers to 999
	Identify and solve subtraction problems
	Identify and use subtraction strategies for basic facts
Measurement	Use non-standard units to measure, estimate and order the size of objects
	Use standard units to measure, estimate and order the size of objects

Term or semester plan

- Identifies how the yearly plan is broken down into units of work which can be completed in a particular number of weeks.
- Contains details relating to sequential considerations and building on prior knowledge.
- May be changed and modified following assessment of the students and evaluation of the teaching program.

Semester program

Strand	Semester one	Semester two
Number	Review two-digit numbers	Read, write, compare and order all three-digit numbers
	Read, write, compare and order 100s	Round three-digit numbers to nearest 10 and 100
	Read, write, compare and order regular three-digit numbers, e.g. 463	Complete two-digit with two-digit addition with renaming
	Read, write, compare and order three-digit numbers with teens, e.g. 317	Compare and order ordinal numbers to 999
	Round numbers to nearest 100	Complete addition basic facts with automatic recall
	Identify and solve addition problems	
	Identify and use addition strategies for basic facts	Identify and use subtraction strategies for basic facts
	Identify and solve subtraction problems	
Measurement	Use non-standard units to measure, estimate and order the size of objects	Use standard units to measure, estimate and order the size of objects

Weekly or daily plan

- Records the finer details of a program including details of specific student outcomes, learning experiences, materials and resources to be used.
- Details how the program is organised, that is, different strands each day, 2 or 3 strands per day, which may vary from week to week.
- Shows variations in class and group organisation and furniture arrangements and assessment procedures to be used.

Daily/weekly program

Monday	Tuesday
Whole class	*Whole class*
Read three-digit numbers	Review problem solving plan
Revise count on strategy	*Paired work*
Group rotations	Addition problems
Group 1	Read, discuss and solve problems in pairs
Read and make three-digit numbers using place value charts and Base 10 materials	Write up solutions on charts
	Whole class
Group 2	Discuss the problems solved and how they were solved
Play Number Bugs	Display charts
Group 3	
Count on sheet and count on game	
Morning tea	Morning tea

Ongoing student assessment and evaluation of the teaching program is essential for the teaching and learning cycle. Monitoring of student progress enables teaching programs to be modified and refined to meet the ongoing needs of the students. In the beginning of the school year, the records and observations made by the previous year's class teachers can provide valuable insight for planning learning experiences to best suit the students' needs. As the year progresses, data gathered from ongoing assessment and monitoring should be used to enable teachers to modify the class program to suit the changing needs of the students.

Planning learning

Effective teaching and learning experiences begin with careful and thorough planning for each unit of work. The first step in planning any learning activities is to understand the learner—what prior knowledge do learners have, what cultural or social features may influence learning, what motivates and challenges students, what are the needs, interests and abilities of individuals and groups within the class. Other factors which need to be taken into consideration are the learning process—how do students learn basic facts, how do students learn two-digit numbers and so on—and providing a supportive yet challenging environment.

Inexperienced teachers often marvel at the abilities of their experienced peers to present lessons of high quality with apparent ease and minimal planning. This is a situation where appearances are usually quite deceptive. Competent experienced teachers have usually spent many years carefully planning appropriate learning experiences to the stage where quality planning now appears to occur quickly and effortlessly. To reach this stage of competence, much initial effort involving thorough planning and preparing needs to be expended. The following guideline is offered as general advice, to be adapted to suit specific needs.

Developing a lesson or unit of work

1 **Learning outcomes**—identify the student outcomes you want to achieve from the learning experiences. What understandings will the students gain from the learning activity? Will the learning experience cover one strand or more than one strand?
2 **Lesson steps**—what steps are needed to achieve the stated outcomes, and in what sequence should they be organised?

   a Introduce the lesson
   - Link with previous experiences—specify appropriate strategies to relate the learnings planned for this lesson to children's prior knowledge, learnings and experiences.
   - Prepare—identify experiences to prepare the children for learning by establishing the motivation, focus, setting, direction and dissonance for the new learning.

   b Develop and sequence the lesson steps
   - Present new learnings—specify the implementation and careful sequencing of the activities that will promote the desired learnings.
   - Personalise—specify the activities that will involve the learner in understanding, conceptualising, internalising and owning the learnings developed in the lesson.
   - Apply—specify the experiences that will provide opportunities for the children to use the learning in situations and settings different from those in which it was learned.

   c Conclude the lesson
   - Closure—identify experiences that will enable the children to draw together the ideas, skills, concepts, processes and so on that have been the focus of the lesson.
   - Communicate—specify the experiences that will enable the students to communicate their understandings, that is, talking to peers, group/class presentation
   - If relevant, use these experiences to prepare for the introduction of further learnings that are to follow in subsequent lessons.

Thorough lesson planning does not guarantee that teaching will be successful or that learning will occur. On the other hand, poorly planned lessons are rarely if ever effective in achieving learning outcomes. Therefore, developing competence in planning effective teaching and learning experiences is an essential part of becoming an excellent teacher of mathematics.

Workshop—a trial planning experience for prospective teachers

1 **Preliminary activities**

   a So that you are aware of recent trends, familiarise yourself with the broad details of relevant Australian documents including the following:
   - *A National Statement on Mathematics for Australian Schools* (1991)
   - *Numeracy, A Priority for All: Challenges for Australian Schools* (DETYA, 2000)

- *Numeracy Benchmarks—Years 3, 5 & 7* (Curriculum Corporation, 2000)

  b To obtain a broad perspective on recent international developments familiarise yourself with the general details of relevant documents like the following:
  - *National Curriculum in Mathematics* (HMSO, UK, 2002). Available online at www.nc.uk.net by clicking on Mathematics
  - *Principals and Standards for School Mathematics* (National Council of Teachers of Mathematics, USA, 2000). Available online at http://www.standards.nctm.org/document/

  c Now examine in more detail the relevant curriculum documents for your state, territory, district or school system. These documents typically include syllabus statements, guidelines for interpreting and implementing syllabus recommendations, source books and other resources for teacher use in planning programs of learning experiences and accompanying assessment programs. They also commonly include suggestions for specific kinds of activities, games, group and individual investigations, appropriate materials and other advice to assist teachers in planning experiences for specific strands of the class mathematics program.

2 **Planning experiences**—the preliminary activities suggested above were chosen to provide some familiarisation with relevant sources of advice that may need to be consulted in more detail at this planning stage. To specify a realistic context for the planning experience, consider yourself in the following situation. For a selected primary year level (e.g. Year 4) design a work program and accompanying assessment instruments in the following context.
   - The primary school at which you are to teach Year 4 for the coming year is about to review its mathematics program to ensure that the school program includes at least all the details for each year of the mathematics syllabus or program for your state or school system.
   - You have been appointed chairperson of the Year 4 Mathematics Committee. The other Year 4 teachers on this committee have agreed that, because of your recognised reputation as a mathematics teacher of excellence, you should prepare discussion documents to help expedite discussion and decision making at the next meeting. You have therefore agreed to prepare draft versions of the following.

   a A yearly work program across four terms for the Year 4 mathematics program. This may be in the form of a scope and sequence chart or some other preferred format.

   b A detailed work program for Term 1 (of ten weeks), showing unit planning details including outcomes, learning experiences, resources and organisational details, cross-referenced to any teacher resource materials including the mathematics curriculum documents for your state or school system.

   c A program of assessment for Term 1 (you will need to study Chapter 9 before undertaking this step). Include any of the following that are appropriate:
   - overall checklists and inventories of expected achievement, knowledge, skills, behaviours and attitudes

- simpler checklists to identify specific behaviours, skills, attitudes and the like that teachers should focus attention on at critical points in learning experiences
- summative evaluation instruments with accompanying specification tables to constitute part of the overall school evaluation program
- use of student profiles
- any other appropriate instruments or details

You should indicate the time during the term program when it would be appropriate to use each designated assessment instrument.

d A concise discussion paper identifying any critical judgements you have made about such details as program sequencing, language and setting out to be used, choice of materials, assessment, and summarising reasons to justify your decisions. Alternatively, you may choose to identify arguments for and against critical decisions, for committee consideration and discussion.

Your work programs should also indicate which specific learning experiences should preferably incorporate group activities, problem-solving experiences, specific calculator and information technology use, and any other important experiences that teachers should be emphasising.

The learning environment

In addition to a teacher's knowledge and thorough planning, the learning environment plays a key role in developing mathematics. The *Principles and Standards for School Mathematics* (2000, p. 18) state:

> Teachers establish and nurture an environment conducive to learning mathematics through the decisions they make, the conversations they orchestrate and the physical setting they create . . . more than just a physical setting with desks, bulletin boards, and posters, the classroom environment communicates subtle messages about what is valued in learning and doing mathematics.

Physical constraints such as the arrangement of the classroom and ways of storing and accessing teaching resources impact on teaching decisions and learning outcomes. A teacher needs to consider the learning environment every bit as carefully as he or she considers the content and detailed sequence of activities that will be required to bring an understanding and ability to use new concepts and processes.

While the learning atmosphere that a class teacher establishes is influenced by physical characteristics such as the shape and size of the allocated classroom, the teaching space available outside the classroom, the furniture and floor and wall coverings in the room, the material and person resources available and storage facilities, it is just as strongly influenced by the pervading educational philosophy, beliefs, preferences and attitudes of the teacher operating within the classroom. Often within a school, teachers can have quite opposing philosophies and beliefs. It is necessary for agreement to be reached within the school as to what is the underlying philosophy on how students best learn mathematics. The commonly held policy and philosophy on how students best learn mathematics can be incorporated into the school mathematics program. The following commonly held beliefs could be discussed when establishing a school philosophy:

- Children learn mathematics more readily when they are active rather than passive learners, when they participate in discussion, when the mathematics is presented attractively and with plenty of appropriate manipulative materials and challenging situations.
- Each class consists of individuals with different backgrounds, needs, capabilities, achievements, and levels of mathematical interest and motivation.
- Learning environments that encourage experimentation, activity, discovery, discussion and decision making are important.
- Being able to discuss and communicate about and in mathematics is essential.
- Problem solving, mathematical thinking and number sense underpin mathematics programs.
- A variety of strategies should be experienced to allow previously acquired knowledge to be applied in new and unfamiliar situations, especially those involving everyday situations.
- Available technology should be integrated into teaching and learning activities and appropriate use of calculators should be developed across all year levels.
- Students and teachers should enjoy all mathematics experiences in a happy class environment.

Teachers who attempt to implement these principles need to be flexible in adapting the learning environment to suit different requirements and circumstances. Issues that require ongoing monitoring and flexible adjustment include the management of learning groups, learning modes, children's behaviour, materials, adult resources and homework experiences.

Learning groups

The learning environment should be sufficiently flexible to allow children to be regrouped in different ways to suit particular circumstances. This means that the classroom furniture and materials should be able to be reorganised conveniently in the available space to suit the whole class, small group or individual experiences, and that children should be able to move conveniently to other locations such as a designated mathematics area when desirable.

Some teachers avoid using groups because of the management problems associated with group work. However, group work can be just as easy to manage as whole class activities provided care is taken. When planning group activities it is important to keep in mind that students need to have group skills, that is, take turns in speaking, listen and respect other members, select members to do different jobs, use quiet voices, vote on action if necessary, discuss and then agree/vote on a plan. It is also important that the activity can be managed by the students themselves; they need to know explicitly what is expected of the group, know where to get necessary resources, know what to do when finished.

The following shows two examples of the many ways in which classrooms may be organised to suit group activities. Flexibility in the learning environment is important as no particular instructional mode is universally appropriate for all circumstances. Rather, different modes (whole class, cooperative groups, small groups, individual work) suit different circumstances and needs. The following broad categories illustrate some of the varied approaches to learning needs.

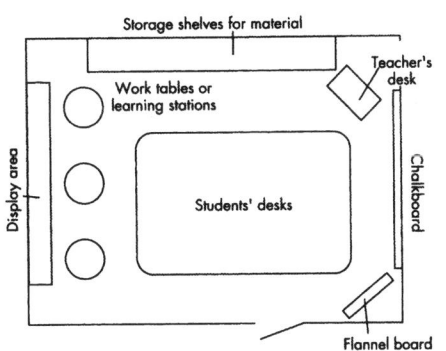
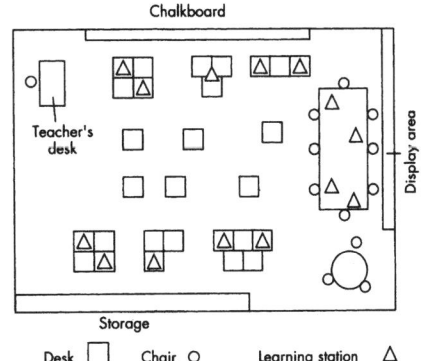

Whole class teaching

There are many circumstances where the teacher may judge it appropriate to take a strong direct leadership role and use a systematically presented sequence of events during a class lesson. This approach may be useful and efficient when clear explicit instruction is desirable to demonstrate the use of new equipment, introduce a new concept, understanding, game or activity, view and discuss a videotape, plan a class activity, summarise the outcomes of a series of group investigations or convey information in a systematic and appropriately paced manner.

Small group cooperative work

Organising children into small teacher-selected groups is often a very effective way to teach. Good group interaction can motivate more persistent effort and a diversity of approaches to demanding tasks with which children may be less inclined to persist individually. Teachers can use this method to:

◆ cater for a wide range of abilities with activities for intervention, consolidation, revision and extension
◆ promote mathematical thinking and problem solving with students working together to solve a problem
◆ develop concepts and skills that can be reinforced in later whole class discussions
◆ provide appropriate settings for cooperative social development

Effective group work requires the same high-quality planning from the teacher as other learning methods. Merely putting children haphazardly into groups and allowing them to proceed with little planned purpose does not ensure that productive work will result. In planning a group activity, the teacher must decide whether groups will be homogeneous or heterogeneous. *Homogeneous* groups consist of members who are comparably capable and are appropriate in those circumstances when very capable children are to be challenged to their limits on one set of tasks while children requiring intervention are grouped together, often with a teacher, teacher aide or parent helper to assist them. This allows the teacher to meet more specific needs of students.

Heterogeneous groups consist of sets of children with abilities ranging across the class spectrum, and are appropriate when children of different capabilities can all make contributions to enhance each other's learning and can participate in different ways in the group effort, with the obvious social benefits that arise when each different individual input can be valued. In such circumstances, the less developed reading or language skills of some children can be minimised by the efforts of the better readers in each group so that all can participate in the mathematical activities without disadvantage. Opportunities for peer teaching—both incidental and deliberate—can also enhance these activities.

Individual work

When planning units of work it is important to consider a balance among whole class, small group and individual work. Some students prefer to work on their own and to set their own goals. Individualisation of learning can motivate such students by allowing them to work at their own pace. Another approach is having children work on a contract system for some learning experiences. It may be helpful for a student experiencing learning difficulties to work individually with a teacher aide or parent helper. Similarly, a student with highly developed abilities may work with a teacher or student from a higher year level or with a mentor from outside the school. Each of these activities can be used when appropriate for extension, review, reinforcement, consolidation and intervention.

Learning styles

Just as there is a wide range of levels of cognitive, physical and social development among the children in any class group, so there is also diversity in the children's preferred modes of working, thinking and learning for different aspects of mathematics. In any situation there may be some who would prefer to operate chiefly in a visual mode where their understanding and learning is enhanced by appropriate visual representations and responses, while others prefer to operate chiefly in a verbal mode involving considerable oral discussion and writing. Others may prefer to work in a concrete mode where the process is developed through much manipulation of concrete materials or through dramatic participation by 'acting it out', as in grouping games and shopping activities. Activities involving all of these modes are clearly appropriate and desirable on most occasions. By incorporating a diversity of modes at different phases of each particular lesson, the teacher accommodates the preferences of different children at some stage while desirably encouraging all to undertake experiences in a variety of modes as well as the ones they might prefer on that occasion. Such variety not only promotes involvement and enjoyment in the learning process, but also assists in developing meaningful learning with understanding instead of relying on mere rote learning.

Managing resources

The approaches to teaching and learning mathematics suggested throughout this book require classrooms to be organised so that children have ready access to materials. Number work may need a variety of collections of discrete objects (shells, counters, buttons, bottle tops and the like) for early sorting, classifying, describing, sharing, matching and counting experiences, as well as appropriate place value materials like Unifix cubes, Bundling sticks, Base 10 blocks and place value charts for learning about two-digit and larger numbers. Spatial experiences require ready access to an assortment of three-dimensional objects, multilink cubes, building blocks, pegboards, geoboards, pattern blocks, geostrips and plenty of spatial puzzles like tangrams, trimension, jigsaw puzzles and the like. Measurement and chance and data activities similarly require a wide variety of everyday materials and appropriate measuring and recording equipment for the many different topics that must be investigated, while a mapping problem may require access to string, counters, grid paper and a calculator.

In preparing for the day's mathematics activities, the class teacher must ensure that all needed materials are readily available at the appropriate time. The ideal situation is one where all the materials relevant to the class mathematics program for the year or at least the current semester are stored in the classroom. These materials must also include any that may be needed for intervention activities as well as extension materials to challenge those who learn more quickly than others in the class.

Materials should be stored so that any child in the class can readily identify what is needed and can pack materials away and return the containers to their allocated locations. Plastic tote trays and cardboard storage boxes can be purchased for this purpose through conventional school supply companies. Plastic containers with fitting lids purchased through supermarkets provide an inexpensive alternative. If the school cannot afford these containers then fruit cartons, shoe boxes and the like can serve the purpose. Appeals to parents usually produce generous responses in supplying such items, which can be decorated appropriately by the children. Open shelving provides ready access to the labelled containers, but school cupboards, trolleys and lockers can also serve this purpose. With a little ingenuity and some community cooperation, any classroom can be organised so that the required materials are neatly stored and adequately labelled to be readily accessible for immediate class use.

Some teachers store all the mathematics equipment in a central 'mathematics area' which might be a set of shelves, a bookcase or even a window sill. Other teachers might organise the room into different activity areas with the various materials in each area. Everyday items such as paper and calculators might be stored in a central area but materials and games for place value might be in one area and materials and games for basic facts in another.

Any worksheets that are planned for use as stimulus materials in concept development or for written reinforcement follow-up activities should also be stored in a convenient systematic manner. A filing cabinet, folders with plastic sleeves, slimpick wallets and tidy trays can all be used for this purpose. Once students are familiar with how and where materials and resources are organised and stored they can be expected to take responsibility for assisting with resource management by keeping areas tidy and restocked as needed.

Managing children's behaviour

Classrooms in which children are encouraged to participate actively in the learning process through the ready manipulation of materials and the use of investigations, discussions, constructions and so on, often in group situations, are generally seen to make greater demands on teachers' management skills than those where children are commonly engaged in silent individual desk work. However, teachers who have developed appropriate strategies to deal with active, motivated children generally report more satisfying and productive learning environments for both teachers and children. Thorough preparation and classroom management strategies, especially those relating to noise, are often called on when effectively managing active group learning situations.

Effective preparation requires learning outcomes that are clearly defined and tasks or resources that are ready before each session begins. Displaying the program of activities for the day or week on a notice board together with materials to be used and group memberships keeps everyone informed so that children know in advance what will happen. These preparatory notice board displays should be used across all year levels as an accepted school policy, so that children from all classes develop the habit of looking for the information in advance, preparing for the activities and making suggestions about them so that they become active participants in their own learning.

Children commonly exhibit negative attitudes and disruptive behaviour when they feel alienated from the system in use. Once children feel secure that they are valued members of a team and responsible for their own learning, they often become too busy and interested to be involved in disruptive behaviour and will exert peer group pressure on class members who disrupt their activities.

Active participatory group learning almost always involves some class noise. Classroom noise can be happy, enthusiastic discovery noise from involved and motivated learners or it can be objectionably loud and disruptive noise that is unproductive and offensive. Loud unproductive noise commonly occurs when group activity learning is undertaken with poor planning of the activities to be used and inadequate preparation of the children and materials. Teachers can prepare the children to cope sensibly with the excitement of group activities by discussing appropriate working behaviour with them and encouraging them to establish the rules that are to apply to disruptive behaviour. As children enjoy learning actively, the associated pleasure can be used as positive reinforcement to encourage appropriate behaviour. As mentioned previously students need to have group work skills and to be able to actively manage the task at hand.

The acceptable noise associated with enthusiastic participation in group activities can worry some teachers to the extent that they even omit desirable experiences merely to avoid noise in the classroom. Teachers should not be embarrassed or concerned when class noise is of the desirable kind associated with good activity learning. Noise can also be minimised by scheduling a mix of quiet and boisterous activities at any one time. Appropriate care at the planning stage can ensure that all groups are not working simultaneously on noisy activities. By considering simple measures like these, any potential school problems arising from the desirable noise associated with good activity learning can be forestalled.

Adult resources

It is often the case that mathematics activities undertaken by children working in groups are more effective and successful when cooperating adults are available to assist the class teacher. Such assistance is usually provided by employed teacher aides and parent volunteers.

Most educational employers have very clear and definite regulations and guidelines to indicate the kinds of duties that may be undertaken by teacher aides. They are usually permitted to prepare materials for teacher use by assisting with typing, word processing, duplicating and photocopying class materials, by setting out and clearing away class equipment and similar helpful activities. Parent volunteers similarly may legitimately assist teachers by helping to prepare materials, rearranging the class furniture for different activities, setting out and clearing away materials, and even by helping children read instructions and sitting with a group and participating in activities when appropriate.

There is a serious obligation on schools to ensure that only qualified teachers are given autonomous teaching responsibilities. It is therefore essential that other assisting adults are not put in any situation that might compromise them or the school. Adult assistance is much too valuable a resource to have its availability jeopardised through poor administration. Teachers therefore must ensure that assisting adults are always welcomed, valued and respected and provided with clear guidelines and exact instructions relating to the limits of their responsibilities, and especially the requirements of avoiding making comparisons between children, discussing children and the like.

Perhaps the most important form of adult assistance available to teachers is the one that can be easily overlooked. As in other school activities, children's performances in mathematics can be strongly influenced by parental attitudes, expectations and support. The quality of the learning climate is enhanced when parents are well informed about classroom activities and expectations, have opportunities to participate in school activities beyond the level of fundraising and are assisted to feel comfortable about discussing matters frankly with teachers. Misunderstandings commonly arise when parents are not given opportunities to be informed about current learning practices that differ markedly from their own school experiences. For example, when a class is about to begin subtraction experiences, it would be opportune to invite parents to attend a meeting at which they could experience the use of Bundling sticks and Base 10 materials in developing the subtraction algorithm with renaming. Encouragement to participate in relevant mathematical games with their children would enhance their involvement in the learning process in a beneficial way and minimise the complaints from children that 'Dad says it's easier to take away by borrowing and paying back'.

Supportive parents are too valuable an asset to risk their loss through insensitivity or inaction. A little effort by teachers to inform and encourage them may well be one of the most effective ways of utilising a valuable freely available resource.

Homework

When used appropriately, homework experiences can provide a valuable adjunct to classroom activities. Homework can provide opportunities to reinforce and consolidate what has happened in class, and can provide time for students to reflect on recent learning. However, excessive use of drill and practice can distort this valuable learning situation.

A far more important goal in using homework activities is to communicate quite effectively to the children and their parents the distinctive characteristics of the classroom learning experiences. For instance, when the class work involves bundling and renaming activities that establish foundations for the development of two-digit place value, the homework should reflect what has been experienced in class. This can be done by having students take home some Bundling sticks to play renaming games with their family, or, to reinforce renaming activities, using worksheets with pictorial representations of the materials currently in use in class and requiring the renaming to be shown by circling, shading or drawing the renamed bundles of tens. Bundling sticks are available in packs of 1000 for just a few dollars. Students could keep a set of Bundling sticks in their tidy trays, which could be taken home on a regular basis for a variety of homework activities.

Similarly, when the homework is to be used to continue the exploration of problem solving undertaken in class, homework problems should be consistent with those under consideration. They should therefore be as interesting, challenging, divergent and process orientated as those used in the class activities. When children are working in groups to solve such problems, they could be invited to continue the group participation by involving other family members in their explorations. Teachers who have tried it report a surprising level of interest and involvement when the children are permitted to invite their participating parents to join them at school in reporting their solutions to the class. Homework problems that involve extensive or difficult computation normally performed on a calculator can serve the additional purpose of demonstrating appropriate uses of calculators in primary classes. Basic facts activities and games similarly provide opportunities for demonstrating to parents the effective use of thinking strategies in preference to traditional rote learning of 'tables' from their own school experiences.

Homework activities should reflect the varied experiences of classroom mathematics. They should therefore utilise the variety of learning experiences used by teachers including the use of materials, writing of explanations, drawing of pictures, finding and correcting of errors and preparing explanations and reports for class.

Used this way, homework is a powerful complement to class work. By contrast, homework that is excessive, punitive, pointless, irrelevant, stereotyped or consisting only of drill and practice activities does not in any way complement a class program. Homework activities have been described as providing windows for parents into their children's classrooms. Used constructively, homework can be an effective means of advising parents about what their children are learning and the effectiveness of the learning experiences being provided.

Beliefs and attitudes in mathematics learning

It has long been recognised (Dungan and Thurlow, 1989; Ellerton and Clements, 1990; McLeod, 1992) that teacher beliefs and attitudes influence how a classroom is organised and what mathematics will be emphasised and valued. The beliefs and attitudes of the students in the class, as well as those of the teacher, also play an important role in teaching mathematics. Students who are generally successful in mathematics often talk about mathematics as enjoyable and challenging while students who dislike mathematics are often those who are less than successful, who talk about it as boring, confusing and something to be avoided. Many people think back on their years of learning school mathematics and remember it as rote memorisation, rule based and lacking in any real sense. These people opted out of mathematics as soon as possible.

When deciding how a concept or process is to be taught, teachers are relaying subtle and at times unintentional beliefs and attitudes about mathematics (Hart & Walker, 1993; Nisbet, 1997). For example, a lesson involving measuring an area of a rectangle could take many different forms. A lesson that takes a 'draw this', 'label this', 'add this' and 'do this' approach with little or no discussion or exploration imparts a message that mathematics is a set of rules that need to be followed. Procedural knowledge rather than understanding is seen to be important.

A different approach is one where the teacher discusses a number of ways to measure an area, has groups of students decide which way they will use, marks out real areas to measure and carry out the measurement. This could be followed by a discussion about the methods chosen and the advantages and disadvantages of each method. The message from this lesson is that mathematics is real and there are different ways problems can be solved to suit different purposes.

In all schools and all classes, students have varying levels of performance in mathematics. How students attribute their performance is also important. Some will attribute their success or failure to their ability and/or effort (an internal factor) while others will attribute it to luck or task difficulty (an external factor). Some of these beliefs are related to gender where males will often attribute their success to ability and/or effort and their failure to an external factor while females will attribute success to luck and failure to themselves (Ellerton and Clements, 1990; Leder, 1992). It is the teacher's role to be aware of these beliefs and to address them within the classroom context. It is important to develop an environment where students see themselves as learners on a path of discovery and that success and understanding comes from effort and persistence and much less from luck and other external factors beyond their control.

Emotional responses to mathematics

In most classrooms mathematics is taught on a daily basis. It can therefore be expected that students may well develop an emotional response to mathematics. For some students this response could be excitement of finding the solution to a problem, satisfaction in completing a complex activity or pleasure at finding it is again 'maths time'. For other students the emotional response could be quite different with feelings of anxiety, frustration or embarrassment. These emotions need to be monitored by the teacher to try and establish the source of these feelings. Is the student worried about keeping up? Do they feel they are not understood by the teacher or other classmates? Do they feel undervalued? A student's perception of themselves as a learner influences their school learning and plays a direct role in their performance in mathematics. Teachers need to assess how students perceive themselves as learners in order to assist them to become confident about their ability to succeed in mathematics (Renga and Dalla, 1993). A classroom environment where all students regardless of ability, background or culture are valued, cared for and supported will assist students to develop positive perceptions, attitudes and emotional responses.

Implications for teaching

Teachers in the primary school, in particular the early years, have an opportunity to develop and influence the attitudes and beliefs of students towards mathematics. The attitudes and beliefs developed in school are usually carried forward into adulthood. Very few people who disliked mathematics at school revise this opinion as an adult. Nisbet, (1997) provided the following suggestions to develop positive attitudes and beliefs in students:

- Take a broad view of mathematics—emphasise patterns, processes and understanding rather than rules and set procedures without meaning.
- Base mathematics learning activities on problems that are the start of the mathematics rather than the application at the end.
- Encourage children to look for different ways of solving problems as well as many solutions.
- Reward all attempts at problems, not just successes. Say 'Good attempt' rather than 'That's wrong'. Say 'That's interesting' or 'Do you think that will work?' rather than 'You're on the wrong track'.
- Provide opportunities for success for all students and give feedback on progress for students own monitoring.
- Have students write about mathematics—what they like or dislike, how they feel about various topics, and what they still don't understand.
- Plan for enjoyment, motivation and fun as well as understanding.

- Establish close links with home—explain the program to parents and let them ask questions about their children's progress. Send puzzles and problems home to convey the message that mathematics is interesting and enjoyable, and that it may have changed since they were at school.

Teaching and mathematics

Teaching mathematics effectively in a primary school classroom is a demanding and challenging undertaking. It requires thorough planning, creative management of the learning environment and ongoing monitoring of the effectiveness of the teaching and learning processes. For teachers who accept the challenge to teach mathematics well, the rewards come in seeing children enjoy the learning experiences and in sharing the pride in their achievement.

Investigate and discuss

1 Compose a letter to be sent by the class teacher to parents either to explain the use of thinking strategies the class is about to use to begin learning addition facts or to invite them to a meeting to show the development of division computation, which is about to begin in class.

2 For the term or semester program developed in the workshop activity on pp. 543–545 construct a list of:
   a all the mathematics materials, formal and informal, that should be obtained for use in that program; and
   b appropriate storage containers that might be requested from parents so that all the materials can be stored for ready access in the classroom.

3 Discuss how you would handle this situation. Mary Smith's parents have sent a note complaining that Tuesday's homework problems allowed the use of calculators. They thought that Mary was at school to learn how to do sums and surely calculators made children lazy. They demand that you not allow calculators in future.

CHAPTER NINE
Assessment in mathematics

> It is imperative that assessment be seen as an integral part of instruction. It provides a window to students' thinking and a compass for instruction. Equally important, what gets assessed — and how it gets assessed — sends clear signals to students about what teachers think is important
>
> *Cooney, Badger and Wilson, 1993, p. 239.*

Assessment is crucial to teaching and learning. It provides teachers with information on a student's developing mathematical understanding; reveals an ability to use particular processes and ways of thinking; shows whether a teaching activity is meeting with success or requires modification; suggests where extensions or building up of underlying ideas might be needed; and provides information to the student, teachers, parents and the wider community on a student's mathematical capabilities. Assessment practices also affect what students learn and the manner in which they learn. In particular, it is important that all the goals of the mathematics curriculum are assessed if students are to develop a range of knowledge, processes and ways of thinking as they build up and use mathematics. If assessment only includes exercises or routine problems that can be answered using standard procedures, a student is unlikely to value originality and the use of personal strategies in mathematical thinking. When there are large numbers of short questions or a limited time frame, a student can come to believe that being 'good at mathematics' means being able to do things quickly. This attitude will then detract from giving thoughtful consideration to a problem or investigation before it is begun and any reflection on the answer or method of solution when it is completed. It may also inhibit the development of any expectation to make sense of the answer and solution process.

When assessing, it is also important to avoid giving the impression that it is only answers that count, for it is the way answers are obtained and the thinking that allows new problems to be tackled that are really at the heart of mathematical learning. Similarly, assessment in which the major focus is to assign marks or grades should also be downplayed as this provides little information on what a student knows and understands. While grades may motivate good students to perform and study for tests, they do not promote learning as an end in itself. Conversely, students who receive poor marks often become discouraged and less motivated to pursue learning. Assessment should no longer be used to deny students the opportunity to learn important mathematics through turning them against the subject or by refusing them entry to particular courses.

Instead, it should become a means of fostering growth in mathematics. Care needs to be taken that there is a balance in the questions that are asked, and that the range of problems reflects due importance to each aspect of the topic under review. Concepts need to be developed in depth because they provide the basis for further generalisations and problem solving. Emphasis should be given to ways of thinking and the understanding of any processes that might be used rather than expecting that all students will solve a given problem using the same method. Practical activities, which make up much of topics such as measurement and chance and data, need to be seen to have an equal weighting with the written tasks associated with computation and problem solving. Indeed, 'if some areas of the curriculum are assessed to the exclusion of other areas, teaching will tend to emphasise those, whether consciously or not' (*A National Statement on Mathematics for Australian Schools*, 1991).

Just as the school curriculum has changed to reflect differing emphases and needs as a consequence of new teaching approaches and the ready availability of technology, so the types and needs of assessment have also had to adapt. It is no longer sufficient to rely solely on particular, school-based tasks related to standardised procedures, a narrow range of memorised facts and straightforward applications to determine proficiency. Rather, the portrayal of a particular student's ability or capacity is required through the use of authentic tasks that resemble activities carried out in the classroom or in real life, together with access to the types of materials and equipments used in the teaching of mathematics. Students will work for what is rewarded and these forms of assessment reward understanding, as opposed to the memorisation encouraged by traditional tests, which may be counterproductive to the learning that is expected. Similarly, if students are routinely encouraged to work cooperatively, they should also be assessed as group members as well as individually. To give credit only to individual performances sends a message that group work is not valued enough to be assessed and students may then be reluctant to work cooperatively and effectively with others.

Testing at the end of a segment of work has usually been referred to as *summative*. The less formal, ongoing assessment used to see how particular individuals were progressing, or how the teaching might need to be adapted, was called *formative* as it was principally used to influence the pattern and direction of instruction. Such distinctions have now largely disappeared, and it is more appropriate to judge assessment practices on the degree to which they inform, the range and depth of information they provide, and the circumstances on which they are able to report.

Assessment has thus become an ongoing aspect of teaching and learning that helps to inform everyday practice, rather than being a one-off 'test' given at the end of a unit of work showing how much or how little an individual learner has acquired, and often simply used to rank one student against another. This means that thought must be given to the form and timing of assessment at the initial planning of the particular unit of work. It is not possible to go back in time to gather evidence after the event, while leaving consideration of assessment until near the end of a teaching segment is likely to leave a formal test as the only avenue to use. Just as importantly, gathering indicators of how a particular student is thinking requires a carefully considered set of tasks or questions that need to be set at aptly chosen times. Decisions regarding students' achievements can then be made on the basis of a convergence of information from a variety of sources gathered during the teaching process and the process of describing what mathematics students know or do not know can now be defined:

> Assessment is the process of gathering evidence about a student's knowledge of, ability to use, and disposition toward, mathematics and of making inferences from that evidence for a variety of purposes (National Council of Teachers of Mathematics, 1995a, p. 3).

The process of assessment

The informal assessment practices that were once valued only as a means of conducting day-to-day teaching are now being seen as paramount in almost all situations. This has led to more systematic recording of significant events and individual responses during the learning cycle using observations, checklists, insightful samples of student work, portfolios selected by teacher and student, student journals and so on. At the same time, any formal testing is becoming more extensive than the traditional pencil and paper test which often reflected exactly what and how something had been taught.

While formal tests have their place in terms of efficiency, as *A National Statement on Mathematics for Australian Schools* observes:

> [p]aper and pencil tests, with or without a time limit, are likely to continue to be one useful and efficient means of gaining feedback about student learning. However, these conventional forms of tests cannot address all areas of the mathematics curriculum. Additional modes of testing must be developed . . . to increase our repertoire of types of questions which can be asked in traditional test settings and develop strategies for judging the response to these questions (1991, p. 22).

More open-ended questions, investigations and projects are being included to provide opportunities for making inferences about a student's learning.

A deeper appraisal of mathematical thinking may also be gained through the use of other methods, such as consultations with individuals, where structured questions or interviews are designed to probe understanding. But if there is any one way in which modern assessment practices might be characterised, it is the shift from simply grading of students on one-off tests to a series of tasks and investigations that are essentially *diagnostic* in nature. No longer is it sufficient to give a score or level. Strengths and weaknesses need to be stated and reasons for them proposed to allow everyone in the educational endeavour to make an informed choice about progress for individual students, directions for particular programs and teaching strategies to follow.

However, assessment has most often been associated with the selection of tasks to be given to students to make judgements on their knowledge and understanding of mathematics. But this is really the third part of an assessment process which draws on initial *observations* about their ways of thinking and acting that lead to *assumptions* about their underlying capacity to understand, use and interpret mathematics that in turn need to be validated by further *probing*. A teacher's assumptions about a particular student's or class's performance is based on day-to-day observations and it is more likely that a teacher will be surprised by a result on a particular test or test item given their expectations from everyday observations than that they will discount their observations in favour of a single formal test result.

Nonetheless, any assumptions about a student's understanding, use and interpretation have to be checked to see whether they can be confirmed. This phase in the overall assessment model should give rise to the more formal probing that is frequently all that assessment is taken to be.

This understanding of the assessment process may then be summarised as:

leading to a three-phase cycle for the assessment process

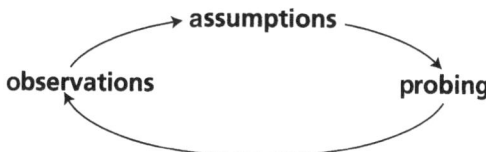

In this model, although observations are a natural starting point, it may be that assessment insights are first generated through discussions with a student or by a student's actions or responses. These will then give rise to particular assumptions that require further probing to confirm or validate the meaning that is taken. This three-phase cycle can then be something that drives a teacher's ability to plan for individual students, to review a proposed sequence of work or activities, and to determine whether to move on to new ideas or applications or to step back to where difficulties are arising for a student or group of students. In this way, assessment is integral to teaching, rather than separate from it, and will no longer be viewed simply as an event that marks the end of a particular topic or segment of work.

Observations

Observations are the main determinant of a teacher's role in the classroom, suggesting adjustments to a program to meet particular needs, identifying those students needing additional support and enabling other students to extend their thinking or move on from a topic to apply themselves to other areas of study when they no longer need to focus on a specific mathematical aspect. However, the observations that teachers use to determine their actions and judgements are often not sufficiently valued by those outside a particular classroom or even by the teachers themselves. New approaches to assessment are focusing on ways to give more validity and comparability to teachers' day-to-day observations through checklists, developmental continua and moderation meetings. These informal assessment tasks and situations can then be more readily shown to stand up over time and take their rightful place as prime determiners of a student's ability and capacity in mathematics.

Used as an assessment tool, observations need to focus on individual students and their behaviour to try to determine thinking and understanding of specific concepts and processes. This focused observation should take place during teaching and may be more easily carried out as students work independently or in small groups. Sometimes observation might only involve examining students' written work or constructions. In all situations a number of questions need to be borne in mind:

- Is the student in control of his or her own thinking or is there an expectation to be told exactly what to do or follow a set procedure without any thought to what is occurring?
- What do the ways of thinking and acting reveal about an individual's mathematical understanding and reasoning?
- Does the student or group of students have an overall plan to solve a problem? Do they consider a variety of approaches using different strategies? Can they articulate their plan and strategies and consider those of other students?
- Are appropriate tools and materials accessed to help in considering the task and used with any working? Are these materials and tools used effectively?
- Is there any organised way of recording the work or does writing about and drawing what has been done seem difficult?

Assumptions

Day-to-day, ongoing observations lead to a teacher making assumptions about a student's understanding and the progress of the class as a whole. The conclusions reached reflect the teacher's knowledge of the underlying mathematical ideas and experience with a range of children in similar situations. In some cases when students are observed, ways of thinking and acting are fairly clear. Sometimes a child's response may be related to several different possibilities. At other times, there may at first be no discernible pattern in what is happening. Of course, there will always be situations where no pattern can be seen at all, either because the responses or behaviours are related to events outside the classroom or because of particular characteristics of the child's social, intellectual or physical make-up.

In making assumptions about the quality of a student's mathematical reasoning or performance, conjectures are formed regarding issues such as:

- the mathematical ways of thinking that are present in any explanations, representations or drawings
- the understanding of any processes that are used
- whether there is appropriate accuracy in measurement, computation or construction
- whether underlying conceptual knowledge is secure and can be brought to bear on non-routine problems or new mathematical developments
- the effects that attitudes, prior experiences and general behaviour might have on the mathematical activities and learning

Any assumption made by the teacher then needs to be confirmed through further probing. In particular, it may be necessary to ask a series of similar questions without indicating any concern with the initial example, in order to see what might be the cause now that there is an awareness of something inappropriate, incomplete or incorrect in the response.

There may also be times when a child's behaviour reveals that there is no understanding of the task at all or that a particular response is given for a totally unrelated consideration. This could even be related to some event outside of the class as when a child asked to name a shape in the form of a regular hexagon simply said that he thought it was called 'savoury or maybe barbeque' from the type of biscuit he remembered eating at home (Clarke, 1997).

Probing

Levels of understanding and proficiency in mathematics cannot always be readily determined and may require close scrutiny of the responses to a series of related tasks or questions. For instance, it may be necessary to identify the precise reason for a child's failure in a learning situation where others have gained competence, or to identify broad achievement levels for a new class member on transfer from another school. It is often the case, for example, that children have difficulties with written algorithms because of undetected misconceptions in numeration concepts or because they use inefficient methods for obtaining number facts. Indeed, any assumption about a student's capacity has to be checked to see if it can be confirmed. In the midst of teaching, it is easy to make quick assumptions with only one possibility being considered and then act by simply telling or re-stating a strategy that seems appropriate from the teacher's perspective. At best, this might make a child very dependent on a teacher interceding to tell him or her just what to do. More likely, it will discourage any self-confidence and induce the child to do the minimum amount of work possible so as to avoid being seen to be having difficulties with work that other children do not find taxing.

Instead, what is needed is to accept a student's response at face value and then provide a similar problem or exercise so that a further observation can be made, this time equipped with an expectation that something is not quite as it ought to be. An observation is now more focused on the particular misconception or faulty process and is likely to provide further information on the source of the difficulty. Armed with one or more possibilities as to the cause, a teacher can then systematically account for each one with well-chosen questions to pinpoint the underlying misconceptions that need attention. Without this in-depth probing, only a surface-level response can be made and neither the teacher nor the student will have any confidence that the same difficulty will not occur again soon after.

Using the assessment cycle

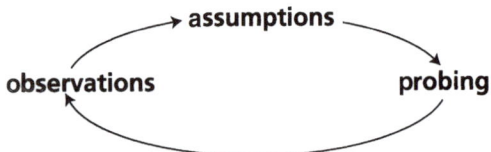

Sometimes when assessment is being conducted, a child's behaviour and thinking may have obvious causes. At other times these may appear obvious, but further observations may reveal a deeper problem:

When a Year 2 class was measuring lengths of common objects with their rulers, one boy obtained answers of 15, 23, 9 and 17 cm for the items that his group measured. The answers given by every other child were 14, 22, 8 and 16 cm. Puzzled, the teacher observed closely as he undertook the next tasks.
 The fact that his answers were 1 cm greater than the other group members gave her a clue to his behaviour. Close attention to his method showed that he aligned the beginning of each item to be measured with the 1 on his ruler, not the 0.

Further probing can identify the thinking more precisely and point to its source:

While the most obvious assumption was that his skills in using a ruler needed addressing, the teacher decided she should probe his understanding of the task more closely. It turned out that he believed that 1 was the first number from his earlier counting experiences and in fact did not always give an appropriate meaning to zero in other aspects of mathematics.

 This example shows how a child's insecurity in one area can affect another seemingly unrelated part of mathematics. On other occasions, a child's behaviour might show that there is little understanding of the task.

A Year 6 boy was given this task:
Here are two (isometric) views of a block of buildings. Indicate in the plan the height of the building in storeys:

His response was simply to write in the number of blocks he could see for the first aspect in the diagram provided, and then draw a similar grid to record what he could see in the second aspect.

It could readily be assumed that he was not able to see that the two pictures could be completed in any way. He also did not see any difficulty with the parts of each picture that he could not visualise.

Assessment task adapted from Olssen (ed.), 1995, p. 59.

At other times the thinking may not be clear, and many possibilities might suggest themselves and require several cycles of observation, assumption and probing to determine the full source and extent of a child's difficulties.

A Year 5 girl who had often struggled with mathematics was feeling very pleased with herself as she had just competed 3 out of 5 additions correctly.

$$\begin{array}{r} \overset{1}{3}4\,6 \\ +2\,7\,9 \\ \hline 6\,2\,5 \end{array} \qquad \begin{array}{r} \overset{1}{5}8\,9 \\ +6\,4\,3 \\ \hline 1\,2\,3\,2 \end{array} \qquad \begin{array}{r} \overset{1}{3}0\,6 \\ +5\,4\,7 \\ \hline 9\,4\,3 \end{array} \qquad \begin{array}{r} \overset{1}{5}8\,9 \\ +8\,7\,0 \\ \hline 1\,4\,5\,9 \end{array} \qquad \begin{array}{r} \overset{1}{7}0\,9 \\ +6\,5\,2 \\ \hline 1\,4\,5\,1 \end{array}$$

The teacher made some assumptions regarding the source of her difficulty.

- It did not appear to be basic facts, as most of the partial sums were correct.
- Since 2 out of the 3 examples with zero were incorrect it could have something to do with facts involving zero.
- There were no errors in the ones place but some errors in both the tens and hundreds places, so it may have to do with renaming.

Her teacher decided to give some other examples to check more closely. At first some without any zeros were given:

```
  ⁷7 2 6         ⁹6 4          ⁴6 8
  +4 7 7         +3 5 8        +3 7 5
  1 2 0 3        1 3 2 2       8 4 3
```

As the girl completed the first three examples, the teacher was able to observe her approach. After she had obtained the correct answers to the first two examples, he asked her to do the third one out loud. He noted that her language was:

> '5 and 8 are 13, put down the 3 and carry one.
> 7 and 7 are 14, put down the 4 and carry one.
> 3 and 5 is eight'.

There was a bottom to top procedure without any meaning related to place value or renaming, and the renaming digit was recorded as a small digit above and to the left of the place being added.

He then presented her with some examples with zeros in the ones place:

```
  ¹7 4 0         ¹8 9 3
  +3 8 9         +4 6 0
  1 1 2 9        1 3 5 3
```

Again she completed them correctly, confidently using the same method, recording and language.

Finally he set her examples with zeros in the tens place:

```
  ¹4 0 6         ¹8 0 9
  +3 7 8         +6 4 3
  8 7 4          1 5 4 2
```

With the first example, she recorded the additional ten in the same manner as a small 1 to the left of the tens digit. But this time, instead of joining this 1 with the other tens digits, she read the top line as 10. The teacher then gave the final example and was able to confirm his observation. Inspection of the initial set of additions showed that this was the method used there.

The problem now was: Why did she only use this method for numbers with an internal zero?

The teacher then recalled that the class had only recently been doing some subtraction with internal zeros and he had been pleased to see the girl's progress there. He decided to pose some subtraction examples to observe how these were done.

```
  ⁷8̶ ¹0 6        ⁵6̶ ¹0 9
  − 3 5 2        − 2 4 8
    4 5 4          3 6 1
```

As she completed the first example, the teacher noted that the language and recording were similar to what she used for addition:

> '6 takeaway 2 is 4.
> 0 takeaway 5, can't do it; borrow one.
> 10 takeaway 5 is 5. 7 takeaway 3 is 4'.

Her answer was correct.

When he observed the second example, which she also completed correctly, he noted that in each case she had interpreted her renaming $^1 0$ as ten, just as she did in the addition examples that caused her problems.

He had found the source of her difficulty—her methods for both addition and subtraction were rotely learned mechanical procedures and she had confused the two because she had no understanding with which to call anything to account.

This last example also shows how an inadequate procedure with one operation can impact on the processes for other operations. Using this cycle of assessment allows a full awareness of a child's understandings, strengths and weaknesses to be determined. Consequently, further work or re-teaching can then build on well-understood strengths rather than surface-level performance.

Alternatives to written tests

While written tests have often been the preferred way of determining the understanding and proficiency of each class member, other possibilities may be more informative. For instance, rather than a test with a number of questions, just one well-chosen example that brings all of the important learning to account could be given to each child in turn. Individual responses would then show each child's proficiency, understanding and, with accompanying observations and questioning, precisely where any aspect was in doubt or in error. Depending on the mathematical and classroom situation, probing can take many forms.

Consultations

The use of structured questions in interviews designed to probe understanding and ability to use particular processes or concepts is usually referred to as a consultation. Students can be asked to illustrate or represent concepts, explain operations or processes, make generalisations, or extend their thinking. Responses, whether oral, written or involving manipulation of materials, should be carefully noted for later analysis or inclusion in a student's portfolio. Prior planning to select appropriate questions and materials along with good rapport in a relaxed, non-threatening atmosphere will also be needed. At the same time, by asking students to give explanations of operations and processes, a consultation can let students know that understanding is valued and they are expected to be able to make sense of any processes they perform.

Consultations are often conducted on a one-to-one basis, but can be even more productive if small groups of students are involved. This allows a teacher to visit each student more frequently if needed. It also allows observations of students making sense to each other and among themselves rather than trying to second guess just what it is that a teacher wants. Even when a student can answer a question appropriately, it requires a further level of thinking to be able to present the reasons for this clearly. Inferring reasons from discussions among students will often be more productive than relying on direct questioning of individuals. Similarly, it may be helpful to have students show how something is done with materials. Their manipulation may give insight into the way ideas are formed and manoeuvred inside their heads in order to direct and control the materials that represent the ideas at a concrete and external level.

Just as observations need to be made in conjunction with classroom activities on a daily basis, so too consultations should be an integral part of teaching and are usually feasible while other children are engaged in work that requires little direct teacher interaction. Left to a special time or place, a student is more likely to treat it as an occasion warranting a search for behaviours and explanations of which the teacher might approve rather than revealing ways of acting akin to those used in less peculiar times. Nonetheless, if an extended period of time is likely to be needed or if a child finds it difficult to concentrate or provide responses in class, it may be necessary to conduct the consultation out of class hours or in another situation.

Classroom observations

An assessment folder can be used to keep together all the observations related to each child. The teacher can simply use an A4 sheet of paper divided into spaces for observing 6 to 8 students which can later be cut out and glued onto a page for the particular child (sticky note pads could also be used).

Two-digit numeration Semester 1, 2004

Mia	James	Alison	Mirlya
• 2 digit counting sound • confident crossing tens • developing number sense • very quick to make numbers with bundling sticks • very sound concept of ten	• looks at person next to him before making 2 digit numbers with materials • counts on accurately but slowly — not overly confident • needs a number board to count back	• has no difficulties with any aspect of 2 digit numbers • able to count on and back from any number in ones and tens • constructs and records numbers quickly • counts in her mind — needs no counting materials • excellent 2 digit place value	• has difficulty counting on and back in tens from any number — needs bundling sticks and a 2 digit number board • makes numbers accurately • still coming to terms with concept of 10
Royce	Anthony	Rebecca	Jonathon
• no problems with 2 digit numbers • ready to move onto 3 digit counting • ready to move from bundling sticks to Base10 materials • very confident in all 2 digit concepts	• makes 2 digit numbers quickly and accurately with bundling sticks • counts on and back in his mind • starting to show confidence in all areas • beginning to verbalise what he is doing	• starting to count on and back but always checks with others • enjoys the work with bundling sticks • able to make and record numbers quickly	• makes 2 digit numbers accurately but always checks with others • 2 digit place value sound • concept of 10 well developed • counts back in his mind — slow but accurate • not confident

An A4 sheet used for recording observations during class

When the time comes for writing reports, most of the information on which these will be based is then in one readily accessible place. At the end of each term or semester, the page or pages with the observations can be easily put into the student's portfolio along with other selected assessment pieces.

An assessment folder labelled for each child in the class

Other procedures which have been developed that are helpful in structuring observations and keeping records to allow a teacher to maintain an up-to-date picture of each student's developing mathematical competence include the following.

Annotated class lists
One of the simplest recording forms is to use the list of students in the class and leave a space for brief comments alongside each name. This ensures that every student is observed over the duration of the unit of work with a brief note that describes the situation and the observed behaviour for any event considered significant. Such events might involve clear evidence of newly acquired mastery or understanding, evidence of incorrect processes or lack of understanding, or examples of atypical behaviour. Observations can be made unobtrusively during class time while children are working, or as soon as possible afterwards and can be accompanied by examples of class work and project reports.

Topic	Two-digit subtraction with renaming using bundling sticks and place value charts in pairs
....	
Wendy Quinn	Uses the material appropriately but needs prompting when verbalising what is happening.
Mustapha Khan	Very confident, able to use materials and verbalise process accurately.
Joel Robbins	Able to use materials and verbalise process but relies on partner for confirmation.
Tim Shanahan	Very confident and accurate and was able to work out how to record process.
Lisa Shiu	Worked with Tim and started to use whiteboard pens to record process.
Donna Waters	Used materials and verbalised process confidently. Ready to start recording.
....	

Checklists

The preparation of an inventory of expected achievements, knowledge, skills, behaviours and attitudes for a particular time of the school year helps class teachers to identify clearly their expectations with respect to the outcomes for the class program. Checklists of relevant parts of these inventories can then be prepared so that some structure is provided for the observations made of students at different times. This structure can improve the quality of the information recorded, can minimise time wasted on irrelevant or redundant assessment, and can readily identify the low-profile students for whom nothing was observed.

A checklist of expected prior achievements is especially useful at the beginning of a school year. It can help to determine the readiness of children for new work and can identify those in need of re-teaching of essential concepts and processes not yet mastered. At other times, simpler checklists can be used to identify the specific behaviours, skills or attitudes appropriate to a particular unit of work. Several textbook series now available in Australia provide class management materials that include checklists and inventories for mathematical topics for each class level. The following general checklist has been adapted from *Making Numbers Make Sense* (Ritchhart, 1994, p. 137).

Task observed .... **Date** ....

Strong Developing Weak Not Observed

Communication	....	....	....
Listens to and follows other students' reasoning			
Incorporates other students' ideas in planning a solution strategy			
Explains own ideas and supports them with evidence			
Asks questions of others to facilitate own understanding			
Cooperation			
Becomes actively involved			
Supports and encourages others			
Assumes personal responsibility with the group			

Checklist for observations on communication and cooperation

Note that it is necessary to be a little wary when using checklists as they appear to be a closed list of all that will matter. There will always be important observations to be made in addition to those that relate to the items listed.

Work samples

Examples of student work chosen to reflect the manner in which a student is working, whether selected by the teacher or nominated by the student, are invaluable records to support observations. Any samples should be dated and annotated when appropriate to provide a means of assessing changes in thinking and working over time. It may also be feasible to photograph or photocopy particular pieces of work or to video or audio tape work in progress. These samples can then be part of the evidence selected to make a profile or portfolio of a student's mathematical achievements.

Portfolios

Information gathered consistently over time from a variety of assessment sources can provide a balanced, descriptive picture of the learning that has taken place. A portfolio might include examples of a student's best mathematical products as well as representative samples of work to show progress in understanding and thinking. Materials from both in-class and out-of-class activities have a place, but conventional worksheets or activities that simply focus on practice examples do not belong. In order to reflect what is valued in class activities, examples from group work can be included as well as a student's reflections in a mathematics journal.

Both students and teachers can be involved in selecting the work that goes into a portfolio, addressing issues such as:

- What does the activity, outcome or investigation disclose about the student's developing mathematical knowledge?
- How does the information reflect change and/or growth in the student's mathematical thinking and understanding?
- Will the information add to what is already known?

To serve as good assessment tools, portfolios need not be standardised and work should not be produced specifically for inclusion. Rather, the curriculum should be rich enough to provide ample examples from which meaningful samples of authentic mathematical tasks can be chosen.

A one-to-one conference with the student to review the portfolio can be useful in allowing both teacher and student to gain insight into how the student operates mathematically, and can serve as a vehicle to support progress, build confidence and identify goals to be addressed in the future. The work samples can also be used to explain to parents, other teachers or schools the sort of mathematics a child is doing and how he or she is making sense of it.

Student journals

Student self-assessment provides an important contribution to the range of understanding a teacher would like to have on a student's developing mathematical ability. At the same time, as students learn to evaluate the activities they are involved in, to measure progress in their learning and to pinpoint the things that give them success or act as barriers to progress, they are able to develop the autonomous learning and assessment that will provide them with success in later life. A five-minute session at the end of each day's mathematics activity can be used to have each child write about:

- what has been learned and feelings about progress
- what was enjoyable
- what has been troublesome to understand and what might need working on in the future to overcome this difficulty
- what was important and any ideas or topics that might be explored further

At the end of a unit of work, this could be reviewed and made into a more extensive journal entry on the topic and learning experiences as a whole. Entries in a journal also make good starting points for a class discussion on a particular topic or problem, allowing each child to readily make a contribution based on his or her own reflections. They also provide insights for a teacher as to what teaching emphases are needed, teaching activities that are successful and the manner in which time might need to be allocated to different aspects of the work.

Formal testing

Formal assessment in primary school mathematics may involve: written tests of achievement, usually to help provide a summative evaluation at the end of a unit of work, school semester or year; practical tests of processes used in measurement, geometry, chance and data, the use of materials with number and patterning, or computer packages; or system-wide tests given by outside agencies. Such tests may be constructed by the class teacher shortly before they are to be used, provided by the school for all classes in a particular year level, or purchased commercially.

When formal testing is conducted, one issue that has to be considered is the amount of time to be allowed to complete an individual question or the entire test. When there is a restricted amount of time, some children may not get the opportunity to show what they know or can do. This can also lead to stress about the test and to children giving only cursory attention to the actual question before beginning work on forming a response. As a consequence, misreadings of the question, careless working and incomplete answers can occur. It is also unlikely that an answer will be considered fully before the next question is begun. On the other hand, an unlimited amount of time can also be stressful when students are unable to attempt certain questions. Further, there are occasions, such as assessing basic fact knowledge, when unlimited time will not actually allow a true assessment of what is known and not known. When time is restricted, students are more likely to reveal their actual thinking processes as they try to find an answer. For example, if only a short time is allowed in providing answers, many students are able to answer $12 - 9$ by counting on their fingers, but find that $12 - 3$ is not possible.

Another point to bear in mind is that after primary school almost all tests do have a restricted time and there will be state-wide tests and competitions even in the primary years where time will be limited. When first introduced to formal tests, children might be allowed as much time as they need, but part of the process of reviewing the solutions should involve showing how an answer can be expressed quickly and succinctly. In this way, children can learn how to make maximum use of the time they have available as well. This review should also assist them to see how a full understanding of the material to be assessed will allow them to understand more quickly a question and the response that is required. Indeed, an important aspect of formal tests in the upper primary school is to prepare children for the process of test taking.

Teacher-constructed tests

Written and practical tests are usually used to measure the achievement of individual children as well as the progress of a whole group. For this to be done reliably and validly, the test must be carefully planned in terms of the outcomes which need to be assessed and the types of questions or activities that can measure these outcomes and to provide a balance of questions and activities that address all the crucial aspects in the unit of work. Conversely, a cursory attempt to write questions or activities usually results in a poorly designed test that may not reflect accurately what was taught or what is important. Constructing a valid and reliable test involves:

1 *Listing the outcomes to be assessed*—the test outcomes should correspond to the learning outcomes of the work to be assessed, although they will tend to be more condensed, relating to the achievement expected rather than the smaller steps required to build up understanding. If the predominant learning experiences have involved learning new concepts, then the test outcomes should focus on these. If, however, learning basic facts for a number operation has dominated recent class time, then the test outcomes should be directed to measuring correct recall of these number facts through efficient use of the learned thinking strategies. When the unit has addressed aspects of three-dimensional shapes in geometry, a practical task to identify, sort or construct particular shapes would be more appropriate. Thus the learning outcomes will suggest the type of assessment procedures and test items to be used, and also indicate when written or practical tests are appropriate. The extent of the learning outcomes will also dictate the type and number of test items. While understanding of a new concept might be measured by questions directed at using and interpreting the concept, achievement of many outcomes, as measured by an end-of-year test, might require selective sampling of major critical issues if the test is not to be excessively long.

2 *Identifying the types of test items to use*—if it has been decided that an outcome can be assessed appropriately using a written test, a further decision is needed on the kinds of items to be used. Knowledge and comprehension levels can often be measured efficiently by the use of short-answer questions where an appropriate word, phrase or short sentence is supplied. Statements that need to be matched to correct responses or which can be marked true or false and multiple choice items may also be used. These types of questions are called outcome items because they can be marked to give identical results from independent assessors. They may be useful in allowing an extensive range of content to be assessed reliably in a single test that can be completed in reasonable time. More lengthy questions that require several steps for each response would be preferable for assessing proficiency with written algorithms, measurement applications and problem-solving processes. A detailed marking scheme that allows for all possible types of responses and errors will be needed and this can also allow independent assessors to give comparable marks if required. Similarly, on a practical test, the activities need to be carefully chosen to allow each student to demonstrate the behaviours that the assessment is seeking to determine without the activity itself being so complex that a student can become tangled up in the practicalities and then be denied the opportunity to reveal what he or she knows.

3 *Deciding the number of questions for each outcome*—if a test is to provide a valid measure of achievement on a set of outcomes, the testing process should relate to the relative amounts of emphasis placed on each outcome during teaching. The number of test items used, their degrees of difficulty and their relative weighting should be in proportion to the amount of emphasis placed on the outcomes to which they refer. At the same time, the number of questions should be appropriate to the time available for completing the test.

In order to ensure that the completed test accurately reflects the design intentions, a table of specifications should be developed by listing the outcomes or topics, the relative emphasis intended for each outcome, and the numbers of items of different difficulty levels that refer to each. Such a table allows for adjustments to be made to ensure that the final set of questions samples appropriately from all the learning outcomes and content included in the unit of work.

4 *Selecting or constructing the test items and collating the test*—each test item should provide evidence of whether a student has achieved an outcome or not. It should be written in clear simple statements, using language that the students understand. Vocabulary and sentence structure should be appropriate for the class level. If each item is to provide evidence that an outcome has been achieved, then it should avoid unimportant details, unrelated bits of information or irrelevant material, unless the issue of selecting what is important, as with problem solving, is a part of the outcome being assessed. At the same time, the structure and presentation of each item should not be likely to cause those who know the correct solution to get it wrong or to allow students to obtain a correct answer by chance or some incorrect process. It is helpful to have a colleague check the validity, relevance and clarity of each item. This will assist in deleting any questions of doubtful relevance and rewriting any that might be ambiguous.

Instructions to students should be clear, concise and complete so that each student knows what is expected even if he or she is unable to complete the activity or answer the question. It may be helpful to begin with some easier items with later questions increasing in difficulty or involving mixed levels of difficulty. It is important to place together all items of one type of structure, such as short answers, multiple choice or questions with several steps. Finally, an answer key should be prepared and the distribution of marks for particular parts determined. This too could be referred to a colleague for feedback as assumptions may have been built into the question without being made clear on the test.

Another option that has been explored is to have children review a unit of work and propose questions, either individually or in groups, that can be considered by the teacher for inclusion in the formal test (Clarke, 1988a, p. 58). The teacher would have the right to choose from the questions submitted, make any editorial changes to rephrase ambiguous or poorly worded questions while retaining their essence, and to add further questions if necessary to obtain a balanced assessment of the unit of work. Not only does this approach heighten students' interest in the outcomes and give them a sense of involvement and control over the assessment process, it also provides a motivating way to review what is really at the heart of the work just covered.

Discussion of the results of the test and the solutions that apply is heightened when students have a part in preparing test items. But however the test is constructed, it is essential for time to be spent in analysing the responses for strengths and weaknesses and for these to be discussed with the class as a whole and individual students if this is warranted. The ways in which this can be done and the outcomes that can be drawn from it are discussed in the section on using and communicating assessment information (p. 581).

Tests prepared by outside agencies

There are many commercially prepared achievement tests designed to help teachers measure whether each child has achieved specific outcomes. Sets of test items that assess proficiency with particular concepts, processes or applications are provided. The results can then be used to determine whether a child has learned from a period of instruction and each child's performance can be compared with his or her own previous performance. This form of test measures achievement with reference to previously specified criteria and so is called a *criterion-referenced* test. Other tests are designed to compare proficiency with other students who have taken the same test. For example, a class test result can be reported to indicate whether each student performed very well, averagely or poorly according to a scale determined by the teacher or relative to the criteria for mastery of the topic.

However, most commercially prepared tests are *standardised* tests as they include carefully prescribed procedures for administering and scoring to ensure that standard procedures are always used, and statistical data which indicate how validly and reliably the test actually measures performances for stated outcomes. Often results are provided to show the scores called *norms* achieved by large specified groups of students in many different schools and classrooms. These norms can be used by class teachers to compare the performances of children in their classes with those of large groups of similar age or experience. The term standardised then refers to procedures for administering, scoring and interpreting results rather than the purposes for which the test may be used or the kinds of achievements it may be designed to measure. The details of a wide range of standardised tests suitable for use in Australian schools are available from ACER Press (Australian Council for Educational Research) at 347 Camberwell Rd (Private Bag 55), Camberwell, Victoria 3124.

Standardised tests do not necessarily provide much useful information and may not reveal a student's real capabilities or may even mask misunderstandings. They cannot be used to provide a full picture of a student's mathematical ability and should be used as just one part of an overall assessment program. Further, norm-referenced tests that seek to compare individuals can be misleading and distort the learning process. In producing the norms, a distribution of scores is created that places 50 per cent above average and 50 per cent below, regardless of their true understanding. Time is usually introduced as a factor in order to produce this normal distribution of scores, ensuring that some students who know the material may not be able to finish the test and may therefore score more poorly. Standardised tests do not usually require students to show their thinking, and do not assign a higher mark for better strategies. Questions are often devoid of context or have contexts that are inaccessible to some students. They also allow for only one correct response to a question, while there may be several ways of perceiving a problem.

While the norming group serves as the standard to which future students are compared, the comparison takes no account of individual differences in the pace of learning. Further, in an effort to prepare students for the test, some teachers rush through the curriculum at a pace that can only allow for superficial understanding. Students may be able to hold on to this for just long enough to do satisfactorily on the test, but the gains may be short-lived and may soon evaporate.

Normed data can be useful in providing information about general trends in performance compared with those of larger groups, for example all Year 5 children in a particular Australian state. However, the tests given across systems by education departments are usually standardised tests and are often criticised for the distorting effects credited to standardised tests in general. Criterion-referenced tests are likely to be of more general use to class teachers because they can provide a means of determining specific strengths and weaknesses for each child in the class. Tests designed by teachers may also do this and these forms of tests may sometimes be used to diagnose specific errors that need re-teaching.

Diagnostic assessment

An awareness of student strengths and weaknesses in understanding, carrying out and applying mathematical ideas and processes is an essential part of planning and implementing effective teaching of mathematics. Traditional tests might reveal some of what a student can and cannot do, but rarely provide reasons that might explain either success or failure. Diagnostic assessment is designed to show not only what is known and what is not, but also to identify the way in which mathematics is understood and used, to highlight the strengths that allow success and to pinpoint the underlying weaknesses that cause errors or difficulties. It is especially dependent on teacher observation of the way in which a particular learner is working and thinking, inferences made from samples of student work or working, entries made in student journals and comments made in class discussion rather than the result of formal tests.

Designing tasks and framing questions to make such judgements is not easy. A full understanding of the underlying mathematics and the sequences of development required to allow individuals to construct or reconstruct this knowledge is essential. But it must also be noted that the sequence of teaching steps used to develop mathematical ideas will necessarily be different from the series of questions that can be used to assess the development of a full understanding and competency. In particular, questions need to probe knowledge at the highest level expected rather than build from the earliest form to the more complex and investigate fundamental concepts on which other understanding depends. Determining the language and thinking that is used to control mathematical processes, the understanding of these processes and the ability to translate knowledge into new situations is also crucial in diagnostic assessment.

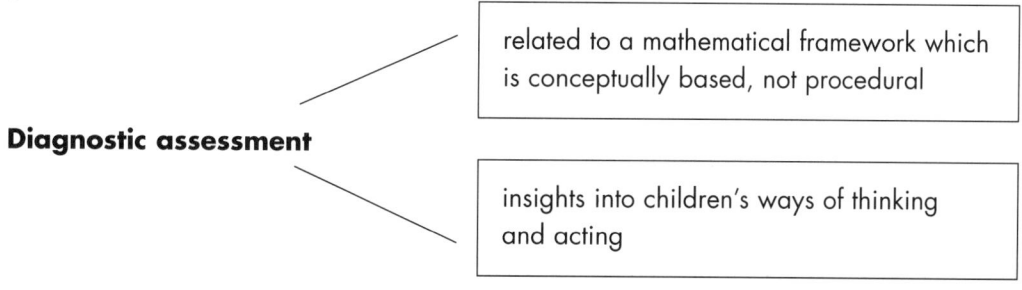

A consideration of what is involved in assessing strengths and weaknesses in division will serve to illustrate what is needed in diagnostic testing in terms of the underlying mathematical framework and provide some insight into the ways of thinking that students might exhibit. The questions have been taken from the *Booker Profiles in Mathematics* (Booker, 1995) and the responses have been gathered at the Mathematics Assistance Clinic at Griffith University in Brisbane.

A well-developed understanding of the division concepts is crucial to an ability to use division in routine situations and in problem solving. If the concepts are available, then all aspects would be known and it is appropriate to ask for an interpretation of the symbolic recording:

Read $3\overline{)18}$ Can you tell a division story using these numbers?
What is the answer for your story?

A lack of understanding in response to these questions is often seen in a left to right reading and literal translation:

Reading   3 divided by 18
Story   3 goes into 18 six times

Other stories simply match words to the symbols without reference to the concept:

There were three groups of eighteen people

There may be an inability to produce any form of story at all, no idea of how to interpret a picture in terms of a particular operation and no notion of how to match materials to the computational situation:

Asked to show with counters the story problem
Five children have enough money to buy 15 apples.
When they share them out, how many will each child get?

A student responded with

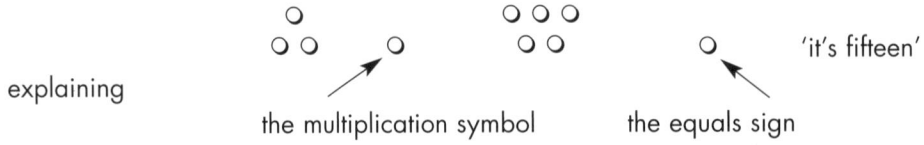

explaining   the multiplication symbol   the equals sign   'it's fifteen'

Finally, there is often a focus on the answer rather than on the thinking that underlies the concept, as in the first response above, almost as if the operation symbols are simply instructions to proceed to an answer instead of a symbolisation of a mathematical process.

Assessment of basic fact knowledge also needs to focus on the levels that would be expected of someone with expertise, requiring immediate, accurate responses. Questions should be organised so as to allow a set amount of time for each individual response, to be representative of the thinking that allows efficient mastery, and to provide opportunities to assess whether rote learning, guessing or inefficient techniques such as counting on fingers are being used.

Read each fact aloud
Allow 3 seconds for an answer before moving to the next fact.

1	18 divided by 2	8	15 divided by 5	15	24 divided by 3
2	12 divided by 6	9	72 divided by 9	16	42 divided by 7
3	16 divided by 8	10	36 divided by 4	17	56 divided by 8
4	14 divided by 7	11	63 divided by 7	18	49 divided by 7
5	25 divided by 5	12	9 divided by 9	19	48 divided by 6
6	15 divided by 3	13	36 divided by 6	20	56 divided by 7
7	45 divided by 9	14	28 divided by 7		

For each question, note whether:
- a response is given immediately
- the answer is given with certainty or tentatively
- the answer is obtained by counting related to addition, multiplication or subtraction
- there is some other form of working out being used explicitly

A sequence of questions to assess knowledge of the division facts

Source: Booker, 1995, *Booker Profiles in Mathematics*, Computation Easel, p. 57, reproduced by permission of the Australian Council for Educational Research.

Assessing ability with the computation of larger numbers is also necessary if a full picture of competency is to be gained. While this must relate to the sequence to be used for the development of each computational process, it also needs to allow for the level that a particular learner might have reached. Thus, rather than aim at a final order skill from the outset, the algorithm needs to be examined for its various plateaus and items arranged within these as well as among them: division by a single-digit number with remainders; division by a single-digit number requiring renaming of hundreds and sharing tens; division with an internal zero in the answer; and division with two-digit divisors. Items can then be arranged within these levels according to specific difficulties related to the underlying multiplication facts, zeros in the computations and so on:

COMPUTATION — DIVISION ALGORITHM

Item		Response	Comments
1	4)96 → 24		2 digit divided by 1 digit, rename tens
2	5)76 → 15r1 or 15.2		2 digit divided by 1 digit, rename tens
3	7)59 → 8r3 or 8.42		2 digit divided by 1 digit, rename hundreds and tens, no tens in answer
4	5)785 → 157		3 digit divided by 1 digit, rename hundreds and tens
5	7)800 → 114r2 or 114.28		3 digit divided by 1 digit, rename hundreds and tens
6	6)394 → 65r4 or 65.66		3 digit divided by 1 digit, no hundreds in answer
7	8)207 → 25r7 or 25.87		3 digit divided by 1 digit, no hundreds in answer
8	6)780 → 130		3 digit divided by 1 digit, zero in answer
9	4)837 → 209r2 or 209.25		3 digit divided by 1 digit, internal zero in answer
10	7)8637 → 1233r6 or 1233.85		4 digit divided by 1 digit

COMPUTATION — DIVISION ALGORITHM continued

Item		Response	Comments
11	8)9007 → 1125r7 or 1125.87		4 digit divided by 1 digit
12	7)9137 → 1305r2 or 1305.28		4 digit divided by 1 digit, internal zero in answer
13	6)3946 → 657r4 or 657.66		4 digit divided by 1 digit, no thousands digit in answer
14	7)1963 → 280r3 or 280.4285		4 digit divided by 1 digit, no thousands digit in answer, zero in answer
15	9)3658 → 406r4 or 406.444		4 digit divided by 1 digit, no thousands digit in answer, internal zero in answer
16	7)59273 → 8467r4 or 8467.571		5 digit divided by 1 digit, no ten thousands digit in answer
17	9)63452 → 7050r2 or 7050.222		5 digit divided by 1 digit, no ten thousands digit in answer, internal zero in answer
18	6)43265 → 7210r5 or 7210.833		5 digit divided by 1 digit, no ten thousands digit in answer, zero in answer
19	13)4256 → 327r5 or 327.3846		4 digit divided by 2 digit
20	34)7026 → 206r22 or 206.647		4 digit divided by 2 digit, no thousands in answer

Items to assess understanding and facility with the division algorithm

Source: Booker, 1995, *Booker Profiles in Mathematics*, Computation Division Algorithm sheets 1 and 2, reproduced by permission of the Australian Council of Educational Research.

Consideration also needs to be given to the relatively routine problems that might be solved with a straightforward use of division. Of course, problem solving depends on more than computational skill. Aspects of the concept can cause difficulties, the working that needs to be examined can be more complex, and the information can be in excess of that needed or presented in an order counter to that in which it might be used. Close observations of an individual's solution to such problems can provide a good deal of information about his or her conceptual knowledge, the skills and understandings of processes that might be used and the meaning that the particular operation has for him or her. While incorrect answers can also be due to inappropriate recording, a lack of understanding of the operations or of the particular problem situation may also mean that an inappropriate response is drawn from a correct computation:

John has 87 marbles. He wants to share them equally with Peter and Michael. How many marbles will each of the three boys get?

$$3\overline{)87} \quad \begin{array}{r} 29 \\ \underline{6} \\ 27 \\ \underline{27} \end{array}$$

The boys will have 27 marbles each

A systematic examination of the concepts, understandings and skills that underpin this complex set of capabilities can help pinpoint where and why difficulties have developed and competencies have emerged. Diagnostic assessment requires an informed set of questions, acute observations of responses and an analysis of the underlying mathematical understandings if it is to provide a basis to plan further learning and develop confident, competent users of mathematics.

Assessing problem solving

An ability to think mathematically and apply this to the solution of problems is crucial in today's world. Yet many people believe that there are those who are born with this ability and others who just do not seem able to obtain it. This, of course, is not the case and problem solving can be built up by judicious teaching as described in Chapter 2. However, a key to assisting children to build on their natural problem solving abilities is to assess their current strengths and weaknesses in coming to terms with problem tasks. Any measure of problem solving ability must focus on the possession of an overall approach to determining the meaning of a problem and ways in which this can be put to use in suggesting means to a solution. Simply noting the presence of a series of isolated techniques, especially when they are just acting on the numbers within a problem or on one of the parts required for a solution, will only give a measure of success or lack of success on particular problems. Since the solution of one problem is no guide to an ability to solve any subsequent problems, no measure of an overall ability to solve problems will be obtained. What is needed is a means of revealing the underlying mathematical thinking by providing a series of problem solving tasks that are similar in structure but have increasing mathematical demands.

Two aspects need to be considered in framing these problems—the balance between the difficulty of the mathematics and the language used to express them, and the complexity of the problems themselves. Initially, problems should use simple language and straightforward mathematics, before involving simple language and more difficult mathematics or more difficult language and simple mathematics. At the highest level, problems would have difficult language and difficult mathematics. Any *operations* involved should be relatively obvious at first before being framed to be less obvious. The *information* required for a solution should move from being fully provided to being more than is needed and then to too little so that more will need to be obtained from understanding the mathematics or from contextual knowledge. The *thinking* applied to the problem should build from straightforward identification and application of known processes to strategic thinking needed to reinterpret existing knowledge in the light of the problem question.

The following questions and means of analysis taken from the *Booker Profiles in Mathematics: Thinking Mathematically* (ACER, 2001) illustrate how these considerations can be used as the basis of an assessment tool designed to assist in building problem solving:

THINKING MATHEMATICALLY — RECORD SHEET — Level 3

Problem	Response	Comments
3.1 Michael delivered 392 newspapers in 6 hours. He can be paid either by the number of newspapers or by the hour. The rate per newspaper is 4.5c and the hourly rate is $3.75 for the first hour and $2.75 for the other hours. Which method of payment is best?	2.75 13.75 × 5 + 3.75 13.75 17.50 392 × 4.5 1764 or $17.64 so payment per paper	
3.2 At the strawberry fest David ate twice as many strawberries as Stephen. Stephen ate 4 strawberries fewer than Jane, and Jane ate 8 more than Craig. If Craig ate 9 strawberries, how many strawberries did David eat?	Work backwards 26 strawberries	
3.3 Judy and her friends drove from Brisbane to Mackay, a distance of 999 km. They left at 9 am and averaged 96 km per hour for the first 3 hours of their trip. They stopped for half an hour for morning tea and an hour for lunch. After lunch they drove for another 4 hours at an average speed of 85 km per hour. How far had they travelled?	96 85 × 3 × 4 288 340 288 + 340 628 They travelled 628 km	
3.4 A farmer wants to fence a rectangular paddock 38 metres longer than it is wide. If she has 1464 metres of fencing, how long and wide should she make the sides?	38 + 38 76 1464 − 76 1388 1388 ÷ 4 = 347 347 m wide 385 m long	
3.5 Bill's car was out of petrol. He pushed it to the service station and put 36 litres of petrol into the tank. As he drove away, he noticed that the tank was still ¼ empty. How many litres does a full tank hold?	36 litres is 3 fourths of a tank 1 fourth is 12 L 4 × 12 is 48 Full tank holds 48 L	

THINKING MATHEMATICALLY — ANALYSIS SHEET — Level 3

Using a highlighter pen, shade cells to indicate questions answered correctly.

Language/Mathematics	Obvious	Less obvious	Too much information	Too little information	Strategic thinking
simple/simple	3.1	3.2	3.3		
simple/more difficult or more difficult/simple	3.6	3.7	3.8	3.4	3.5
difficult/difficult				3.9	3.10

Shade cells to indicate quality of responses (where observed).

	Yes	Sometimes	No
Able to read questions?			
Were questions read and analysed?			
Was all information considered?			
Was irrelevant information discounted?			
Were numbers simply manipulated?			
Was order of steps considered?			
Were possible solutions explored?			
Was conceptual understanding used?			
Was a diagram used?			
Was a table or list used?			
Was the mathematics carried out systematically?			
Were answers expressed in a sentence?			
Were answers examined for reasonableness?			

Further comments:

Poor performance at this level indicates a need to:
- analyse problems;
- anticipate the form of the answer;
- explore a variety of possible solutions;
- check that an answer fits the problem conditions.

Items to assess and analyse higher level problem solving abilities

Source: Booker and Bond, (2001), *Booker Profiles in Mathematics: Thinking Mathematically*, problems and analysis sheet from level 3, reproduced by permission of the Australian Council of Educational Research.

Using and communicating assessment information

The aim of any assessment should be to improve learning and teaching. Designing and administering it correctly is an important part of the process, but using the results effectively is equally significant. Assessment results, including test scores, merely provide information that is of little use unless teacher and student then do something effective as a result of analysing it. For instance, while there is a tendency to conclude that a student has not developed understanding and skill when his or her results on a test are low, it could just as readily indicate that there is something amiss with the test itself, with the mathematics program or with the teaching approaches, especially if several students do poorly.

Assessment results, then, might lead a teacher to re-examine the whole unit of work, although at other times they might indicate whether the unit should be continued or whether it is appropriate to move on to the next phase in the overall mathematics program. For the students, the results might confirm their approaches to learning mathematics or might lead them to question the way they enter into the activities or complete work in their own time. They might also lead to an awareness that there are aspects of earlier work that might need revisiting, either by the student alone or with help from the teacher. Discussions with a student's parent or care-giver might reveal differences in expectations and support between home and school. It is this capacity for informing all of the participants in learning that is, after all, the major reason that assessment is carried out.

Informing students

The results of any assessment should be discussed with individuals and the class to bring out the implications for the topic under review and for further learning. Unfortunately, many students simply look to their mark or comment to see how they compare with other students rather than try to interpret what it might mean for their mathematical knowledge. For the students, a test should be able to show what they know, what they need to know and, with the help of a teacher or peers, how to better come to terms with the material and be able to communicate that knowledge to others through the tasks and investigations that they are able to complete. By comparing their own responses with those of others in the class as well as a teacher's summary of the expected responses and any errors made, students can see how to make sense of the questions they are asked and how to present their knowledge accurately, completely and concisely.

Informing the teacher

When the *observe, assume, probe* cycle is used at all times, it should enable a teacher to intervene in the learning before any misconception or inadequate way of thinking becomes entrenched. However, there will be times when individual children will develop inappropriate ways of acting and thinking, and full remedial activities will be needed. Rich assessment information provides a teacher with full details of the nature of any difficulty so that re-teaching can be tailored to a child's specific needs. For example, knowing only that a child has scored 40 per cent on a numeration test provides little advice for improving learning. Knowing that all items referring to two-digit numbers were correct except those relating to teen numbers does, however, signal the most appropriate starting point for remedial activities.

Informing teaching

Assessment results can also provide valuable information to help teachers evaluate various aspects of the teaching process when results of a class or group are considered as a whole. For example, when the majority of a group perform poorly on the same test items, it may indicate poor learning on the part of the students, but it could also indicate that problems exist in the instructional processes used. Particular learning outcomes may have been unrealistic, ineffective methods may have been followed, or materials may not have been used adequately to produce the desired learning. This is commonly the case when children are introduced to new concepts and processes in a formal symbolic mode without first having extensive experience using materials to develop an appropriate language to talk about what is happening.

In other cases, mediocre performance by the class might indicate that otherwise good learning experiences were poorly organised or badly implemented, or that time was misallocated for the learning activities. Often, planning for a unit of work over a five-week period results in the material to be learned being broken into five equal pieces. Yet it may be better to allocate half of the time to the initial concepts, and then build more quickly on this secure foundation. In all cases, a careful analysis of common student errors is useful in providing valuable evidence to assist class teachers to identify any aspects of their planning or implementation of class learning experiences which should be reviewed. Post-test discussions of results may also provide valuable clues to sources of difficulties which may be corrected by making changes to the instructional processes.

Informing parents, care-givers and other teachers

One further use of assessment results is in reporting to parents or care-givers at the end of a semester or year. Most schools have their own formats for this procedure, ranging from only giving a comment on mathematics as a whole to a more insightful breakdown of the subject into the various processes and concepts involved. It is also usual to write a brief comment on a child's overall learning and ability to use mathematics. To be helpful to both parent and child, this should provide an overview of strengths and weaknesses drawn from a range of assessments and should relate to the specific learning and teaching undertaken in that time period. Subsequent discussions with the parent or care-giver can then show specific examples from a portfolio of the child's achievements to clarify what has been reported and the basis for any comments made, and can lead to an understanding of what is needed in future learning. A report backed up with evidence of this nature can also be invaluable for a teacher in a succeeding year or in another school should the student be transferring.

Assessment, learning and teaching

Assessment is important in understanding the knowledge that students are constructing, the meanings they give to mathematical ideas and their developing mathematical ability. It needs to focus on conceptual understanding rather than simply relate to procedural knowledge and should provide a full profile of a student's ability or capacity. In order to do this, it needs to draw on multiple sources of information and reflect the activities promoted and valued in the classroom as well as in the child's own life. Assessment needs to be authentic, resembling real learning tasks and actively involving students. After all, assessment is in many ways a contract between teachers and students in which it is the student's responsibility to demonstrate understanding and proficiency and the teacher's responsibility to provide the opportunity and the means for that demonstration (Clarke, 1997).

Assessment must also be aligned with teaching so that it supports the overall goals for the learner, the teacher and the mathematics. If the focus and form of assessment are different from those of teaching and learning, then assessment can subvert students' learning by sending conflicting messages about what is valued. When instruction pursues one set of goals and assessment another, students are faced with a dilemma and must assume that the goals of assessment are the ones that count (National Council of Teachers of Mathematics, 1995a, p. 13). On the other hand, assessment that enhances mathematics learning can become an integral part of ongoing classroom activity, encouraging and supporting further learning. Indeed, continuous assessment of students' work and working not only facilitates their learning of mathematics but also enhances their confidence and understanding. As they learn to monitor their own development, they can reflect on their progress, be confident in their understanding and proficiency and ascertain what they have yet to learn.

References

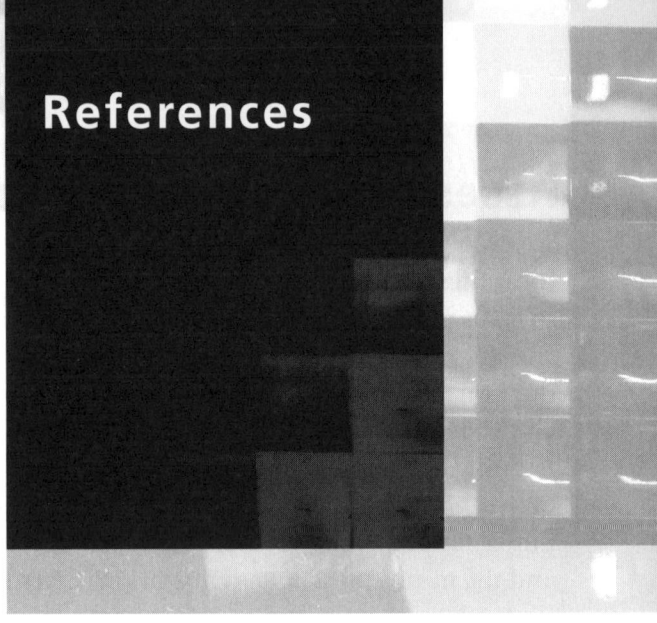

Ainley, J., 1990, 'Playing Games and Learning Mathematics' in L. Steffe and T. Wood (eds), *Transforming Children's Mathematics Education: International Perspectives*, Hillsdale, NJ: Lawrence Erlbaum.

Allende, I., 2001, *Portrait in Sepia* London: HarperCollins p. 151.

Artigue, M., 1992, 'The Importance and Limits of Epistemological Work in Didactics' in W. Geeslin (ed.), *Proceedings of the 15th International Conference of the Psychology of Mathematics Education*, Durham: University of New Hampshire.

Artzt, A. and Armour-Thomas, E., 1990, 'Protocol Analysis of Group Problem Solving in Mathematics: A Cognitive–Metacognitive Framework for Assessment', paper presented at the Annual Meeting of the American Educational Research Association, Boston.

Atweh, B, Owens, K. and Sullivan, P., 1996, *Research in Mathematics Education in Australasia 1992–1995*, Campbelltown: MERGA.

Australian Association of Mathematics Teachers, 1990, *A National Statement on Girls and Mathematics*, Canberra: GEMS.

Australian Association of Mathematics Teachers, 1997, *Numeracy = Everyone's Business, Report of the Numeracy Education Steering Committee Development Conference*, Adelaide: AAMT.

Barry, B., Booker, G., Perry, B. and Siemon D., 1996, *HBJ Mathematics Series*, Sydney: Harcourt Brace.

Battista, M., 2002, 'Learning Geometry in a Dynamic Computer Environment', *Teaching Children Mathematics*, February, pp. 333–339.

Bay-Williams, J., 2001, 'What is Algebra in Elementary School?', *Teaching Children Mathematics*, December, pp. 196–200.

Beesey, C., Clarke, B., Clarke, D., Stephens, M. and Sullivan, P., 1998, *Effective Assessment for Mathematics*, Melbourne: Longman.

Bishop, A.J., 1983, 'Space and Geometry' in R. Lesh and M. Landau (eds), *The Acquisition of Mathematical Concepts and Processes*, New York: Academic Press.

Bishop, A., 1997, 'Imagining, Locating and Mapping: The Geographical Possibilities' in D. Clarke, P. Clarkson, D. Gronn, M. Horne, L. Lowe, M. Mackinley and A. McDonough (eds), *Mathematics: Imagine the Possibilities* (pp. 75 - 79), Melbourne: Mathematical Association of Victoria.

Bishop, A., Clarke, B. and Ocean, J., 2002, 'Highly Accomplished Teachers in Mathematics' in C. Vale, J. Roumeliotis and J. Horwood (eds), *Valuing Mathematics in Society* (pp. 1-10), Melbourne: Mathematical Association of Victoria.

Blum, H. and Yocom, D., 1996, 'Using Instructional Games to Foster Student Learning', *Teaching Exceptional Children*, November/December, pp. 60-63.

Bond, D., 1996, 'Developing Children's Mathematical Thinking' in H. Forgasz, T. Jones, G. Leder, J. Lynch, K. Maguire and C. Pearn (eds), *Mathematics: Making Connections*, Proceedings of the 1996 Annual Conference, Melbourne: Mathematics Association of Victoria.

Booker, G., 1995, *Booker Profiles in Mathematics: Numeration and Computation*, Melbourne: Australian Council for Educational Research (ACER).

Booker, G., 2000, *The Maths Game: Using Games to Teach Mathematics*, Wellington: New Zealand Council for Educational Research.

Booker, G., 2002, *Win With Maths!: A CD-ROM of Early Numeracy Games*, Wellington: New Zealand Council for Educational Research.

Booker, G. and Bond, D., 2001, *Booker Profiles in Mathematic: Thinking Mathematically*, Melbourne: Australian Council for Educational Research (ACER).

Brousseau, G., 1997, *Theory of Didactical Situations in Mathematics*, Dordrecht: Kluwer.

Brownell, W., 1935, 'Psychological Considerations in the Learning and the Teaching of Mathematics' in W. Reeve (ed.), *The Teaching of Arithmetic*, 10th Yearbook of the National Council of Teachers of Mathematics, New York: Teachers College, Columbia University.

Brownell, W., 1938, in G. Buswell, W. Brownell and L. John, *Living Arithmetic*, Boston: Ginn.

Brownell, W., 1945, 'When is Arithmetic Meaningful?', *Journal of Educational Research*, Vol. 38, No. 7, pp. 481-498.

Burton, L., 2001, 'Fables: The Tortoise? The Hare? The Mathematically Underachieving Male' in B. Atweh, H. Forgasz and B. Nebres (eds), *Sociocultural Research on Mathematics Education: An International Perspective*, Mahwah, NJ: Lawrence Erlbaum.

Buswell, G., Brownell, W. and John, L., *Living Arithmetic*, Boston: Ginn.

Clarke, D., 1988a, *Assessment Alternatives in Mathematics*, Melbourne: Curriculum Corporation.

Clarke, D., 1988b, *The Mathematics Curriculum and Teaching Project*, Canberra: Curriculum Development Centre.

Clarke, D., 1997, *Constructive Assessment in Mathematics*, Berkeley: Key Curriculum Press.

Clements, D.S., 1997, '(Mis?)Constructing Constructivism', *Teaching Children Mathematics*, December, pp. 198-200.

Clements, D., 2003, in D. Clements & G. Bright (eds), *Learning and Teaching Measurement: 2003 Yearbook*, Reston, VA: National Council of Teachers of Mathematics.

Clements. D. and Battista, M., 1992, 'Geometry and Spatial Reasoning' in D. Grouws (ed.), *Handbook of Research on Mathematics Teaching and Learning* (pp. 420 - 464), New York: Macmillan.

Clements, D. and Callahan, L.G., 1983, 'Number or Prenumber Foundational Experiences for Young Children: Must we choose?', *Arithmetic Teacher*, November, pp. 34-37.

Clements, D. and McMillen, S., 1996, 'Rethinking Concrete Manipulatives', *Teaching Children Mathematics*, January, pp. 270-279.

Cobb, P., 1987, 'An Analysis of Three Models of Early Number Development', *Journal for Research in Mathematics Education*, Vol. 18, pp. 163-179.

Cobb, P., 1988, 'The Tension Between Theories of Learning and Instruction in Mathematics Education', *Educational Psychologist*, Vol. 23, No. 2, pp. 87–103.

Cobb, P., 1989, 'Experimental, Cognitive and Anthropological Perspectives in Mathematics Education', *For the Learning of Mathematics*, Vol. 9, No. 2, pp. 32–42.

Cobb, P. and Bauersfeld, H., 1995, *The Emergence of Mathematical Meaning: Interaction in Classroom Cultures*, Mahwah, NJ: Lawrence Erlbaum.

Cobb, P., Yackel, E. and Wood, T., 1992, 'A Constructivist Alternative to the Representational View of Mind in Mathematics Education', *Journal for Research in Mathematics Education*, Vol. 23, No. 1, pp. 2–33.

Cockcroft, W., 1982, *Mathematics Counts: Report of the Committee of Inquiry into the Teaching of Mathematics in Schools*, London: Her Majesty's Stationery Office.

Confrey, J., 1985, 'Toward a Framework for Constructivist Education', Proceedings of the Ninth International Conference on the Psychology of Mathematics Education, London, pp. 477–483.

Cooney, T., Badger, E. and Wilson, M., 1993, 'Assessment, Understanding Mathematics, and Distinguishing Visions from Mirages' in N. Webb and A. Coxford, *Assessment in the Mathematics Classroom* (pp. 239–247), Reston, Virginia: National Council of Teachers of Mathematics.

Coupland, M. and Wood, L., 1998, 'What Happens When the Girls Beat the Boys? Community Reactions to the Improved Performance of Girls in Final Examinations' in C. Keitel (ed.), *Social Justice and Mathematics Education. Gender, Class, Ethnicity and the Politics of Schooling* (p. 243), Berlin: Freie University.

Crowther, G., 1959, *Report to the Central Advisory Council for Education*, London: HMSO.

Cundy, H.M. and Rollett, A.R., 1981, *Mathematical Models* (3rd edn), Norfolk: Tarquin Publications.

Curcio, F., 2001, *Developing Data-graph comprehension in Grades K-8* (2nd edn), Reston, Virginia: National Council of Teachers of Mathematics.

Curriculum Corporation, 1991, *A National Statement on Mathematics for Australian Schools*, Carlton, Victoria: Curriculum Corporation.

Curriculum Corporation, 2000, *Numeracy Benchmarks Years 3, 5 & 7*, Melbourne: Curriculum Corporation.

Curriculum Council, 1998, *Curriculum Framework for Kindergarten to Year 12 Education in Western Australia*, Perth: Author.

d'Ambrosio, U., 1985, 'Ethnomathematics and its Place in the History and Pedagogy of Mathematics', *For the Learning of Mathematics*, Vol. 5, No. 1, pp. 44–48.

Datta, B. and Singh, A., 1962, *History of Hindu Mathematics* (2 volumes), Bombay: Asia Publishing House.

Davey, G., 1998, 'Space and Geometry' in G. Booker, D. Bond, J. Briggs and G. Davey, *Teaching Primary Mathematics* (2nd edn), Melbourne: Longman.

Davey, G. and Pegg, J., 1992, 'Research in Geometry and Measurement' in B. Atweh and J. Watson (eds), *Research in Mathematics Education in Australia 1988–1991* (pp. 231–248), Kelvin Grove: Mathematics Education Research Group of Australasia (MERGA).

Davis, R., 1993, 'Reality of Negative Numbers' in R. Davis and C. Maher (eds), *Schools Mathematics and the World of Reality*, New York: Allyn and Bacon.

Delaney, K., 1979, 'A Place for Space', *Mathematics Teaching* 86, pp. xvii.

Delaney, K. and Dichmont, J., 1979, 'Do it Yourself Islamic Patterns', *Mathematics Teaching*, 86, pp. xvii–xix.

Department for Education and Employment, 1998, *The Implementation of the National Numeracy Strategy: The Final Report of the Numeracy Task Force*, London: DfEE.

Department of Education, 1994, *Literacy and Numeracy Strategy 1994-98*, Brisbane: Queensland Department of Education.

Department of Education, Training and Youth Affairs, 2000, *Numeracy: A Priority For All: Challenges for Australian Schools*, Canberra: DETYA.

Dick, T., 1988, 'The Continuing Calculator Controversy', *Arithmetic Teacher*, April, pp. 37-41.

Dungan, J. and Thurlow, G., 1989, Students' Attitudes to Mathematics: A Review of the Literature, *The Australian Mathematics Teacher*, Vol. 45, No. 2, October, pp. 8-11.

Education Department of Western Australia, 1998, *Outcomes and Standards Framework: Mathematics Student Outcome Statements*, Perth: Education Department of Western Australia.

Ellerton, N.F. and Clements, M.A., 1990, *Attitudes and Appreciations of Mathematics*, Adelaide: Australian Association of Mathematics Teachers (Teacher's Handbook).

Ernest, P., 1986, 'Games—a Rationale for their Use in the Teaching of Mathematics', *Mathematics in School*, Vol. 15, No. 1, pp. 2-5.

Fernandez, M., 1994, 'Problem Solving: Managing It All', *The Mathematics Teacher*, Vol. 87, No. 3, March, pp. 195-199.

Finlay, E. and Lowe, I., 1993, *Chance and Data: Exploring Real Data*, Melbourne: Curriculum Corporation.

Flegg, G., 1984, *Numbers: Their history and meaning*, Harmondsworth: Penguin.

Forgasz, H., and Leder, G., 2001, 'A+ for Girls, B for Boys: Changing Perspectives on Gender Equity and Mathematics' in B. Atweh, H. Forgasz and B. Nebres (eds), *Sociocultural Research on Mathematics Education: An International Perspective*, Mahwah, NJ: Lawrence Erlbaum.

Friel, S., Mokros, J. and Russell, S., 1992, *Used Numbers—Statistics: Middles, Means and In-Betweens, Grades 5-6*, Palo Alto: Dale Seymour.

Fuson, K., 1988, *Children's Counting and Concepts of Number*, New York: Springer-Verlag.

Fuys, D. Geddes, D. and Tischler, R., 1998, *The van Hiele Model of Thinking in Geometry Among Adolescents*, JRME Monograph 3, Reston, Virginia: NCTM.

Gillings, R., 1982, *Mathematics in the Times of the Pharaohs*, New York: Dover.

Goldin, G., 2002, 'Representation in Mathematical Learning and Problem Solving' in L. English (ed.), *Handbook of International Research in Mathematics Education*, Mahwah, NJ: Erlbaum, pp.197-218.

Gonzales, N., Mitchell, M. and Stone, A., 2002, *Mathematical History: Activities, Puzzles, Stories and Games* (2nd edn), Reston, Virginia: National Council of Teachers of Mathematics.

Hanbury, L., 1999, *Geometry and Measurement in the Primary School*, Griffith University: Faculty of Education.

Hanbury, L., 2000, 'Student Problems Inhibiting the Successful Answering of Geometrical Questions' in J. Wakefield (ed.), *Mathematics: Shaping the Future* (pp. 340-347), Melbourne: MAV.

Hart, L.E. and Walker, J., 1993, 'The Role of Affect in Teaching and Learning Mathematics' in D.T. Owens (ed.), *Research Ideas for the Classroom: Middle Grades Mathematics* (pp. 22-38), Reston, Virginia: National Council of Teachers of Mathematics.

Herscovics, N. and Bergeron, J., 1984, 'A Constructivist vs A Formalist Approach in the Teaching of Mathematics', Proceedings of the Eighth International Conference on the Psychology of Mathematics Education, Sydney, pp. 190-196.

Hoffer, A., 1981, 'Geometry is More Than Proof', *Mathematics Teacher*, No. 74, pp. 11-18.

Hoffer, A., 1983, 'Van Hiele Based Research' in R. Lesh and M. Landau (eds), *The Acquisition of Mathematical Concepts and Processes*, New York: Academic Press.

Horne, M., 1998, 'An Angle on Geometry' in J. Gough and J. Mousley (eds), *Mathematics: Exploring All Angles* (pp. 201–209), Melbourne: Mathematics Association of Victoria (MAV).

Huinker, D., 2002, 'Calculators as Learning Tools for Young Children's Explorations of Number', *Teaching Children Mathematics*, February, pp. 316–321.

Ifrah, G., 1985, *From one to zero: A universal history of numbers*, New York: Viking Penguin.

Irons, C., Burnett, J. and Fong, S., 1994, *Mathematics from Many Cultures*, Hawthorne, Victoria: Mimosa Publications.

Isaacs, A.C. and Carroll, W.M., 1999, 'Strategies for Basic-facts Instruction', *Teaching Children Mathematics*, Vol. 5, No. 9, pp. 508–515.

Jones, G., Thornton, C., Langrall, C. and Tarr, J., 1999, 'Understanding Children's Probabilistic Reasoning' in L. Stiff and F. Curcio (eds), *Developing Mathematical Reasoning in Grades K–12*, 1999 Yearbook, Reston, Virginia: National Council of Teachers of Mathematics.

Joseph, G., 1991, *The Crest of the Peacock*, Harmondsworth, UK: Penguin.

Kamii, C. and Housman, L., 2000, *Young Children Reinvent Arithmetic: Implications of Piaget's Theory* (2nd edn), New York: Teachers College Press.

Kaplan, R., 2000, *The Nothing That Is*, Harmondsworth, UK: Penguin.

Katz, V., 1993, *A History of Mathematics: An Introduction*, New York: Harper Collins.

Kerrigan, J.,2002, 'Powerful Software to Enhance the Elementary School Mathematics Program', *Teaching Children Mathematics*, February, pp. 364–370.

Kilpatrick, J., 1992, 'A History of Research in Mathematics Education' in D. Grouws (ed.), *Handbook of Research on Mathematics Teaching and Learning*, Reston, Virginia: NCTM.

Klaebe, K., 1986, *Dictionary of Mathematics*, Sydney: Harcourt Brace.

Labinowicz, E., 1980, *The Piaget Primer: Thinking, Learning, Teaching*, Menlo Park: Addison-Wesley.

Larouche, C., Bergeron, J. and Herscovics, N., 1984, 'Research on the Role of Play and Games in the Learning of Mathematics', *Proceedings of the Sixth Annual Meeting of the North American Chapter of the International Group for the Psychology of Mathematics Education*, Madison, Wisconsin, pp. 203–212.

Leder, G.C., 1992, 'Attitudes Towards Mathematics', *Mathematics Education Research Journal*, 4(3), pp.1–7.

Lester, F., 1985, 'Methodological Considerations in Research on Mathematical Problem Solving Instruction' in E.A. Silver (ed.), *Teaching and Learning Mathematical Problem Solving: Multiple Research Perspectives* (pp. 41–69), Hillsdale, NJ: Lawrence Erlbaum Associates.

Lindquist, M.M. and Shulte, A.P. (eds), 1987, *Learning and Teaching Geometry K–12*, Reston, Virginia: National Council of Teachers of Mathematics.

Lovitt, C. and Lowe, I., 1991, *Chance and Data*, Melbourne: Curriculum Corporation.

Ma, Liping, 1999, *Knowing and Teaching Elementary Mathematics*, Mahwah, NJ: Lawrence Erlbaum.

Malone, J. and Taylor, P., 1993, *Constructivist Interpretations of Teaching and Learning Mathematics*, Perth: Curtin University.

Mankiewicz, R., 2000, *The Story of Mathematics*, London: Cassell.

Mansfield, H., and Happs, J., 1991, 'Concept Maps', *The Australian Mathematics Teacher*, 47(3) pp. 30–33.

McIntosh, A., Reys, B. and Reys, R., 1992, 'A Proposed Framework for Examining Basic Number Sense', *For the Learning of Mathematics*, Vol. 12, No. 3, pp. 2–8, 44.

McIntosh, A., Reys, B. and Reys, R., 1998, *Number SENSE*, Palo Alto, CA: Dale Seymour Publications.

McLeod, D.B., 1992, 'Research on Affect in Mathematics Education: A Reconceptualisation' in D.A. Grouws (ed.), *Handbook of Research on Mathematics Teaching and Learning*, A project of the National Council of Teachers of Mathematics, New York: Macmillan.

Moon, B., 1986, *The 'New Maths' Curriculum Controversy*, London: The Falmer Press.

National Council of Teachers of Mathematics, 1980, *Agenda for Action: Recommendations for School Mathematics of the 1980s*, Reston, Virginia: NCTM.

National Council of Teachers of Mathematics, 1980, *Problem Solving in School Mathematics*, 1980 NCTM Yearbook, Reston, Virginia: NCTM.

National Council of Teachers of Mathematics, 1989, *Curriculum and Evaluation Standards*, Reston, Virginia: NCTM.

National Council of Teachers of Mathematics, 1989, *Historical Topics for the Mathematics Classroom*, Reston, Virginia: NCTM.

National Council of Teachers of Mathematics, 1991, *Assessment Standards for Teaching Mathematics*, Reston, Virginia: NCTM.

National Council of Teachers of Mathematics, 1995a, *Assessment Standards for School Mathematics*, Reston, Virginia: NCTM.

National Council of Teachers of Mathematics, 1995b, *Professional Standards for School Mathematics*, Reston, Virginia: NCTM.

National Council of Teachers of Mathematics, 2000, *Principles and Standards for School Mathematics*, Reston, Virginia: NCTM.

National Council of Teachers of Mathematics, 2003, *Principles and Standards for School Mathematics—Overview*, http://www.nctm.org/standards/overview.htm

A National Statement on Mathematics for Australian Schools, 1991, Melbourne: Curriculum Corporation.

Needham, J., 1959, *Science and Civilisation in China*, Cambridge: University Press.

Nisbet, S., 1997, 'Beliefs and Attitudes in Mathematics Learning' in G. Booker, D. Bond, J. Briggs and G. Davey, *Teaching Primary Mathematics* (2nd edn, pp. 410–411), Melbourne: Longman.

Noddings, N., 1993, in R. Davis and C. Maher (eds), *Schools, Mathematics and the World of Reality*, New York: Allyn and Bacon.

O'Brien, T., 1980, *Wollygoggles and Other Creatures*, New York: Cuisenaire.

Olssen, K.(ed.), 1995, *Working Mathematically: Investigations*, Carlton, Victoria: Curriculum Corporation.

Organisation for Economic Co-operation and Development, 1999, *Measuring Student Knowledge and Skills: A New Framework for Assessment*, Paris: OECD.

Orrill, O., 2001, 'Mathematics, Numeracy and Democracy' in L. Steen (ed.), Preface to *Mathematics and Democracy*, Washington DC: National Council on Education and the Disciplines.

Owens, K. and Mousley, J., 2000, *Research in Mathematics Education in Australasia 1996–1999*, Kelvin Grove: MERGA.

Piaget, J., 1972, *To Understand is to Invent*, New York: Grossman.

Pinel, A., 2002, '90 Minutes on a Sunday Afternoon', *Mathematics Teaching*, 179, pp. 8–10.

Pirie, S., 1998, 'Crossing the Gulf Between Thought and Symbol: Language as (Slippery) Stepping-Stones' in H. Steinbring, M. Bussi and A. Sierpinska (eds), *Language and Communication in the Mathematics Classroom*, Reston, Virginia: National Council of Teachers of Mathematics.

Polya, G., 1965, *Mathematical Discovery* (Vol. 1, p. 117), New York: Wiley.
Polya, G., 1990, *How to Solve It*, Melbourne: Penguin.
Principles and Standards for School Mathematics, 2000, Reston, VA: National Council of Teachers of Mathematics.
Principles and Standards for School Mathematics—Overview, 2003, http:/www.nctm.org/standards/overview.htm: National Council of Teachers of Mathematics.
Putnam, J., Lampert, M. and Peterson, P., 1990, 'Alternative Perspectives on Knowing Mathematics in Elementary Schools', *Review of Research in Education*, Vol. 16, pp. 57–150.
Rathmell, E., 1978, 'Using Thinking Strategies to Teach the Basic Facts' in *Developing Computational Skills*, 1978 NCTM Yearbook (pp. 13–38), Reston, Virginia: NCTM.
Renga, S. and Dalla, L., 1993, 'Affect: A Critical Component of Mathematical Learning in Early Childhood' in R. Jensen (ed.), *Research Ideas for the Classroom: Early Childhood Mathematics* (pp. 22–39), New York: Macmillan.
Reys, B., 1995, *Mission Statement*, Number Sense Research Group, University of Missouri, http://tiger.coe.missouri.edu/~barb/number.html
Rich, W. and Jotner, J., 2002, 'Using Interactive Web Sites to Enhance Mathematics Learning', *Teaching Children Mathematics*, February, pp. 380–383.
Ritchhart, R., 1994, *Making Numbers Make Sense*, Menlo Park, California: Addison-Wesley.
Rubenstein, R., 1996, 'Strategies to Support the Learning of the Language of Mathematics' in *Communication in Mathematics, K–12 and Beyond*, 1996 NCTM Yearbook (pp. 214–218), Reston, Virginia: NCTM.
Schoenfeld, A., 1985, *Mathematical Problem Solving*, Orlando: Academic Press.
Schoenfeld, A., 1987, 'Cognitive Science and Mathematics Education: An Overview', in A.H. Schoenfeld (ed.), *Cognitive Science and Mathematics Education* (pp. 1–31), Hillsdale, New Jersey: Lawrence Erlbaum Associates.
Schoenfeld, A., 1989, 'Teaching Mathematical Thinking and Problem Solving' in L. Resnick and L. Klopfer (eds), *Toward the Thinking Curriculum: Current Cognitive Research*, 1989 Yearbook of the Association for Supervision and Curriculum Development.
Schoenfeld, A (1992), 'Learning to think mathematically: problem solving, metacognition, and sense-making in mathematics', in D. Grouws, (ed.), *Handbook for Research on Mathematics Teaching and Learning* (Chapter 15, pp. 334–370), New York: MacMillan.
Schoenfeld, A., 2001, 'Reflections on an Impoverished Education' in L. Steen (ed.), *Mathematics and Democracy* (pp. 37–48), Washington DC: National Council on Education and the Disciplines.
Schwartzman, S., 1994, *The Words of Mathematics: An Etymological Dictionary of Mathematical Terms Used in English*, Washington DC: Mathematical Association of America.
Sharan, S., 1980, Cooperative Learning in Small Groups: Recent Methods and Effects on Achievement, Attitude, and Ethnic Relations, *Review of Education Research*, No. 50, pp. 241–271.
Siemon, D., 1998, 'Problem Solving: Addressing the Myths' in G. Booker, D. Bond, J. Briggs and G. Davey, *Teaching Primary Mathematics* (2nd edn, pp. 46–48), Melbourne: Longman.
Siemon, D. and Booker, G., 1990, 'Teaching and Learning FOR, ABOUT and THROUGH Problem Solving', *Vinculum*, Vol. 27, No. 2, June, pp. 4–12.
Silver, E., 1994, 'On Mathematical Problem Posing', *For the Learning of Mathematics*, Vol. 14, No. 1, pp. 19–28.

Slavin, R.E., 1983, *Cooperative Learning*, New York: Longman.
Smith, D. E., (Dover reprint – originally published 1925), *History of Mathematics, vols 1 & 2*, New York: Dover.
Smith, J. P., 2002, 'The Development of Students' Knowledge of Fractions and Ratios', in B. Litwiller and G. Bright (eds), *Making Sense of Fractions, Ratios and Proportions*, 2002 NCTM Year Book (p. 3), Reston Virginia: NCTM.
Stanic, G. and Kilpatrick, J., 1989, *Historical perspectives on problem solving in the mathematics curriculum*, in R. Charles & E. Silver (eds), The teaching and assessing of mathematical problem solving (pp. 1-22), Reston, VA: National Council of Teachers of Mathematics.
Steen, L. (ed.), 1997, *Why Numbers Count: Quantitative Literacy for Tomorrow's America*, New York, NY: The College Board.
Steen, L., 2001, 'Embracing Numeracy', Epilogue to L. Steen (ed.), *Mathematics and Democracy*, Washington DC: National Council on Education and the Disciplines.
Steen, L. (ed.), 2001, *Mathematics and Democracy*, Washington DC: National Council on Education and the Disciplines.
Steffe, L., Cobb, P. and von Glasserfeld, E., 1988, *Young Children's Construction of Arithmetical Meanings and Strategies*, New York: Springer-Verlag.
Steffe, L. and Wiegel, H., 1994, 'Cognitive Play and Mathematical Learning in Computer Microworlds', *Journal of Research in Childhood Education*, 8(2), pp. 117–131.
Streefland, L., 1985, 'Search for the Roots of Ratio: Some Thoughts on the Long Term Learning Process', *Educational Studies in Mathematics*, Vol. 10, No. 1, pp. 75–94.
Sullivan, P. and Lillburn, P., 1997, *Open Ended Mathematics Activities*, Melbourne: Oxford University Press.
Swan, P. and Sparrow, L., 1998, 'Calculators in the Primary School: Challenging the Myths', *Australian Primary Mathematics Classroom*, Vol. 3, No. 4, pp. 28–32.
Swetz, F., 1996, 'Problem Solving From the History of Mathematics' in *Proceedings of the History and Pedagogy of Mathematics Conference* (pp. 201–208), Braga, Portugal, 24–30 July.
Tharoor, S., 1984, *The Great Indian Novel*, London: Picador.
Thorndike, E.L., 1922, *The Psychology of Arithmetic*, New York: Macmillan.
Thornton, C., 1990, 'Strategies for the Basic Facts' in J. Payne (ed.), *Mathematics for the Young Child* (pp. 133–151), Reston, Virginia: NCTM.
Usiskin, Z., 1996, 'Mathematics as a Language' in *Communication in Mathematics, K–12 and Beyond* (pp. 231–243), 1996 NCTM Yearbook, Reston, Virginia: NCTM.
van Hiele, P.M., 1986, *Structure and Insight*, New York: Academic Press.
Vygotsky, L., 1978, *Mind in Society*, Cambridge Mass: MIT Press.
Wheatley, G., 1995, 'Calculators and School Mathematics' in R. Reys and N. Nohda (eds), *Computational Alternatives for the Twenty-first Century* (p. 116), Reston, Virginia: NCTM.
Willis, S., 1990, *Being Numerate: What Counts?*, Melbourne: Australian Council for Educational Research (ACER).
Willis, S., 1989, *Real Girls Don't Do Maths*, Geelong: Deakin University Press.
Willoughby, S., 2000, *Perspectives on Mathematics Education*, Reston, VA: National Council of Teachers of Mathematics.
Wilson, P. and Rowland, R., 1993, Chapter 8, in R.J. Jensen (ed.), *Research Ideas for the Classroom*, NCTM: Macmillan.

Yackel, E., Cobb, P., Wood, T., Wheatley, G. and Merkel, G., 1990, 'The Importance of Social Interaction in Children's Construction of Mathematical Knowledge', *Teaching and Learning Mathematics in the 1990s*, 1990, NCTM Yearbook, Reston, Virginia: NCTM.

Zaslavsky, C., 1996, *The Multicultural Mathematics Classroom: Bringing in the World*, Portsmouth, New Hampshire: Heinemann.

Index

0–99 boards 104, 121
2D shapes 397–417, 422–7
3D shapes 408, 417–27

achievements, minimum *see Numeracy Benchmarks*
addition
   algorithm for 182–4, 204–13
   basic facts 180, 195–204, 533–4
   calculators in 217–20, 222
   concept of 191–5
   with decimal fractions 220–5
   difficulties with 203
   estimation with 213–16
   with like common fractions 220–5
   mental computations 216–17
   overview 190
   sequence for developing 190–1, 192–3
   in subtraction basic facts 235–6
   with unlike common fractions 223–4, 254–8
   see also computation
algebra
   development of patterning skills for 170–4
   history of 363–4
   introduction of 364–6
algorithms
   addition 182–4, 204–13
   defined 175
   division 186, 316–31
   in estimation and mental computation 186–90
   multiplication 184–5, 275–88
   subtraction 184, 238–46

Allende, Isabel 12
angles
   in 2D shapes 410–16
   in measurement 463
area 462, 480–6
arrangement *see* location and arrangement strand
arrays 258, 260, 302–3 *see also* division; multiplication
arrowhead-shaped figures 406, 407
assessment
   alternatives to written tests 565–70
   defined 558
   diagnostic 575–9
   formal testing 558, 571–9
   by observation of play activities 27
   overview 556–8, 583
   of problem solving skills 579–80
   process of 558–65
   standardised tests 574–5
   by task 393–4, 442–4
   teacher-constructed tests 571–3
   use of results 581–2
assessment folders 566–7
average 22, 510–13

bar graphs 507, 515, 520
Base 10 blocks 105
basic fact wheel 235–6, 312
basic facts
   addition 180, 195–204, 533–4
   for computation 180–2
   learning steps for 237

multiplication 180, 266–75, 534–5
subtraction 180, 234–8, 533–4
behaviour, managing children's 550–1, 554
benchmarks *see Numeracy Benchmarks*
billions 126–7
'birds-eye view' sketches 425–7
Bishop, A. 368
BODMAS 363
BOMDAS 363
Booker, G. 43, 56
box plots 517–19
boys, ability of 28–30, 553–4
Brackets, Of, Divide, Multiply, Add, Subtract (BODMAS) 363
bringing down, in division 302, 319, 327
Brownell, William 10
Burton, L. 30

calculators
in addition 217–20, 222
in data analysis 508, 522
in division 309, 335–7, 338–9
in multiplication 282, 292–4, 297–8
operation of 33, 218
policy for 32, 69
spaces in four-digit numbers 116
in subtraction 249–51
use of 31–4
capacity and volume 455, 462–3, 486–8
Celsius, Anders 467
chance and data
concept development 501–2
in curriculum 67, 69, 468, 500–1
data analysis 509–21
data handling 506–9
probability 467–8, 502–6
teaching process 521–30
Chance and Data (Curriculum Corp) 524
checklists, in assessment process 568–9
chevron-shaped figures 406, 407
circle graphs 519
circles 408–10, 436, 484
Clarke, D. 393
class lists, in assessment process 567–8
classroom environment 25–6, 545–52, 554
clocks 269, 412, 464–6 *see also* time
Cobb, P. 14
column graphs 507, 515, 520
commas, conventions for 126, 147
common fractions
addition and subtraction with unlike 254–8

addition with 220–5
division with 317, 339–40
equivalent fractions 162–7, 254, 255–7, 258, 340–4
introduction of 136, 138
like and unlike 161, 223
multiplication with 298–9, 302
naming 156–9
renaming 159–62
subtraction with 251–8
communication *see* language
commutative property 264
compasses 408
comprehension *see* language; understanding
computation
algorithms for larger numbers 180, 182–90
basic facts for 180–2
concepts for 179
defined 175
errors and misconceptions with 348–59
learning process 178–9
rules for mixed 362
skill development 68–70
standard process for 349
teaching of 175–8, 359–66
see also addition; division; multiplication; subtraction
computers
in data analysis 508, 522
instructional games 27, 91, 446, 447
use of 31–2, 34, 69
concept maps 394
concepts, constructivist teaching approach 10–14
cones 420
congruent shapes 370
conservation of number 87
constructivism
defined 10–14
learning process in 537–8
and measurement 456
and teaching shape and space 384–5, 386
consultations, in assessment process 565–6
contingency tables 521
cooperative learning 25–6
counting
one-digit numbers 86, 88–9
two-digit numbers 104
three-digit numbers 112
four, five and six-digit numbers 116
Coupland, M. 30
Crest of the Peacock, The (Joseph) 35, 36

criterion-referenced tests 574
Cuisenaire rods 433
curriculum
   content and processes in 67–70, 531–6
   homework 552–3
   learning environment 545–52, 554
   planning mathematics program 1–2, 4–7, 450, 538–42
   planning the learning 542–5
   resources for developing 450, 543–4
   students' attitudes and beliefs 553–5
cylinders 420

data analysis
   average 510–13
   graphs and plots 507, 514–21
   mean 510–12
   median 512–13
   mode 513
   overview 509–10
   quartiles 513–14
   range 513
   standard deviation 514
   see also chance and data
data handling 468, 506–9 *see also* chance and data
databases 508
Davey, G. 374, 375
Davis, R. 24
decimal fractions
   addition with 220–5
   decimal points 126, 147, 221–2
   in division 317, 329–31, 337–9
   history of 134
   introduction of 135–7
   multiplication with 221
   in probability measurement 503–6
   recording and reading of 145–53
   renaming 165
   subtraction with 251–4
decimal points
   conventional symbols 126, 147
   'lining up' rule 221–2
Delaney, K. 432
Descartes, René 364
diagnostic assessment 575–9
diagonals, in 2D shapes 415–16
diamond-shaped figures 406
Dichmont, J. 432
Dick, T. 33–4
Dienes, Zoltan 105

dilation, in transformational geometry 430
direction *see* location and arrangement strand
distributive law 264
division
   algorithm for 186, 316–31
   basic facts 534–5
   calculators in 309, 335–7, 338–9
   concept of 302, 304–7, 309–10
   diagnostic assessment 576–8
   difficulties with 310, 315, 327–8, 331
   estimation with 331–3
   with fractions 317, 337–48
   invert and multiply rule 339
   linking to multiplication 311–15
   mental computations 333–5
   overview 302–3
   remainders 307–8, 329–31
   in renaming fractions 165
   sequence for developing 303–4, 307, 317
   see also computation
dot paper 424–5, 439
drawing
   3D shapes 424–7
   to develop spatial awareness 376

early-mathematical thinking 86–9
ellipses 408–10
enlargement, in transformational geometry 430–1
equilateral triangles 401
equipment *see* materials and equipment
equivalent fractions 162–7, 254, 255–7, 258, 340–4
errors and misconceptions
   intervention to correct 356–9
   in measurement 495
   origins of 14–15, 350–2
   overview 348–50
   patterns of 6, 11
   probing for reasons 561–5
   in problem solving 51–6
   reversal errors 228
   in shape and space learning 381–4
   sources of learned difficulties 352–6
Escher, M. C. 433, 436
Escher tessellations 433, 436–8
estimation
   in addition 213–16
   in computational skills development 186–90
   in division 177, 331–3
   in measurement 476

in multiplication 177, 288–90
  pre-requisites for 181
  in subtraction 246–8
etymology, of mathematical names 21–2
Euler, Leonhard 409
exams *see* tests
exponential notation 128–32, 353–4

factor trees 129
fathoms 458
females, ability of 28–30, 553–4
Fibonacci series 19
figures *see* two-dimensional (2D) shapes
five-digit numbers 117–19, 124–5
formulas
  introduction of 459
  teaching of 480, 482, 488
four-digit numbers 115–16, 118–19, 124–5
fraction ideas
  defined 133
  history of 133–4, 145
  language of 134
fractions
  addition with decimal and common 220–5
  addition with unlike common 254–8
  common 136, 138, 156–67
  concept development 134–5, 139–45
  decimals 135–7, 145–53, 165
  in division 317, 329–31, 337–48
  like and unlike 161, 223
  multiplication with common 298–9, 302
  multiplication with decimal 294–8, 301–2
  multiplication with per cents 299–302
  overview 133–9
  per cents 138–9, 153–6, 165
  in probability 503–6
  proper and improper 135, 160, 162
  ratios 167–70
  renaming as equivalent 162–7, 254, 255–7, 258, 340–4
  subtraction with common 251–8
  use of models 15–16, 17
frequency tables 520

games
  2D—3D transformations 423–4
  addition using calculators 218–19, 220
  algebraic number puzzles 366
  basic fact wheel 235–6
  birds-eye view sketches 426
  for chance and probability 522–30
  to compare and sequence numbers 90
  for counting skills 88–9
  for division 305–6, 308, 313
  for fractions 142–3, 157–8
  and learning process 26–8
  for one-digit number naming 83–5
  renaming fractions 160–1
  rounding numbers 122
  for shape and space 387, 389–90, 391, 395, 401
  shape and structure strand 399, 402–4, 414–15
  for three-digit numbers 108–9
  for two-digit numbers 95, 96, 97, 98–9, 102, 105
  using arrays 263–4
gender, and ability 28–30, 553–4
geoboards 407, 481
geometry
  applications of 377
  in curriculum 67, 69, 371–2
  defined 368
  forms of 370–1
  ways of working in 374–8, 385
  see also shape and space
Giles, Geoff 443
girls, ability of 28–30, 553–4
golden ratio 19, 368
graphs and plots 507, 514–21
Great Indian Novel, The (Tharoor) 82
grids 447–9

Hanbury, L. 370, 375
Hiele, Dina van 378, 380–1 *see also* van Hiele theory
Hiele, Pierre van 378, 380–1 *see also* van Hiele theory
high school mathematics 364, 366, 367–8
histograms 520
history of mathematics
  algebra 363–4
  fraction ideas 133–4, 145
  geometrical ideas 368–9
  mathematical education 35–6
  measurement ideas 457–9
  number words 21–2
  numbering systems 81–2
  pattern recognition 18, 19
Hoffer, A. 374, 385–6
homework 552–3
horizontal recording
  in addition 194
  introduction of 361–2

in multiplication 265
in subtraction 228, 232
Horne, M. 378

imaginary numbers 18
information and communication technology *see* technology
inquiry teaching-model 386
instructional games *see* computers; games
Internet 31, 34, 127
Internet Assigned Numbers Authority 127
intervention, for error correction 356–9
inventories, in assessment process 568–9
'invert and multiply' rule 339
irrational numbers 344
Islamic patterns 432–3
isometric sketches 424–5
isosceles triangles 401

Jones, G. 501
Joseph, G. 35, 36
journals, students' 570

kite-shaped figures 406, 407

language
  for computational skills 179
  conventions for measurement units 496–9
  of direction 446–7
  in learning process 20–3, 70
  in learning shape and space 369–70
  'length' words 471–2
  of location 446
  'location' words 471–2
  of measurement 455–6
  in shape and space 375–6, 387, 392, 418
learned difficulties
  remedial teaching 581–2
  sources of 352–6
learning difficulties
  in computation 348–52
  remedial teaching 581–2
  see also errors and misconceptions
learning process
  beliefs and attitudes to 553–5
  constructivist approach to 10–14, 537–8
  in cooperative framework 25–6
  environment for 25–6, 545–52, 554
  homework 552–3
  planning for 542–5
  understanding critical to 1–4, 68, 350

see also models (theoretical)
Leibniz, Gottfried 364
length 462, 471–80
Lesseps, Ferdinand De 443
lessons
  planning individual 542–3
  planning mathematics program 1–2, 4–7, 450, 538–42
letters, as numeric symbols 172, 364–6
line graphs 515–16
line plots 516–17
lists, in assessment process 567–8
literacy, mathematical 7–10
location and arrangement strand (shape and space)
  concepts development 445–6
  language of direction 446–7
  language of distance 471–2
  language of location 446, 471–2
  maps 447–9
  Numeracy Benchmarks minimum achievements 444–5
logical thinking, development of 377
long division *see* division
Lotto 500

Ma, Liping 10
Making Numbers Make Sense (Ritchhart) 568–9
males, ability of 28–30, 553–4
maps 447–9
mass 463–4, 488–90
materials and equipment
  in problem solving 61–2
  for shape and space 391
  storage and management of 549–50
  use of 13–17, 20
Mathematical Activity Tiles (MATs) 414
mathematical literacy 7–10
'mathematical thinking', concept of 24–5, 37–40, 42
mathematics, defined 431
mathematics program
  components of 37
  planning individual lessons 542–3
  planning of 1–2, 4–7, 450, 538–42
measurement
  constructivism and 456
  in curriculum 69, 451–2, 496
  defined 452–3
  difficulties with 495
  history of 457–9

language for 455–6
learning sequence for 452, 453–5, 496
materials and equipment for 456–7
non-SI units 498
Numeracy Benchmarks minimum achievements 459–61
principles of 454
standard units 454–5, 458, 496
syllabus topics 461–8
teaching process for 469–94
measurement topics
area 480–6
capacity and volume 455, 462–3, 486–8
geometric measures 462–3
length 471–80
mass 463–4, 488–90
money and value 467, 493–4
overview 461–2
probability 467–8, 502–6
statistics 468
temperature 466–7, 492–3
time 269, 455, 464–6, 490–2
value and money 467, 493–4
volume and capacity 455, 462–3, 486–8
weight 463–4
median 512–13
mental arithmetic *see* mental computation
mental computation
addition 216–17
in computational skills development 187–90
division 333–5
multiplication 290–2
subtraction 248–9
metric system 458, 496–9
millions 125–7
minus 227, 228 *see also* subtraction
mixed numbers 135, 160, 162
mode, in data analysis 513
models (3D)
to represent concepts 376–7
in teaching fractions 15–16, 17
models (theoretical)
Hoffer's theory 374–8, 385–6
inquiry model 386
of Polya 42–3, 44, 56
of Siemon & Booker 43, 48–9, 56–9, 62
van Hiele theory 378–81, 385–6
see also constructivism
money and value 467, 493–4
Mousley, J. 378
Multi-base Arithmetic Blocks (MABs) 105

multi-link patterns 431–2
multiplication
algorithm for 184–5, 275–88
basic facts 180, 266–75, 534–5
calculators in 282, 292–4, 297–8
with common fractions 298–9, 302
concepts for 259–66
with decimal fractions 221, 294–8, 301–2
difficulties with 265, 274–5, 280, 285–6, 288
estimating addition using 216
estimation with 288–90
linking division to 311–15
mental computations 290–2
overview 258
with per cents 299–302
sequence for developing 259, 262–3, 276
see also computation
multiplication tables 266, 267–8, 274

National Statement on Mathematics for Australian Schools, A (AEC)
assessment of students 557, 558
chance and data 468
communication ability 65
data collection and analysis 507, 521
definition of mathematics 431
estimation 247
framework for teaching 538
learning process 538
measurement 451
patterning 171
place value 78
problem solving 69
purpose of 45
shape and space 371, 372–3, 387
negative numbers 18, 466
nets, of 3D solids 422–4
Newton, Isaac 364
Nisbet, S. 554
noise, in the classroom 550–1
norms, student 574–5
nth term 172–3, 365
number, conservation of 87
number fact wheel 235–6, 312
number lines 18, 135
number patterns *see* patterning
number processes 75–6
number sense 8, 38–40, 42
numbers
mixed 135, 160, 162
naming 72–4, 78, 82–5

naming large 125–32
naming three-digit 107–9
rational and irrational 344
renaming 74–5, 77
renaming three-digit 109–11
written words for 83
see also numeration
numeracy
  in curriculum 68
  data sense and 530
  defined 2, 7–10, 371
Numeracy, a Priority for All (DETYA) 538–9
Numeracy Benchmarks (Curriculum Corp)
  basic facts 180–1
  location and arrangement strand 444–5
  for measurement 459–61
  numeracy levels 9–10
  program organisation 538, 539
  shape and space 371–2, 373
  shape and structure strand 396–7
  transformation and symmetry strand 427–8
numeration
  concept of ten 92
  defined 71–2
  exponential and scientific notation 128–32
  five- and six-digit numbers 117–19
  four-digit numbers 115–16, 118–19
  large numbers 125–8
  learning process 77–8, 533–5
  naming numbers 72–4, 78
  number processes 75–6
  one-digit numbers 81–94
  place value 78–80
  renaming numbers 74–5, 77
  rounding 120–5
  sequence for developing 80
  teaching of 76–7, 79, 80
  three-digit numbers 107–14
  two-digit numbers 94–106

oblongs 401, 406
O'Brien, T. 399
one-digit numbers
  comparing, sequencing and ordering 87, 89–90
  counting 86, 88–9
  difficulties with 93
  early-mathematical thinking 86–9
  history of 81–2
  naming of 82–5
  ordinal numbers 91
  pre-school skill acquisition 81

sequence for developing 92–3
order of operations 362
ordinal numbers 91
origins of mathematics *see* history of mathematics
Orrill, O. 7
ovals 408–10
Owens, K. 378

palindromes 443
parallelism 416–17
parallelograms 405–7, 483
parents
  assessment reports for 582
  assistance from 549, 551–2
partition, defined 304
patterning
  for algebraic skills 170–4
  basis of tuition 17–20
  in early-mathematical thinking 87
  in transformational geometry 428–33
Pegg, J. 374, 378
per cents
  history of 134
  introduction of 138–9
  multiplication 299–302
  in probability 503–6
  recording and reading 153–6
  renaming 165
perspective sketches 424–5
pi (π) 408, 409, 484
Piaget, Jean 86
picture graphs 514
pie graphs 519
Pinel, A. 402
place value
  concept of ten 92
  exponential and scientific notation 128–32
  four, five and six-digit numbers 115–19
  for large numbers 125–8
  in numeration 78–80
  three-digit numbers 107–14
  two-digit numbers 92, 94–106
plane (2D) shapes 397–417
Platonic solids 418–19
play, role of 26
plots and graphs 507, 514–21
plus, use of 23, 192, 193
Polya, G. 24, 42–3, 44, 56
polygons
  classification of 397, 399–408

learning difficulties with 382
naming 369
symmetry in 441–2
polyhedrons 418–20, 421–3, 424–7
portfolios, students' 565, 569–70
powers of ten 128–32
pre-number skills 87
prime factors 129–31
prime numbers 129
Principles and Standards for School Mathematics (NCTM) 545
prisms 419–20, 424
probability 467–8, 502–6 *see also* chance and data
problem solving
   assessment of student ability 579–80
   computation strategies for 360–3
   in curriculum 2, 4, 532
   difficulties with 51–6
   number sense and 38–40, 42
   organising a program for 47–9
   Polya's model 42–3, 44, 56
   problem selection 46–7
   problem writing to assist 65
   process for 56–65
   Siemon & Booker's model 43–4, 48–9, 56–9, 62
   spatial sense and 40
   standard process for 349
   strategies for 44–6, 60–2, 63–4
   worked example 49–51
problem structure tables 47–8
problem writing 65
program organisation *see* mathematics program
protractors 384, 412, 463
pyramid graphs 520
pyramids 419–20

Q-plots 521
quadrilaterals 400, 404–8, 435–6
quartiles 513–14
quotition, defined 304

randomness *see* chance and data
range, in data analysis 513
Rangoli patterns 439
rational numbers 344
ratios 167–70, 503–6
rectangles 401, 405–6, 407
reflection, in transformational geometry 429
remainders *see* division
remedial teaching 581–2
reports, assessment 582

resources, classroom management of 549–50
   see also materials and equipment
reversal errors 228
Reys, B. 38
rhombuses 405, 406
Ritchhart, R. 2, 568–9
rotagrams 410–11, 412, 447, 463
rotation, in transformational geometry 429
rote learning 196, 266, 538
rounding
   in division 332–3
   of division remainders 330–1
   in estimation 188, 214–16
   in measurement 477
   in multiplication 289
   of numbers 120–5
   in subtraction 247
Rowland, R. 453

samples, of students' work 569
scalene triangles 401
scatter plots 521
Schlafi coding system 421–2, 434
Schoenfeld, A. 43
school reports 582
scientific notation 128–32
self-assessment, by students 570
shape and space
   constructivist teaching approach 384–5, 386
   in curriculum 67, 69, 371–2
   development of teaching method 392–6
   difficulties with 381–4
   forms of geometry 370–1
   geometrical ways of working 374–8, 385
   history of geometrical ideas 368–9
   inquiry teaching-model 386
   language and vocabulary 369–70, 375, 387, 392
   learning process for 372–4
   syllabus strands 396, 450
   teaching goals 367–8, 449
   teaching principles 386–92
   van Hiele theory of learning 378–81, 385–6
   see also location and arrangement strand; shape and structure strand; transformation and symmetry strand
shape and structure strand (shape and space)
   2D—3D transformations 422–7
   2D shapes 397–417
   3D shapes 408, 417–27
   Numeracy Benchmarks minimum achievements 396–7

short division 316, 319, 327 *see also* division
SI (Système International d'Unités) 458, 496
Siemon, D. 43, 48–9, 56
signs, defined 23
six-digit numbers 117–19, 124–5
social context, of mathematical development 25–9
software programs 31, 34, 91, 446, 447
solids *see* three-dimensional (3D) shapes
space *see* shape and space
spatial awareness *see* shape and space
spheres 420
squares 401, 405–6, 407
standard deviation 514
standard units of measurement
   history of 458
   metric system 454–5, 496–9
statistics, defined 468 *see also* chance and data
Steen, L. 7
stem-and-leaf plots 517, 520
Stevin, Simon 145
subitisation 88
subtraction
   algorithm for 184, 238–46
   basic facts 180, 234–8, 533–4
   calculators in 249–51
   concepts of 226–34
   with decimal fractions 251–5
   difficulties with 237, 245–6
   estimation with 246–8
   with like common fractions 251–5
   mental computations 248–9
   overview 225
   sequence for developing 226, 228–9
   with unlike common fractions 254–8
   see also computation
syllabus *see* curriculum
symbols
   addition (+) 23, 192, 193
   algebraic 364–6
   concept of 23–5
   for decimal points 126, 147
   division 303, 309
   equals (=) 194, 228, 309
   greater than (>) 89, 103, 111
   introduction of 89, 103, 111, 351
   learned difficulties associated with 353–4
   less than (<) 103, 111
   letters as 172
   multiplication (×) 261, 276
   *n*th term 172–3, 365

   per cent (%) 154
   pi (π) 408, 409, 484
   subtraction (–) 227–8
symmetry
   assessment activities 442–4
   learning difficulties with 383
   Numeracy Benchmarks minimum achievements 427–8
   pattern blocks 438–9
   in polygons 441
   Rangoli patterns 439
   testing for 440–1
   types of 438
Système International d'Unités (SI) 458, 496

tables, multiplication 266, 267–8, 274
takeaway *see* subtraction
tangrams 390, 391, 481
teachers
   in constructivist learning 11–13
   in cooperative learning 26
   role of 4–7, 359
   traditional approach of 13
teachers' assistants 551–2
teaching standards 538
technology, and mathematics curriculum 31–4, 68–9
teen numbers 79, 96–7, 100–3, 115, 369
temperature 466–7, 492–3
ten
   concept of 92
   powers of 128–32
tessellation 383, 433–8, 481
tests
   alternatives to written 565–70
   assessment by task 393–4, 442–4
   evaluation of results 581–2
   formal 558, 571–9 *see also* assessment
   standardised 574–5
   teacher-constructed 571–3
Tharoor, S. 82
'thinking mathematically', concept of 24–5, 37–40, 42
Thorndike, E. L. 181
thousands 115–19
three-digit numbers
   comparing, sequencing and ordering 110, 111
   counting 112
   difficulties with 112–13
   naming 107–9
   renaming 109–11

rounding 123
sequence for developing 112
three-dimensional (3D) shapes 408, 417–27
time
    learning to tell 269, 490–2
    measurement of 455, 464–6
times, concept of 261–2, 353–4 *see also* multiplication
transformation and symmetry strand (shape and space)
    Numeracy Benchmarks minimum achievements 427–8
    patterns 431–3
    symmetry 438–44
    tessellation 433–8
    transformation 428–38
transformational geometry 428–38
translation, in transformational geometry 428–9
trapeziums 405, 483
triangles 401–4, 435–6, 482–3
two-digit numbers
    20–99 numbers 94–100
    comparing, sequencing and ordering 103–5
    concept of ten 92
    counting 104
    difficulties with 106
    rounding 121–2
    sequence for developing 105
    teen numbers 79, 96–7, 100–3, 369
two-dimensional (2D) shapes 397–417, 422–7

understanding
    curricular emphasis on 1–4
    difficulties due to inadequate 68, 350
Used Numbers—Statistics (Friel *et al.*) 511

value and money 467, 493–4
van Hiele theory 378–81, 385–6
vertical recording
    in addition 194
    in multiplication 258, 265
    in subtraction 228, 232
Viete, François 364
visual skills 40, 370, 371, 375
vocabulary *see* language; words
volume and capacity 455, 462–3, 486–8
Vygotsky, L. 26

websites 450, 544
weight 463–4
Wheatley, G. 31
wheel, for basic facts 235–6, 312
Willis, S. 29
Wilson, P. 453
Win with Maths! (software) 34, 84, 91
Wollygoggles and other creatures (O'Brien) 399
Wood, L. 30
word problems 360
words
    for length 471–2
    for location 471–2
    origins of 21–2
    for space and shape activities 388–9
    specificity of 22–3
workshops, for teachers 543–5
writing, number words and symbols 83

zero
    concept of 24
    in division 313–15, 324–6
    history of 24, 82
    in multiplication 272–3
    in numeration 74, 76, 79–80
    in subtraction 236, 244–6
    in thousands 115